BASIC CHEMICAL KINETICS

BASIC CHEMICAL KINETICS

H. EYRING

Department of Chemistry
University of Utah

S. H. LIN

Department of Chemistry
Arizona State University

S. M. LIN

National Chung-Shan Institute of Sciences and Technology

A Wiley-Interscience Publication

JOHN WILEY & SONS

New York · Chichester · Brisbane · Toronto

Library of Congress Cataloging in Publication Data:

Eyring, Henry, 1901–
  Basic chemical kinetics.
  "A Wiley–Interscience publication."
  Includes index.
  1. Chemical reaction, Rate of. I. Lin, Sheng
Hsien, 1937–   joint author. II. Lin, S. M., joint
author. III. Title.
QD502.E97   541′.39   79-26280
ISBN 0-471-05496-8

Printed in the United States of America

10 9 8 7 6 5 4 3 2 1

Preface

This book is the outgrowth of the course Chemical Kinetics taught by S. H. Lin and H. Eyring and is intended for undergraduate seniors, graduate students, and researchers. The presentation is elementary and self-contained wherever possible. Both classical and quantal treatments are presented for comparison. Problems are presented to help the readers understand the material covered in this book. In view of the recent, rapid progress in photochemistry, photophysics, and photobiology, two chapters are used to present the elementary processes in these areas. Wherever derivations were too long to be included in detail, references are given to smooth the way.

Two of us, H. E. and S. H. L., would like to thank their wives, Winifred and Pearl, for their understanding, patience, and moral support; S. M. L. would like to thank Dr. R. E. Weston for his hospitality.

We would like to thank the *Journal of Chemical Physics* for allowing us to reproduce material previously published there. We further acknowledge The Chemical Society for permission to reproduce figures published in the *Journal of the Chemical Society, Faraday Transactions*

H. EYRING*
S. H. LIN†
S. M. LIN‡

* *Salt Lake City, Utah*
† *Tempe, Arizona*
‡ *Taipei, Taiwan*
  January 1980

v

Contents

Contents

BASIC
CHEMICAL
KINETICS

One

Introduction

CONTENTS

In this chapter we briefly present some elementary mathematical methods that are frequently used in solving chemical kinetics problems. The application of these methods is demonstrated by solving some particular reactions. We also discuss the validity and limitations of the applications of steady-state approximation and equilibrium approximation to reaction kinetics.

1.1 METHOD OF SEPARATION OF VARIABLES

According to the law of mass action for the chemical reaction

$$aA + bB \longrightarrow cC + dD \tag{1.1}$$

the rate of reaction is proportional to a product of concentrations of reactants,

$$\text{rate} = kC_A^a C_B^b \tag{1.2}$$

1

where k is called the *rate constant*. The rate of reaction can be represented by $-(dC_A/dt)$ or $-(dC_B/dt)$ or (dC_C/dt) or (dC_D/dt). There exists a relation among these representations

$$-\frac{1}{a}\frac{dC_A}{dt} = -\frac{1}{b}\frac{dC_B}{dt} = \frac{1}{c}\frac{dC_C}{dt} = \frac{1}{d}\frac{dC_D}{dt} \tag{1.3}$$

assuming that the stoichiometry of reactions is also given by eq. 1.1.

Next we consider the simplest case

$$aA \longrightarrow \text{products} \tag{1.4}$$

In this case the rate of reaction is given by

$$-\frac{dC_A}{dt} = kC_A^a \tag{1.5}$$

This equation can be solved by using the method of separation of variables

$$-\frac{dC_A}{C_A^a} = k\,dt \tag{1.6}$$

If $a \neq 1$, we can carry out the integrations of both sides of eq. 1.6 independently to obtain

$$\frac{1}{a-1}\cdot\frac{1}{C_A^{a-1}} = kt + \frac{1}{a-1}\cdot\frac{1}{C_{A0}^{a-1}} \tag{1.7}$$

Here we have assumed that the initial concentration of A is C_{A0} (i.e., at $t = 0, C_A = C_{A0}$). Similarly, for $a = 1$, we have

$$\log C_A = -kt + \log C_{A0} \tag{1.8}$$

or

$$C_A = C_{A0}\exp(-kt) \tag{1.9}$$

It should be noticed that eq. 1.8 can be obtained from eq. 1.7 by using the l'Hospital rule.

Equation 1.7 indicates that the plot of $1/C_A^{a-1}$ versus t is linear with the slope of $(a-1)k$ and the intercept of $1/C_{A0}^{a-1}$. Similarly, eq. 1.8 shows that $\log C_A$ versus t is linear; the slope is $-k$ and the intercept is $\log C_{A0}$. These features provide us a method to determine the order of reactions and the rate constant. For reactions like those given by eq. 1.4, the half-life $t_{1/2}$, which is defined as the time required for the reactant concentration to reduce to one-half of its value, exists and can easily be found by setting $C_A = \frac{1}{2}C_{A0}$ in eqs. 1.7 and 1.8 to yield

$$t_{1/2} = \frac{2^{a-1}-1}{(a-1)kC_{A0}^{a-1}} \tag{1.10}$$

for $a \neq 1$ and

$$t_{1/2} = \frac{\log 2}{k} \tag{1.11}$$

for $a = 1$, respectively.

Now we consider another simple case of eq. 1.1.

$$A + B \rightarrow \text{products} \tag{1.12}$$

If the initial concentrations of A and B are C_{A0} and C_{B0}, then at a particular instant $C_A = C_{A0} - \chi$ and $C_B = C_{B0} - \chi$. Thus the rate of reaction in this case can be written as

$$-\frac{dC_A}{dt} = -\frac{dC_B}{dt} = \frac{d\chi}{dt} = kC_A C_B = k(C_{A0} - \chi)(C_{B0} - \chi) \tag{1.13}$$

Separating the variables in eq. 1.13 and carrying out the partial fraction of the resulting expression yields

$$\frac{d\chi}{(C_{A0} - \chi)(C_{B0} - \chi)} = \frac{d\chi}{C_{B0} - C_{A0}} \left(\frac{1}{C_{A0} - \chi} - \frac{1}{C_{B0} - \chi} \right) = k\, dt \tag{1.14}$$

which can easily be integrated

$$\frac{1}{C_{B0} - C_{A0}} \log \frac{C_{B0} - \chi}{C_{A0} - \chi} = kt + \frac{1}{C_{B0} - C_{A0}} \log \frac{C_{B0}}{C_{A0}} \tag{1.15}$$

or

$$\frac{1}{C_{B0} - C_{A0}} \log \frac{C_B}{C_A} = kt + \frac{1}{C_{B0} - C_{A0}} \log \frac{C_{B0}}{C_{A0}} \tag{1.16}$$

Equation 1.13 can also be integrated by using the relation $C_B = C_{B0} - C_{A0} + C_A$

$$-\frac{dC_A}{dt} = kC_A(C_{B0} - C_{A0} + C_A) \tag{1.17}$$

and by carrying out the partial fraction of $1/C_A(C_{B0} - C_{A0} + C_A)$. Equation 1.16 indicates that for the reaction mechanism given by eq. 1.12 the plot of $\log(C_B/C_A)$ versus t is linear with the slope $k(C_{B0} - C_{A0})$ and the intercept $\log(C_{B0}/C_{A0})$. Notice that when $C_{A0} = C_{B0}$, eq. 1.13 reduces to eq. 1.15 for $a = 2$ and eq. 1.16 can be reduced to eq. 1.7 by using the l'Hospital rule. Other cases of eq. 1.1 can be discussed similarly (Benson, 1960; Cappellos and Bielski, 1972; Laidler, 1965).

1.2   DETERMINANT METHOD (EIGENVALUE METHOD)

To demonstrate this method let us consider the reaction

$$A \underset{k_b}{\overset{k_f}{\rightleftharpoons}} B \tag{1.18}$$

The rate equations are given by

$$\frac{dC_A}{dt} = -k_f C_A + k_b C_B \tag{1.19}$$

and

$$\frac{dC_B}{dt} = k_f C_A - k_b C_B \tag{1.20}$$

To solve these equations, we let

$$C_A = A_1 e^{-\lambda t}, \qquad C_B = A_2 e^{-\lambda t} \tag{1.21}$$

It follows that

$$\begin{aligned}(\lambda - k_f)A_1 + k_b A_2 &= 0 \\ k_f A_1 + (\lambda - k_b)A_2 &= 0\end{aligned} \tag{1.22}$$

In eq. 1.21, A_1, A_2, and λ are to be determined. For A_1 and A_2 to have non-trivial solutions, we must have

$$\begin{vmatrix} \lambda - k_f & k_b \\ k_f & \lambda - k_b \end{vmatrix} = 0 \tag{1.23}$$

This is usually called the secular determinant and λ is called the eigenvalue of this determinant. The two roots of eq. 1.23 are given by

$$\lambda_1 = 0, \qquad \lambda_2 = k_f + k_b \tag{1.24}$$

Thus the solutions of eq. 1.21 become

$$\begin{aligned}C_A &= A_{11} e^{-\lambda_1 t} + A_{12} e^{-\lambda_2 t} \\ C_B &= A_{21} e^{-\lambda_1 t} + A_{22} e^{-\lambda_2 t}\end{aligned} \tag{1.25}$$

Notice that for $\lambda = \lambda_1$, we have

$$(\lambda_1 + k_f)A_{11} + k_b A_{21} = 0 \tag{1.26}$$

or

$$A_{21} = \frac{k_f}{k_b} A_{11} \tag{1.27}$$

Similarly, for $\lambda = \lambda_2$

$$(\lambda_2 - k_f)A_{12} + k_b A_{22} = 0 \tag{1.28}$$

or

$$A_{22} = -A_{12} \tag{1.29}$$

Substituting eqs. 1.27 and 1.29 into eq. 1.28 yields

$$C_A = A_{11}e^{-\lambda_1 t} + A_{12}e^{-\lambda_2 t}$$

$$C_B = \frac{k_f}{k_b}A_{11}e^{-\lambda_1 t} - A_{12}e^{-\lambda_2 t} \tag{1.30}$$

Suppose at $t = 0$, $C_A = C_{A0}$ and $C_B = C_{B0}$,

$$C_{A0} = A_{11} + A_{12}$$

$$C_{B0} = \frac{k_f}{k_b}A_{11} - A_{12} = KA_{11} - A_{12} \tag{1.31}$$

where $K = k_f/k_b$, the equilibrium constant. Solving for A_{11} and A_{12}, we obtain

$$A_{11} = \frac{C_{A0} + C_{B0}}{1 + K}, \qquad A_{12} = \frac{KC_{A0} - C_{B0}}{1 + K} \tag{1.32}$$

Therefore the solutions given by

$$C_A = \frac{(C_{A0} + C_{B0})}{1 + K} + \frac{(KC_{A0} + C_{B0})}{1 + K}e^{-(k_f + k_b)t}$$

$$C_B = \frac{K(C_{A0} + C_{B0})}{1 + K} - \frac{(KC_{A0} - C_{B0})}{1 + K}e^{-(k_f + k_b)t} \tag{1.33}$$

Let $t \to \infty$. Equation 1.33 then reduces to

$$C_{Ae} = \frac{C_{A0} + C_{B0}}{1 + K}, \qquad C_{Be} = \frac{(K(C_{A0} + C_{B0}))}{1 + K} \tag{1.34}$$

which of course represent the equilibrium concentrations of A and B.

One of the main purposes in chemical kinetics is to determine the rate constants. In the above discussed reversible reaction, two rate constants k_f amd k_b are involved and to be determined. Suppose that $k_f C_{A0} > k_b C_{B0}$. Then the reaction proceeds from the left to the right until the equilibrium is reached. In this case the concentration of A decreases with t, and the concentration of B increases with t. If we follow the reaction by measuring C_A

as a function of time t, then from the limiting value of $C_A = C_{Ae}$ as $t \to \infty$, we obtain one condition of $C_{Ae} = (C_{A0} + C_{B0})/(1 + K)$ (i.e., we determine the equilibrium constant) and from the slope of the plot of $\log(C_A - C_{Ae})$ versus t, we obtain $k_f + k_b$. These two conditions provide us enough information to determine k_f and k_b.

Although we have applied the determinant method to solve eqs. 1.19 and 1.20, actually eqs. 1.19 and 1.20 can be solved by using the method of separation of variables. Notice that

$$\frac{dC_A}{dt} + \frac{dC_B}{dt} = 0 \tag{1.35}$$

which indicates that $C_A + C_B$ is a constant, that is,

$$C_A + C_B = C_{A0} + C_{B0} \tag{1.36}$$

Using eq. 1.36, we can eliminate C_B from eq. 1.19 to obtain

$$\frac{dC_A}{dt} = -(k_f - k_b)C_A + k_b(C_{A0} + C_{B0}) \tag{1.37}$$

which can of course be easily solved by using the method of separation of variables.

The determinant method can be applied to solve a set of first-order equations or pseudo-first-order equations. If the set of reactions is first order, for example,

then among the three eigenvalues obtained, one of them will be zero; this eigenvalue corresponds to the equilibrium situation. In this case the concentration of, say A, can be expressed as

$$C_A = A_{11}e^{-\lambda_1 t} + A_{12}e^{-\lambda_2 t} + A_{13}e^{-\lambda_3 t} \tag{1.38}$$

where $\lambda_1 = 0$. Here $A_{11}, A_{12}, A_{13}, \lambda_2$, and λ_3 are related to the initial concentrations and the four rate constants. Now if we follow the reactions by measuring C_A versus t, we can determine $A_{11}, A_{12}, A_{13}, \lambda_2$, and λ_3, which can in turn be used to determine the four rate constants if we know the initial concentrations of at least two reactants. On the other hand, for the set of reactions

there are six rate constants to be determined and C_A is still given by eq. 1.38. In this case, if we measure C_A versus t and even if we know the initial concentrations of all the three reactants, we have only five relations A_{11}, A_{12}, A_{13}, λ_2, and λ_3 for six unknown rate constants. In other words, in this case we need another independent measurement to determine the six rate constants completely.

The determinant method has been applied to study the effect of vibrational relaxation on unimolecular reactions (Lin *et al.*, 1972; Lin and Eyring, 1974) and on molecular luminescence (Lin, 1972), the stochastic models of reaction kinetics (Widom, 1974; Bartis and Widom, 1974), and the kinetic Ising model (Lacombe and Simha, 1974).

1.3 LAPLACE TRANSFORM METHOD

In this section we discuss how to employ the Laplace transform method (cf. Appendix One) to solve chemical kinetic problems.

To begin with, we consider the reaction

$$A \rightarrow \text{products}$$

for which we have

$$\frac{dC_A}{dt} = -kC_A \tag{1.39}$$

Applying the Laplace transformation to eq. 1.39, we obtain

$$-C_{A0} + P\bar{C}_A = -k\bar{C}_A, \qquad \bar{C}_A = \frac{C_{A0}}{k + P} \tag{1.40}$$

Carrying out the inverse Laplace transformation of eq. 1.40 yields eq. 1.9 (see Appendix One).

Next we consider the reaction

$$A \underset{k_b}{\overset{k_f}{\rightleftharpoons}} B \tag{1.41}$$

The rate equations for this reaction are given by eqs. 1.19 and 1.20 and have been solved by using the determinant method. We solve these same equations by using the Laplace transform method.

Carrying out the Laplace transformation of eqs. 1.19 and 1.20, we obtain

$$P\bar{C}_A - C_{A0} = -k_f\bar{C}_A + k_b\bar{C}_B \tag{1.42}$$

and

$$P\bar{C}_B - C_{B0} = k_f\bar{C}_A - k_b\bar{C}_B \tag{1.43}$$

where $\bar{C}_A$ and $\bar{C}_B$ represent the Laplace transforms of C_A and C_B,

$$\bar{C}_A = \int_0^\infty e^{-Pt} C_A \, dt \tag{1.44}$$

and

$$\bar{C}_B = \int_0^\infty e^{-Pt} C_B \, dt \tag{1.45}$$

Solving for $\bar{C}_A$ and $\bar{C}_B$ from eqs. 1.42 and 1.43, we find

$$\bar{C}_A = \frac{C_{A0}(P + k_b) + k_b C_{B0}}{P(P + k_f + k_b)} \tag{1.46}$$

and

$$\bar{C}_B = \frac{C_{B0}(P + k_f) + k_f C_{A0}}{P(P + k_f + k_b)} \tag{1.47}$$

Noticing that

$$\frac{1}{P(P + k_f + k_b)} = \left(\frac{1}{P} - \frac{1}{P + k_f + k_b}\right)\frac{1}{k_f + k_b} \tag{1.48}$$

we can carry out the inverse transformation of eqs. 1.46 and 1.47 (see Appendix One)

$$C_A = \frac{C_{A0} + C_{B0}}{1 + K} + \frac{KC_{A0} - C_{B0}}{1 + K} \exp[-(k_f + k_b)t] \tag{1.49}$$

and

$$C_B = \frac{K(C_{A0} + C_{B0})}{1 + K} - \frac{(KC_{A0} - C_{B0})}{1 + K} \exp[-(k_f + k_b)t] \tag{1.50}$$

Next we show the application of the Laplace transform method to the diffusion problem (Crank, 1957). For this purpose we first derive the diffusion equation for the one-dimensional case. Let us consider the points χ and $\chi + d\chi$. Then at a particular instant t if we let $J(\chi)$ and $J(\chi + d\chi)$ be the flux at χ and $\chi + d\chi$, respectively, the accumulation of mass between χ and $\chi + d\chi$ in the time interval dt causes the change in concentration dC,

$$[J(\chi) - J(\chi + d\chi)]\, dt = dC\, d\chi \tag{1.51}$$

Using the Taylor expansion, eq. 1.51 can be written as

$$\frac{\partial C}{\partial t} = -\frac{\partial J}{\partial \chi} \tag{1.52}$$

Now according to Fick's empirical law, we have $J(\chi) = -D(\partial C/\partial \chi)$, where D represents the diffusion coefficient. Inserting this relation into eq. 1.52, we obtain

$$\frac{\partial C}{\partial t} = \frac{\partial}{\partial \chi}\left(D\,\frac{\partial C}{\partial \chi}\right) \tag{1.53}$$

If D is independent of C, then eq. 1.53 reduces to

$$\frac{\partial C}{\partial t} = D\,\frac{\partial^2 C}{\partial \chi^2} \tag{1.54}$$

For the three-dimensional case, it can easily be shown that the corresponding diffusion equation is given by

$$\frac{\partial C}{\partial t} = D\nabla^2 C \tag{1.55}$$

where

$$\nabla^2 = \frac{\partial^2}{\partial x^2} + \frac{\partial^2}{\partial y^2} + \frac{\partial^2}{\partial z^2}$$

As an example of the application of the Laplace transform method, we consider the problem of diffusion in a semiinfinite medium, $\chi > 0$, in which the boundary is kept at a constant concentration C_0, the initial concentration being zero throughout the medium. Thus we have

$$C = C_0, \qquad \chi = 0, \qquad t > 0 \tag{1.56}$$

and

$$C = 0, \qquad \chi > 0, \qquad t = 0 \tag{1.57}$$

Carrying out the Laplace transformation of eq. 1.54 yields,

$$P\bar{C}(p) = D\frac{d^2\bar{C}(P)}{d\chi^2} \tag{1.58}$$

which can easily be solved

$$\bar{C}(P) = A_1 \exp\left(-\chi\sqrt{\frac{P}{D}}\right) + A_2 \exp\left(\chi\sqrt{\frac{P}{D}}\right) \tag{1.59}$$

where A_1 and A_2 are the integration constants. As $C(P)$ is finite as $\chi \to \infty$, we find $A_2 = 0$. From eq. 1.56, we know that at $\chi = 0$,

$$\bar{C}(P) = \frac{C_0}{P} \tag{1.60}$$

by carrying out the Laplace transformation of eq. 1.56, that is, $A_1 = C_0/P$. It follows

$$\bar{C}(P) = \frac{C_0}{P} \exp\left(-\chi\sqrt{\frac{P}{D}}\right) \tag{1.61}$$

Inverting the transformation (see Appendix One) yields

$$C(\chi, t) = C_0 \frac{2}{\sqrt{\pi}} \int_{\frac{\chi}{2\sqrt{Dt}}}^{\infty} e^{-u^2}\, du = C_0 \operatorname{erfc} \frac{\chi}{2\sqrt{Dt}} \tag{1.62}$$

From eq. 1.62 we can see that in order to determine D, we need only measure the concentration at a particular χ and particular instant t.

To conclude this section of the Laplace transform method, it should be noted that as long as the rate equations are linear with respect to concentrations of reactants, this method can be used. Nonlinear rate equations can be made linear by making the concentrations of certain reactants in large excess in comparison with others. We return to the application of the Laplace transform method in unimolecular reactions.

1.4 MISCELLANEOUS METHODS

In this connection we discuss several methods of considerable generality. For other particular methods the literature may be consulted (Benson, 1952; Mowery, 1974; Kremer and Baer, 1974; Donohue, 1974). As usual we illustrate these methods by examples. First we consider

$$A \xrightarrow{k_1} B \xrightarrow{k_2} C \tag{1.63}$$

The rate equations in this case are given by

$$\frac{dC_A}{dt} = -k_1 C_A \tag{1.64}$$

$$\frac{dC_B}{dt} = k_1 C_A - k_2 C_B \tag{1.65}$$

$$\frac{dC_C}{dt} = k_2 C_B \tag{1.66}$$

These equations can of course be solved by using the determinant method and Laplace transform method. We can also solve these equations by starting with eq. 1.64. The solution of eq. 1.64 can easily be carried out

$$C_A = C_{A0} e^{-k_1 t} \tag{1.67}$$

where C_{A0} is the initial concentration. Substituting eq. 1.67 into eq. 1.65 yields

$$\frac{dC_B}{dt} + k_2 C_B = k_1 C_{A0} e^{-k_1 t} \tag{1.68}$$

This can be solved by noticing that the integration factor is $e^{k_2 t}$ (Frost and Pearson, 1961), that is,

$$\frac{d}{dt}(C_B e^{k_2 t}) = k_1 C_{A0} e^{t(k_2 - k_1)} \tag{1.69}$$

It follows that

$$C_B = \frac{k_1 C_{A0}}{k_1 - k_1}(e^{-k_1 t} - e^{-k_2 t}) \tag{1.70}$$

Here it has been assumed that $C_B = 0$ at $t = 0$. Equation 1.68 can also be solved by first finding the auxillary solution $C_B^{(a)}$ from

$$\frac{dC_B^{(a)}}{dt} + k_2 C_B^{(a)} = 0 \tag{1.71}$$

to yield

$$C_B^{(a)} = \alpha e^{-k_2 t} \tag{1.72}$$

where α is the integration constant. Next we have to find the particular solution $C_B^{(P)}$ from eq. 1.68 by letting $C_B^{(P)} = \beta e^{-k_1 t}$ to obtain

$$\beta = \frac{k_1 C_{A0}}{k_2 - k_1} \tag{1.73}$$

The general solution of eq. 1.68 is then

$$C_B = C_B^{(a)} + C_B^{(P)} = \alpha e^{-k_2 t} + \frac{k_1 C_{A0}}{k_2 - k_1} e^{-k_1 t} \qquad (1.74)$$

The integration constant α can be determined by using the initial condition. This method can be applied to linear high-order differential equations. The term C_C can be found by noticing that $(dC_A/dt) + (dC_B/dt) + (dC/dt) = 0$ or $C_A + C_B + C_C$ is a constant.

Another method for solving the linear differential equations often appears in the literature. The essential point of the method is to replace simultaneous differential equations by a higher-order differential equation. For example, C_A can be eliminated by using eqs. 1.64 and 1.65 to yield

$$\frac{d^2 C_B}{dt^2} + (k_1 + k_2) \frac{dC_B}{dt} + k_1 k_2 C_B = 0 \qquad (1.75)$$

which can easily be solved by letting $C_B = \beta e^{-\lambda t}$ to find $\lambda = k_1, k_2$, that is,

$$C_B = \beta_1 e^{-k_1 t} + \beta_2 e^{-k_2 t} \qquad (1.76)$$

The integration constants β_1 and β_2 can be determined by using the initial conditions and by putting eq. 1.76 back into the original differential equations.

Next we consider (Weston and Schwarz, 1972)

$$A + C \xrightarrow{\;k_a\;} P \qquad (1.77)$$

$$B + C \xrightarrow{\;k_b\;} Q \qquad (1.78)$$

The corresponding rate equations are given by

$$\frac{dC_A}{dt} = -k_a C_A C_c \qquad (1.79)$$

$$\frac{dC_B}{dt} = -k_b C_B C_c \qquad (1.80)$$

$$\frac{dC_c}{dt} = -k_a C_A C_c - k_b C_B C_c \qquad (1.81)$$

To solve these equations we take the ratio between eqs. 1.79 and 1.80

$$\frac{dC_A}{dC_B} = \frac{k_a C_A}{k_b C_B} \qquad (1.82)$$

which can easily be integrated

$$\log \frac{C_A}{C_{A0}} = \frac{k_a}{k_b} \log \frac{C_B}{C_{B0}}, \qquad C_B = C_{B0} \left(\frac{C_A}{C_{A0}} \right)^r \qquad (1.83)$$

where C_{A0} and C_{B0} are the initial concentrations and $r = k_b/k_a$. Since $dC_c/dt - dC_A/dt - dC_B/dt = 0$, we obtain

$$\dot{C}_C = C_{C0} + (C_A - C_{A0}) + (C_B - C_{B0}) \tag{1.84}$$

or

$$C_C = C_{C0} + C_A - C_{A0} - C_{B0}\left[1 - \left(\frac{C_A}{C_{A0}}\right)^r\right] \tag{1.85}$$

Substituting eq. 1.85 into eq. 1.79 and carrying out the separation of variables, we find

$$\int_{C_A}^{C_{A0}} \frac{dC_A}{C_A\left[C_{C0} + C_A - C_{A0} - C_{B0}\left\{1 - \left(\frac{C_A}{C_{A0}}\right)^r\right\}\right]} = k_a t \tag{1.86}$$

Analytical expressions of eq. 1.86 can be obtained only when r is $\frac{1}{2}$ or an integer. Otherwise the numerical calculation should be used.

1.5 STEADY-STATE AND EQUILIBRIUM APPROXIMATION

Although we have presented a number of mathematical methods to solve rate equations in chemical kinetics, very often one may run across rate equations so complicated that they cannot be solved analytically by using any of the conventional methods. In that case the steady-state approximation or equilibrium approximation are often used. To illustrate these approximations we consider the reactions given by eq. 1.63 and the Lindemann scheme for unimolecular decompositions.

$$A + M \underset{k_{-1}}{\overset{k_1'}{\rightleftharpoons}} A^* + M \tag{1.87}$$

$$A^* \xrightarrow{k_2} P \tag{1.88}$$

We shall assume that the concentration of M is much bigger than that of A. In that case we can conveniently rewrite eqs. 1.87 and 1.88 as

$$A \underset{k_{-1}}{\overset{k_1}{\rightleftharpoons}} B, \quad B \xrightarrow{k_2} P \tag{1.89}$$

where $k_1 = k_1'M$, $k_{-1} = k_{-1}'M$ and $B = A^*$.

Let us first consider the reactions given by eq. 1.63. To test the validity of the steady-state approximation as applied to B, we compare the exact expression for the rate equation for C with the approximate rate equation for C obtained from the use of the steady-state approximation, that is,

$$D(k_i, t) = \frac{k_2 B}{k_1 A} = \frac{k_2}{k_2 - k_1} \{1 - \exp[-(k_2 - k_1)t]\} \qquad (1.90)$$

The factor D may be regarded as representing the extent of deviation for the steady-state approximation. In other words, the factor D may be used to test the performance of the steady-state approximation. However, this is not the only way to study the validity of the steady-state approximation. It is useful to consider the time required for the $D = 1$ state to be reached. If we let this time be denoted by t_s, then an explicit expression for t_s can be obtained from setting $dB/dt = 0$, that is,

$$t_s = \frac{1}{k_2 - k_2} \log \frac{k_2}{k_1} \qquad (1.91)$$

For the purpose of numerical comparison it is convenient to use a set of dimensionless variables defined by $k_2^* = k_2/k_1$, $t^* = k_1 t$, and $B^* = B/A_0$. In terms of these dimensionless variables we have

$$D(k_2^*, t^*) = \frac{k_2^*}{k_2^* - 1} \left\{ 1 - \frac{1}{k_2^*} \exp[-(k_2^* - 1)\Delta t^*] \right\} \qquad (1.92)$$

and

$$t_s^* = \frac{1}{k_2^* - 1} \log k_2^* \qquad (1.93)$$

where $\Delta t^* = t^* - t_s^*$. When $t^* \to \infty$, eq. 1.92 reduces to

$$D(k_2^*, \infty) = \frac{-k_2^*}{k_2^* - 1} \qquad (1.94)$$

Both eqs. 1.92 and 1.94 indicate that the steady-state approximation is valid when $k_2^* \gg 1$ or $k_2 \gg k_1$. Under this condition eq. 1.93 indicates that the steady state is reached quickly. In other words, in this case the steady-state approximation holds under the condition in which the intermediate B disappears quickly once it is formed.

Figure 1.1 is a plot of D versus Δt^* for various k_2^*, where Δt^* denotes the time after which t_s^* has been reached (Volk et al., 1977). The upper limit for each curve is $k_2^*/(k_2^* - 1)$, and the lower limit is $D = 1$. The distance from the $D = 1$ line to each D versus Δt^* curve gives the deviation involved in the steady-state approximation for a particular k_2^* value. For example, for the steady-state approximation to hold within 5%, it is required that $k_2^* > 20$ (see Fig. 1.1). Figure 1.2 shows a plot of t_s^* versus k_2^*; that is, it shows how

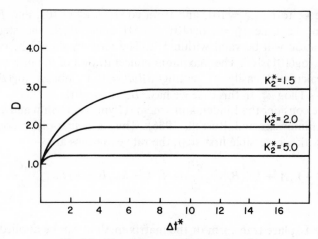

Fig. 1.1 Plot of D versus Δt^* for various k_2^*.

rapidly the true steady state is reached under various conditions. Sometimes it is important to estimate the relative magnitude of the intermediate B relative to A, that is, to see if the reaction intermediate is detectable. For this purpose it should be noted that $B/A = D/k_2^*$, which indicates that under the condition in which the steady-state approximation is valid, the concentration of B is very small.

Suppose we require the steady-state approximation to be valid within 1 % (i.e., $0.99 \le D \le 1.01$). In this case, from eq. 1.94 and $D = 1.01$, we can

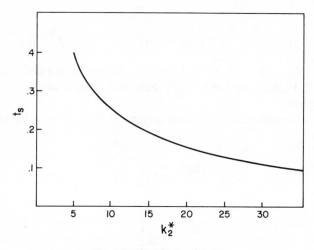

Fig. 1.2 Plot of t_s^* versus k_2^*.

determine k_2^* to be $k_2^* = 101$, and from eq. 1.92, $k_2^* = 101$, and $D = 0.99$, we obtain the time $t_M^* = [\log(101/2)/100]$ after which the steady-state approximation will be valid within 1%. The term t_M^* should be compared with $t_s^* = \log(101)/100$. The maximum concentration of the intermediate B can be calculated easily by setting $dB/dt = 0$ to obtain $\log(B_m/A_0) = -k_2^*/(k_2^* - 1)\log k_2^*$. In this case we have $B_m/A_o = (101)^{-1.01}$.

Next we consider the Lindemann scheme (Pyun 1971; Shin and Giddings, 1961; Jachimowski and Russell, 1966). This is a first-order consecutive reaction with a reversible first step; the rate equations are

$$\frac{dA}{dt} = -k_1 A + k_{-1}B, \qquad \frac{dB}{dt} = k_1 A - k_{-1}B - k_2 B, \qquad \frac{dC}{dt} = k_2 B$$

$$(1.95)$$

Either the Laplace transform or the matrix method can be applied to solve eq. 1.95 exactly; the resulting dimensionless solutions are given by

$$A^* = \frac{1}{\lambda_1^* - \lambda_2^*} [(1 - \lambda_2^*)e^{-\lambda_1^* t^*} + (\lambda_1^* - 1)e^{-\lambda_2^* t^*}] \qquad (1.96)$$

$$B^* = \frac{(\lambda_1^* - 1)(1 - \lambda_2^*)}{k_{-1}^*(\lambda_1^* - \lambda_2^*)} (e^{-\lambda_2^* t^*} - e^{-\lambda_1^* t^*}) = \frac{1}{\lambda_1^* - \lambda_2^*} (e^{-\lambda_2^* t^*} - e^{-\lambda_1^* t^*}) \qquad$$

$$(1.97)$$

where

$$\lambda_1^* = \tfrac{1}{2}(1 + k_{-1}^* + k_2^* + \sqrt{(1 + k_{-1}^* + k_2^*)^2 - 4k_2^*}) \qquad (1.98)$$

and

$$\lambda_2^* = \tfrac{1}{2}(1 + k_{-1}^* + k_2^* - \sqrt{(1 + k_{-1}^* + k_2^*)^2 - 4k_2^*}) \qquad (1.99)$$

with $A^* = A/A_o$, $B^* = B/A0$, and $t^* = k_1 t$. Here it is assumed that at $t = 0$, $B = C = 0$. In this case both steady-state and equilibrium approximations can be applied.

Applying steady-state approximation to B (i.e., $dB/dt = 0$), eq. 1.95 is reduced to

$$B_{ss}^* = \frac{A^*}{k_{-1}^* + k_2^*} \qquad (1.100)$$

and dC/dt becomes

$$\frac{dC^*}{dt^*} = \frac{k_2^*}{k_{-1}^* + k_2^*} A^* \qquad (1.101)$$

Following the discussion for the first case the deviation expression D_{ss} can be obtained from eq. 1.101 and eq. 1.100 as

$$D_{ss}(k_i^*, t^*) = (k_{-1}^* + k_2^*)\frac{B^*}{A^*} \tag{1.102}$$

where B^* and A^* are given in eqs. 1.96 and 1.97, respectively. When t^* approaches infinity, eq. 1.102 reduces to

$$D_{ss}(k_i^*, \infty) = \left(1 + \frac{k_2^*}{k_{-1}^*}\right)(1 - \lambda_2^*) \tag{1.103}$$

The expression for t_s^* can be obtained from applying the steady-state condition to eq. 1.97

$$t_s^* = \log\left(\frac{\lambda_2^*/\lambda_1^*}{\lambda_2^* - \lambda_1^*}\right) \tag{1.104}$$

Assuming equilibrium between the reactant A and the intermediate B that is, the equilibrium approximation, we have

$$\frac{B_{eq}}{A} = \frac{k_1}{k_{-1}} \tag{1.105}$$

$$\left(\frac{dC^*}{dt^*}\right)_{eq} = \frac{k_2^*}{k_{-1}} A^* \tag{1.106}$$

and

$$D_{eq}(k_i^* t^*) = \frac{k_{-1}^* B^*}{A^*} \tag{1.107}$$

From eq. 1.107 and eq. 1.102, D_{eq} is related to D_{ss} by

$$D_{eq}(k_i^*, t^*) = \frac{k_{-1}^*}{k_{-1}^* + k_2^*} D_{ss}(k_i^*, t^*) \tag{1.108}$$

The limiting form of D_{eq} is given by

$$D_{eq}(k_i^*, \infty) = (1 - \lambda_2^*) \tag{1.109}$$

Plots of D_{ss} versus t^* and D_{eq} versus t^* are shown in Figs. 1.3 and 1.4, respectively (Volk et al., 1977). Both plots are based on values of the ratio k_2^*/k_{-1}^* larger than and smaller than one. From Fig. 1.3 we see that for $k_2^*/k_{-1}^* < 1$, the smaller the value of k_2^*/k_{-1}^*, the better the steady-state approximation and for $k_2^*/k_{-1}^* > 1$, the larger value of k_2^*/k_{-1}^* gives a better approximation. Figure 1.4 shows that for $k_2^*/k_{-1}^* < 1$ the equilibrium approximation improves with decreasing k_2^*/k_{-1}^* and for $k_2^*/k_{-1}^* > 1$, the

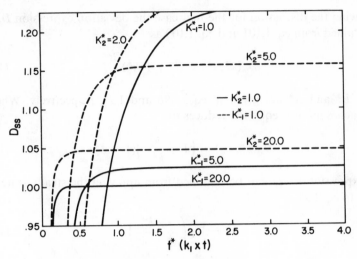

Fig. 1.3 Plot of D_{ss} versus t^*.

equilibrium approximation does not apply. (See also eq. 1.108.) The plot of B^*/A^* versus t^* is shown in Fig. 1.5. Recently Richardson et al. have applied the singular perturbation method to solve the Lindemann scheme and investigate the validity of the steady-state and equilibrium approximations analytically (Richardson et al., 1973).

The plot of t_s^* versus k_2^* with k_{-1}^* varied from 0.1 to 35 is shown in Fig. 1.6.

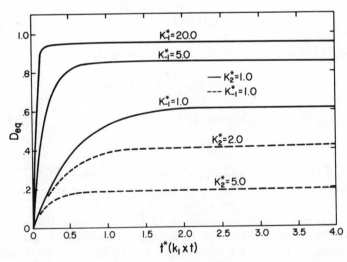

Fig. 1.4 Plot of D_{eq} versus t^*.

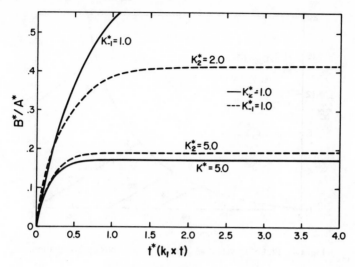

Fig. 1.5 Plot of B^*/A^* versus t^*.

From this plot one can see that at small values of k^*_{-1} and k^*_2 the system needs more time to reach the steady state. But, in general, the larger the value of k^*_2, the shorter the time for the system to reach the steady state. In Fig. 1.7 the plot of k^*_2 versus k^*_{-1} at D_{ss} $(k^*_i, \infty) = 1.01$ (eq. 1.103 is obtained as a limiting curve. Any combination of k^*_2 and k^*_{-1} in the region bounded by this curve will give a D_{ss} value in the range of 1.01 and 0.99. For practical use the

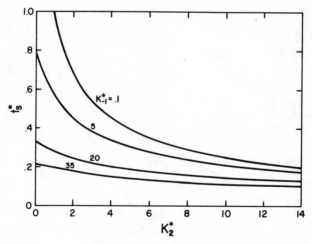

Fig. 1.6 Plot of t^*_s versus k^*_2.

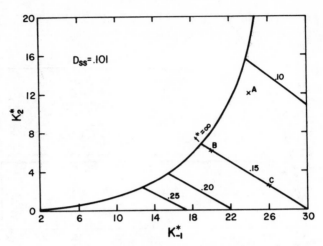

Fig. 1.7   Plot of k_2^* versus k_{-1}^* for the steady-state approximation.

time contour lines at $t^* = 0.10, 0.15, 0.2,$ and 0.25 are calculated from eq. 1.102 for the values of $D_{ss}(k_i^*, t^*)$ that are in the range of 0.99 and 1.01 and plotted in the same figure. With these contour lines the minimum time required for the system to reach the steady state within $\pm 1\%$ approximation can be determined from this figure. For example, points B and C give a D_{ss} value of 0.99 at $t^* = 0.15$, but as t^* approaches infinity, they give D_{ss} values of 1.009 and 1.003, respectively (see eq. 1.103). All combination of k_2^* and k_{-1}^* above

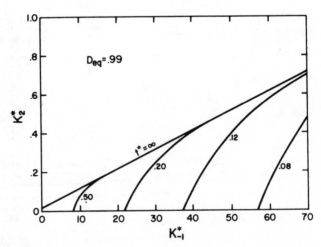

Fig. 1.8   Plot of k_2^* versus k_{-1}^* for the equilibrium approximation.

the region bounded by the limiting curve and contour t^* lines represent the cases in which the system will reach the steady state within $\pm 1\%$ approximation.

A similar plot for the equilibrium approximation is shown in Fig. 1.8. The contour t^* lines are calculated from eq. (1.108) at the values of $t^* = 0.08$, 0.12, 0.20, and 0.50 and plotted in the same figure. The behavior of the plot shown in Fig. 1.8 is similar to that mentioned in the steady-state approximation case. It should be noted that D_{ss} at time infinity cannot take values smaller than unity (see eq. 1.103), whereas D_{eq} at time infinity cannot exceed unity.

The validity of the application of the steady-state and equilibrium approximations to other complex reactions has been investigated by Volk et al. (Volk et al, 1977). They have also studied the effect of the order of reactions and of the reversibility of reactions on the applicability of the steady-state approximation and/or the equilibrium approximation.

REFERENCES

Bartis, J. T. and B. Widom (1974), *J. Chem. Phys.*, **60**, 3474.

Benson, S. W. (1952), *J. Chem. Phys.*, **20**, 1605.

Benson, S. W. (1960), *The Foundations of Chemical Kinetics*, McGraw-Hill.

Cappellos, C. and B. H. J. Bielski (1972). *Kinetic Systems*, Wiley-Interscience.

Crank, J. (1957), *Mathematics of Diffusion*, Oxford University Press.

Donohue, E. (1974), *J. Chem. Phys.*, **61**, 2180.

Frost, A. A. and R. G. Pearson (1961), *Kinetics and Mechanisms*, Wiley-Interscience.

Jachimowski, C. J. and M. E. Russell (1966), *Z. Phys. Chem.*, **48**, 102.

Kremer, M. L. and S. Baer (1974), *J. Phys. Chem.*, **78**, 1919.

Lacombe, R. H. and R. Simha (1974), *J. Chem. Phys.*, **61**, 1899.

Laidler, K. J. (1965), *Chemical Kinetics*, McGraw-Hill.

Lin, S. H. (1972), *J. Chem. Phys.*, **56**, 4155.

Lin, S. H., K. H. Lau, W. Richardson, L. Volk, and H. Eyring (1972). *Proc. Natl. Acad. Sci. U.S.*, **69**, 2778.

Lin, S. H., and H. Eyring (1974), *Ann. Rev. Phys. Chem.*, **25**, 39.

Mowery, D. W., Jr., *J. Phys. Chem.*, **78**, 1918 (1974).

Pyun, C. W. (1971), *J. Chem. Educ.*, **48**, 194.

Richardson, W., L. Volk, K. H. Lau, S. H. Lin, and H. Eyring (1973), *Proc. Natl. Acad. Sci. U.S.*, **70**, 588.

Shin, H. K. and J. C. Giddings (1961), *J. Phys. Chem.*, **65**, 1164 (1961).

Volk, L., W. Richardson, K. H. Lau, M. Hall, and S. H. Lin, *J. Chem. Educ.*, **54**, 95.

Weston, R. E., Jr., and H. A. Schwarz (1972), *Chemical Kinetics*, Prentice-Hall.

Widom, B. (1974), *J. Chem. Phys.*, **61**, 672.

Two

Potential
Energy Surfaces

CONTENTS

The theoretical approach to the calculation of chemical reactions consists of three separate parts: (1) the determination of the potential energy surfaces for the interacting species, (2) the evaluation of the reaction cross section (or reaction probability) as a function of the initial states of the reactants and the final states of the products, and (3) the determination of the total reaction rate by integrating the reaction cross section over the initial state distributions of the reactants. Thus we can see that the construction of reliable potential energy surfaces is of primary importance in the discussion of basic chemical kinetics. In this chapter we describe the theoretical methods that have been commonly used in the construction of these energy surfaces and of the properties associated with them; in particular, we choose one of the simplest cases, the $H + H_2$ exchange reaction, as an example for detailed discussion of the results obtained from various methods. The semiempirical methods and a priori methods used in the construction of potential energy surfaces can also be applied to other areas such as hydrogen bonding and molecular energy transfer.

2.1 THE SEPARATION OF ELECTRONIC AND NUCLEAR MOTION

The study of the atomic and molecular collisions leading to chemical reactions requires, in principle, the solution of the Schrödinger equation,

$$\hat{H}\psi(r, R) = E\psi(r, R) \tag{2.1}$$

with the full Hamiltonian of the system $\hat{H}$ given as

$$\hat{H} = \hat{T}_e + \hat{T}_n + V(r, R) \tag{2.2}$$

where $\hat{T}_e$ denotes the kinetic energy operator of the electrons and $\hat{T}_n$, that of the nuclei; $V(r, R)$ represents the total potential energy of nuclei and electrons, including the potential energy of electron-electron, electron-nucleus, and nucleus-nucleus interactions: $r = (r_i)$ is the set of electronic coordinates and $R = (R_\alpha)$ the set of nuclear coordinates. The solution of the wave equation (2.1) is, in general, a very difficult problem. However, making use of the fact that the mass of every atomic nucleus is at least a couple of thousand times as great as the mass of an electron, Born and Oppenheimer (Born, 1951; Born and Huang, 1954) demonstrated the separation of electronic and nuclear motion of the wave equation by carrying out a systematic expansion of the wave function and other quantities entering in eq. 2.1 as a power series in $(m/M)^{1/4}$ in which M is an average nuclear mass and m an electron mass. Here for simplicity we present the derivation due to Born for discussing the separation of electronic and nuclear motion of eq. 2.1 (Born, 1951).

From eq. 2.2, the Hamiltonian corresponding to fixed nuclei can be written as

$$\hat{H}_e = \hat{T}_e + V(r, R) \tag{2.3}$$

Suppose now that the energy and the wave function of the electron in the state n for a fixed nuclear configuration R are known,

$$\hat{H}_e \phi_n(r, R) = U_n(R)\phi_n(r, R) \tag{2.4}$$

Then the wave equation of the system can be solved by introducing the expansion

$$\psi(r, R) = \sum_n H_n(R)\phi_n(r, R) \tag{2.5}$$

Substituting eq. 2.5 into eq. 2.1, multiplying the result by $\phi_n^*(r, R)$, and integrating over r yields

$$[\hat{T}_n + U_n(R) - E]H_n(R) + \sum_{n_1} C_{nn_1}(R, P)H_{n_1}(R) = 0 \tag{2.6}$$

where

$$C_{nn_1}(R, P) = \sum_\alpha \frac{1}{M_\alpha}(\mathbf{A}_{nn_1}^{(\alpha)} \cdot \mathbf{P}_\alpha + B_{nn_1}^{(\alpha)}) \tag{2.7}$$

with $\mathbf{P}_\alpha = -i\hbar\nabla_\alpha \cdot \mathbf{A}_{nn_1}^{(\alpha)}$ and $B_{nn_1}^{(\alpha)}$ in eq. 2.7 are defined by

$$\mathbf{A}_{nn_1}^{(\alpha)}(R) = \int \phi_n^*(r, R)\mathbf{P}_\alpha\phi_{n_1}(r, R) \, dr \tag{2.8}$$

and

$$B_{nn_1}^{(\alpha)}(R) = \frac{1}{2}\int \phi_n^*(r, R)P_\alpha^2\phi_{n_1}(r_1 R) \, dr \tag{2.9}$$

For stationary states the electronic wave function $\phi_n(r, R)$ can be chosen as a real function so that it can be normalized to the same constant value for all values of R. Thus

$$\mathbf{A}_{nn}^{(\alpha)}(R) = -\tfrac{1}{2}in\nabla_\alpha \int \phi_n^2(r, R) \, dr = 0 \tag{2.10}$$

and eq. 2.6 reduces to

$$[\hat{T}_n + U_n(R) - E]H_n(R) + \sum_n{}' C_{nn_1}(R, P)H_{n_1}(R) = 0 \tag{2.11}$$

with

$$U_n^{(c)}(R) = U_n(R) + C_{nn}(R) = U_n(R) + \sum \frac{1}{M_\alpha} B_{nn}^{(\alpha)}(R) \tag{2.12}$$

Hence one can see that when the coupling of different electronic states C_{nn_1} is neglected, eq. 2.11 reduces to

$$[\hat{T}_n + U_n^{(c)}(R)]H_n(R) = EH_n(R) \tag{2.13}$$

Equation 2.13 is the result due to the Born-Oppenheimer adiabatic approximation. It implies that on account of the disparity of masses of electrons and nuclei the electrons can carry out many cycles of their motion in the time required for the nuclear configuration to change appreciably, and that in consequence it is allowed to quantize their motion for fixed nuclear configuration by solving the electronic wave equation, and then to use the electronic energy functions as potential energy functions in determining the motion of nuclei. Such a physical argument is supported by the experimental evidence of molecular spectroscopy, which shows that the internal energy of a molecule in many cases can be obtained by summing up the electron energy, the vibrational energy, and the rotational energy.

It is the quantity $U_n^{(c)}(R)$, but not $U_n(R)$, that plays the part of the potential energy of the nuclei when the coupling of different electronic states C_{nn_1} can be neglected. The difference between $U_n^{(c)}(R)$ and $U_n(R)$ can be determined when the electronic eigenfunctions for fixed nuclear configurations are known (Kolos, 1970).

The question of under what conditions the coupling parameters C_{nn_1} will be small so that the Born-Oppenheimer adiabatic approximation is valid cannot be answered in general. In the regime of the collision energies of most chemical interest, however, the nuclear velocities are in general sufficiently small relative to those of the electrons that the Born-Oppenheimer separations of nuclear and electronic motion are applicable. This implies that the nuclear motion may be regarded as adiabatic and a single electronic eigenfunction may be used to represent the state of the electrons throughout the reactive encounter. Thus in the Born-Oppenheimer approximation, the determination of a potential energy surface involved in a chemical reaction requires solution of the electronic Schrödinger equation, eq. 2.4, with the Hamiltonian given by

$$\hat{H}_e = -\frac{\hbar^2}{2m} \sum_i \nabla_i^2 - \sum_i \sum_\alpha \frac{Z_\alpha e^2}{r_{i\alpha}} + \sum_{i<j} \frac{e^2}{r_{ij}} + \sum_\alpha \sum_\beta \frac{Z_\alpha Z_\beta e^2}{R_{\alpha\beta}} \quad (2.14)$$

where the second term on the right-hand side of eq. 2.14 represents the nucleus-electron interaction, the third term, the electron-electron repulsion, and the last term, the nucleus-nucleus repulsion. Hence by solving a series of energy surfaces corresponding to the different electronic energy eigenvalues, the reacting system may be regarded as proceeding across these well-defined potential energy surfaces.

We should also be aware that even if the coupling parameters C_{nn_1} in eq. 2.11 are not small, their influence will be negligible if the electronic state n under discussion is separated from all others by a large energy gap. This can be shown from the perturbation theory (Lin and Eyring, 1974),

$$\mathbf{A}_{nn_1}^{(\alpha)}(R) = -\frac{i\hbar \langle \phi_{n_1}^0 | (\nabla_\alpha V)_0 | \phi_n^0 \rangle}{U_{n_1}^0 - U_n^0} + \cdots \quad (2.15)$$

and

$$B_{nn_1}^{(\alpha)}(R) = -\frac{\hbar^2}{2} \frac{\langle \phi_n^0 | (\nabla_\alpha^2 V)_0 | \phi_{n_1}^0 \rangle}{U_{n_1}^0 - U_n^0} \tag{2.16}$$

$$- \hbar^2 \sum_m{}' \frac{\langle \phi_n^0 | (\nabla_\alpha V)_0 | \phi_m^0 \rangle \langle \phi_m^0 | (\nabla_\alpha V)_0 | \phi_{n_1}^0 \rangle}{(U_{n_1}^0 - U_n^0)(U_{n_1}^0 - U_m^0)} + \cdots$$

where

$$\langle \phi_{n_1}^0 | (\nabla_\alpha V)_0 | \phi_n^0 \rangle = \int \phi_{n_1}^*(r, R_0)(\nabla_\alpha V)_0 \, \phi_n(r, R_0) \, dr$$

and so on. This will be the case for the ground state of many molecules and nonconducting metals. Here the zero-order approximation is a nonharmonic nuclear vibration with the potential energy $U_n^{(c)}(R)$, and the coupling with higher electronic states can be calculated from eq. 2.11 by the perturbation method. For metals in which the electronic states form a quasi-continuum, the summation in eq. 2.11 cannot be regarded as a small perturbation and will become an integral-differential equation such that the coupling of electronic and nuclear motion may be found in a rigorous way. On the other hand, in the case of close-lying vibronic states belonging to different electronic configurations, it is shown that the adiabatic approximation will completely fail. Such a breakdown of the Born-Oppenheimer approximation is well known for the degenerate electronic state, as the Jahn–Teller effect (Jahn and Teller, 1930), for the nearly degenerate state, as the pseudo Jahn–Teller effect, and for widely separated electronic states giving rise to vibrationally induced electronic transitions, as Hertzberg–Teller effects (Herzberg and Teller, 1933).

Recently, many investigations on the approximation of the Born–Oppenheimer separation have been carried out. Fisk and Kistman (1964) and Jepsen and Hirschfelder (1960), respectively, evaluated the energy corrections to the adiabatic approximations for the H_2 and H_2^+ molecules. Chiu (1964) discussed the rotation-electronic interactions of diatomic molecules from the nonadiabatic viewpoint of the Born–Oppenheimer approximation. Wu and Bhatia (1956) and Dalgarno and McCorrol (1957) respectively studied the interaction of hydrogen and helium atoms in the ground and excited states and found that the diagonal terms of the coupling between electronic and nuclear motion are not negligible at large separations. The nonstationary character of the adiabatic approximation has been discussed by Lin (1966, 1967) and the transition rate for the system to oscillate from one electronic state to another accompanied by a change in the quantum states of nuclear motion to conserve energy has been derived. The viewpoint of the breakdown of the adiabatic approximation has been adopted by Lin and Bersohn (1968), Jortner et al. (1969), Robinson and Frosch (1963, 1964), and Siebrand (1967) in discussing the radiationless transitions of molecular luminescence.

2.2 VALENCE BOND METHOD

As discussed in the previous section, the potential energy of nuclear motion of a molecule can be obtained by solving the electronic wave equation for any fixed nuclear configuration if the Born–Oppenheimer adiabatic approximation is valid. In this section we discuss the solution of the electronic wave equation by the valence bond method, and for this purpose we use the three-electron system as an example. The derivation, however, can be generalized for more complicated systems (Eyring et al, 1944).

2.2.1 The London Equation

For a system of three atoms, each with one valence electron, we denote the wave functions of the valence orbitals of the atom by $a(x, y, z)$, $b(x, y, z)$, and $c(x, y, z)$. If the atoms are divided into pairs (a, b), (b, c), and (c, a), the most stable configuration would be the one that corresponds to the maximum number of bonds. To the approximation in which spin interactions are neglected, the spin operators $\hat{S}^2$ and $\hat{S}_z$ commute with the Hamiltonian and may be used to reduce the order of the secular determinant. For the eigenvalue of $\hat{S}_z$ to be $\frac{1}{2}$, we have

$$
\begin{array}{cccl}
c & b & a & \\
\alpha & \alpha & \beta & \phi_1 = |(a\beta)_1(b\alpha)_2(c\alpha)_3| \\
\alpha & \beta & \alpha & \phi_2 = |(a\alpha)_1(b\beta)_2(c\alpha)_3| \\
\beta & \alpha & \alpha & \phi_3 = |(a\alpha)_1(b\alpha)_2(c\beta)_3|
\end{array}
$$

where α and β represents the spin wave functions and ϕ_i are the Slater determinants,

$$
\phi_3 = \frac{1}{\sqrt{3}}
\begin{vmatrix}
(a\alpha)_1 & (b\alpha)_1 & (c\beta)_1 \\
(a\alpha)_2 & (b\alpha)_2 & (c\beta)_2 \\
(a\alpha)_3 & (b\alpha)_3 & (c\beta)_3
\end{vmatrix}
\tag{2.17}
$$

For a linear combination of the ϕ_i's that corresponds to a bond between a and b, it is necessary for a and b to have opposite spins such that we are limited to the functions ϕ_1 and ϕ_2. The combination is therefore of the form

$$
\psi_{ab} = a_1\phi_1 + a_2\phi_2
\tag{2.18}
$$

If we interchange the spins on a and b, the function ψ_{ab} must change sign since the spin function associated with a stable bond is antisymmetric in the electrons,

$$\psi_{ab} = -a_2\phi_1 - a_1\phi_2 \tag{2.19}$$

Equations 2.18 and 2.19 are consistent only if $a_2 = -a_1$. We have, therefore, the unnormalized eigenfunction representing the bond a—b as

$$\psi_{ab} = \phi_1 - \phi_2 \tag{2.20}$$

Similarly, we find the eigenfunction for the bond b—c as

$$\psi_{bc} = \phi_2 - \phi_3 \tag{2.21}$$

The variational wave function can then be expressed in terms of these two independent canonical bond eigenfunctions

$$\psi = A\psi_{ab} + B\psi_{bc} \tag{2.22}$$

and the corresponding secular equation is given by

$$A(H_{AA} - S_{AA}U) + B(H_{AB} - S_{AB}U) = 0 \tag{2.23}$$
$$A(H_{BA} - S_{BA}U) + B(H_{BB} - S_{BB}U) = 0$$

where $H_{AA} = \langle\psi_{ab}|\hat{H}_e|\phi_{ab}\rangle$, $H_{AB} = \langle\psi_{ab}|\hat{H}_e|\psi_{bc}\rangle$, $H_{BB} = \langle\psi_{bc}|\hat{H}_e|\psi_{bc}\rangle$. $S_{AA} = \langle\psi_{ab}|\psi_{ab}\rangle$, $S_{AB} = \langle\psi_{ab}|\psi_{bc}\rangle$, and $S_{BB} = \langle\psi_{bc}|\Psi_{bc}\rangle$. Elimination of A and B gives

$$U^2(S_{AA}S_{BB} - S_{AB}^2) - U(H_{AA}S_{BB} + H_{BB}S_{AA} - 2H_{AB}S_{AB}) \\ + H_{AA}H_{BB} - H_{AB}^2 = 0 \tag{2.24}$$

or

$$U_\pm = \{(H_{AA}S_{BB} + H_{BB}S_{AA} - 2H_{AB}S_{AB}) \\ \pm [(H_{AA}S_{BB} + H_{BB}S_{AA} - 2H_{AB}S_{AB})^2 \\ - 4(H_{AA}H_{BB} - H_{AB}^2)(S_{AA}S_{BB} - S_{AB}^2)]^{1/2}\}/2(S_{AA}S_{BB} - S_{AB}^2) \tag{2.25}$$

This is the general expression of the potential energy for a system of three atoms. The normalization and overlap matrix elements in eq. 2.25 can be evaluated as follows:

$$S_{AA} = \langle\phi_1 - \phi_2|\phi_1 - \phi_2\rangle = \langle\phi_1|\phi_1\rangle + \langle\phi_2|\phi_2\rangle - 2\langle\phi_1|\phi_2\rangle \\ = 2 - \Delta_{bc}^2 - \Delta_{ac}^2 + 2\Delta_{ab}^2 - 2\Delta_{ab}\Delta_{bc}\Delta_{ac} \tag{2.26}$$

where $\Delta_{ab} = \langle a|b\rangle$ and so on. Similarly,

$$S_{BB} = 2 - \Delta_{ab}^2 - \Delta_{ac}^2 + 2\Delta_{bc}^2 - 2\Delta_{ab}\Delta_{ac}\Delta_{bc} \tag{2.27}$$

and

$$S_{AB} = 1 - \Delta_{ab}^2 - \Delta_{bc}^2 + 2\Delta_{ac}^2 + \Delta_{ab}\Delta_{ac}\Delta_{bc} \qquad (2.28)$$

Defining the Coulomb integral $Q = \langle (a)_1(b)_2(c)_3 | \hat{H}_e | (a_1)_1(b)_2(c)_3 \rangle$, the exchange integral $(ab) = \langle (a)_1(b)_2(c)_3 | \hat{H}_e | (b)_1(a)_2(c)_3 \rangle$, and so on, the Hamiltonian matrix elements can be expressed as

$$
\begin{aligned}
H_{AA} &= \langle \psi_{ab} | \hat{H}_e | \psi_{ab} \rangle \\
&= \langle \phi_1 | \hat{H}_e | \phi_1 \rangle + \langle \phi_2 | \hat{H}_e | \phi_2 \rangle - 2\langle \phi_1 | \hat{H}_e | \phi_2 \rangle \\
&= [Q - (bc)] + [Q - (ac)] - 2[-(ab) + (bca)] \\
&= 2Q - (bc) - (ac) + 2(ab) - 2(bca)
\end{aligned} \qquad (2.29)
$$

Similarly, H_{BB} and H_{AB} are given by

$$H_{BB} = 2Q - (ab) - (ac) + 2(bc) - 2(bca) \qquad (2.30)$$

and

$$H_{AB} = -Q - (ab) - (bc) + 2(ac) + (bca) \qquad (2.31)$$

If we assume that the orbitals a, b, and c are mutually orthogonal, we have eqs. 2.26–2.28 reduced to

$$S_{AA} = 2, \qquad S_{BB} = 2, \qquad S_{AB} = -1$$

Substituting these values into eq. 2.25 yields

$$
\begin{aligned}
U_\pm &= \tfrac{1}{3}(H_{AA} + H_{BB} + H_{AB}) \pm \tfrac{1}{3}\{\tfrac{1}{2}[CH_{AA} - H_{BB})^2 + (H_{AA} + 2H_{AB})^2 \\
&\quad + (H_{BB} + 2H_{AB})^2]\}^{1/2} \\
&= Q \pm [\tfrac{1}{2}\{[(ab) - (bc)]^2 + [(bc) - (ac)]^2 \\
&\quad + [(ac) + (ab)]^2\}]^{1/2} - (bca)
\end{aligned} \qquad (2.32)
$$

If the double-exchange integral (bca) equals zero, eq. 2.32 reduces to the London formula

$$U_\pm = Q \pm [\tfrac{1}{2}\{[(ab) - (bc)]^2 + [(bc) - (ac)]^2 + [(ac) - (ab)]^2\}]^{1/2} \qquad (2.33)$$

From the preceding derivation we can see that the London equation is only applicable to homopolar compounds with the overlap integrals and the double-exchange integrals neglected.

Now suppose that we remove the atom C to infinity; we have the case of a diatomic system and

$$S_{AA} = 2 + 2\Delta_{ab}^2, \qquad S_{BB} = 2 - \Delta_{ab}^2, \qquad S_{AB} = -1 - \Delta_{ab}^2$$

$$H_{AA} = 2Q + 2(ab), \qquad H_{BB} = 2Q - (ab), \qquad H_{AB} = -Q - (ab)$$

Substituting these equations into eq. 2.25, we obtain

$$U_{\pm} = \frac{[Q \mp (ab)]}{1 \mp \Delta_{ab}^2} \qquad (2.34)$$

This is the Heitler–London equation for homopolar bonding (Heitler and London, 1927). The lower sign gives rise to the lower energy, since the exchange integral is always negative in value and corresponds to the bound state $^1\Sigma$ of the molecule while the upper sign refers to the repulsive state. The Heitler–London equation leads to a minimum potential energy of 72.4 kcal for $^1\Sigma$ of H_2 and it accounts for only 66% of the experimental value, 109.4 kcal. If the overlap integral is ignored, eq. 2.34 then reduces to

$$U_{\pm} = Q \mp \alpha \qquad (2.35)$$

where $\alpha = (ab)$. The calculated energy at the minimum of the $^1\Sigma$ curve is now 107.5 kcal, in better agreement with the experimental value than that given by eq. 2.34. There is, however, no justification for neglecting the overlap integral, the better agreement being considered due to a cancellation of errors.

The London equation is good only for arriving at the general form of the surfaces, but is not very accurate. There are a number of reasons why the London equation cannot give reliable potential energies. In the first place, the Heitler–London equation gives for the H_2 binding energy an error that is much greater than the activation energy of the $H + H_2$ reaction. Although the London equation is not directly related to the Heitler–London equation (2.34) but rather the simplified equation (2.35), a trend of similar energy error may be expected in the London equation for the triatomic system. Second, the double-exchange integral neglected in the London equation has been shown to be quite important. Third, London (1929) in fact wrote down the following equation without proof

$$U_{\pm} = Q \pm \{\tfrac{1}{2}[(\alpha_{ab} - \alpha_{bc})^2 + (\alpha_{bc} - \alpha_{ac})^2 + (\alpha_{ac} - \alpha_{ab})^2]\}^{1/2} \qquad (2.36)$$

where $Q = Q_{ab} + Q_{bc} + Q_{ac}$ and Q_{ab}, α_{ab}, and so on, are the Coulombic and exchange integrals for diatomic molecules. In other words, eqs 2.33 and 2.36 are not equivalent since the integrals Q_{ab}, α_{ab}, and so on, used in the original London equation (2.36) are obtained by assuming the third atom being removed to infinity. Thus there is the limitation of its application and it has been critically discussed by James and Coolidge (1934) and others. At any rate, even though the London equation is not quantitatively correct, it was useful in the earlier days in the construction of the potential surfaces within which it was possible to understand the significance of the activation energy (Slater, 1931; Cashion and Herschbach, 1964).

2.2.2 The London–Eyring–Polanyi Method

Making use of the London equation (2.36), Eyring and Polanyi (1931) in 1931 developed a semiempirical method in computing the energy of the three-electron system; energy surfaces based on their method are now known as the London–Eyring–Polanyi (LEP) surfaces. Though the LEP method is not accurate, it has proved useful in making rough estimates of energies of activation.

Though the binding energy of a three-electron system can be evaluated from eq. 2.25 by calculating all of one-, two-, and three-centered integrals, the binding energy obtained thereby would be no more accurate than what is given from the Heitler–London equation (2.34) where the correct diatomic limits cannot be obtained when one atom is removed to infinity. Recognizing that the errors thus introduced would be too great in performing these calculations, the experimental bond dissociation energy is then used in the semiempirical approach. From an inspection of Sugiura's (1927) calculation of the Coulombic and exchange integrals for the H_2 molecules, Eyring and Polyanyi concluded that over a range of interatomic distances (in particular, for $R > 0.8$ Å) the fraction

$$\rho = \frac{Q}{Q + \alpha} \tag{2.37}$$

is roughly constant at 10–15 %. For any triatomic configuration it is therefore possible to compute the Coulombic and exchange energies for each pair of atoms on the basis of the spectroscopic value for the total energy, and by substituting these quantities into eq. 2.36 we can obtain the required potential energy for the three-atom system. For the bonding energy of a diatomic molecule $A - B$ as a function of internuclear distance, Eyring and Polanyi use the Morse potential,

$$U_{ab} = D_{ab}[\exp(-2\beta_{ab} X_{ab}) - 2\exp(-\beta_{ab} X_{ab})] \tag{2.38}$$

where X_{ab} is the displacement from the equilibrium internuclear distance, D_{ab} the classical dissociation energy, and β_{ab} a spectroscopic constant. The spectroscopic constant can be determined by differentiating U_{ab} with respect to X_{ab} twice,

$$\beta_{ab} = \pi C \omega_{ab}^0 \left(\frac{2\mu_{ab}}{D_{ab}}\right)^{1/2} \tag{2.39}$$

where u_{ab} is the reduced mass and ω_{ab}^0 the ground state vibrational frequency expressed in wave members.

The LEP method in general gives rise to a surface that has a basin at the activated state; the depth of the basin increases as ρ in eq. 2.37 is increased.

Moreover, the activation barrier also varies rather strongly with the fraction ρ. For the $H + H_2$ system the LEP method cannot predict a correct energy of activation without giving a basin. Since the presence of a basin is in contrast to the experimental evidence and results from other theoretical calculations, the LEP method is good only in making rough estimates of activation energy.

2.2.3 The London–Eyring–Polanyi–Sato Method (LEPS)

To eliminate the unrealistic basin present in the LEP method, Sato (1955) has proposed an alternative method to construct the potential energy surface of a triatomic system by making use of the shape of the repulsive state curve, which is well known for the H_2 molecule. To determine the analytical expression for the repulsive curve, Sato modifies the Morse potential by changing the sign between the two exponential terms from minus to plus and dividing the resulting expression by 2,

$$U_+ = \frac{D}{2}\left[\exp(-2\beta^1 x) + 2\exp(-\beta^1 x)\right] \tag{2.40}$$

According to the simplified Heitler–London treatment, eq. (2.35), we thus have

$$Q - \alpha = \frac{D}{2}\left[\exp(-2\beta^1 x) + 2\exp(-\beta^1 x)\right] \tag{2.41}$$

and

$$Q + \alpha = D\left[\exp(-2\beta x) - 2\exp(-\beta x)\right] \tag{2.42}$$

for the repulsive and ground states, respectively. Hence Q and α can be calculated as a function of internuclear distance without the assumption of a constant ρ. Moreover, Sato employs a modified expression to replace the original London equation (2.36)

$$U_\pm = \frac{1}{1 + S^2}\left[Q \pm \{\tfrac{1}{2}[(\alpha_{ab} - \alpha_{bc})^2 + (\alpha_{bc} - \alpha_{ac})^2 + (\alpha_{ac} - \alpha_{ab})^2]\}^{1/2}\right]$$

$$\tag{2.43}$$

where S is the overlap integral. Sato concludes on intuitive ground that this equation is valid only if the overlap integrals for the diatomic species are equal. The validity of the Sato equation has been discussed in detail by Eyring and Lin (1974).

Though the LEPS method is as arbitrary and empirical as the LEP method, the former is preferable to the latter since it leads to surfaces free of basins.

In a comparison of these two methods, however, Weston (1959) observes that the LEPS method gives barriers that are too thin such that more tunneling may come into effect for the $H + H_2$ reaction.

2.2.4 The Modified London–Eyring–Polanyi Method

A few methods have been developed that basically follow the LEP approach but avoid some of the approximations pertaining to the London equation. Porter and Karplus (1964) applied the complete expression for the three-atom system to obtain the potential energy surfaces of the $H + H_2$ exchange reaction. To parametrize the energy expression, they first decompose the Coulombic integral Q in eq. 2.32 into diatomic contributions by choosing the zero energy as that of the separated atoms

$$Q = Q_{ab} + Q_{bc} + Q_{ac} \tag{2.44}$$

where

$$Q_{ab} = -2\left\langle b \left| \frac{e^2}{r_{ai}} \right| b \right\rangle + \left\langle ab \left| \frac{e^2}{r_{ij}} \right| ab \right\rangle + \frac{e^2}{R_{ab}}$$

$$Q_{bc} = -2\left\langle c \left| \frac{e^2}{r_{bi}} \right| c \right\rangle + \left\langle bc \left| \frac{e^2}{r_{ij}} \right| bc \right\rangle + \frac{e^2}{R_{ac}}$$

and

$$Q_{ac} = -2\left\langle a \left| \frac{e^2}{r_{ci}} \right| a \right\rangle + \left\langle ac \left| \frac{e^2}{r_{ij}} \right| ac \right\rangle + \frac{e^2}{R_{ac}}$$

The diatomic Coulombic integral, say, Q_{ab}, represents the Coulomb integral for an H_2 molecule with internuclear distance R_{ab}. For the single-exchange integrals, they use

$$(ab) = \alpha_{ab} + \Delta\alpha_{ab} \tag{2.45}$$

where α_{ab} is the exchange integral of the $A - B$ molecule

$$\alpha_{ab} = -2\,\Delta_{ab}\left\langle a \left| \frac{e^2}{r_{ai}} \right| b \right\rangle + \left\langle ab \left| \frac{e^2}{r_{ij}} \right| ba \right\rangle + \Delta_{ab}^2 \cdot \frac{e^2}{R_{ab}} \tag{2.46}$$

and $\Delta\alpha_{ab}$ represents the residual terms composed of triatomic interactions,

$$\Delta\alpha_{ab} = 2\,\Delta_{ab}\left[-\left\langle a \left| \frac{e^2}{r_{ci}} \right| b \right\rangle + \left\langle ac \left| \frac{e^2}{r_{ij}} \right| bc \right\rangle \right]$$

$$+ \Delta_{ab}^2\left[-\left\langle c \left| \frac{e^2}{r_{ai}} \right| c \right\rangle - \left\langle c \left| \frac{e^2}{r_{bi}} \right| c \right\rangle + \frac{e^2}{R_{bc}} + \frac{e^2}{R_{ca}} \right] \tag{2.47}$$

Expanding the double-exchange integral (*cab*) similarly yields

$$
(cab) = -\Delta_{ab}\Delta_{bc}\left[\left\langle a\left|\frac{e^2}{r_{bi}}\right|c\right\rangle + \left\langle a\left|\frac{e^2}{r_{ai}}\right|c\right\rangle\right]
$$

$$
-\Delta_{ac}\Delta_{bc}\left[\left\langle b\left|\frac{e^2}{r_{ci}}\right|a\right\rangle + \left\langle b\left|\frac{e^2}{r_{bi}}\right|a\right\rangle\right]
$$

$$
-\Delta_{ab}\Delta_{ca}\left[\left\langle c\left|\frac{e^2}{r_{ai}}\right|b\right\rangle + \left\langle c\left|\frac{e^2}{r_{ci}}\right|b\right\rangle\right]
$$

$$
+\Delta_{ab}\left\langle ac\left|\frac{e^2}{r_{ij}}\right|cb\right\rangle + \Delta_{bc}\left\langle ab\left|\frac{e^2}{r_{ij}}\right|ca\right\rangle
$$

$$
+\Delta_{ca}\left\langle bc\left|\frac{e^2}{r_{ij}}\right|ab\right\rangle + \Delta_{ab}\Delta_{bc}\Delta_{ca}\sum_{\alpha>\beta}\sum\frac{e^2}{R_{\alpha\beta}} \tag{2.48}
$$

To determine Q_{ab} and α_{ab}, and so on, Porter and Karplus make use of the Heitler–London equation for the ground state and the triplet excited state, eq. 2.34, to obtain

$$
Q_{ab} = \tfrac{1}{2}[U_{+ab} + U_{-ab} + \Delta_{ab}^2(U_{-ab} - U_{+ab})]
$$

$$
\alpha_{ab} = \tfrac{1}{2}[U_{-ab} - U_{+ab} + \Delta_{ab}^2(U_{-ab} + U_{+ab})] \tag{2.49}
$$

For the analytical expressions of empirical energies U_{+ab} and U_{-ab}, they employ the Morse potential and Sato potential,

$$
U_- = {}^1D\{\exp[-2\beta(R - R^0)] - 2\exp[-\beta(R - R^0)]\} \tag{2.50}
$$

and

$$
U_+ = {}^3D\{\exp[-2\beta^1(R - R^0)] + 2\exp[-\beta^1(R - R^0)]\} \tag{2.51}
$$

The quantities ${}^1D$, ${}^3D$, β, β^1, and R^0 have been determined by fitting eqs. 2.50 and 2.51 to the calculated results of Kolos and Roothaan (1960) for the singlet and triplet states of the H_2 molecules; they are given in Table 2.1. The overlap integrals Δ_{ab}, and so on, are, however, calculated directly from the $1s$ orbital of the exponential screening constant,

$$
S = 1 + \chi\exp(-\lambda R) \tag{2.52}
$$

Table 2.1
Potential Energy Constants for $H + H_2$

${}^1D = 4.7466$ eV	$\lambda = 0.65$	$\chi = 0.60$
$R^0 = 1.40083$ au	${}^3D = 1.9668$ eV	$\delta = 1.12$
$\beta^1 = 1.000122$ au	$\beta = 1.04435$ au	$\varepsilon = -0.616$

with χ and λ are chosen to approximate the Wang results for H_2 (1928); for instance, they are evaluated analytically by

$$\Delta_{ab} = \left[1 + \frac{SR_{ab}}{a_0} + \frac{1}{3}\frac{S^2 R_{ab}^2}{a_0^2}\right]\exp\left(-\frac{SR_{ab}}{a_0}\right) \tag{2.53}$$

where S is defined in eq. 2.52.

For the evaluation of $\Delta\alpha_{ab}$, and so on, the unscaled $1s$ orbital has been used, and the two-center integrals in eq. 2.47 have been evaluated by using the elliptic coordinate system. The approximate expression for $\Delta\alpha_{ab}$ is given by

$$\Delta\alpha_{ab} = \delta\,\Delta_{ab}^2\left[\frac{e^2}{R_{ac}}\left(1 + \frac{R_{ac}}{a_0}\right)\exp\left(-\frac{2R_{ab}}{a_0}\right)\right.$$
$$\left. + \frac{e^2}{R_{bc}}\left(1 + \frac{R_{ac}}{a_0}\right)\exp\left(-\frac{2R_{bc}}{a_0}\right)\right] \tag{2.54}$$

A correction factor δ is introduced by Porter and Karplus in eq. 2.54 to adjust the empirical values for α_{ab} to match the calculated Heitler–London values. Over the significant range of internuclear distances, the average value of the correction factor δ is found to be 1.12. The double-exchange integral (cab) is written as

$$(cab) = \varepsilon\,\Delta_{ab}\,\Delta_{bc}\,\Delta_{ca} \tag{2.55}$$

where ε, given in Table 2.1, is assumed to be constant. Actually the ratio $(cab)/\Delta_{ab}\,\Delta_{bc}\,\Delta_{ca}$ has been shown to be nearly constant except for the smaller internuclear distances, which are of less importance for thermal reactions. Thus one is now ready to make the calculation of the potential energy surfaces of H_3 from eq. 2.25 or

$$U_- = -\frac{C_2 - C_2^2 - C_1 C_3)^{1/2}}{C_1} \tag{2.56}$$

where U_- is chosen because it corresponds to the lower energy state in the region of interest. Here C_1, C_2, and C_3 are defined as follows:

$$C_1 = 3(1 - \Delta_{ac}\,\Delta_{bc}\,\Delta_{ca})^2 - \tfrac{3}{2}[(\Delta_{ab}^2 - \Delta_{bc}^2)^2 + (\Delta_{bc}^2 - \Delta_{ca}^2)^2 + (\Delta_{ca}^2 - \Delta_{ab}^2)^2]$$
$$C_2 = -3[Q - (bca)](1 - \Delta_{ab}\,\Delta_{bc}\,\Delta_{ca}) + \tfrac{3}{2}\{(\Delta_{ab}^2 - \Delta_{bc}^2)[(ab) - (bc)]$$
$$+ (\Delta_{bc}^2 - \Delta_{ca}^2)[(bc) - (ca)] + (\Delta_{ca}^2 - \Delta_{ab}^2)[(ca) - (ab)]\}$$

and

$$C_3 = 3[Q - (bca)]^3 - \tfrac{3}{2}\{[(ab) - (bc)]^2 + [(bc) - (ca)]^2 + [(ca) - (ab)]^2\}$$

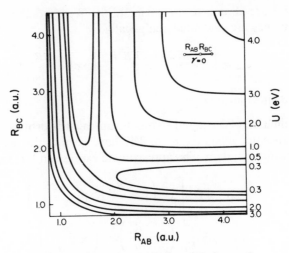

Fig. 2.1   Potential energy surface of H_3 for $\gamma = 0$.

Employing eq. 2.56, the potential energy surfaces of the reaction $H + H_2$, as shown in Figs. 2.1–2.4, are constructed by Porter and Karplus as a function of the distances R_{ab} and R_{ac} for a fixed bending angle γ. As shown for these figures, the contour maps for various γ's are qualitatively similar except for $\gamma = \frac{2}{3}\pi$. For $\gamma = 2\pi/3$, which corresponds to the equilateral triangle configuration along the line $R_{ab} = R_{bc}$, there exist cusps along the line $R_{ab} = R_{bc}$

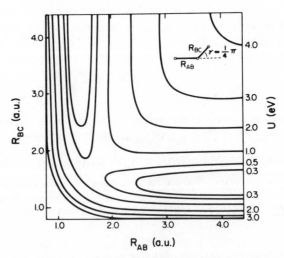

Fig. 2.2   Potential energy surface of H_3 for $\gamma = \pi/4$.

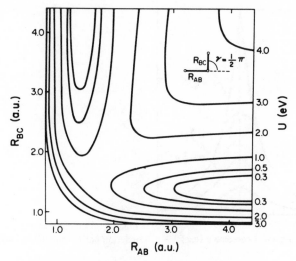

Fig. 2.3 Potential energy surface of H_3 for $\gamma = \pi/2$.

that arise from the degeneracy of the equilateral triangle and the resulting Jahn–Teller instability (Karplus, 1970). The significance of various bending angles is clearly illustrated in Fig. 2.5, from which one can see that the path of collinear collision requires the least energy, whereas for $\gamma = \frac{2}{3}\pi$, the potential energy curve along the path of minimum energy is considerably steeper and

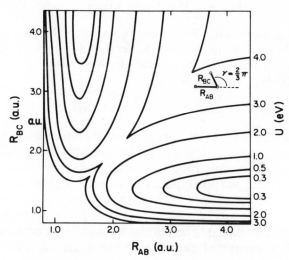

Fig. 2.4 Potential energy surface of H_3 for $\gamma = 2\pi/3$.

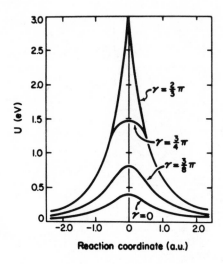

Fig. 2.5 Potential energy of H_3 along the reaction path of minimum energy.

has a cusp at its maximum corresponding to an equilateral triangle configuration. For γ angles other than $\frac{2}{3}\pi$, the potential energy curves are well behaved with maxima considerably lower than that of $\gamma = \frac{2}{3}\pi$.

Another procedure adopted by Cashion and Herschbach (1964) for the calculation of the potential energy surfaces for the $H + H_2$ reaction is based on the London equation originally employed by Eyring and Polanyi:

$$U_- = Q_{ab} + Q_{bc} + Q_{ca}$$
$$- \{\tfrac{1}{2}[(\alpha_{ab} - \alpha_{bc})^2 + (\alpha_{bc} - \alpha_{ca})^2 + (\alpha_{ca} - \alpha_{ab})^2]\}^{1/2} \qquad (2.57)$$

where Q_{ab}, α_{ab}, and so on, refer to diatomic integrals. To determine Q_{ab}, α_{ab}, and so on, they utilize the simplest Heitler–London equation (2.35) for the potential energy curves of the ground electronic state $^1\Sigma_g^+$, and the first repulsive state $^3\Sigma_u^+$, for example,

$$Q_{ab} = \tfrac{1}{2}(U_{-ab} + U_{+ab}), \qquad \alpha_{ab} = \tfrac{1}{2}(U_{-ab} - U_{+ab}) \qquad (2.58)$$

The accurate potential energy curves of the H_2 molecule are then used to evaluate U_{-ab} and U_{+ab}. This permits the potential energy surface for the $H + H_2$ reaction to be constructed over a wide range of interatomic distances without introducing further empirical adjustments. For the potential energy surface obtained by using eqs. 2.57 and 2.58 to be accurate to within 1 kcal/mole over the range $R_{ab} \geq 0.5$ Å, $R_{bc} \leq 2.5$ Å, the potential energy curves of diatomic molecules should be accurate to within 0.2 kcal/mole over the range $R = 0.5$–5 Å. The potential energy curve to such accuracy is available for the $^1\Sigma_g^+$ state of H_2 but not for the triplet state.

For the $^1\Sigma_g^+$ state of H_2 the potential energy curve has been derived from the spectroscopic data of Herzberg and Howe (1959) by means of the Rydberg–Klein–Rees (Tobias and Vanderslice, 1961) (RKR) method. The classical turning points thus obtained span the region of $R = 0.411–3.284$ Å. The perturbation calculation of Dalgarno and Lynn (1956), which extends from $R = 2.1$ Å to beyond 6 Å, agrees very closely with the RKR points in the region of their overlap (0.15 kcal/mole disparity at worst). In the calculation of the potential energy surface of $H + H_2$, Cashion and Herschbach use the potential energy curve of the $^1\Sigma_g^+$ state of H_2 obtained by seventh-order Lagrangian interpolation of the RKR points up to $R = 3.2$ Å and the points given by Dalgarno and Lynn outside this range. They point out that the variational calculation of Kolos and Roothaan (1960) agrees with the RKR results within 0.2 kcal/mole in the region of 0.5–1.3 Å, but is found to be too high by 2.4 kcal/mole at $R = 2.2$ Å.

For the potential energy curve of the $^3\Sigma_u^+$ state of H_2, several theoretical calculations have been carried out; these results are shown in Fig. 2.6 for comparison. The potential curves of Kolos and Roothaan (1960) and Hirschfelder and Linnett (1950) are obtained by a variational method. The perturbation calculation by Dalgarno and Lynn (1956) for $R \geq 2.1$ Å seems

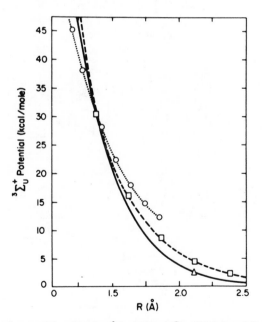

Fig. 2.6 Calculated potential curves for $^3\Sigma$ of H_2: $\cdots\bigcirc\cdots$ Kolos and Roothaan; $--\square--$ Hirschfelder and Linnett; $-\triangle-$ Dalgarno and Lynn.

to be the best in view of the excellent agreement their procedure gives for the $^1\Sigma_g^+$ state, as mentioned earlier. In a later paper Dalgarno (1961) gives an empirical equation for calculating the energy separation of the triplet and singlet states:

$$U_- - U_+ = 7696.8R^2 \exp(-3.730R) \tag{2.59}$$

where the energy is in kcal/mole and R is in angstroms. Cashion and Herschbach choose the perturbation results of Dalgarno given by eq. 2.59 for the potential curve of the $^3\Sigma_u^+$ state down to $R = 1.4$ Å, and then join it smoothly to the potential curve of Kolos and Roothaan for $R < 1.3$ Å.

Figure 2.7 shows a contour map of the potential energy surface obtained from eqs. 2.57 and 2.58 for a linear complex of the hydrogen exchange reaction by Cashion and Herschbach. They have compared the Coulomb fraction of the binding energy of H_2, ρ, defined by eq. 2.37, as a function of internuclear distance obtained by various methods. As shown in Fig. 2.8, these curves are fairly close together near the bond distance of the activated complex, $R = 0.96$ Å; however, neither Sato's $\rho(R)$ nor that calculated from the Heitler–London–Sugiura integrals (Hirschfelder and Linnett, 1941) approach the proper limit at large distances where $\rho(R)$ should approach unity. This type of variation of $\rho(R)$ seems to be required if the washbowl or basin is to be eliminated from the potential surfaces (see Fig. 2.7). It is noted that in the LEP method, the ρ value of 0.14 has often been used for each diatomic pair in the calculation of the potential energy surface with

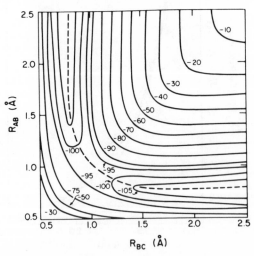

Fig. 2.7   Potential surface for $H + H_2$. (Energies are in kcal/mole relative to a zero of energy at infinity separation of three atoms.)

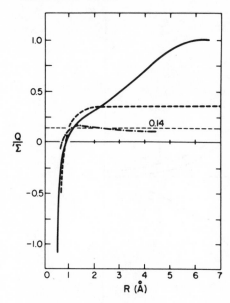

$\frac{Q}{\Sigma|}$

R (Å)

0.14

Fig. 2.8 Coulomb fraction of the binding energy of H_2 as a function of internuclear distance: ——— Cashion and Herschbach; ———Sato; —·— Heitler–London–Sugiura.

good results in some cases even though the London equation is applicable only to s electrons and ρ may become much larger for bonds involving higher orbitals (Fraga and Mulliken, 1960). Cashion and Herschbach have also compared their $^1\Sigma_g^+$ and $^3\Sigma_u^+$ potentials of H_2 with Sato's approximations (cf. eqs 2.41 and 2.42). The results are shown in Fig. 2.9; the potential curve of the triplet state given by Sato is too high at $R = 0.96$ Å by 14.6 cal/mole (a factor of 1.18) and at $R = 1.93$ Å by 6.0 kcal/mole (a factor of 2.35).

It is interesting that when the calculations of Porter and Karplus have been repeated with the use of the best potential curves of H_2 adopted by Cashion and Herschbach, the classical activation energy is reduced to 4.6 kcal compared with their original value of 9.1 kcal and with the 9.7 kcal of Cashion and Herschbach. Further comparisons of the hydrogen exchange reaction are given in the next section.

2.3 POTENTIAL ENERGY SURFACES FOR VARIOUS SYSTEMS

2.3.1 H + H₂ Reaction

The potential energy surfaces of the $H + H_2$ reaction discussed thus far are derived from the London equation (2.36). Although these methods seem to have some successful calculations on the $H + H_2$ system, we present here several purely quantum-mechanical treatments that have been employed in

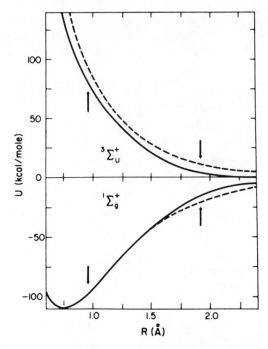

Fig. 2.9   Comparison of the "best" potential curves for H_2: ——— "best"; ––– Sato.

the construction of potential energy surfaces for this system. Results of these calculations are then compared with those obtained by the semiempirical methods.

Shavitt et al. (1968) employ the variational method to minimize the energy of the system by writing the wave function of the system ϕ as a linear combination of the real configuration Φ_r:

$$\phi = \sum C_r \Phi_r \tag{2.60}$$

It follows that

$$\sum_s (H_{rs} - U \, \delta_{rs}) C_s = 0, \qquad r = 1, 2, 3, \ldots \tag{2.61}$$

where $H_{rs} = \langle \Phi_r | \hat{H}_e | \Phi_s \rangle$ and δ_{rs} is the Kronecker delta. The configurational wave function Φ_r consists of sums of Slater determinants constructed from one-electron exponential functions centered on the nuclei. Calculations are carried out for optimized exponent basis sets consisting of 6 orbitals ($1s$, $1s'$ on each nucleus of H_3) and 15 orbitals of $1s$, $1s'$, $2P_x$, $2P_y$, and $2P_z$ on each nucleus of H_3. To achieve comparable accuracy for different nuclear geometries (i.e., symmetric and asymmetric, linear and nonlinear), all possible determinants that can be formed from a given basis set are included in the configuration interaction wave function ϕ (e.g., for the 15 orbital set, there are 200 configurations for the linear symmetric case and 402 for the linear asymmetric case). The basis orbitals used to form the configurational determinants Φ_r are not the atomic orbitals, but arbitrary symmetry orbitals defined as linear combinations of the atomic orbitals. The only restriction on the symmetry orbitals is that they belong to the proper irreducible representations of the symmetry group corresponding to the nuclear geometry. A Schmidt orthogonalization routine is used to convert the symmetry orbitals to an orthogonal set.

To determine the barrier height of the H_3 potential energy surface, the corresponding results for the H_2 molecules are required. Shavitt et al. carry out a full set of calculations for H_2 with a 4 orbital basis set ($1s$, $1s'$ on each center of H_2) and a 10 orbital basis set ($1s$, $1s'$, $2P_x$, $2P_y$, and $2P_z$ on each center of H_2). They include all configurations in the optimized configuration interaction calculations for H_2:6 configurations for the ($1s$, $1s'$) set and 16 configurations for the ($1s$, $1s'$, $2p$) set. Theoretical results for the equilibrium distance, force constant, and dissociation energy are given in Table 2.2 for comparison with experimental values. Table 2.2 shows that although the equilibrium distance and force constant obtained with the 10 orbital basis set are satisfactory, the calculated dissociation energy is still 3.0 kcal/mole above the experimental value.

The potential energy surfaces of H_3 have been obtained from the 6 orbital calculation and from the 15 orbital calculation, and the contour maps of the potential energy surface in the saddle point region of linear H_3 are given in Fig. 2.10. In Fig. 2.11 are shown the locations of the minimum energy paths as a function of apex angle for comparison. Shavitt et al. have also made some calculations for the linear case to test whether the correct $H_2 + H$ limit

Table 2.2
Results of Optimized CI Calculations for H_2

Basis set	R_e (au)	$U(R_e)$ (au)	D_e (kcal)	K (au)	Optimized exponents (in order of basis set)
$1s$	1.42	-1.1479	93	0.43	1.193
$1s$, $1s'$	1.4148	-1.1528	96	0.35	1.122, 1.386
$1s$, $1s'$, $2p$	1.4018	-1.16959	106	0.36	1.078, 1.426, 1.800
$1s$, $1s'$, $2s$, $2p$	1.4013	-1.16696	104	—	0.965, 1.43, 1.16, 1.87(σ), 1.71(π)
Experiment	1.4008	-1.17445	109	0.365	

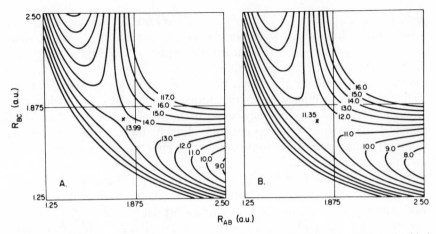

Fig. 2.10   Potential energy surface in the saddle point region for linear H_3: (*a*) from 6-orbital (*b*) from 15-orbital calculation.

is approached as one H atom is removed to infinity; with the R_{ab} distance equal to the H_2 theoretical value and the other atom at distance greater than 11 au, full agreement with the separately calculated H_2 result is obtained. In these calculations they also find a possible van der Waals minimum occurs at a distance of about 6 au. Since the depth of the well is very small (approximately, 4×10^{-4} au), more effort in terms of exponent optimizations and higher integral accuracy would be required to obtain reliable results.

For the $H + H_2$ system another calculation scheme is due to Conroy and Bruner (1967). To describe their procedure, one has to consider first the one-electron problem (Conroy, 1970). In this case the Schrödinger equation

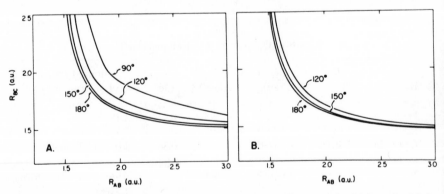

Fig. 2.11   Position of minimum-energy path as a function of apex angle: (*a*) from 6-orbital calculation; (*b*) from 15-orbital calculation.

in atomic units for a single electron moving in the field of a number of fixed nuclei with charges $Z_a, Z_b, \ldots$, is given by

$$\hat{H}_e \phi = -\tfrac{1}{2} \nabla^2 \phi + V\phi = U\phi \qquad (2.62)$$

where

$$V = -\sum_\alpha \frac{Z_\alpha}{r_\alpha} \qquad (2.63)$$

The potential energy V reaches $-\infty$ whenever the coordinates of the electron coincide with those of any nucleus, and since U is always finite it follows that for any exact solution ϕ, the quantity $\nabla^2 \phi$ must become infinite at the nuclei. To construct an eigenfunction with the desired property, a new variable d is introduced,

$$\gamma d = \sum_\alpha Z_\alpha r_\alpha, \qquad \gamma = \sum_\alpha Z_\alpha \qquad (2.64)$$

It can easily be shown that

$$\gamma \nabla^2 d = 2 \sum \frac{Z_\alpha}{r_\alpha} = -2V \qquad (2.65)$$

Thus a trial function containing d,

$$F_1 = \exp(-\gamma d) \qquad (2.66)$$

will give a complete cancellation of V including the accompanying singularities at the nuclei from eq. 2.62,

$$-\tfrac{1}{2} \nabla^2 F_1 = -[V + \tfrac{1}{2}\gamma^2 (\nabla d)^2] F_1$$

$$F_1^{-1} \hat{H}_e F_1 = -\tfrac{1}{2}\gamma^2 (\nabla d)^2 \qquad (2.67)$$

The quantity on the right side of the second equation in eq. 2.67 lies in a numerical value between $-\tfrac{1}{2}\gamma^2$ and zero; it obviously never reaches infinity, but neither is it everywhere equal to the eigenvalue U. Thus the trial function F_1 is only a crude approximation. Choosing the origin of the spherical coordinates (r, θ, ϕ) at the center of nuclear charges, that is, $\sum_\alpha Z_\alpha \mathbf{r}_\alpha = 0$, it can be shown that $\lim_{r \to \infty} (d/r) = 1$, from which it follows that

$$\lim_{r \to \infty} \gamma^2 (\nabla d)^2 = \gamma^2 (\nabla r)^2 = \gamma^2 \qquad (2.68)$$

The asymptotic behavior at large r is therefore not correctly given by F_1 in eq. 2.66, since usually $-\tfrac{1}{2}\gamma^2 < U$. This can be remedied by using

$$F = \sigma^{-\alpha} \exp[-\gamma d + (\gamma - \varepsilon)\sigma^{-1}] \qquad (2.69)$$

where $\sigma = (r^2 + s^2)^{1/2}$, $\varepsilon = (-2U)^{1/2}$, and $\alpha = \gamma/\varepsilon - 1$. The term s is a parameter that is chosen as an average radius, specifically as that value of r

at which half of the integrated radial density is accumulated. Operationally this procedure for determining s requires the cumbersome three-dimensional integrations. It is found that the utility of F is not critically dependent on s and that the deficiency due to the choice of some nonoptimal value is rather quickly compensated by the radial series expansion. Thus Conroy (1967) introduces

$$F = \sigma_2^{-\alpha} \exp[-\gamma d + (\gamma + \varepsilon)\sigma_1^{-1}] \qquad (2.70)$$

where $\sigma_1 = (r^2 + d_o^2)^{-1/2}$, $\sigma_2 = (r^2 + d_o^2 + \varepsilon^{-2})^{-1/2}$, and d_o is the value of d at the origin. To add flexibility the final form of the one-electron molecular wave function is chosen to be a series expansion

$$\phi = \sum_n C_n X_n = \sum_n C_n A_n(r, \theta, \phi) F(r, \theta, \phi) \qquad (2.71)$$

where $F(r, \theta, \phi)$ is defined by eq. 2.70 and the functions $A_n(r, \theta, \phi)$ are members of a single-center basis set

$$A_n(r, \theta, \phi) = r^j \sigma_2^{i-1} Y_j^k(\theta, \phi) L_{i+j}^{2j+1}(q) \qquad (2.72)$$

with $q = 2\varepsilon(\sigma_1^{-1} - d_o)$. The terms $Y_j^k(\theta, \phi)$ and $L_{i+j}^{2j+1}(q)$ represent the spherical harmonics and associated Laguerre polynomials, respectively.

For the two-electron problem, Conroy writes the wave function as a product of two factors, an electron correlation factor ϕ_c and a molecular shape factor ϕ_s

$$\phi = \phi_c \phi_s \qquad (2.73)$$

The functional form of the shape factor is restricted to a linear superposition of configurations; each configuration is taken as a symmetrized or antisymmetrized product of one-electron terms

$$\phi_s = \sum_{nm} [X_n(1)X_m(2) \pm X_n(2)X_m(1)]C_{nm} \qquad (2.74)$$

where the positive sign and the negative sign are used for the singlet and triplet spin states, respectively. The electron correlation factor ϕ_c consists of two parts,

$$\phi_c = a(r_{12})b(r_{12}, \sigma_1, \sigma_2) \qquad (2.75)$$

The first part, $a(r_{12})$, is introduced to improve convergence by explicitly taking care of the singularity $1/r_{12}$ in the Hamiltonian and is defined by

$$a(r_{12}) = \sum_{k=0}^{\infty} \frac{1}{k!(k+1)!} \left(\frac{r_{12}}{t}\right)^k \qquad (2.76)$$

where $t = 1 - (1/r)(-2U/N)^{1/2}$, N being the number of electrons; $b(r_{12}, \sigma_\perp, \sigma_2)$ is introduced to provide more flexibility (mostly for angular correlation),

$$b(r_{12}, \sigma_1, \sigma_2) = \sum_{ijk} C_{ijk} W_{12}^i \lambda_{12}^j \mu_{12}^k \qquad (2.77)$$

with $W_{12} = (r_{12}^2 + t^2)^{1/2}$, $\lambda_{12} = \sigma_1^{-1}(1) + \sigma_1^{-1}(2)$, and $\mu_{12} = \sigma_1^{-1}(1) - \sigma_1^{-1}(2)$. This procedure can be expanded to more than two electrons by including all singlet and triplet pair correlations and proper antisymmetrization (Conroy, 1967).

To find the optimal wave function and energy, Conroy proposes to minimize the energy variance

$$W^2 = \frac{\int (\hat{H}_e \phi - U\phi)^2 \, d\tau}{\int \phi^2 \, d\tau} \qquad (2.78)$$

For example, for ϕ given by eq. 2.71, W^2 is given by

$$W^2 = \frac{\sum_n \sum_m (H_{nm} - U V_{nm} + U^2 S_{nm}) C_n C_m}{\sum_n \sum_m C_n C_m S_{nm}} \qquad (2.79)$$

where $H_{mm} = \langle \hat{H}_e X_n | \hat{H}_e X_m \rangle$, $V_{nm} = \langle X_n | \hat{H}_e | X_m \rangle + \langle X_m | \hat{H}_e | X_n \rangle$, and $S_{nm} = \langle X_n | X_m \rangle$. The best possible ϕ is the one that gives the lowest numerical value of W^2. Minimization of eq. 2.79 gives a secular equation of the type

$$|H_{nm} - U V_{nm} + (U^2 - W^2) S_{nm}| = 0 \qquad (2.80)$$

The main advantage of this procedure is the reduced sensitivity of the energy obtained with respect to errors in the matrix elements, and this energy variance calculation permits extrapolation to a more accurate approximation of the true energy than can be obtained solely from a knowledge of the ordinary upper and lower bounds. It is well known that the energy E obtained from the minimization of

$$E = \frac{\langle \phi | \hat{H}_e | \phi \rangle}{\langle \phi | \phi \rangle} \qquad (2.81)$$

gives the upper bound to the true lowest energy U_0. On the other hand, Temple (1928) derives a formula allowing the simultaneous calculation of a lower bound to U_0,

$$U_0 \geq \lambda_T = E - \frac{W^2}{U_1 - E} \qquad (2.82)$$

where U_1 is the energy of the first excited state. The quantity λ_T is the Temple lower bound to the energy. The range of uncertainty in U_0, $E - \lambda_T = W^2(U_1 - E)^{-1}$ can be made to vanish if the energy variance W^2 vanishes.

Thus by plotting E versus W obtained from the use of a number of different wave functions and extrapolating to zero W, one obtains the best energy.

The method described above has been applied to H_3 by Conroy and Bruner (1967). The energy contour maps for linear and isosceles H_3 arrangements are given in Figs. 2.12 and 2.13. In agreement with other treatments the surface of linear configuration lies lower than that of the triangular ones. The saddle point is at $R_{ab} = R_{bc} = 1.76$ au with an activation energy of 7.74 kcal/mole above $H + H_2$.

The calculated results of the potential energy surfaces of H_3 at the saddle point region obtained by various theoretical approaches are summarized in Table 2.3 for comparison. To permit a valid comparison Shavitt et al. (1968) have repeated the semiempirical calculations of Cashion and Herschbach and of Eyring and Polanyi by using the potential energy curves of $^1\Sigma_g^+$ and $^3\Sigma_u^+$ for H_2 adopted by Porter and Karplus. The energy profile along the minimum energy path for the linear H_3 obtained from various treatments is shown in Fig. 2.14 and the location of the minimum energy path for various computed linear H_3 surfaces is given in Fig. 2.15. In these figures and Table 2.3, "E" refers to the original version of Eyring and Polanyi, "ES," the modification of the Sato type by Cashion and Herschbach, "PK," the semi-

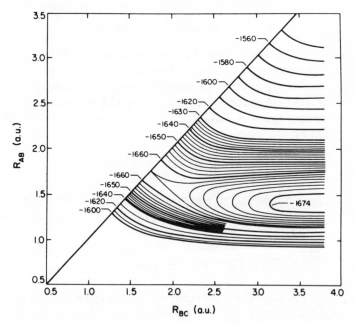

Fig. 2.12 Potential energy surface for linear H_3.

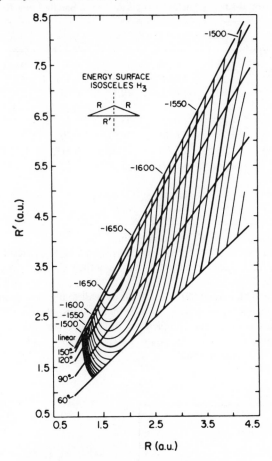

Fig. 2.13 Potential energy surface for isosceles H_3.

empirical calculation by Porter and Karplus, "CB," the a priori treatment of Conroy and Bruner, and "I" and "II," the theoretical calculation by Shavitt et al. for the 6 orbital basis set and 15 orbital basis set, respectively. Figure 2.14 shows that the original LEP method with $\rho = 0.14$ gives a curve having a very broad and pronounced well, and other semiempirical calculations and a priori treatments show no minimum. Also, as shown in Fig. 2.15, all the semiempirical curves have some "kinks" in the saddle point region in contrast to the a priori curves of Shavitt et al. which are completely smooth; the curve of Conroy and Bruner gives some indication of a slight kink corresponding to that of Cashion and Herschbach. Curves I and II are almost parallel but they differ from all other curves in approaching the H_2 equilibrium

Table 2.3
Potential-Energy Surface Properties in the Saddle Point Region

Surface	R_{sp} (au)	U_{sp} (au)[a]	ε_0^+ (kcal)[b]	K_{11} (au)	K_{22} (au)	K_{33} (au)
E[c]	1.614	-1.6565	11.3	0.323	0.041	$(+)0.136$
ES	1.781	-1.6497	15.6	0.331	0.028	-0.137
PK	1.701	-1.6600	9.1	0.36	0.024	-0.124
BS($1s, 1s'$)[d]	1.779	-1.6119	15.4	0.29	0.023	-0.047
$1s$[e]	1.883	-1.6106	23.4	—	—	—
I($1s, 1s'$)	1.788	-1.6305	14.0	0.30	—	—
II($1s, 1s', 2p$)	1.764	-1.6521	11.0	0.31	0.024	-0.061
CB	1.76	-1.6621	7.7	0.32	0.026	~0.00
One-center expansion[f]	~1.8	-1.6358	—	—	—	—
Gaussian set[g]	~1.8	-1.6493	13.5	—	—	—

[a] U_{sp} is the total energy at the saddle point.
[b] ε_0^+ is the barrier height referred to the corresponding H_2, H result (see text).
[c] All results for surface E refer to the bottom of the well ($R_{AB} = R_{BC}$) and not its rim.
[d] The Boys and Shavitt (1959) calculations used all configurations for the ($1s, 1s'$) set with exponents $\alpha_{1s} = 1.0\lambda$, $\alpha_{1s'} = 1.5\lambda$ for all atoms, where λ is a scale factor.
[e] This calculation used the complete set of four configurations and optimized exponents $\alpha_{1sA} = \alpha_{1sC} = 1.058$, $\alpha_{1sB} = 1.202$ (unpublished calculation).
[f] Edmiston and Krauss (1965).
[g] Hayes and Parr (1967).

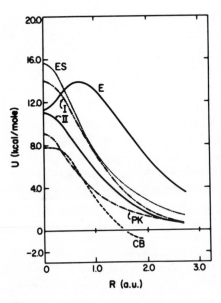

Fig. 2.14 Energy profile along the minimum-energy path for linear H_3 from various computations: E, Eyring; ES, Eyring–Sato; PK, Porter–Karplus; CB, Conroy–Bruner.

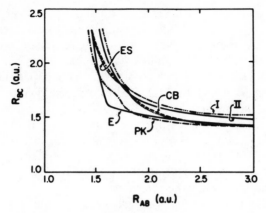

Fig. 2. 15 Location of the minimum-energy path for linear H_3 from various computations.

distance much more slowly as the third atom moves away. Curves II and CB are in excellent agreement on the saddle point region, but diverge rapidly in the intermediate region.

The potential energy of H_3 as a function of internuclear distance for the linear-symmetric geometry is plotted in Fig. 2.16 for the surface II, CB, and PK. The curves represent the potential function for the symmetric stretching vibration at the saddle point. Aside from the difference in the height and position of the saddle point, all three curves in Fig. 2.16 are similar and curves CB and II are almost parallel to each other.

For the semiempirical surfaces and the CB surface the classical activation energies, as given in Table 2.3, are computed relative to the Kolos–Roothaan energy for the H_2 molecule, whereas in other cases (BS, 1s basis set, I and II), the values of classical activation energies are relative to a H_2 energy computed

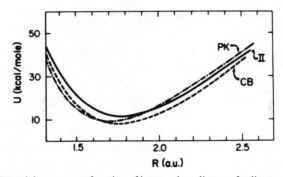

Fig. 2.16 Potential energy as a function of internuclear distance for linear symmetric H_3.

with a wave function constructed from a comparable basis set. Table 2.3 shows that the calculation II and the Gaussian calculation (Edmiston and Krauss, 1965) are about 1.9 and 4.4 kcal/mole, respectively, above the PK result whereas the CB value is about 1.4 kcal/mole below. The calculation II value is certainly somewhat too high but it does not appear likely to be too high by as much as the 3.3 kcal required to bring it down to the CB result. This is because the method used by Conroy and Bruner includes an extrapolation procedure that does not necessarily yield an upper bound. In addition, in Fig. 2.14, the CB energy curve drops to more than 1 kcal/mole below the H_2 + H limit. This somewhat surprising result has been attributed by Conroy and Bruner to the fact that their calculation at any point could be 0.6–1.2 kcal above or below the true value.

Table 2.3 also indicates that the calculations II and CB for the internuclear distance at the saddle point are in close agreement ($R_0 = 1.765$ au) while the PK value ($R_0 = 1.701$ au) is significantly smaller. The symmetric stretching force constant K_{11} and the bending force constant K_{22} vary only by relatively small amounts among all the calculations. The third force constant K_{33} changes more markedly for the different surfaces; in particular, it is positive for surface E evaluated at the center of the well. In general, K_{33} is significantly larger for the semiempirical surfaces ES and PK than for the a priori surfaces. Thus considerable differences in the tunneling would be predicted from the simple models, which fit the barrier to the force constant K_{33}.

From comparisons with the earlier a priori studies of the H_3 potential surface (Laidler, 1969), it is clear that a significant improvement in accuracy has been achieved by the 15 orbital multiconfiguration calculations by Shavitt et al. and the CB calculation. It is very likely that the general features of the H_3 surfaces are essentially correct, and that in the near future more accurate theoretical calculations can be done to reduce the uncertainty in the energy of potential surfaces to the acceptable value of 0.3 kcal/mole.

2.3.2   K + NaCl Reaction

The calculation of the potential energy surface of the reaction K + NaCl carried out by Roach and Child (1968) is based on a model in which a single valence electron moves in the fields of Na^+, K^+, and Cl^- with the ions retaining their essential structural identity at all configurations of interest. Their approach is suggested by Rittner's successful calculation of the potential curves of isolated alkali halide molecules. Rittner's (1951) calculation is based on the assumption of classical interactions between two mutually polarized charge spheres, together with supplementary van der Waal's forces and close range repulsions. If μ_+ and μ_- are the point dipoles induced at each ion, the potential energy of an isolated alkali halide molecule in atomic units can be written as

$$U(R) = -\frac{1}{R} - \frac{\mu_+ + \mu_-}{R^2} - \frac{2\mu_+\mu_-}{R^3} + \frac{\mu_+^2}{2\alpha_+} + \frac{\mu_-^2}{2\alpha_-} - \frac{C}{R^6} + Ae^{-R/\rho} \quad (2.83)$$

where the successive terms represent the charge-charge, charge-dipole, dipole-dipole interactions, the quasi-elastic energies of dipole formation, the van der Waals energy, and the close range overlap repulsion. The terms α_+ and α_- are the ion polarizabilities. Thus the Hamiltonian of the reaction K + NaCl can be written in the form

$$\hat{H}_e = \hat{H}_e^0 + V(\text{core})$$
$$\hat{H}_e^0 = -\tfrac{1}{2}\nabla^2 + V(\text{Na}^+) + V(\text{K}^+) + V(\text{Cl}^-) \quad (2.84)$$

The term $\hat{H}_e^0$ includes the potential energy of its interaction with the isolated ions, all polarization terms being neglected. In eq. 2.84 $V(\text{core})$ is composed mainly of the interactions between the ion cores, a generalization of eq. 2.83, but also includes the contribution of the valence electron to the polarization energy, which is treated as a first-order correction to the electronic energy U_n^0 obtained from the Schrödinger equation,

$$\hat{H}_e^0 \phi_n = U_n^0 \phi_n \quad (2.85)$$

The valence electron that occupies a K 4s orbital in the ground state of the reactants is transferred in the course of reaction to a Na 3s orbital. Thus Roach and Child expand ϕ_n at all configurations as a linear combination of Na and K valence atomic orbitals. The inclusion of the four p orbitals directed in the plane of the nuclei in addition to Na 3s and K 4s orbitals is found to contribute substantially to the stability of the ground state potential surface and is necessary to obtain the correct form of long-range dipole-induced dipole interactions. Strictly speaking, ϕ_n must be orthogonal to all the occupied orbitals of Na$^+$, K$^+$, and Cl$^-$; however, the Na valence orbitals, for instance, although orthogonal to the occupied Na$^+$ core orbitals, have small nonzero overlaps with the K$^+$ and Cl$^-$ core functions. To avoid complete and rigorous orthogonalization of the atomic orbitals to all foreign core orbitals, the problem is treated by using suitably modified expressions for $V(\text{Na}^+)$, $V(\text{K}^+)$, and $V(\text{Cl}^-)$ such that their interaction with the undisturbed atomic functions reproduces the energetic effects of orthogonalization (Austin et al, 1962). The expressions for Na$^+$ and K$^+$ adopted by Roach and Child are

$$V(M^+) = 0 \qquad r < \sigma$$
$$= \frac{-1}{r} \qquad r > \sigma \quad (2.86)$$

where σ is the gaseous ionic radius proposed by Rittner, and for Cl^-

$$V(Cl^-) = \frac{1}{r} \tag{2.87}$$

In calculating the matrix elements of $\hat{H}_e$, ionization potentials I are used to estimate all one-center and kinetic energy integrals; for example,

$$\langle X_{Na,s} | -\tfrac{1}{2}\nabla^2 + V(Na^+) | X_{Na,s} \rangle$$
$$= I(Na, s); \langle X_{Na,s} | -\tfrac{1}{2}\nabla^2 + \tfrac{1}{2}V(Na^+) + \tfrac{1}{2}V(K^+) | X_{K,p\sigma} \rangle$$
$$= \tfrac{1}{2}[I(Na, s) + I(K, P)]\langle X_{Na,s} | X_{K,p\sigma} \rangle \tag{2.88}$$

and so on. The two-center integrals are calculated analytically in elliptic coordinates and the nine three-center integrals of the type $\langle X_{Na} | V(Cl^-) | X_K \rangle$ are evaluated numerically. For this purpose, single-term Slater-type orbitals are chosen (Clementi, 1963, 1964):

$$X_{Na,s} = N_{Na,s} r^2 e^{-0.85r}, \qquad X_{K,s} = N_{K,s} r^3 e^{-0.736r} \tag{2.89}$$

It is assumed that the radial parts of the corresponding p orbitals are given by the same expressions. The electronic energy also contains the contributions from the term $V(core)$. These interactions include the charge-charge electrostatic energies, the van der Waals' attraction for each pair of ions, the close-range overlap repulsions between the ions, the polarization of ion cores, and so on (cf. eq. 2.83). The calculation of terms involving the electron is simplified to the interaction of point charges and dipoles.

Using the approach described earlier, Roach and Child carry out the test calculations of potential curves for the ions Na_2^+ and K_2^+; results of these calculations are found to be in very satisfactory agreement with the experimental binding energies, bond lengths, and spectroscopic constants. For the reaction between K and NaCl, the calculated reaction exothermicity is 6 kcal/mole compared with an experimental value of 4 kcal/mole (Brewer and Brackett, 1961). The ground state potential energy surface is found to be of the attractive type and there is no activation barrier. In addition, the calculations of Roach and Child also predict a potential well with maximum depth, $D_o^0 = 13.5$ kcal/mole, in a triangular configuration at $R_{NaCl} = 1.7$ Å, $R_{KCl} = 2.9$ Å, and $<NaClK = 75°$. The contour maps of the reaction NaCl + K as a function of the angle $\theta = <NaClK$ are shown in Fig. 2.17. They conclude that an adiabatic model adequately describes reactive collisions at thermal energies but that a nonadiabatic mechanism in producing electronically excited products could become important for highly energetic collisions.

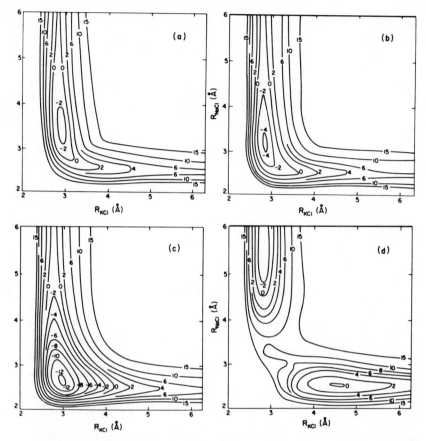

Fig. 2.17 The ground state potential surface: (a) 180°, (b) 135°, (c) 90°, (d) 45°.

2.3.3 Formaldehyde Surface

The photochemistry of formaldehyde has been widely studied as a prototype carbonyl system. By the $n \to \pi^*$ excitation (2800–3550 Å), two elementary processes result

$$H_2CO \longrightarrow H + HCO \tag{I}$$

$$H_2CO \longrightarrow H_2 + CO \tag{II}$$

the radical and molecular mechanisms, respectively. It was found that I predominates at shorter wavelengths and II at longer wavelengths. Recently there has been a concerted effort to understand the detailed mechanisms of the H_2CO $n \to \pi^*$ photochemistry. Yeung and Moore(1973) measured the fluorescence from individual vibronic levels of the $n \to \pi^*$ singlet state (S_1) and found that nonradiative transitions back to the ground electronic state (S_0) predominate over intersystem crossing to the $n \to \pi^*$ triplet (T_1). However, Brand and co-workers (Brand and Stevens, 1973, Brand and Liu, 1974) have observed singlet-triplet perturbations in the S_1 state believed to be caused by intersystem crossing $(S_1 \to T_1)$. Miller and Lee (1974) and Luntz and Maxson (1974) have found the $T_1 \to S_0$ intersystem crossing to be an efficient process. Miller and Lee (1975) have also determined that the out-of-plane bending mode (v_4') is the promoting mode for S_1 nonradiative transitions (to an unspecified final state) as predicted by the model calculations of Yeung and Moore. The evidence is still inconclusive as to the fate of the S_1 molecules.

Hayes and Morokuma (1972) have reported *ab initio* Self Consistent Field (SCF) MO-CI calculations of the three lowest potential energy surfaces of the unimolecular process (I). They find that nonradiative transitions to either T_1 or S_0 must occur during the decomposition, but cannot rule out either of the two pathways. For the molecular mechanism, recent *ab initio* potential surface calculations by Jaffe et al. (1974) have not found any low-energy reaction pathway. The S_1 and T_1 states of H_2CO correlate to energetically inaccessible product states implying that dissociation must occur through the ground electronic state. However, the minimum energy requirement for reaction on the ground state potential surface was found to be 4.30 eV. This is considerably in excess of the experimental threshold of 3.66 eV (McQuigg and Calvert, 1969). As the reaction is nearly thermoneutral ($\Delta H = 0.3$ eV), the existence of a large barrier to reaction would mean that the product must be formed with considerable translational or internal energy. Klimek and Berry (1973) have measured high-frequency laser emission from the photolysis of HFCO; the photochemistry of this system is expected to be similar to that of H_2CO. They find that less than 10% of the available energy ends up in high-frequency vibration. Similarly, Moore and co-workers (private communication) have ascertained that the CO molecules formed by process II in formaldehyde are mostly in the ground vibrational level. Thus the details of the molecular dissociation are not well understood. Questions on the height of the barrier to molecular dissociation and the total resolution of the energy distribution among the products remain unanswered.

Alternative mechanisms have been suggested by Moore and Luntz (private communication). They suggest that many of the previous experiments have not been performed at sufficiently low pressures to ensure unimolecularity. Luntz has estimated that the collision-induced electronic transitions $(S_1 \to T_1$ and $S_1 \to S_0)$ and dissociation predominate above pressures of 100 mtorr. Moore has observed a sizable time delay between the disappearance of S_1 and the appearance of the product CO molecules, suggesting the presence of an intermediate species and collisional phenomena. More experiments are in progress that may shed more light on this process.

Recently Jaffe and Morokuma (1976) have carried out the *ab initio* (MCSCF) calculations for the unimolecular process II; the ground state potential energy surface for the dissociation of formaldehyde is calculated with an extended (4–31G) basis set. Their emphasis has been on locating the minimum energy barrier between H_2CO and $H_2 + CO$. In addition the force constants for the S_0 equilibrium geometry and transition state have been calculated and the normal vibrational modes determined. At the equilibrium H_2CO geometry, the Hartree–Fock configuration is

$$^1A_1 = (1a_1)^2(2a_1)^2(3a_1)^3(4a_1)^2(1b_2)^2(5a_1)^2(1b_1)^2(2b_2)^2$$

Eighteen configurations were included in the MCSCF wave function; these were selected to describe properly the H_2CO, $H_2 + CO$ and $H + HCO$ limits. Calculations of the ground state electronic energy were run at numerous geometries to determine the C_{2v} equilibrium geometry and the C_s saddle point geometry for the molecular dissociation pathway.

Jaffe and Morokuma (1976) have found the height of the barrier for process I to be 4.55 eV. In addition, they have calculated several force constants for the equilibrium geometry by numerical second differentiation of the MCSCF energy with respect to small displacements in the positions of the atomic centers. Their results are in good agreement with experimental data (Takagi and Oka, 1963) and the more accurate theoretical values reported by Pulay and Meyer (1974). The geometry of the saddle point is planar with the C_s symmetry. If we let R_{CO} represent the C—O distance, R_{HH} the H—H distance, D^1 the distance between the C atom and center of H—H, ϕ the angle between R_{CO}, and D^1 and θ the angle between R_{CO} and R_{HH}, then the geometry of the saddle point for process I can be described by

$$D^1 = 1.293 \text{ Å}, \quad R_{HH} = 1.336 \text{ Å}, \quad R_{CO} = 1.176 \text{ Å}$$
$$\phi = 49°, \quad \theta = 108°$$

Jaffe and Morokuma (1976) have also determined a full set of force constants for the saddle point geometry by numerical differentiation of the calculated energies. These force constants follow:

$$K_{D'D'} = 0.213 \qquad K_{R_{HH}R_{HH}} = -0.116$$
$$K_{R_{CO}R_{CO}} = 14.73 \qquad K_{\phi\phi} = 3.106, \quad K_{\theta\theta} = 2.529$$
$$K_{D'R_{HH}} = 1.802 \qquad K_{D'R_{CO}} = 1.337$$
$$K_{D'\phi} = -3.397 \qquad K_{D'\theta} = 2.015$$
$$K_{R_{CO}R_{HH}} = -0.116 \qquad K\phi_{R_{HH}} = 0.300, \quad K\phi_{R_{CO}} = 0.500$$
$$K_{\theta R_{HH}} = 0.383 \qquad K_{\theta R_{CO}} = -0.624, \quad K_{\theta\phi} = -2.461$$

where stretching constants are in mdyn/Å, stretching-bending force constants are in mdyn, and bending force constants are in mdyn Å. The force constant for the out-of-plane angle (angle of rotation of R_{CO} around the D' axis) has been calculated to be 0.179 mdyn Å.

2.3.4 Other Systems

In the previous sections we have discussed in some detail the calculation of the potential surfaces of $H + H_2$, $K + NaCl$, and formaldehyde. Because of the rapid progress in recent years, it is clearly impractical to describe or review all the surfaces published in the literature. The surfaces of several other systems of interest are, however, briefly discussed in the following presentation.

Because of its importance in modeling the performance of the HF chemical laser, there has been considerable theoretical (Thompson, 1972; Wilkins, 1975) and experimental (Kwok and Wilkins, 1974; Bott and Heidner, 1977) interest in the deactivation of vibrationally excited HF by H atoms. In addition to V-T and V-R inelastic collisions, the chemical reactions

$$H^1 + HF(v) \longrightarrow H^1F(v^1) + H \qquad (1)$$

and

$$H + HF(v) \longrightarrow H_2 + F \qquad (2)$$

may be important, depending on the level of vibrational excitation and its role in overcoming the reaction barriers. The reverse of reaction 2,

$$F + H_2 \longrightarrow HF(v' < 3) + H \qquad (3)$$

is one of the most thoroughly studied reactions. The fact that reactions 1–3 possess a common potential energy surface led dynamicists to employ for reaction 1 the semiempirical surfaces that had been adjusted to agree with the known experimental data for reaction 3 (Thompson, 1972; Wilkins, 1975). However, recent *ab initio* calculations on the collinear reaction 1 have shown that the energy barrier for these semiempirical surfaces are all too small (Bender et al., 1975; Botschwina and Meyer, 1977); they predict barriers ranging from 42.4 to 49.0 kcal/mole, whereas the semiempirical surfaces possess barriers from -5.2 to 28.6 kcal/mole. The *ab initio* results are supported by the recent experiments of Heidner and Bott (Bott and Heidner, 1977; Bott, 1976), which have set an upper bound on the rate of exchange reaction. In particular, the discharge experiments (Bott, 1976) appear to show that the barrier for exchange reaction 1 is comparable to or greater than that for abstraction reaction 2, that is, 33.0 kcal/mole. Extensive *ab initio* calculations on reactions 2 and 3 have shown that they proceed via a collinear transition state (O'Neil et al., 1972; Jaffe et al., 1975; Lester and Rebentrost, 1976). For reaction 1 the published calculations (Bender et al., 1975; Botschwina and Meyer, 1977) have considered only collinear geometries. The semiempirical surfaces assume a collinear transition state geometry for all these reactions. Wadt and Winter (1977) have recently performed the first calculations on the $HF + H^1 \rightarrow H^1F + H$ reaction at nonlinear geometries with various multiconfiguration Self Consistent Field Configuration Interaction (SCF CI) wave functions. Their calculations demonstrate the importance of diffuse functions on the fluorine for describing nonlinear geometries, and the energy barrier for the exchange reaction is calculated to be ~ 45 kcal/mole, which is comparable to the values obtained in previous *ab initio* calculations on the collinear reaction surface. More importantly, their calculations show that the saddle point region is very flat, the barrier changing by only 1–2 kcal/mole between collinear (180°) and perpendicular (90°) geometries; the optimum angle for the transition state geometry is calculated to be 106°.

Recent molecular beam experiments of elastic and reactive scattering of the ground state (Wilcomb et al., 1977a) and electronically excited (Johnson et al., 1975) Hg atoms with halogens have stimulated an interest in the potential energy surfaces for these systems. In addition, electronic state correlations have provided qualitative understanding of the results of the excited Hg reactions. Semiquantitative features of the ground state adiabatic potential hypersurface for the $Hg + I_2$ system have recently been deduced by Wilcomb et al. (1977b) using recent molecular beam scattering experiments, spectroscopic and structural data, and electronic state correlation diagrams. The key element of the potential energy surface is a deep attractive "basin" implied from the reactive scattering data, which provided evidence for the existence of a long-lived intermediate complex, believed to be IHgI. This empirical surface is characterized by the following features: (1) a shallow well in the entrance valley corresponding to the weakly bound (~ 0.06 eV) van der Waals adduct $Hg \cdot I_2$; (2) a subsequent barrier of about 0.7 eV in the entrance valley due to avoided crossings of diabatic potential curves, followed by a "fall-off" leading to insertion of the $Hg(^1S_0)$ into the I_2 ($^1\Sigma_g^+$) molecule; (3) a deep potential well (-1.45 eV) corresponding to the stable IHgI complex, taken to be gaseous mercuric iodide in its ground electronic state $^1A_1(^1\Sigma_g^+)$; (4) an exit valley with a minimum energy path rising essentially monotonically, with a negligible intrinsic barrier (< 0.03 eV) from -1.45 to $+1.15$ eV to yield HgI ($X^2\Sigma^+$) + $I(^2P_{3/2})$; (5) an essentially monotonic exit path rising to the threshold ($+1.54$ eV) for collision-induced dissociation; (6) a barrier to reaction in the collinear configuration (Hg-I-I) in which IHgI complex formation is sterically precluded.

Recently the London–Eyring–Polanyi–Sato method has been widely used to study the interaction of a diatomic molecule with a solid surface (Wolken and McCreery, 1975, 1976a, 1976b); McCreery and Wolken (1975, 1976a, b) has given details of the application of the model to the absorption of H_2 and HD on a variety of surfaces. Wolken and McCreery (1977) have extended the LEPS method to the interaction of three atoms with each other and with a solid surface to study the collision of H_2 with a surface containing an adsorbed hydrogen atom; they have also examined the effects of the adsorbed atoms on various aspects of the molecule-surface potential and on the molecule-surface adsorption dynamics and calculated the sticking probability. It decreases with increasing coverage, which is consistent with the experimental findings (Rye and Barford, 1974); this is due to two effects: (1) at higher coverage there are fewer available adsorption sites and (2) at higher coverage more gas phase molecules tend to collide with occupied sites.

There are a number of studies on the water dimer, an important system for the understanding of the hydrogen bond; however, such studies have been restricted either in the method adopted (SCF) (Popkie et al., 1975; Allen and Kollman, 1972) or in the extent to which the potential has been scanned (Dierksen et al., 1975). Although in the SCF approximation the electron correlation effects are neglected, these studies successfully provided a qualitative understanding of the water dimer and even of liquid water. However, the correlation effect is important for quantitative prediction of the dimerization energy. Recently Matsuoka et al., (1976) have reported an *ab initio* calculation of the potential surface of the water dimer based on a configuration-interaction method; the main purpose of their work is to obtain a quantitatively accurate description of the pair potential function for two water molecules, and hence they have carried out calculations for the water dimer, not only in the vicinity of the equilibrium geometrical configuration, but also for a wide range of geometrical configurations. The computed dimerization binding energies corresponding to the potential minima for the linear, cyclic, and bifurcated configurations are -5.6, -4.9, and -4.2. kcal/mole, respectively; the correlation effects account for -1.1, -1.2, and -0.9 kcal/mole, respectively, of the total binding energy for these three dimeric forms. Matsuoka et al., have also obtained the analytical expressions for the water dimer potential surface by fitting the calculated energies; these expressions are useful for determining the structure of liquid water in the pairwise approximation and with Monte Carlo techniques.

2.4 SYMMETRY RULES IN REACTION KINETICS

We present, in this section, the subject of symmetry rules. This subject, which is derived from the molecular orbitals (MO) theory with chemical reaction considered as a perturbation along the reaction coordinate, has been widely used recently in predicting the course of chemical reactions. As we know, all MO's of a system correspond to the binding together of certain atoms, the antibonding of other atoms, and the nonbonding of the remaining atoms. Notice that in a chemical reaction certain MO's must be vacated of electrons and others must be filled to create the new bonding situation; that is, any detailed analysis of a reaction mechanism requires a knowledge of how the electron distribution changes for various modes of the system. The most important of these changes is a flow of electrons from the highest occupied MO to the lowest unoccupied MO (Fukui, 1965). Electron movement between two orbitals cannot occur unless the orbitals meet the symmetry requirement (Bader, 1962). The subject of orbital symmetry in reaction kinetics has been reviewed recently by Woodward and Hoffmann (1965–1970) and by Pearson (1976).

2.4.1 General Theory

The chemical reaction may be regarded as a perturbation on the reactant system and any arbitrary small motion of the nuclei away from the original configuration can be analyzed as a sum of displacements corresponding to the normal modes of the system representing the reactants. When the nuclei are displaced from their equilibrium positions, the electron distributions relax in such a way as to follow the motion of nuclei. By determining which nuclear motion allows for the most favorable relaxation of the electron density, we can in fact determine the reaction coordinate. Now we expand the Hamiltonian of electronic motion in power series of the reaction coordinate Q about the original configuration with Hamiltonian $\hat{H}_e^0$,

$$\hat{H}_e = \hat{H}_e^0 + \left(\frac{\partial V}{\partial Q}\right)_0 Q + \frac{1}{2}\left(\frac{\partial^2 V}{\partial Q^2}\right)_0 Q^2 + \cdots \tag{2.90}$$

If the last two terms in eq. 2.90 are regarded as perturbation, then using the perturbation theory we can solve for the wave functions and energies as

$$\phi^0 = \phi_0^0 + \sum_k{}^1 \frac{\langle\phi_0^0|(\partial V/\partial Q)_0|\phi_k^0\rangle Q}{U_0^0 - U_k^0} \phi_k^0 + \cdots \tag{2.91}$$

and

$$U_0 = U_0^0 + \left\langle\phi_0^0\left|\left(\frac{\partial V}{\partial Q}\right)_0\right|\phi_0^0\right\rangle Q + \frac{Q^2}{2}\left\langle\phi_0^0\left|\left(\frac{\partial^2 V}{\partial Q^2}\right)_0\right|\phi_0^0\right\rangle$$

$$+ Q^2 \sum_k{}^1 \frac{|\langle\phi_0^0|(\partial V/\partial Q)_0|\phi_k^0\rangle|^2}{U_0^0 - U_k^0} + \cdots \tag{2.92}$$

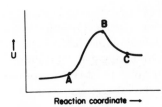

Fig. 2.18   Potential energy curve of a chemical reaction.

Although eqs. 2.91 and 2.92 are valid only for small Q, they are applicable for our discussion here since we can select any suitable configuration to carry out the expansion of eq. 2.90.

Figure 2.18 shows the usual adiabatic plot of potential energy along the reaction coordinate. At any maximum or minimum in the potential energy curve $\partial U_0/\partial Q = 0$, and eq. 2.92 reduces to

$$U_0 = U_0^0 + \frac{Q^2}{2}\left\langle \phi_0^0 \left| \left(\frac{\partial^2 V}{\partial Q^2}\right)_0 \right| \phi_0^0 \right\rangle + Q^2 \sum_k{}^1 \frac{\left|\left\langle \phi_0^0 \left| \left(\frac{\partial V}{\partial Q}\right)_0 \right| \phi_k^0 \right\rangle\right|^2}{U_0^0 - U_k^0} + \cdots$$

(2.93)

The physical meaning of eqs. 2.91–2.93 may be conveniently interpreted in terms of the electron density functions $L_{00} = |\phi_0^0|^2$ and $L_{0k} = \phi_0^{0*}\phi_k^0$. If the undistorted position is chosen to be the equilibrium configuration of the system, then the quantity L_{00} is the electron density for the undistorted system, and the second term of eq. 2.93 determines the increase in the energy of the system when the nuclei are displaced from their equilibrium positions with the electron distribution remaining unchanged. In eq. 2.91, the electron density function L_{0k} in $\langle \phi_0^0|(\partial V/\partial Q)_0|Q_k^0 \rangle$ is often termed the transition density (Bader, 1960). It is a measure of the amount of charge that is transferred within the molecule when the nuclei are displaced from their equilibrium positions and represents the correction to the electron density of the undistorted system. As the integral of the transition density overall space is equal to zero, L_{0k} does not represent any absolute amounts of charge but instead gives a three-dimensional representation of the movements of charge density within the system. Thus we may now interpret the matrix element $\langle \phi_0^0|(\partial V/\partial Q)_0|\phi_k^0 \rangle$ as the force (often called the transition force exerted on the nuclei that are displaced in the mode Q by the displaced charge density, or, alternatively, the matrix element may be regarded as the interaction of the point dipoles centered on the nuclei with the displaced charge interaction.

It is evident from the denominator in the expression U_0 that the lowering in energy due to the relaxation effect is greatest for that mode of vibration that permits an interaction with the lowest excited state. Furthermore, the transition force $\langle \phi_0^0 | (\partial V/\partial Q)_0 | \phi_k^0 \rangle$ is different from zero only if the transition density L_{0k} and Q have identical symmetries. We then assume that for the most favorable motion, the one that leads to the smallest increase in potential energy will be that one whose symmetry allows for an interaction with the lowest excited state. The principal change brought about in the electron distribution for any of the nuclear motions is assumed to be determined by the admixture of the lowest excited state of the proper symmetry. For convenience we assume that ϕ_0 is nondegenerate. At all points in the potential energy curve other than at maxima or minima, the linear term in eq. 2.92 dominates. In this case the reaction coordinate belongs to the totally symmetric representation, because the direct product of a nondegenerate symmetric representation with itself is always totally symmetric. That is, since $\phi_0^{0^2}$ is totally symmetric, $(\partial V/\partial Q)$ and also Q must be totally symmetric; otherwise the product of $\phi_0^{0^2}$ and $(\partial V/\partial Q)_0$ will not be totally symmetric. This means that once a reaction embarks on a particular reaction path it must stay within the same point group until it reaches an energy maximum or minimum.

As mentioned earlier, each excited state wave function is mixed into the ground state wave function with the amount of mixing given in eq. 2.91. The wave function is changed only because the resulting electron distribution ϕ_0^2 is better suited to the new nuclear positions. Salem (1969) calls the resulting decrease in energy the relaxability of the system along the coordinate Q. Now since $(\partial V/\partial Q)_0$ is totally symmetric, we can easily show that excited state wave functions ϕ_k^0 that have the same symmetry as ϕ_0^0 can mix in and lower the potential barrier. This can be easily verified from eqs. 2.92 or 2.93, because U_0 decreases with increasing $|\langle \phi_0^0 | (\partial V/\partial Q)_0 | \phi_k^0 \rangle|^2$. The last term in eqs. 2.92 or 2.93 represents the change in energy that results from changing the electron distribution to one more suited to the new nuclear positions determined by Q. Its value is always negative since $U_0^0 - U_k^0$ is a negative number. Thus for a reaction to occur with a reasonable activation energy, there must be low-lying excited states of the same symmetry as the ground state. In other words, during the chemical reaction the symmetry of the wave function of the system is preserved. Such a reaction is said to be symmetry allowed. A symmetry-forbidden reaction is simply one having a very high activation energy because of the absence of suitable excited states.

For practical applications some other assumptions must be made. One is that LCAO-MO theory will be used in place of the exact wave functions, ϕ_0^0, ϕ_k^0, and so on, since MO theory has the great advantage of accurately showing the symmetries of various electronic states. This creates no serious error as we are only concerned with the symmetry properties. The second assumption is that we replace the infinite sum of excited states in eqs. 2.92 and 2.93 by only a few lowest-lying states by recognizing that the various states contributing to eqs. 2.91–2.93 fall off very rapidly as the difference $|U_0^0 - U_k^0|$ becomes large. Thus the symmetry of $\phi_0^0 \phi_k^0$ is replaced by $X_i X_f$, where X_i is the occupied MO in the ground state and X_f is the MO in the excited state. Since $(\partial V/\partial Q)$ is a one-electron operator, the matrix element $\langle \phi_0^0 | (\partial V/\partial Q)_0 | \phi_k^0 \rangle$ in MO theory reduces to

$$\left\langle \phi_0^0 \left| \left(\frac{\partial V}{\partial Q} \right)_0 \right| \phi_k^0 \right\rangle = C \left\langle X_i \left| \left(\frac{\partial V_j}{\partial Q} \right)_0 \right| X_f \right\rangle$$

where C is a constant. Since excitation of an electron from the highest occupied MO (HOMO) to the lowest unoccupied MO (LUMO) defines the lowest excited state, X_i and X_f refer to HOMO and LUMO, respectively, From the preceding discussion, we can see that the orbitals X_i and X_f are of the same symmetry, which implies that they have a nonzero overlap.

2.4.2 Bimolecular Reactions

Suppose that two molecules approach each other with a definite orientation. They have started to interact with each other, but the interaction energy is still small. This means that the MO's of the two separate molecules are still a good starting point for considering the combined system (point A in Fig. 2.18). Those of the same symmetry interact more and more strongly as the reaction coordinate is traversed, and at the transition state (point B in Fig. 2.18) quite different MO's are produced. The transition state of a bimolecular reaction represents the configuration of maximum potential energy along the reaction coordinate and, therefore, as for a stable molecule, the energy is independent of the reaction coordinate to the first order. In this case we have

$$\frac{|\langle \phi_0^0 | (\partial V/\partial Q)_0 | \phi_k^0 \rangle|^2}{U_k^0 - U_0^0} > \frac{1}{2} \left\langle \phi_0^0 \left| \frac{\partial^2 V}{\partial Q^2} \right| \phi_0^0 \right\rangle$$

Thus we should expect to find that at least one excited state is relatively low-lying for a transition state molecule in order to meet the preceding requirement. It is in general the nature of this first excited state that determines the course of reaction. Now we can add an additional requirement on X_i and X_f by using chemical knowledge rather than quantum-mechanical arguments; for their bonding parts, X_i must represent bonds that are broken and X_f bonds that are made during the reaction. The reverse statement holds for their antibonding parts. The requirement for a bimolecular reaction is simply that the two have a net overlap. Since some atoms are much more electronegative than other atoms, electrons move more easily from X_i to X_f when they move toward the more electronegative atoms. In such cases

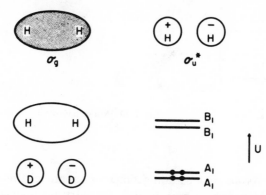

Fig. 2.19 Interaction between H_2 and D_2: $C_2(Z)$; $\sigma_v(xz)$; $\sigma'_v(yz)$.

$|U_0^0 - U_k^0|$ will be small and the stabilizing effect of electron excitation will be large.

To illustrate these principles, we consider one of the simplest chemical reactions:

$$H_2 + D_2 \longrightarrow 2HD \tag{2.94}$$

by assuming that the reaction occurs via a bimolecular mechanism in which H_2 and D_2 collide broadside, giving rise to a four-center transition state (see Fig. 2.19). The point group of this transition state is C_{2v}, and the character table of C_{2v} is given in Table 2.4. The MO's of H_2 and D_2 can now be classified as A_1 for the bonding σ_g and B_1 for the antibonding σ_u^*. As Fig. 2.20 shows, there is no empty MO of the same symmetry as any of the filled MO's. Hence the reaction is forbidden by orbital symmetry.

The same conclusion can be obtained by using the orbital correlation method due to Woodward and Hoffmann (1970). To prepare a correlation diagram for a chemical reaction, one writes down the approximately known energy levels of the reactants on one side and those of the products on the

Table 2.4
Character Table for C_{2v}

C_{2v}		E	C_2	σ_v	σ'_v
z	A_1	1	1	1	1
	A_2	1	1	-1	-1
x	B_1	1	-1	1	-1
y	B_2	1	-1	-1	1

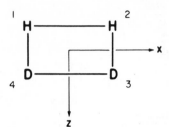

Fig. 2.20 Molecular orbitals and energy levels of H_2 and D_2.

other side. Assuming a certain geometry of approach one can classify levels on both sides with respect to the symmetry maintained throughout the approach, and then connect levels of like symmetry. Such a molecular correlation diagram yields valuable information about the intermediate region in a chemical reaction, that is, the transition state of the reaction. In this case we choose the plane occupied by H_2 and D_2 as the XZ plane (see Fig. 2.19) and use the following notations for the molecular orbitals of reactants and products,

	Bonding	Antibonding
$H_2(1.2)$	σ	σ^*
$D_2(3.4)$	σ^1	σ^{1*}
$HD(1.4)$	σ_1	σ_1^*
$HD(2.3)$	σ	σ_2^*

It can then easily be shown that the character $X(R)$ of the reducible representation results from the molecular orbitals of both reactants $(H_2 + D_2)$ and products $(2\,HD)$ are given by

$$X(E) = 4, \qquad X(C_2) = 0, \qquad X(\sigma_v) = 4, \qquad X(\sigma_v^1) = 0 \qquad (2.95)$$

Making use of the relation (Eyring et al., 1944)

$$a_i = \frac{1}{h} \sum_R X(R) X_i(R) \qquad (2.96)$$

we can find the number of times the irreducible representation Γ_i occurring in the reducible representation. The term h in eq. 2.96 represents the number of elements in the symmetry group. In this case we obtain $a_{A_1} = 2$ and $a_{B_1} = 2$; in other words, we expect that there will be two MO's belonging to the A_1

symmetry and two MO's for the B_1 symmetry. The symmetry orbitals can easily be found as

$$X_1(A_1) = \sigma, \qquad X_2(A_1) = \sigma^1, \qquad X_1(B_1) = \sigma^*, \qquad X_2(B_1) = \sigma^{1*} \qquad (2.97)$$

for reactants and

$$X_1^1(A_1) = \sigma_1 + \sigma_2, \qquad X_2^1(A_1) = \sigma_1^* + \sigma_2^*$$

$$X_1^1(B_1) = \sigma_1 - \sigma_2, \qquad X_2^1(B_2) = \sigma_1^* - \sigma_2^* \qquad (2.98)$$

for products. Based on this information we can prepare a correlation diagram for the reaction (eq. 2.94) as shown in Fig. 2.21. From this correlation diagram we can see that if orbital symmetry is to be conserved, there is a very large symmetry-imposed barrier to the reaction under discussion.

Four-center reactions of diatomic molecules almost always turn out to be forbidden. The statement that the reaction is forbidden, as stated earlier, simply indicates that the assumed mechanism has an excessive activation. Indeed, the energy of the transition state in this case can be calculated quite accurately by *ab initio* quantum-mechanical methods (Conroy and Malli, 1969); it lies 123 kcal above the energy of the reactants H_2 and D_2. Thus the mechanism of a direct reaction is impossible for all practical purposes and for reaction of eq. 2.94, it is considered to proceed via the atom-molecule mechanisms since reactions of free atoms and radicals rarely have serious symmetry restrictions.

To further illustrate the construction of a molecular correlation diagram we choose the maximum symmetry approach of two ethylene molecules leading to a product of cyclobutane as another example (see Fig. 2.22). As usual in theoretical investigations, maximum insight into the problem is gained by simplifying the case as much as possible, while maintaining the essential physical features. We need only the four π orbitals of the two ethylene molecules in the correlation diagram since they are the only ones that are

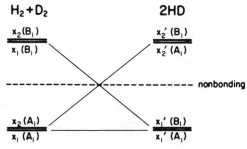

Fig. 2.21 Orbital correlation for $H_2 + D_2 \rightarrow 2\,HD$.

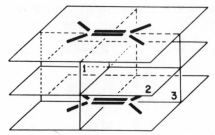

Fig. 2.22 Parallel approach of two ethylene molecules: $3 = x - z$; $2 = x - y$; $1 = y - z$.

transformed into four σ orbitals of cyclobutane in the course of reaction. The C—H and the C—C σ bonds of the ethylene skeleton may be omitted from the correlation diagram because, although they undergo hybridization changes in the course of reaction, their number, their approximate positions in energy, and, in particular, their symmetry properties are unchanged. The first step in the construction of a correlation diagram involves isolating the essential bonds and placing them at their approximate energy levels in reactants and products; the result is shown for the case under discussion in Fig. 2.23, in which the dashed horizontal line is the nonbonding level. In the next step the proper molecular orbitals for the reactants and products are written down. Molecular orbitals must be symmetric or antisymmetric with respect to any molecular symmetry element that may be present. To simplify

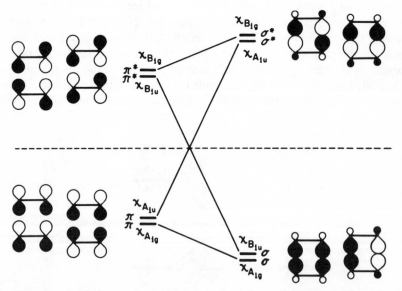

Fig. 2.23 Correlation diagram for the formation of cyclobutane from two ethylene molecules.

the discussion, we deliberately choose C_{2v} as the point group of this system although we could use the appropriate D_{2h} symmetry. Let π_1, π_2, π_1^*, and π_2^* be the four π orbitals of the two ethylene molecules, the characters of the reducible representation of these four π orbitals can be shown to be

$$X(E) = X(\sigma_v) = 4, \qquad X(C_2) = X(\sigma_v^1) = 0$$

Thus by using eq. 2.96 we find $a_{A_1} = 2$ and $a_{B_1} = 2$; that is,

$$A_1 = \pi_1, \pi_2, \qquad B_1 = \pi_1^*, \pi_2^* \tag{2.99}$$

Choosing the proper linear combination between π_1 and π_2, and between π_1^* and π_2^* so that the resulting orbitals will be either symmetric or antisymmetric with respect to inversion, we obtain

$$X_{A_{1g}} = \pi_1 + \pi_2, \qquad X_{A_{1u}} = \pi_1 - \pi_2$$
$$X_{B_{1g}} = \pi_1^* - \pi_2^*, \qquad X_{B_{1u}} = \pi_1^* + \pi_2^* \tag{2.100}$$

We must analyze next the situation in cyclobutane in an entirely analogous way. The results are as follows:

$$X_{A_{1g}}^1 = \sigma_1 + \sigma_2, \qquad X_{B_{1u}}^1 = \sigma_1 - \sigma_2$$
$$X_{A_{1u}}^1 = \sigma_1^* + \sigma_2^*, \qquad X_{B_{1g}}^1 = \sigma_1^* - \sigma_2^* \tag{2.101}$$

where σ_1, σ_2, σ_1^*, and σ_2^* represent the four σ orbitals of the cyclobutane molecule. Now we are ready to examine the correlation of the orbitals of reactants with those of the product (see Fig. 2.23). The direction in which the various levels will move may be obtained without detailed calculation by examining in each case whether any level is bonding or antibonding along the reaction coordinate. It should be emphasized that electrons placed in a bonding orbital bring the nuclei closer together (i.e., $\partial U/\partial R > 0$), whereas electrons put into an antibonding orbital pull the nuclei apart (i.e., $\partial U/\partial R < 0$). Also, in bonding orbitals the electrons must be lying in the region between the nuclei and are shared by them, whereas in antiboding orbitals there usually exist nodes (or a node) between the nuclei, which isolate electrons populating the orbitals in the regions of individual terminal nuclei.

From Fig. 2.23 we can see that the lowest level $X_{A_{1g}}$ of two ethylenes is bonding in the region of approach of the two molecules and thus will be stabilized by interaction. The $X_{A_{1u}}$ level has a node and consequently is antibonding in the region of approach. At large distances the interaction is inconsequential, but as the distance between the reacting molecule diminishes, this orbital is destabilized and moves to higher energy. Similarly, the antibonding $\pi^* X_{B_{1u}}$ orbital becomes bonding in the region of approach. It will thus be stabilized as the reaction proceeds whereas the antibonding $\pi^* X_{B_{1g}}$ orbital will be destablized. On the cyclobutane side both the σ levels are bonding in the region where the cyclobutane is being pulled apart. Thus they resist the motion; that is, they are destabilized along the reaction coordinate. On the other hand, both the σ^* levels are antibonding along the reaction coordinate. On the other hand, both the σ^* levels are antibonding along the reaction coordinate and thus move to lower energy as the cyclobutane is pulled apart. These qualitative conclusions can also be arrived at from a completed correlation diagram in which levels of like symmetry are connected. The most obvious and striking features of this diagram is the correlation of a bonding reactant level with an antibonding product level, and vice versa. Clearly, if orbital symmetry is to be conserved, two ground state ethylene molecules cannot combine in a concerted reaction to give a ground state cyclobutane, and vice versa, through a transition state having the geometry assumed here. In other words, there exists a symmetry-imposed barrier to the reaction in both directions. On the other hand, there is no such symmetry-imposed barrier to the reaction of one molecule of ethylene with another if one of these electrons had been promoted, say, by photochemical excitation, to the lowest antibonding orbital shown in Fig. 2.24. For these reasons reactions of the first type are designated as symmetry-forbidden and those of the second type as symmetry allowed.

The argument may be further illustrated by the inspection of the corresponding electronic state correlation diagram for the reaction under discussion. The ground state electronic configuration of two ethylene molecules correlates with a very high-energy doubly excited state of cyclobutane; conversely, the ground state of cyclobutane correlates with a doubly excited

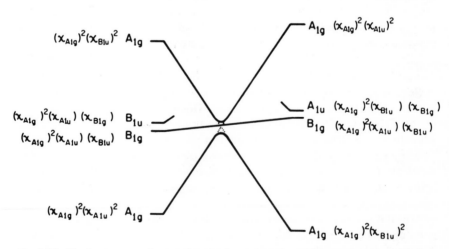

Fig. 2.24 Electronic state diagram for the formation of cyclobutane from two ethylene molecules.

state of two ethylenes. Electron interaction will prevent the resulting crossing and force a correlation of ground state with ground state. But in the actual physical situation the reaction has to overcome the activation energy for the intended but avoided crossing.

The lowest excited state of two ethylenes, the configuration $(X_{A_{1g}})^2(X_{A_{1u}})^1$ $(X_{B_{1u}})^1$ or B_{1g} correlates directly with the first excited state of cyclobutane. Consequently, there is no symmetry-imposed barrier to this formation. This represents the course actually occurring in many photochemical transformations. However, there are ambiguities in excited state reactions that do not exist in their simpler thermal counterparts. Thus it may happen that the chemically reactive excited state is not the one reached in the initial excitation. In fact, radiationless decay may be so efficient that the chemical changes subsequent to irradiation may be those of a vibrationally excited ground state or lower excited electronic state and the formation of a transition state for a given concerted reaction may be competitive with the relaxation of the excited state component to an equilibrium geometry that renders the reaction geometrically impossible. None of these details in any way vitiates the consequences of the orbital symmetry rule. The principle of conservation of orbital symmetry remains applicable, provided that the chemically reactive excited state is identified.

The conclusions obtained by using the principle of conservation of orbital symmetry for the reaction $2C_2H_4 \rightarrow C_4H_8$ can also be obtained by using eqs. 2.90–2.93.

2.4.3 Unimolecular Reactions

Consider point B in Fig. 2.18. This point refers to the activated complex for which the term linear in Q eq. 2.92 vanishes. The first quadriatic term is positive and the second term is negative. Clearly, at the maximum point B the second quadratic term is larger than the first, but the magnitude of the first term determines whether we have a high or low potential barrier. Again the existence of low-lying state ϕ_k of the correct symmetry to match with ϕ_0 is important, ϕ_0, $(\partial V/\partial Q)_0$, and ϕ_k are bound by the symmetry requirement that their direct product must contain the totally symmetric representation. In MO theory the product $\phi_0 \phi_k$ is again replaced by $X_i X_f$, where both the occupied and empty MO's must be in the same molecule. Electron transfer from X_i to X_f results in a shift in charge density in the molecule and its density increases in the regions where they have the same signs. The positively charged nuclei then move in the direction of increased electron density, which defines a reaction coordinate.

The size of the energy gap between X_i and X_f is critical. A small gap means an unstable structure unless no vibrational mode of the right symmetry exists for the molecule capable of changing its structure. A large energy gap between the HOMO and LUMO implies a stable molecular structure. In this case reaction may occur but only with a high activation energy. For an activated complex there must necessarily be at least one excited state of low energy. The symmetry of this state and the ground state then determines the mode of decomposition of the activated complex. Now suppose the unimolecular reaction proceeds from point C to point B in Fig. 2.18. When a molecule lies in a shallow potential well at point C, the activation energy for unimolecular change is small, and again we expect a low-lying excited state. The symmetry of this state and the ground state determine the preferred reaction of the unstable molecule. Thus for a series of similar molecules we expect a correlation between the position of the absorption bands in the visible-UV spectrum and the stability (Pearson, 1976).

As an example, consider the two molecules O_3 and SO_2 (Bader, 1962). The former is blue and is highly unstable whereas the latter is colorless and is much more stable toward dissociation into SO and O. An *ab initio* calculation gives the MO sequences, ... $(3b_1)^2(4b_1)^2(6a_1)^2(1a_2)^2(2b_2)^0$ for O_3 with SO_2 probably having the same sequence. Both molecules have an angular structure and the point group is C_{2v}. The a_2 and b_2 orbitals are π orbitals whereas the a_1 and b_1 orbitals are σ orbitals. The lowest-energy transition is expected to be between the nonbonding a_2 orbital and the antibonding b_2 orbital (Peyerimhoff and Buenker, 1967). The symmetry of the transition is $a_2 \times b_2 = b_1$, which indicates that Q belongs to b_1. The b_1 vibration is the unsymmetric stretch in which one O—O bond shortens and the other lengthens. This corresponds to the dissociation of the following:

$$O_3 \longrightarrow O_2 + O, \qquad SO_2 \longrightarrow SO + O \qquad (2.102)$$

The first absorption bands for O_3 are at 1.5 and 2.1 eV, and those for SO_2 are at 3.2 and 3.7 eV (Maria et al., 1970). Unfortunately, for the simple interpretation those correspond to triplet and singlet excitations from the $6a_1$ MO to the $2b_2$ MO. Hence the symmetry of $X_i \times X_f$ is b_2, which does not correspond to any vibration of these molecules. A higher pair of bands at 2.2 and 4.7 eV for O_3 and 3.7 and 5.3 eV for SO_2 does correspond to the required a_2 to $2b_2$ transition. For most molecules there is still considerable uncertainty in assigning the observed absorption bands to the definite MO transitions.

For molecules that lie in deep potential wells, the LUMO is not so important as higher-lying states since a high activation energy is required. This leads to a difficult problem in placing the higher excited states of a molecule in correct order. Nevertheless, the symmetry rules may still be of great help in selecting the reaction path. Suppose we know that a certain unimolecular reaction occurs, but without knowing the detailed mechanism. Thus certain bonds must be broken and made during the reaction; they are given as X_i and X_f, respectively. These MO's will in turn fix the symmetry of the reaction coordinate Q. The only requirement is a knowledge of the symmetry of the MO's that relates to the bonds affected.

Next let us discuss the application of the principle of molecular orbital symmetry to unimolecular reactions. For this purpose we consider electrocyclic reactions and use the conversion of butadiene to cyclobutene as an example (Woodword and Hoffmann, 1965). An electrocyclic reaction is defined as the formation of a single bond between the termini of a linear system containing $k\pi$ electrons, and the converse process. It can be disrotatory as shown in Fig. 2.25. In the former case the transition state is characterized by a plane of symmetry and in the latter a twofold axis of symmetry is preserved. The essential molecular orbitals are the four π orbitals of the butadiene, the π and π^* levels of the cyclobutene double bond, and the σ and σ^* orbitals of the simple bond to be broken (see Fig. 2.26). We assume that the system belongs to the point group C_{2v} and let π_{1b}, π_{2b}, π_{3a}, and π_{4a} be the four π orbitals of butadiene. The characters of the reducible representation of these four π orbitals are

$$X(E) = 4, \qquad X(\sigma_v) = -4, \qquad X(C_2) = X(\sigma_v') = 0 \qquad (2.103)$$

thus $a_{A_2} = 2$ and $a_{B_2} = 2$. In other words, we expect to have two molecular orbitals belonging to A_2 and two belonging to B_2,

$$X_{A_2} = \pi_{2b}, \qquad X_{A_2}^* = \pi_{4a}, \qquad X_{B_2} = \pi_{1b}, \qquad X_{B_2}^* = \pi_{3a} \qquad (2.104)$$

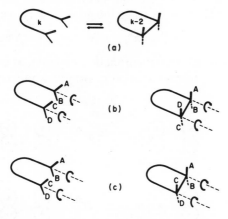

Fig. 2.25 Diagram for electrocyclic reactions. (*a*) Electrocyclic reaction, (*b*) disrotatory, (*c*) conrotatory.

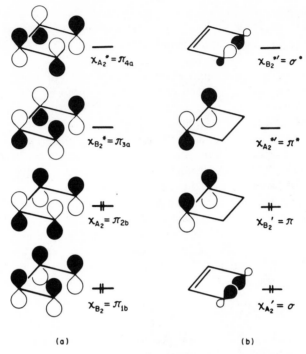

Fig. 2.26 Molecular orbitals of (*a*) butadiene and (*b*) cyclobutene.

First let us consider the conrotatory electrocyclic reaction. Carrying out a conrotatory motion of the σ orbitals of cyclobutene, the resulting orbitals are shown in Fig. 2.27. Thus by using the group theory, we find

$$X'_{A_2} = \sigma, \qquad X^{*\prime}_{A_2} = \pi^*, \qquad X'_{B_2} = \pi, \qquad X^{*\prime}_{B_2} = \sigma^* \qquad (2.105)$$

Now we can prepare an orbital correlation diagram, as shown in Fig. 2.28 by using the principle of conservation of orbital symmetry. From Fig. 2.28

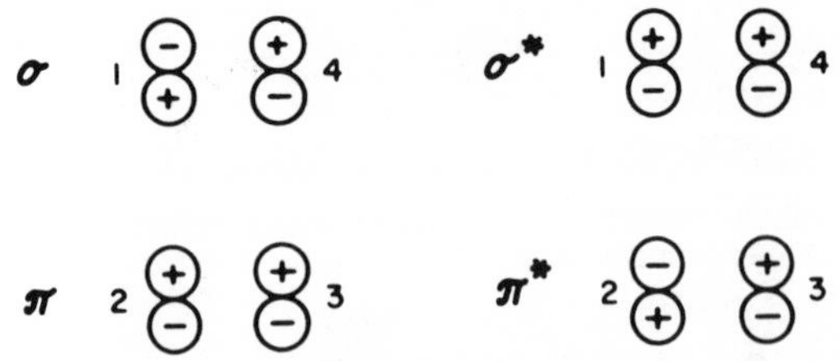

Fig. 2.27 Correlation of orbitals.

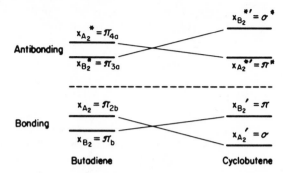

Fig. 2.28 Correlation diagrams for the conrotatory conversion of cyclobutenes to butadienes.

we can see that there is no symmetry-imposed barrier to the conrotatory reaction. In other words the thermal reaction should be a facile one.

Next we consider the disrotatory reaction. We find

$$X'_{B_2} = \sigma, \qquad X''_{B_2} = \pi, \qquad X^{*'}_{A_2} = \pi^*, \qquad X^{*''}_{A_2} = \sigma^* \qquad (2.106)$$

The orbital correlation diagram in this case is given in Fig. 2.29. Conservation of orbital symmetry requires in this case a high-lying transition state and the thermal reaction is symmetry-forbidden. It is clear that in the conrotatory process a twofold rotation axis is maintained at all times, whereas in the disrotatory motion an invariant plane of symmetry is maintained. Obviously, the correlation diagram for the conrotatory process is

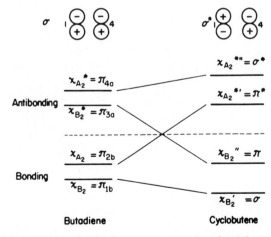

Fig. 2.29 Correlation diagram for the disrotatory conversion of cyclobutenes to butadienes.

characteristic of a symmetry-allowed reaction, whereas the pattern for the disrotatory process is a symmetry-forbidden reaction.

REFERENCES

General

Born, M. and K. Huang (1954), *Dynamical Theory of Crystal Lattices*, Oxford University Press.

Eyring, H., J. Walter, and G. E. Kimball (1944), *Quantum Chemistry*, Wiley.

Glasstone, S., K. J. Laidler, and H. Eyring (1941), *The Theory of Rate Processes*, McGraw-Hill.

Laidler, K. J. (1969), *Theories of Chemical Reaction Rates*, McGraw-Hill.

Laidler, K. J. and J. C. Polanyi (1965), *Prog. React. Kinet.*, **3**, 1.

Nitkin, E. E. (1970), *Adv. Quantum Chem.*, **5**, 135.

Schlier, C., Ed. (1970). *Molecular Beams and Reaction Kinetics*, Academic.

Woodward, R. B. and R. Hoffman (1970), *The Conservation of Orbital Symmetry*, Academic.

Special

Albrecht, A. C. (1960), *J. Chem. Phys.*, **33**, 156, 169.

Allen, L. C. and P. A. Kollman (1972), *Chem. Rev.*, **72**, 283.

Austin, B., V. Heine, and L. J. Sham (1962), *Phys. Rev.*, **127**, 276.

Bader, R. F. W. (1960), *Mol. Phys.*, **3**, 137.

Bader, R. F. W. (1962), *Can. J. Chem.*, **40**, 1164.

Bender, C. F., B. J. Garrison, and H. F. Schaefer (1975), *J. Chem. Phys.*, **62**, 1188.

Born, M. (1951), *Göttingen Nachrichten Math. Phys. Kl.*, **1**.

Botschwina, P. and W. Meyer (1977), *Chem. Phys.*, **20**, 43.

Bott, J. F. (1976), *J. Chem. Phys.*, **65**, 1976.

Bott, J. F. and R. F. Heidener II (1977), *J. Chem. Phys.*, **66**, 2878.

Boys, S. F. and I. Shavitt (1959), Univ. Wisc. Naval Res. Lab. Tech. Rep. WIS-AF-13.

Brand, J. C. D. and C. G. Stevens (1973), *J. Chem. Phys.*, **58**, 3331.

Brand, J. C. D. and D. S. Liu (1974), *J. Phys. Chem.*, **78**, 2270.

Brewer, L. and E. Brackett (1961), *Chem. Rev.*, **61**, 425.

Cashion, J. K. and D. R. Herschbach (1964), *J. Chem. Phys.*, **40**, 2358.

Chiu, Y. N. (1964), *J. Chem. Phys.*, **41**, 3235.

Clementi, E. (1963), *J. Chem. Phys.*, **38**, 1001.

Clementi, E. (1964), *J. Chem. Phys.*, **41**, 295.

Clementi, E., O. Matsuoka, and M. Yoshimine (1976), *J. Chem. Phys.*, **64**, 1351.

Conroy, H. (1967), *J. Chem. Phys.*, **47**, 912.

Conroy, H., and B. L. Bruner (1967), *J. Chem. Phys.*, **47**, 921.

Conroy, H., and G. Malli (1969), *J. Chem. Phys.*, **50**, 5049.

Conroy, H. (1970), in *Molecular Beams and Reaction Kinetics*, C. H. Schlier, Ed., Academic, p. 349.

Dalgarno, A. and R. McCorrol (1956), *Proc. Roy. Soc. (Lon.)*, *237A*, 383.

Dalgarno, A. and N. Lynn (1956), *Proc. Phys. Soc. (Lon.)*, **A69**, 821.

Dalgarno, A. and R. McCorrol (1957), *Proc. Roy. Soc. (Lon.)*, **239A**, 413.

Dalgarno, A. (1961), *Proc. Roy. Soc. (Lon.)*, **A262**, 132.

Dierksen, G. H. F., W. P. Kramer, and B. O. Ross (1975), *Theor. Chim. Acta.*, **36**, 249.

Edmiston, D., and M. Krauss (1965), *J. Chem. Phys.*, **42**, 1119.

Eyring, H. and M. Polanyi (1931), *Z. Phys. Chem.*, **B12**, 279.

Eyring, H., J. Walter, and G. E. Kimball (1944), *Quantum Chemistry*, Wiley.

Eyring, H. and S. H. Lin (1974), in *Physical Chemistry*, Vol. 6A, W. Jost, Ed., Academic, pp. 121–186.

Fisk, G. A. and B. Kirtman (1964), *J. Chem. Phys.*, **41**, 3516.

Fraga, S. and R. S. Mulliken (1960), *Rev. Mod. Phys.*, **32**, 254.

Fukui, K. (1965), in *Modern Quantum Chemistry*, O. Sinanogulu, Ed., Academic, p. 49.

Glasstone, S., K. J. Laidler, and H. Eyring (1940), *The Theory of Rate Processes*, McGraw-Hill.

Hayes, D. M. and K. Morokuma (1972), *Chem. Phys. Lett.*, **12**, 539.

Hayes, E. F. and R. G. Parr (1967), *J. Chem. Phys.*, **47**, 3961.

Heitler, W. and F. London (1927), *Z. Phys.*, **44**, 455.

Herzberg, G. and E. Teller (1933), *Z. Phys. Chem.* (Leipzig), **218**, 410.

Herzberg, G. and L. L. Howe (1959), *Can. J. Phys.*, **37**, 636.

Hirschfelder, J. O. and J. W. Linnett (1941), *J. Chem. Phys.*, **9**, 645.

Hirschfelder, J. O. and J. W. Linnett (1950), *J. Chem. Phys.*, **18**, 130.

Jaffe, R. L., D. M. Hayes, and K. Morokuma (1974), *J. Chem. Phys.*, **60**, 5108.

Jaffe, R. L., K. Morokuma, and T. F. George (1975), *J. Chem. Phys.*, **63**, 3417.

Jaffe, R. L. and K. Morokuma (1976), *J. Chem. Phys.*, **64**, 4881.

Jahn, J. A. and A. Teller (1930), *Proc. Roy. Soc. (Lond.)*, **161A**, 220.

James, A. S. and H. M. Coolidge (1934), *J. Chem. Phys.*, **2**, 811.

Jepsen, D. W. and J. O. Hirschfelder (1960), *J. Chem. Phys.*, **32**, 1323.

Johnson, S. G., H. F. Krause, S. Datz, and F. K. Schmidt-Bleek (1975), *Chem. Phys. Lett.*, **31**, 577.

Jortner, J., S. A. Rise, and R. M. Hochstrasser (1969), *Adv. Photochem.*, **7**, 149.

Karplus, J. (1970), in *Molecular Beams and Reaction Kinetics*, C. H. Schlier, Ed., Academic, p. 320.

Klimek, D. E. and M. J. Berry (1973), *Chem. Phys. Lett.*, **20**, 141.

Kolos, W. (1970), *Adv. Quantum Chem.*, **5**, 99.

Kolos, W. and C. C. J. Roothaan (1960), *Rev. Mod. Phys.*, **32**, 219.

Kwok, M. A. and R. L. Wilkins (1974), *J. Chem. Phys.*, **60**, 2189.

Laidler, K. J., and J. C. Polanyi (1965), *Prog. React. Kinet.*, **3**, 1.

Laidler, K. J. (1969), *Theories of Chemical Reaction Rates*, McGraw-Hill.

Lester, W. A. and F. Rebentrost (1976), *J. Chem. Phys.*, **63**, 3739, 3879.

Lin, S. H. (1966), *J. Chem. Phys.*, **44**, 3759.

Lin, S. H. (1967), *Theoret. Chim. Acta*, **8**, 1.

Lin, S. H. and R. Bersohn (1968), *J. Chem. Phys.*, **48**, 2732.

Lin, S. H. and H. Eyring (1974), *Proc. Natl. Acad. Sci. U.S.*, **71**, 3415, 3802.

London, F. (1929), *Z. Elektrochem.*, **35**, 552.

Luntz, A. C. and V. T. Maxon (1974), *Chem. Phys. Lett.*, **26**, 553.

Maria, H. J., P. Larson, M. E. McCarville, and S. P. McGlynn (1970), *Acc. Chem. Res.*, **3**, 368.

Matsuoka, O., E. Clementi, and M. Yoshimine (1976), *J. Chem. Phys.*, **64**, 1351.

McQuigg, R. D. and J. G. Calvert (1969), *J. Am. Chem. Soc.*, **91**, 1590.

Miller, R. G. and E. K. C. Lee (1974), *Chem. Phys. Lett.*, **27**, 475.

Miller, R. G. and E. K. C. Lee (1975), *Chem. Phys. Lett.*, **33**, 104.

Moffit, W. and W. Thorson (1957), *Phys. Rev.*, **108**, 1251.

Pearson, R. G. (1976), *Symmetry Rules for Chemical Reactions: Orbital Topology and Elementary Processes*, Wiley.

Peyerimhoff, S. D. and R. J. Buenker (1967), *J. Chem. Phys.*, **47**, 1953.

Popkie, H., J. Kistenmacher, and E. Clementi (1975), *J. Chem. Phys.*, **59**, 1325.

Porter, R. N. and M. Karplus (1964), *J. Chem. Phys.*, **40**, 1105.

Pulay, P. and W. Meyer (1974), *Theor. Chim. Acta.*, **32**, 253.

Rittner, E. S. (1951), *J. Chem. Phys.*, **19**, 1030.

Roach, A. C. and M. S. Child (1968), *Mol. Phys.*, **14**, 1.

Robinson, G. W. and R. P. Frosch (1963), *J. Chem. Phys.*, **38**, 1187.

Robinson, G. W. and R. P. Frosch (1964), *J. Chem. Phys.*, **41**, 357.

Rollefson, G. K. and H. Eyring (1932), *J. Am. Chem. Soc.*, **54**, 170.

Rye, R. R. and B. D. Barford (1974), *J. Chem. Phys.*, **60**, 1046.

Salem, L. (1969), *Chem. Phys. Lett.*, **3**, 99.

Sato, S. (1955a), *Bull Chem. Soc. Jap.*, **28**, 450.

Sato, S. (1955b), *J. Chem. Phys.*, **23**, 592.

Sato, S. (1955c), *J. Chem. Phys.*, **23**, 2465.

Shavitt, I., R. M. Stevens, F. L. Minn, and M. Karplus (1968), *J. Chem. Phys.*, **48**, 2700.

Siebrand, W. (1967), *J. Chem. Phys.*, **46**, 440.

Slater, J. C. (1931), *Phys. Rev.*, **38**, 1109.

Sugiura, Y. (1927), *Z. Phys.*, **45**, 484.

Takagi, K. and T. Oka (1963), *J. Phys. Soc. Jap.*, **18**, 1174.

Temple, G. (1928), *Proc. Roy. Soc. (Lond.)*, **A119**, 276; cf. A. Froman and G. G. Hall (1963), *J. Chem. Phys.*, **38**, 1104.

Thompson, D. L. (1972), *J. Chem. Phys.*, **57**, 4170.

Tobias, I. and J. T. Vanderslice (1961), *J. Chem. Phys.*, **35**, 1852.

Wadt, W. R. and N. W. Winter (1977), *J. Chem. Phys.*, **60**, 3068.

Wang, S. C. (1928), *Phys. Rev.*, **31**, 579.

Weston, R. E. (1959), *J. Chem. Phys.*, **31**, 892.

Wilcomb, B. E., T. M. Mayer, and R. B. Bernstein (1977a), *J. Chem. Phys.*, **60**, 3507.

Wilcomb, B. E., T. M. Mayer, J. T. Mackerman, and R. B. Bernstein (1977b), *J. Chem. Phys.*, **60**, 3555.

Wilkins, R. L. (1975), *Mol. Phys.*, **29**, 555.

Wolken, G. and J. H. McCreery (1975), *J. Chem. Phys.*, **63**, 2340.

Wolken, G. and J. H. McCreery (1976a), *J. Chem. Phys.*, **65**, 3510.

Wolken, G. and J. H. McCreery (1976b), *Chem. Phys. Lett.*, **39**, 478.

Woodward, R. B. and R. Hoffman (1965a), *J. Am. Chem. Soc.*, **87**, 395.

Woodward, R. B. and R. Hoffman (1965b), *J. Am. Chem. Soc.*, **87**, 2046.

Woodward, R. B. and R. Hoffman (1965c), *J. Am. Chem. Soc.*, **87**, 2511.

Woodward, R. B. and R. Hoffman (1970), *The Conservation of Orbital Symmetry*, Academic.

Wu, T. Y. and A. B. Ghatia (1956), *J. Chem. Phys.*, **24**, 48.

Yeung, E. S. and C. B. Moore (1973), *J. Chem. Phys.*, **58**, 3988.

Three

Collision Dynamics

CONTENTS

The subject of molecular collision dynamics is of great interest to many physical chemists today. It is concerned with both intermolecular motions and intermolecular forces, which together are of fundamental importance for understanding elementary physical and chemical reactions. With the advanced development of molecular beam technique, molecular collisions can be directly studied under the conditions of single collisions with velocity-selected reactants and in some cases in specified quantum states.

Molecular collisions, in general, are of three types: elastic, inelastic, and reactive; each of these three events occurs with a finite probability under a specified set of initial conditions. In an elastic collision the internal quantum states of collision pairs remain unchanged. In an inelastic collision the quantum numbers of collision pairs are changed, whereas in a reactive collision the chemical identity of the reactants is changed. In this chapter we discuss the features of elastic scattering and show how they are probed experimentally to derive the intermolecular forces governing this simple process.

3.1 CLASSICAL MECHANICS OF TWO-BODY COLLISIONS

3.1.1 Motion in a Central Force Field (Goldstein 1959)

We begin our discussion of elastic scattering by assuming a central force problem which states that the collision particles are considered to be structureless and that there are no external forces other than the intermolecular force acting on the particles. Thus the potential governing the collision process is considered to be a simple function of the relative position of collision pairs; that is, $V = V(r)$.

Since the intermolecular force depends only on the relative separation r between the collision particles, it is advantageous to describe the collision in terms of not the motion of each individual particle but rather their relative motion and the motion of their center of mass (CM). Figure 3.1 shows the position, vector $\mathbf{r}_1$ and $\mathbf{r}_2$ that locate the two particles in a fixed laboratory (LAB) coordinate system. Also shown are the relative position vector $\mathbf{r}$ and the CM vector $\mathbf{R}$, defined as

$$\mathbf{r} = \mathbf{r}_1 - \mathbf{r}_2$$

$$\mathbf{R} = \frac{m_1\mathbf{r}_1 + m_2\mathbf{r}_2}{M} \tag{3.1}$$

where $M = m_1 + m_2$. The linear momentum, angular momentum, and kinetic energy are given by (Landau and Lifschitz, 1960)

$$\mathbf{P} = m_1\dot{\mathbf{r}}_1 + m_2\dot{\mathbf{r}}_2 = M\dot{\mathbf{R}}$$

$$\mathbf{L} = m_1(\mathbf{r}_1 \times \dot{\mathbf{r}}_1) + m_2(\mathbf{r}_2 \times \dot{\mathbf{r}}_2) = M(\mathbf{R} \times \dot{\mathbf{R}}) + \mu(\mathbf{r} \times \dot{\mathbf{r}}) \tag{3.2}$$

$$T = \tfrac{1}{2}m_1\dot{\mathbf{r}}_1 \cdot \dot{\mathbf{r}}_1 + \tfrac{1}{2}m_2\dot{\mathbf{r}}_2 \cdot \dot{\mathbf{r}}_2 = \tfrac{1}{2}M\dot{\mathbf{R}} \cdot \dot{\mathbf{R}} + \tfrac{1}{2}\mu\dot{\mathbf{r}} \cdot \dot{\mathbf{r}}$$

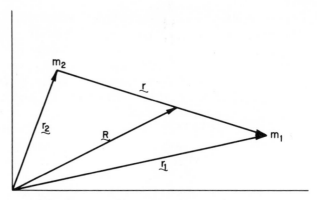

Fig. 3.1 Coordinate vectors for a two-body system.

where $\mu = m_1 m_2/M$, the reduced mass of the two particles. By differentiating eq. 3.1 twice with respect to time, we get

$$\ddot{\mathbf{R}} = \frac{m_1\ddot{\mathbf{r}}_1 + m_2\ddot{\mathbf{r}}_2}{M} = \frac{\mathbf{F}_1 + \mathbf{F}_2}{M}$$

Since no external force acts on the system, $\mathbf{F}_1 = -\mathbf{F}_2$ and $\ddot{\mathbf{R}} = 0$, that is; the CM motion remains unchanged before and after the collision and the linear momentum is a constant vector. If we define a new coordinate system whose origin always travels along the CM motion, $\mathbf{R}$ and $\dot{\mathbf{R}}$ vanish, and

$$\mathbf{L} = \mu(\mathbf{r} \times \dot{\mathbf{r}})$$

$$T = \tfrac{1}{2}\mu\dot{\mathbf{r}} \cdot \dot{\mathbf{r}} \tag{3.3}$$

Thus the original two-body problem is reduced to a motion of a hypothetical particle of mass μ. Given a spherical potential, $V(r)$, the interaction force, $\mu\ddot{\mathbf{r}}$, is always along $\mathbf{r}$ and

$$\dot{\mathbf{L}} = \mu\frac{d}{dt}(\mathbf{r} \times \dot{\mathbf{r}}) = \mu(\mathbf{r} \times \ddot{\mathbf{r}}) = 0$$

That is, $\mathbf{L}$ is also a constant vector. Since $\mathbf{r}$ and $\dot{\mathbf{r}}$ are always perpendicular to the fixed direction of $\mathbf{L}$, the motion must be confined to a plane perpendicular to $\mathbf{L}$ but containing $\mathbf{r}$ and $\dot{\mathbf{r}}$. Using the polar coordinate r and θ in that plane, the complete solution of the motion of a particle in a central force field is solved from

$$E = T + V(r) = \tfrac{1}{2}\mu(\dot{r}^2 + r^2\dot{\theta}^2) + V(r)$$

$$L = \mu r^2\dot{\theta} \tag{3.4}$$

3.1.2 Deflection Function and Collision Trajectories

For a complete description of classical trajectories in the CM system, we introduce here two important parameters:

g = the asymptotic relative velocity of the particles when they are at infinite separation

b = impact parameter, the closest distance of approach of the two particles when no deflection occurs

Making use of these two parameters, the radial motion of eq. 3.4 may be rewritten as

$$\frac{1}{2} \mu g^2 = \frac{1}{2} \mu \dot{r}^2 + \frac{1}{2} \frac{L^2}{\mu r^2} + V(r)$$

or

$$E = \frac{1}{2} \mu \dot{r}^2 + \frac{Eb^2}{r^2} + V(r) \tag{3.5}$$

During the course of elastic collision, the asymptotic relative velocity and impact parameter must be constant and the only variable after collision is that the velocity vector **g** be rotated through an angle $\chi = \pi - 2\theta_c$ (Fig. 3.2), where θ_c can be found by eliminating the variable t from eq. 3.4 (Hirschfelder et al., 1954):

$$\theta_c = \int_c^{\theta_c} d\theta = -b \int_\infty^{r_c} \frac{dr}{r^2(1 - b^2/r^2 - V/E)^{1/2}}$$

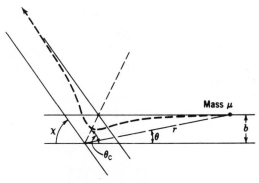

Fig. 3.2 A classical trajectory in the CM system.

Hence the angle of deflection during the collision is given by

$$\chi(E, b) = \pi - 2b \int_{r_c}^{\infty} \frac{dr}{r^2(1 - b^2/r^2 - V/E)^{1/2}} \tag{3.6}$$

where the radial distance of closest approach r_c is determined by the condition that at r_c, $\dot{r} = 0$. From eq. 3.5 we find

$$b^2 = \frac{r_c^2(1 - V(r_c))}{E} \tag{3.7}$$

If $V(r)$ is known, eq. 3.6 may be integrated to determine the deflection function $\chi(E, b)$. Note that, from eq. 3.7, $r_c < b$ if $V(r_c) < 0$ (an attractive interaction at the classical turning point) and $r_c > b$ if $V(r_c) > 0$ (a repulsive potential at the classical turning point). Alternatively, we may state that for an elastic scattering characterized by a large-impact parameter and low collision energy, the classical turning point moves outward and the deflection function is solely determined by the long-range potential, which is typically attractive for atomic and molecular systems. In this case we anticipate $b > r_c$ and the deflection angle is negative. On the other hand, at high energies the collision time is short and the molecules may not have time enough to feel the existence of long-range attraction; thus they may strike the repulsive part of the potential and bounce back. In this case the impact parameter is smaller than the closest approach. However, the deflection function for a collision process only samples the potential function along the trajectory specified by the initial condition E and b but gives no information for potential energy $V(r)$ at $r < r_c(E, b)$.

Returning to eq. 3.5, the equation of radial motion may be regarded as the one-dimensional motion of a particle governed by a total collision energy E and an effective potential $U(r, L)$ made up of true potential $V(r)$ and the centrifugal repulsion:

$$U(r, L) = V(r) + \frac{L^2}{2\mu r^2} \tag{3.8}$$

This effective potential is of fundamental importance for the molecular collision process. In contrast to molecular spectroscopy, where the angular momentum is in general thermally equilibrated and fairly small so that $U(r, L)$ is practically the same as $V(r)$, in the process of molecular collision very large values of b (and L) may be important and the centrifugal repulsion term may become the dominant part of $U(r, L)$. To illustrate the detailed features of molecular trajectories undergoing an elastic collision, we consider here molecules that obey the Lennard–Jones (6, 12) potential,

$$V(r) = 4\varepsilon\left[\left(\frac{\sigma}{r}\right)^{12} - \left(\frac{\sigma}{r}\right)^6\right] \tag{3.9}$$

Here ε is the depth of the potential well and σ is the radius at which $V(\sigma) = 0$. Hence we have an effective potential

$$U(r, L) = 4\varepsilon\left[\left(\frac{\sigma}{r}\right)^{12} - \left(\frac{\sigma}{r}\right)^{6}\right] + \frac{Eb^2}{r^2}$$

Detailed numerical results for elastic collision governed by this effective potential can be conveniently evaluated in terms of reduced dimensionless variables (Hirschfelder et al., 1954):

$$r^* = \frac{r}{\sigma} \qquad E^* = \frac{E}{\varepsilon}$$

$$b^* = \frac{b}{\sigma} \qquad V^* = \frac{V}{\varepsilon}$$

$$L^* = \frac{L}{(2\mu\varepsilon\sigma^2)^{1/2}}$$

$$U^* = \frac{U}{\varepsilon} = V^* + E^*\left(\frac{b^*}{r^*}\right)^2$$

Figure 3.3 shows the effective potential curves for three values of L^*. For $L^* = 1.569$, a point of inflection occurs at $U^* = 0.8$. For higher values of L^* the effective potential is repulsive everywhere, whereas for lower L^*, the effective potential has zones of repulsion at large and small r^*, separated by a zone of attraction at intermediate r^*. Therefore, for $0 < L^* < 1.569$, there appears a potential well in addition to a centrifugal barrier. It is at these lower values of L^* that numerous important features of molecular collision have been observed, making it possible to extract the detailed intermolecular potential.

We recall here that the energy is related to the impact parameter by $E^* = (L^*/b^*)^2$. For a given $L^*(<1.569)$, trajectories for molecular collision vary for different combination of E^* and b^*; they are shown in Fig. 3.4. At relatively high energies $E^* > U^*(r^*_{max})$ the particle feels an attractive force at a large distance and experiences a small deflection as it approaches the scattering center. As time passes, however, it eventually reaches the repulsive wall and thus bounces off. Trajectories for this high incident collision energy (and thus small impact parameter) have a large but positive deflection angle (case c). At $E^* = U^*(r^*_{max})$ the particle reaches the maximum of the centrifugal barrier with zero radial velocity but has a finite angular velocity $\dot{\theta} = L/\mu r^2_{max}$. The particle thus orbits indefinitely and the deflection angle becomes arbitrarily large. For the third feature of lower energy scattering at $E^* < U^*(r^*_{max})$, the particle feels an acceleration initially and then strikes the

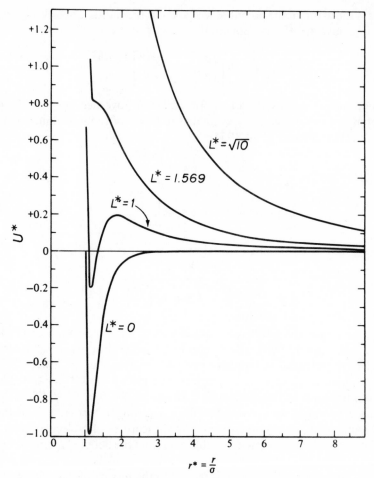

Fig. 3.3 Reduced effective potential as a function of L^* for Lennard–Jones (6,12) potential.

repulsive wall at the foot of the centrifugal barrier. Since the particle cannot penetrate the barrier classically (quantum tunneling effect makes it accessible, however), the particle thus rebounds back without feeling the existence of the potential well at small r^* (Weston and Schwartz, 1972).

For the Lennard–Jones (6, 12) potential, the general features of the deflection function as a function of the reduced impact parameter is summarized in Fig. 3.5 for three incident energies. For large impact parameters, the deflection angle is small and negative because of a weak long-range attraction. As b^* decreases, the deflection becomes large. If $E^* \leq 0.8$, a centrifugal

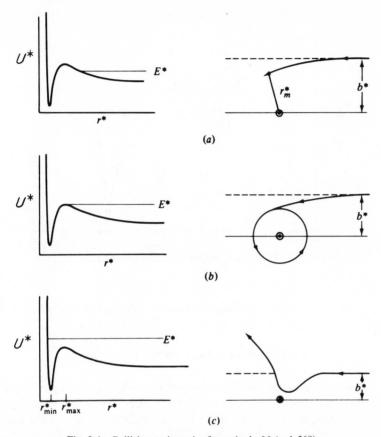

Fig. 3.4 Collision trajectories for a single L^* (<1.569).

barrier may appear in U^* and as b^* is decreased, the condition for orbiting may be achieved and $\chi \to -\infty$. If $E^* > 0.8$, χ reaches a finite minimum (negative) and then increases as b^* further decreases. A head-on collision at $x = \pi$ will eventually occur as b^* approaches zero. Thus it is clear that classical trajectories of molecular collision can be fully discussed using the initial collision energy and the impact parameter for a given potential. At this juncture the three special features that arise are as follows:

$$b^* = b_r^* \quad \text{where} \quad \frac{d\chi}{db^*} = 0 \quad \text{(rainbow)}$$

$$b^* = b_g^* \quad \text{where} \quad \chi = 0 \quad \text{(glory)}$$

$$b^* = b_0^* \quad \text{where} \quad \chi = -\infty \quad \text{(orbiting)} \tag{3.10}$$

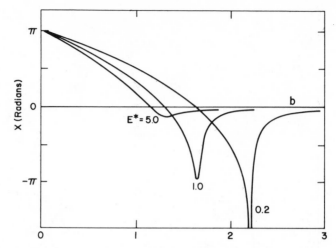

Fig. 3.5   The classical deflection functions for the Lennard–Jones (6, 12) potential.

These special features, however, cannot be adequately described without quantum consideration; they are examined in detail in Sec. 3.3.

3.1.3   Differential and Total Cross Sections

In physical applications we measure not the deflection of a single particle, but that of a beam of finite flux density at which all possible impact parameters are present. Therefore there is a probability function for particles scattered in various direction. In the CM system this probability function is called the differential cross section. The differential cross section $I(\chi)$ is defined physically as the number of particles scattered into a unit solid angle dw in unit time divided by the incident flux density. Because of spherical symmetry of the intermolecular force, the deflection pattern has an axis of symmetry along the line of approach through the center of force. Consequently, the angular distribution is independent of azimuthal angle and depends only on the polar angle of deflection χ; that is,

$$dw = 2\pi \sin \chi \, d\chi$$

and the number of particles scattered per unit time with deflection between χ and $\chi + d\chi$ is

$$dQ = I(\chi)2\pi \sin \chi \, d\chi$$

For a given velocity of the beam, the number of molecules deflected dQ between χ and $\chi + d\chi$ is equal to the number in the incident beam whose

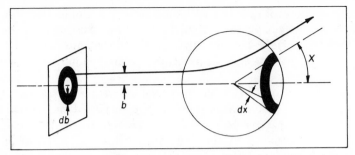

Fig. 3.6 Scattering for an incident beam of particles by a central force.

lines of approach pass through an annulus width db and area $2\pi b\ db$ (Fig. 3.6). Hence we have

$$dQ = 2\pi b\ db = 2\pi I(E, \chi)\sin \chi\ d\chi$$

or

$$I(E, \chi) = \frac{b}{(\sin \chi\,|d\chi/db|)} \qquad (3.11a)$$

We recall from Fig. 3.5 that the relation between deflection angle and impact parameter is not always a one-to-one correspondence at a single energy, a general expression of the differential cross section

$$I(E, \chi) = \sum_i \left(\frac{b_i}{(\sin \chi\,|d\chi/db_i|)}\right) \qquad (3.11b)$$

should replace eq. 3.11a whenever applicable. The total cross section for elastic scattering is obtained by integrating the differential cross section through all solid angles

$$Q(E) = 2\pi \int_0^\pi I(E, \chi)\sin \chi\ d\chi$$

or

$$= 2\pi \int_0^{b_{\max}} b\ db \qquad (3.12)$$

Thus, once the angle of deflection is known for a given E and b, the differential cross section and the total elastic cross section may be readily calculated. However, the classical expression for the differential cross section $I(E, \chi)$ at $x = 0$ is not finite if the scattering is governed by the long-range potential that may extend to infinity, and the total cross section is also infinity. This can be readily seen from Fig. 3.5 since there is no cutoff on the impact parameter and $b_{\max} \to \infty$. In addition, since the variation of the deflection function with the impact parameter is not a monotonic function for a molecular system, physical singularities for the differential cross section are also found at (1) the rainbow angle, (2) the glory scattering, and (3) the orbiting. These features can be remedied when quantum mechanics is considered, however.

We should point out that when molecular scattering is governed by the long-range attractive force, the molecule may be deflected into a small angle with a large impact parameter. Thus the total cross section for elastic scattering is mainly contributed by the small-angle scattering corresponding to large impact parameter. Moreover, the physical singularities of $I(E, \chi)$ resulting from the glory scattering and the orbiting offer limited information on the intermolecular force, classically, whereas the singularity appearing at the rainbow angle, $\chi = \chi_r$, has physical significance since it is directly related to the magnitude of the potential well. Details of this effect are discussed in the next section.

3.1.4   Approximation for Small-Angle Scattering

The exact integration of the deflection function for any realistic potential of a molecular system is often found practically impossible, and thus some sort of approximations are usually employed to discuss the physical significance of the collision cross sections. The approximate method described here for small-angle scattering is extremely useful in leading to physical insight into molecular collisions.

The simplest approximation applied to the small-angle trajectories is called the *impulse-momentum method*; it assumes that the deflection is due to the perturbation of a small interaction force on a straight-line trajectory. We assume that the direction of the initial momentum $P = \mu g$ lies along the z axis and the motion of the collision process is confined to the yz plane, hence $\sin \chi = P'_y/P'$. For small deflection angles, $\sin \chi$ may be replaced by χ and P', the momentum after scattering, may be approximated as P by assuming that the particle moves along a straight line trajectory with a constant velocity $\mathbf{g}$. Hence we have

$$\chi \approx \frac{P'_y}{\mu g} \tag{3.13}$$

Now, since $P'_y = F_y$, the total increment of momentum in the y direction is

$$P'_y = \int_{-\infty}^{\infty} F_y \, dt$$

$$F_y = \frac{F_y}{r} = \frac{Fb}{(b^2 + g^2 t^2)^{1/2}} \tag{3.14}$$

and the deflection angle is approximated as

$$\chi = \frac{b}{\mu g} \int_{-\infty}^{\infty} \frac{F\,dt}{(b^2 + g^2 t^2)^{1/2}} = \frac{b}{E} \int_{b}^{\infty} \frac{F\,dt}{(r^2 - b^2)^{1/2}} \tag{3.15}$$

Compared with eq. 3.6, this equation represents the classical approximation that the term V/E in the integrand of eq. 3.6 is small and the exact solution for a straight-line trajectory is used to evaluate the approximate effect on the trajectory by a small interaction force. Thus we anticipate that this expression provides valid answers for large impact parameter or high collision energy such that $(b/r)^2 \gg V(r)/E$ for all $r > b$. This is probably true for a repulsive force where r is much larger than b for most of the trajectory; for attractive forces, however, r may be equal to or close to b for a considerable range of the trajectory such that a weak force may produce a large angle of deflection and this small angle approximation fails.

To illustrate its physical implication, we assume a power potential of the form $V(r) = Cr^{-n}$, where $n > 0$. Equation 3.15 can then be integrated by changing to the variable $a = b/r$

$$\chi = \frac{nc}{Eb^n} \int_{0}^{1} a^n (1 - a^2)^{-1/2}\,da$$

and the approximate result for small-angle scattering can be explicitly expressed in terms of a gamma function

$$\chi = \frac{CK_n}{Eb^n}$$

with

$$Kn = \pi^{1/2} \frac{\Gamma[(n + 1)/2]}{\Gamma(n + 2)} \tag{3.16}$$

This equation shows that the scattering angle is very sensitive to the magnitude of the impact parameter for large values of n. The differential cross section is found to be

$$I(E, \chi) = \frac{1}{n} \left(\frac{CKn}{E} \right)^{2/n} \chi^{-(2 + 2/n)} \tag{3.17}$$

Thus for an inverse sixth-power attractive potential, the small-angle scattering should obey the relation $I(E, \chi) \alpha \chi^{-7/3}$. Similarly, we may also find an expression for the total elastic cross section. We recall that eq. 3.12 diverges classically since $b_{\max} \to \infty$ as $\chi \to 0$. However, an effective total cross section determined by the experimental detector resolution may be expressed by

$$Q(E) = 2\pi \int_{\chi_0}^{\pi} I(E, \chi) \sin \chi\,d\chi \tag{3.12a}$$

where χ_0, the smallest angle of deflection, can be measured experimentally, Rearrangement of eq. 3.16 gives

$$b_{max} = \left(\frac{CK_n}{\chi_0 E}\right)^{1/n}$$

and

$$Q(E) = 2\pi \int_0^{b_{max}} b\, db = \pi\left(\frac{CK_n}{\chi_0 E}\right)^{2/n} \tag{3.18}$$

This result predicts that a plot of $\ln Q(E)$ versus $\ln E$ should give a straight line with slope equal to $-2/n$ or $-\frac{1}{3}$ for an inverse sixth-power potential.

We may now return to the Lennard–Jones (6, 12) potential and show how its potential well depth ε can be derived from the approximate method at high energies. Analogous to the derivation for eq. 3.16, the general expression for the deflection angle at the high-energy limit is given by

$$\chi = \frac{\pi^{1/2}}{E}\left[\frac{C_m \Gamma[(m+1)/2]}{b^m \Gamma(m/2)} + \frac{C_n \Gamma[(n+1)/2]}{b^n \Gamma(n/2)}\right]$$

for

$$V(r) = C_m r^{-m} + C_n r^{-n} \tag{3.19}$$

where $C_m > 0 > C_n$ and $m > n > 1$. If molecules are subject to the Lennard–Jones (6, 12) potential, the deflection angle is found to be

$$\chi = \frac{15\pi\varepsilon}{4E}\left[\frac{231}{160}\left(\frac{\sigma}{b}\right)^{12} - \left(\frac{\sigma}{b}\right)^6\right] \tag{3.20}$$

At the rainbow angle where $d\chi/db = 0$, we obtain

$$b_r = 1.193\sigma \qquad \text{and} \qquad \chi_r = -\frac{2.04\varepsilon}{E} \tag{3.21}$$

Thus for a single incident energy E, the potential well depth can be extracted from the observed rainbow angle.

3.1.5 Transformation Between the CM-LAB Coordinate Systems

As we have shown, the two-body central force problem may be reduced to the motion of a fictitious particle in the CM coordinates. In practice, however, the scattering processes always involve two particles under the experimental conditions and thus the measurable parameters in the LAB system are different from those we have described in the CM system; for instance, the speeds of collision pairs is not necessarily unchanged during the course of

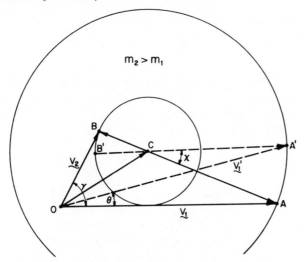

Fig. 3.7 Velocity diagram for a two-body collision.

elastic collision in the LAB system. This can be easily demonstrated for the collision between two billiard balls. When a billiard ball is initially at rest, it carries away a finite velocity and kinetic energy as a result of collision. To maintain the constant total kinetic energy, the cue ball therefore must reduce by an equivalent amount of energy.

Experimentally, one can arrange two beams of particles intercepting at an angle of γ and then measure the LAB angle θ, with respect to one of the incident beams. Figure 3.7 shows the velocity diagram for a two-body collision; this figure is drawn for $m_2 > m_1$. Here the CM motion lies along the line OC and the magnitude of the initial relative velocity g is represented by the length AB. Note that since the total linear momentum in the CM system is conserved, the initial relative velocity of particles 1 and 2 is retained in the following equation,

$$\frac{\overline{AC}}{\overline{BC}} = \frac{v_{1c}}{v_{2c}} = \frac{m_2}{m_1} \tag{3.22}$$

Therefore the relative motion of two particles, after the collision, has been rotated through an angle χ with its relative velocity unchanged $g = g'$ (or $AB = A'B'$). By applying the cosine law, we get the LAB velocity v_1' after collision,

$$v_1'^2 - 2v_1'v_1 \cos\theta + v_1^2 - \frac{2m_2^2(v_1^2 + v_2^2 - 2v_1v_2 \cos\gamma)(1 - \cos\chi)}{M^2} = 0 \tag{3.23}$$

This equation can be simplified to

$$v_1' = \frac{v_1(2m_1m_2 \cos \chi + m_1^2 + m_2^2)^{1/2}}{M} \tag{3.24}$$

if one assumes a target molecule is at rest initially; that is $v_2 = 0$ and $\gamma = 0$. From eq. 3.24, one can further show that the maximum energy is obtained by a particle originally at rest when the two particles move in opposite directions ($\chi = \pi$) after the collision. In this case we find

$$E_{2'\max}' = \frac{4m_1m_2E_1}{M^2} = \frac{4\mu E_1}{M} \tag{3.25}$$

It shows that the most efficient energy transfer occurs when $m_1 \approx m_2$.

Thus far our attention has largely focused on the effects of the velocity parameter in the CM and LAB coordinate systems. The actual transformations of the angular distributions are in fact the probability functions for particles scattered in the various directions. If $d\Omega$ and dw are the elements of solid angle in the LAB and CM systems, respectively, the observed LAB intensity $I(\theta)$ can be readily transformed into the CM intensity $I(\chi)$, as shown in Fig. 3.7, by

$$I(\chi) = I(\theta) \left| \frac{d\Omega}{dw} \right| \tag{3.26}$$

Here the ratio of solid angles is essentially the Jacobian of the transformation.

Finally, in most scattering experiments, measurements are made of the angular distribution of only one of the two particles in the LAB system. Since the scattered angle in the CM system is the same for both species, the derived CM angular distribution from one of the species could be used to back-calculate a LAB angular distribution of the other species.

3.1.6 Limitations of Classical Mechanics

The classical treatment of molecular collision theory suffers a fundamental defect—the constraint of the uncertainty principle since the classical trajectories specified by the deflection function and impact parameter at a given energy can not be accurately used to represent the collision events through the entire process. Consider a wave packet with a breadth Δy traveling in the z direction; the uncertainty of its tranverse momentum ΔP_y can be prescribed as $\Delta Py > \hbar/\Delta y$. This means that the impact parameter from the classical theory is subject to an uncertainty $\Delta b \approx \hbar/\Delta P_y$. Hence we have an equal amount of uncertainty for the angle of deflection

$$\Delta \chi = \frac{\Delta P_y}{P_z} \geq \frac{\hbar}{P_z \Delta b} = \frac{\hbar}{\Delta L} \tag{3.27}$$

That is, the uncertainty principle prevents both angular momentum and the deflection angle from being defined accurately at any given time. This inequality would eventually break down when χ approaches zero. In this case classical trajectories are misleading and the differential cross section at $\chi = 0$ becomes unjustified classically.

In addition to the limitation of the uncertainty principle, the interference effect and the quantum tunneling in the classical forbidden region also restrict the validity of special features described by classical mechanics (eq. 3.10). Owing to the interference effect arising from the different branches of the classical trajectories near the rainbow angle (eq. 3.11b), the differential cross section derived from quantum mechanics has a rapid oscillatory structure as it crosses the classical rainbow angle. On the other hand, the quantum tunneling effect at the centrifugal barrier would also complicate the classical orbiting phenomenon; here a possible quasibound molecular state may be found when resolution of collision energy is high enough. We thus anticipate that in order to interpret the quantitative results of molecular scattering adequately, one has to employ quantum theory rather than classical mechanics.

Despite these constraints, classical collision theory has its own merit: It offers the perceptible physical pictures of particle trajectories when collision events take place. In addition, since the experimental energy resolution is far from being monoergic, most quantum features have rarely been observed as yet; in particular, aside from the rainbow phenomenon in the angular distribution, the detailed interference patterns of $I(E, \chi)$ and of $Q(E)$ are seldom found in low-energy molecular collision. For this reason most molecular collision processes (especially the heavy systems) can be explained well qualitatively and semiquantitatively by classical mechanics alone.

3.2 QUANTUM MECHANICS

3.2.1 Introduction

In quantum mechanics the trajectory of a single particle can not be well defined since the uncertainty principle restricts an exact description of the position vector and momentum vector simultaneously. Instead of having the initial conditions of $\mathbf{r}(t)$ and $\mathbf{P}(t)$, we have to consider a wave function $\psi(r, t)$ that satisfies the equation of motion

$$\hat{H}\,|\,\psi(r, t)\,\rangle = i\hbar\,\frac{\partial}{\partial t}\,|\,\psi(r, t)\rangle \tag{3.28}$$

where $\hat{H} = -(\hbar^2/2\mu)\nabla^2 + V$. In a scattering experiment the incident beam is generally steady and large, that is, for a constant flux at any time interval that is much longer than $\hbar/\Delta E$ where ΔE is the energy spread of the beam and a beam dimension larger than the extent of the scattering field, the initial wave packet, is well approximated by a plane wave extending infinitely in space and time. Hence the wave function $\psi(r, t)$ may be constructed as

$$\psi(r, t) = \exp\left[\frac{i(\mathbf{P} \cdot \mathbf{r} - Et)}{\hbar}\right] \tag{3.29}$$

Assuming that the scattering is governed by a central force field $V = V(r)$, the Hamiltonian $\hat{H}$ is independent of time and a separation of time and space is possible. Equation 3.29 may then be rewritten as (Burhop 1961)

$$\psi(r, t) = \phi_k(r)\exp\left(\frac{-iEt}{\hbar}\right) \tag{3.30}$$

Substituting eq. 3.30 into eq. 3.28 yields a time-independent Schrödinger equation

$$\nabla^2\phi_k(r) + [k^2 - U(r)]\phi_k(r) = 0 \tag{3.31}$$

$$k^2 = \frac{2\mu E}{\hbar^2}, \qquad U(r) = \frac{2\mu V(r)}{\hbar^2}$$

and all relevant information about the scattering process is contained in $\phi_k(r)$. If for the interaction field $V(r) = 0$, we have a plane wave $\phi_k(r) = \exp(i\mathbf{P} \cdot \mathbf{r}/\hbar)$.

For a scattering problem governed by a central force $V(r)$, the solution of eq. 3.31 in the asymptotic region $r \rightarrow \infty$ is

$$\phi_k(r) \sim e^{ikz} + \frac{f(\theta)e^{ikr}}{r} \tag{3.32}$$

The first term represents a plane wave advancing in the positive direction along the z axis with the second term an outgoing wave deviating from the z axis by an angle θ. Making use of the conservation of flux density between incoming and outgoing wave packets, the scattering amplitude $f(\theta)$ has a direct relation to the differential cross section. The current density $\mathbf{j}_s$ in the spherical scattering wave at large r can be readily evaluated:

$$\mathbf{j}_s = \frac{\hbar}{2i\mu}(\phi^*\nabla\phi - \phi\nabla\phi^*)$$

$$= \frac{\hbar k}{\mu}\frac{|f(\theta)|^2}{r^2}\hat{r} \tag{3.33}$$

where $\hat{r}$ is a unit vector in the radial direction. Thus the current density scattered into solid angle dw is then the product $g|f(\theta)|^2\, dw$ whereas the incident current density is simply g. Hence we have the differential cross section

$$I(\theta)\, dw = |f(\theta)|^2\, dw \tag{3.34}$$

and the total cross section is obtained by integrating

$$Q = 2\pi \int_0^\pi |f(\theta)|^2 \sin\theta\, d\theta \tag{3.35}$$

The scattering problem in quantum mechanics is therefore to determine the scattering amplitude $f(\theta)$ for a given interaction potential $V(r)$ (Mott and Massey 1965).

3.2.2 Partial Wave Treatment

Following the partial wave treatment, the plane wave of a free particle, $\exp(ikz)$, in eq. 3.32 is expanded in the form

$$e^{ikz} = \sum_{l=0}^\infty (2l+1)i^l \frac{u_l(r)}{kr} P_l(\cos\theta) \tag{3.36}$$

where the radial wave function $u_l(r)$ satisfies

$$\nabla^2 u_l + \left[\frac{k^2 - l(l+1)}{r^2}\right] u_l = 0 \tag{3.37}$$

and $P_l(\cos\theta)$ are the Legendre polynomials. At $r \to \infty$, $u_l(r)$ has an asymptotic form

$$u_l(r) \sim \sin\left(kr - \frac{l\pi}{2}\right)$$

Analogous to this partial wave expansion, the asymptotic wave function for the scattering process becomes

$$\phi_k(r) \sim \sum_l a_l \frac{v_l(r)}{kr} P_l(\cos\theta) \tag{3.38}$$

and the radial wave function $v_l(r)$ is the solution of

$$\nabla^2 v_l + [k^2 - U_l(r)]v_l = 0$$
$$U_l(r) = U(r) + \frac{l(l+1)}{r^2} \tag{3.39}$$

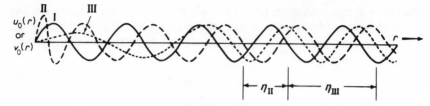

Fig. 3.8 The radial wave functions.

At large r where $U(r) \to 0$, both u_l and v_l represent free particles and can differ at most by a phase factor η_l and v_l can be expressed by

$$v_l(r) \sim \sin\left(kr - \frac{l\pi}{2} + \eta_l\right)$$

where l is the orbiting angular momentum quantum number and η_l, the lth order phase shift.

This phase change is illustrated in Fig. 3.8, which shows the form of $u_0(r)$ (curve I where $V(r) = 0$) and of $v_0(r)$ for an attractive potential (curve II) and a repulsive potential (curve III). An attractive potential implies an acceleration of the particle at small r so that the localized de Broglie wavelength is decreased and η_l is positive. On the other hand, a negative η_l results from a repulsive scattering potential. Substituting eq. 3.36 and 3.38 into eq. 3.32 with the condition that the second term of eq. 3.32 only represents an outgoing wave, we find the coefficient for eq. 3.38:

$$a_l = (2l + 1)i^l e^{i\eta_l} \tag{3.40}$$

and the scattering amplitude

$$f(\theta) = \frac{1}{2ik} \sum_i (2l + 1)(e^{2i\eta_l} - 1)P_l(\cos\theta) \tag{3.41}$$

The differential cross section is then simply given by

$$I(\theta) = |f(\theta)|^2 = A^2 + B^2$$

$$A = \text{Re } f(\theta) = \frac{1}{2k} \sum_l (2l + 1)\sin 2\eta_l P_l(\cos\theta) \tag{3.42}$$

$$B = \text{Im } f(\theta) = \frac{1}{k} \sum_l (2l + 1)\sin^2 \eta_l P_l(\cos\theta)$$

and the total scattering cross section

$$Q = 2\pi \int_0^\pi I(\theta)\sin\theta \, d\theta = \frac{4\pi}{k^2} \sum_l (2l + 1)\sin^2 \eta_l \tag{3.43}$$

It is clear that all relevant information about molecular scattering is contained in the scattering amplitude (eq. 3.41) in which the Legendre polynomial $P_l(\cos \theta)$ determines the angular distribution of scattering particles and the term containing the phase shift η_l is due to the interaction potential $V(r)$. Moreover, because of the interference between the neighboring wave packets we anticipate an oscillatory structure of the differential cross section quantum mechanically. It can be further shown that the total cross section is directly related to the forward scattering amplitude $f(\theta = 0)$ by the relation

$$Q = \frac{4\pi}{k} \operatorname{Im} f(0) \tag{3.44}$$

which is the so-called optical theorem.

Although the formalism of quantum theory appears to be simple, its application to the molecular scattering problem may involve a large number of phase shifts, and the numerical calculation is almost impossible except for a simple scattering system. To see how many phases are involved in the calculation, we start from eq. 3.39; if the interaction potential $U(r)$ is small for $r > [l(l + 1)]^{1/2}/k$, the incident wave will not be seriously distorted and the phase shift η_l will be small. The number of phases that influence the scattering for a short-range potential can then be obtained:

$$l(l + 1) = k^2 d^2 \qquad \text{or} \qquad l \approx kd \tag{3.45}$$

where d is a radius at which the scattering potential becomes negligibly small. For the potential of the form $V(r) = Cr^{-n}$, the condition is

$$d \gg \left(\frac{C}{E}\right)^{1/n} \qquad \text{or} \qquad l \gg C^{1/n}\left(\frac{2\mu}{\hbar^2}\right)^{1/2} E^{(1/2)-(1/n)} \tag{3.46}$$

Therefore if $n > 2$, most phases must vanish as E approaches zero. This means that for low-energy scattering by a short-range potential perhaps only one or a few phases need be considered and direct computation of all phases contributing to the scattering process may be feasible. For high energies or heavy masses, on the other hand, the number of phases that contribute to the scattering may become very large and a suitable approximation may be necessary not only to ease the computation but also to get a physical insight into the molecular collision process. For instance, we consider the molecular scattering at the thermal energy where the de Broglie wavelength may be as small as 0.01 Å but the effective range of the interaction potential may be extended to 20 Å. Thus the summation for the scattering amplitude will contain nearly 2000 terms, making it impractical to make a direct computation. In the next two sections we discuss how the scattering problem may be simplified without losing the detailed physical picture of molecular scattering.

3.2.3   The Born Approximation

The Born approximation, which is regarded as a perturbation method by assuming that the scattered wave is a small addition to the unperturbed incident wave, may be applicable to molecular scattering when the kinetic energy of the collision pairs is much larger than the interaction potential. There are several methods to obtain the Born approximation; the most direct and convenient way is by multiplying eq. 3.37 by $v_l(r)$ and eq. 3.39 by $u_l(r)$, subtracting the resulting equations, and integrating over r from 0 to ∞:

$$\int_0^\infty dr \left[v_l \frac{d^2 u_l}{dr^2} - u_l \frac{d^2 v_l}{dr^2} \right] = - \int_0^\infty dr \, u_l \, U(r) v_l \qquad (3.47)$$

Integrating by parts on the left-hand side, we get

$$\sin \eta_l = - \frac{1}{k} \int_0^\infty dr \, u_l \, U(r) v_l \qquad (3.48)$$

If the interaction potential is weak, the distortion of the incident wave will be small. We thus expect that η_l will be small and it is a reasonable approximation to replace v_l with u_l. The latter is simply the spherical Bessel function

$$\eta_l = \frac{\pi \mu}{\hbar^2} \int_0^\infty [J_{l+(1/2)}(kr)]^2 V(r) r \, dr \qquad (3.49)$$

Making use of the harmonic expansion,

$$\frac{\pi}{2kr} \sum_l (2l + 1) P_l(\cos \theta) [j_l(kr)]^2 = \frac{\sin Kr}{Kr}$$

the scattering amplitude becomes

$$f(\theta) = - \frac{2\mu}{\hbar^2} \int_0^\infty \frac{\sin Kr}{Kr} V(r) r^2 \, dr$$

$$K = 2k \sin \frac{\theta}{2} \qquad (3.50)$$

Note that $f(\theta)$ does not explicitly involve the scattering angle θ except through the magnitude of the momentum change. Also in contrast to the general expression for $f(\theta)$ in eq. 3.41, the scattering amplitude given by the first Born approximation is always real.

At high velocities if the scattering angle is not too small, the integrand of eq. 3.50 becomes the product of a rapidly oscillatory function of $\sin Kr$ and a slowly varying function $rV(r)$, and the cancellation tends to make

$f(\theta)$ small. The resultant total cross section decreases and the scattering becomes concentrated at small angles. Therefore at large l in accord with large E (eq. 3.46), the rapidly oscillatory Bessel function may be replaced as follows:

$$\langle J_{l+(1/2)}^2(kr) \rangle = \frac{1}{\pi k r} \left[1 - \frac{l^2}{k^2 r^2} \right]^{-1/2} \qquad \text{for} \quad r > r_0 \qquad (3.51)$$

and r_0 is the first zero of the Bessel function. Substituting eq. 3.51 into eq. 3.49 with $l = kb$, we get

$$\eta_{JB} = - \frac{k}{2E} \int_{r_0}^{\infty} \frac{V(r)\,dr}{(1 - b^2/r^2)^{1/2}} \qquad (3.52)$$

This is the Jeffreys–Born approximation for the higher-order phases. For an inverse power potential, $V(r) = Cr^{-n}$

$$\eta_{JB} = - \frac{kC}{2Eb^{n-1}} \frac{K_n}{n-1} \qquad (3.53)$$

where K_n is defined in eq. 3.16. Thus at the high-energy limit the phase shift is inversely proportional to $E^{1/2}$ and the total cross section, which can be evaluated from eq. 3.43 by making the approximation $\sin^2 \eta_l = \eta_l^2$, becomes inversely proportional to E.

The condition for the validity of the first Born approximation is that the phase shift has to be small. We therefore anticipate that its application must depend strongly on the magnitude of both collision energy E and the strength of the scattering potential $V(r)$. If the phase shift is approximately given by

$$\eta = \left(\frac{2\mu}{\hbar^2} \right)^{1/2} \int_0^{\infty} [(E - V)^{1/2} - E^{1/2}]\,dr \qquad (3.54)$$

where $V/E \ll 1$ and the effective range of the potential is confined to d, we get

$$\eta \doteq \left(\frac{\mu}{2E\hbar^2} \right)^{1/2} \int V\,dr \doteq \left(\frac{\mu}{2E\hbar^2} \right)^{1/2} \bar{v}\,d \ll 1$$

and

$$E \gg \frac{\mu}{2} \left(\frac{\bar{v}d}{\hbar} \right)^2 \qquad (3.55a)$$

is the high-energy requirement for the Born approximation. If the incident energy is small, on the other hand, the de Broglie wavelength is large compared to the effective range of potential $kd \ll 1$, eq. 3.49 implies

$$\eta_l \doteq \frac{\mu}{\hbar^2} \int V\,dr\,r \doteq \frac{u}{\hbar^2} \bar{v}d^2 \ll 1 \qquad (3.55b)$$

In this case sin Kr/Kr remains near unity in the range where the potential is appreciable and therefore $f(\theta)$ becomes only weakly dependent on the angle θ and the information on the interaction potential cannot be extracted from this approximation.

3.2.4 The Semiclassical Approximation

This method applies under the conditions that the localized de Broglie wavelength of the motion changes slowly within the range of the scattering potential. Since the de Broglie wavelength of molecules at thermal energies is small and the variation of the scattering potential within this wavelength is also small, the semiclassical treatment provides a convenient way to introduce quantum effects for describing the molecular scattering at low energies (Child, 1974; Ford and Wheeler 1959).

The derivation of the semiclassical approximation starts with the phase shift calculated from the JWKB approximation (H. Jeffreys, D. Wentzel, A. Kramers, and L. Brillovin),

$$\eta_l = \lim_{r \to \infty} \left\{ \frac{(l + \frac{1}{2})\pi}{2} + \int_{r_0}^r k_l(r)\,dr - k_l r \right\}$$

and

$$k_l(r)^2 = k_l^2 - U(r) - \frac{(l + \frac{1}{2})^2}{r^2} \qquad (3.56)$$

where r_0 is the classical turning point at which $k_l = 0$. Note that in the semi-classical treatment $l(l + 1)$ in the centrifugal term is replaced by $(l + \frac{1}{2})^2$ throughout the derivation. The next step involves the Legendre polynomial; for large l the Legendre polynomial is approximated by the asymptotic expression,

$$P_l(\cos \theta) \doteq \left[(l + \tfrac{1}{2}) \frac{\pi}{2} \sin \theta \right]^{-1/2} \sin \left[(l + \tfrac{1}{2})\theta + \frac{\pi}{4} \right] \qquad \text{for} \quad \sin \theta \geq \frac{1}{l}$$

$$(3.57)$$

In addition, the summation over all l's for the scattering amplitude in eq. 3.41 is replaced by an integral by assuming that many partial waves contribute to the scattering and that the phase shift varies slowly and smoothly with l.

The scattering amplitude of the semiclassical treatment will then have the form

$$f(\theta) = - \left[\frac{\lambda^2}{2\pi \sin \theta} \right]^{1/2} \int_0^\infty (l + \tfrac{1}{2})^{1/2} [e^{i\phi_+} - e^{i\phi_-}]\,dl$$

$$\text{for} \quad \theta \gg 0 \qquad \text{and} \qquad \phi_+ = 2\eta_l \pm (l + \tfrac{1}{2})\theta \pm \frac{\pi}{4}$$

$$(3.58)$$

Since the terms in the integrand of eq. 3.58 are rapidly oscillatory functions, these oscillations are destructive when integrated and lead to a zero contribution to $f(\theta)$ for most l's except when one of the exponents has an extremum; that is the nonzero contributions occur at l's that satisfy

$$\frac{d\phi_{\pm}}{dl} = 2\frac{d\eta_l}{dl} \pm \theta = 0 \qquad (3.59)$$

The relation between the phase shift and the classical deflection function can be found by the direct differentiation of eq. 3.56 and by setting $h(l + \frac{1}{2}) = \mu g b$,

$$2\frac{d\eta_l}{dl} = \pi - 2\int_{r_0}^{\infty} \frac{(l + \frac{1}{2})\,dr}{r^2[k^2 - U - (l + \frac{1}{2})^2/r^2]^{1/2}} = \chi \qquad (3.60a)$$

This implies that the constructive interference only occurs at $\chi = |\theta|$. For the Lennard–Jones (6, 12) potential, we recall that $\chi > 0$ for a repulsive potential whereas for an attractive potential $\chi < 0$. Thus for the attractive potential the condition is that

$$2\frac{d\eta_l}{dl} = \chi = -\theta \qquad (3.60b)$$

and the term containing ϕ_+ is dominant. For the contribution from the repulsive potential, the ϕ_- term dominates.

The relation between the constructive interference and the classical differential cross section can be readily seen for a purely repulsive potential. There is a single stationary phase point at $l = l_\theta$ for the ϕ_- term. On expanding ϕ_- about the l_θ:

$$\phi_- \approx \phi_-(l_\theta) + \frac{1}{2}q(l_\theta)(l - l_\theta)^2$$

with

$$q(l_\theta) = 2\left(\frac{\partial^2\eta_l}{\partial l^2}\right) = \frac{\partial\chi}{\partial l} < 0 \qquad (3.61)$$

and introducing the standard integrals

$$\int_{-\infty}^{\infty} e^{iq\chi^2}\,d\chi = \left(\frac{\pi}{q}\right)^{1/2}\exp\left(\frac{i\pi}{4}\right) \qquad \text{for} \quad q > 0$$

$$= \left(\frac{\pi}{|q|}\right)^{1/2}\exp\left(\frac{-i\pi}{4}\right) \qquad \text{for } q < 0$$

eq. 3.58 reduces to

$$f(\theta) = \left[\frac{\lambda^2(l_\theta + \frac{1}{2})}{\sin\theta|q|}\right]^{1/2}e^{i\alpha} \qquad (3.62a)$$

where

$$\alpha = 2\eta_{l_\theta} - (l_\theta + \tfrac{1}{2})\theta - \frac{\pi}{2}$$

The semiclassical differential cross section is then

$$I(\theta) = |f(\theta)|^2 = \frac{(l_\theta + \tfrac{1}{2})}{(k^2 \sin \theta |q|)} \tag{3.63}$$

which is again identical to the classical expression for $l + \tfrac{1}{2} \approx kb$. Thus the scattering amplitude may be rewritten as

$$f(\theta) = [I(\theta)]^{1/2} e^{i\alpha} \tag{3.62b}$$

For a more typical scattering potential, say the Lennard–Jones potential, the classical relation between impact parameter and the angle of deflection is not in a one-one correspondence; that is, two or more values of the impact parameter may produce the same scattering angle, as shown in Fig. 3.9.

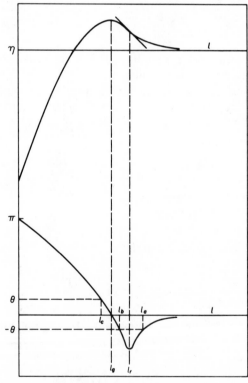

Fig. 3.9 The semiclassical phase shift and the classical deflection angle for a Lennard–Jones (6, 12) potential.

Since the experimental observation cannot distinguish between positive and negative scattering angles, the measured intensity at a given angle is actually a summation over all possible contributions. Thus at a given scattering angle $\chi < \chi_r$, where χ_r is the classical rainbow angle, the classical differential cross section sums over the contributions from three different trajectories, l_a, l_b, and l_c (eq. 3.11b). In semiclassical theory, however, the scattering amplitude leads to the summation

$$I(\theta) = | f_a(\theta) + f_b(\theta) + f_c(\theta)|^2 \qquad (3.64)$$

and because of the interference among these three terms we have an oscillatory structure in the semiclassical differential cross section. Analogous to eq. 3.62, the total scattering amplitude is then

$$f(\theta) = \sum_i f_i(\theta) = \sum_i [I_i(\theta)]^{1/2} e^{i\alpha i} \qquad (3.65)$$

where $I_i(\theta)$ is the classical differential cross section from the ith branch and the phase factors α_i are found to be

$$\alpha_a = 2\eta_{la} + (l_a + \tfrac{1}{2})\theta - \frac{\pi}{2}$$

$$\alpha_b = 2\eta_b + (l_b + \tfrac{1}{2})\theta - \pi \qquad (3.66)$$

$$\alpha_c = 2\eta_{lc} - (l_c + \tfrac{1}{2})\theta - \frac{\pi}{2}$$

Note that these interference effects reduce to two branches of contributions at the rainbow angle where $l_a \approx l_b$ and $\eta_{la} \approx \eta_{lb}$; also, $l_b \approx l_c$ and $\eta_{lb} \approx \eta_{lc}$ for small-angle scattering. It is also important to note that for small-angle scattering, $I(\theta)$ becomes increasingly dominated by the l_a branch and this leads to the same conclusion discussed in the classical mechanics (Sec. 3.1.4).

3.3 LOW-ENERGY ELASTIC SCATTERING OF THE MOLECULAR SYSTEM

3.3.1 Introduction

The interaction of any two atoms or molecules consists of a long-range attractive force and a short-range repulsion. For atoms or molecules that cannot form stable molecules, the weak attractive forces or the van der Waals forces at large intermolecular separation are mainly due to the dispersion forces arising from their polarization. At short distances, repulsive forces appear because of the electron exchange and the Coulomb repulsion when their electron clouds overlap. In the intermediate range of 3–6 Å, there appears a shadow well depth about the order of 0.01 eV. Although there are several mathematical forms designated for this type of intermolecular force, we intend to use the simple two-parameter Lennard–Jones (6, 12) potential throughout this section to illustrate the important features of low-energy molecular scattering and to show how these two parameters can be extracted from the observed elastic scattering cross sections.

For low-energy molecular collisions the localized de Broglie wavelength of the relative motion is in general much smaller than the effective range of the intermolecular potential. Thus classical mechanics alone is often found to be adequate to describe the dynamics of molecular collisions. However, many unique fine structures of molecular scattering observed in the more sophisticated molecular beam experiments require a quantum-mechanical treatment. Since the exact quantum calculation based on the partial wave analysis shows that the variation of phase shift is a slow and smooth function of collision energy for low-energy molecular collision, a semiclassical procedure proves to be a good approximation without losing the physical insight of the collision problem. In semiclassical theory, the motion is described by the wave packets along the classical trajectories and this makes it possible to have a profound interference effect arise where classical mechanics fails (Bernstein, 1966).

For molecular scattering at low energies, the total elastic cross sections are often found to be of the order of 100–1000 Å^2. In terms of classical mechanics, such a large magnitude of cross section is possible only when the collision trajectories governed by the large impact parameter are dominant and thus one would expect to have a strong forward scattering. Accordingly, the differential cross sections at small angles reflect the long-range attractive potential when two particles approach. As energy increases, however, the long-range interaction becomes less important and the collision pairs start to feel the existence of the repulsive wall. In this case, wide-angle scattering due to the repulsive potential becomes significant and the total elastic cross section decreases.

3.3.2 Rainbow Scattering

The rainbow scattering in the semiclassical treatment is a product of the confluence of the stationary phase shift at l_a and l_b, as shown in Fig. 3.9,

$$f(\theta) = f_r(\theta) + f_c(\theta) \tag{3.67}$$

where the scattering amplitude $f_c(\theta)$ due to the repulsive potential is exactly identical to the expression given in eq. 3.62. To evaluate the $f_r(\theta)$ contribution, we note that the rainbow angle appears at $d^2\eta_l/dl^2 = d\chi/dl = 0$. Thus at l near l_r

$$\chi = \chi_r + a_r(l - l_r)^2 \tag{3.68}$$

where $a_r = \frac{1}{2}(\partial^2\chi/\partial l^2) = \frac{1}{2}\lambda^2(\partial^2\chi/\partial b^2) > 0$ measures the curvature of the deflection at the minimum (Fig. 3.9). Integrating over eq. 3.60b and combining the result with eq. 3.68 gives

$$\eta_l = \eta_r - \frac{1}{2}\theta_r(l - l_r) + \frac{1}{6}a_r(l - l_r)^3 \tag{3.69}$$

Note that angle θ is always positive and at the rainbow angle $\theta_r = -\chi_r$. Substituting eq. 3.69 into eq. 3.58, we get

$$\phi_+(l) = \phi_+(l_r) + (\theta - \theta_r)(l - l_r) + \frac{1}{3}a_r(l - l_r)^3 \tag{3.70}$$

where

$$\phi_+(l_r) = 2\eta_r + (l_r + \tfrac{1}{2})\theta + \frac{\pi}{2}$$

Introducing $t = a_r^{1/3}(l - l_r)$, eq. 3.58 becomes

$$f_r(\theta) = \lambda \left[\frac{2\pi(l_r + \tfrac{1}{2})}{\sin\theta_r}\right]^{1/2} a_r^{-1/3} e^{i\delta} \, \text{Ai}\left[\frac{(\theta - \theta_r)}{a_r^{1/3}}\right] \tag{3.71}$$

where $\text{Ai}(y)$ is the Airy function

$$\text{Ai}(y) = \frac{1}{2\pi} \int_{-\infty}^{\infty} \exp\left[i\left(yt + \frac{t^3}{3}\right)\right] dt$$

and

$$\delta = 2\eta_r + (l_r + \tfrac{1}{2})\theta + \frac{5\pi}{4}$$

If there is no significant interference with amplitudes from other branches of the deflection function, the semiclassical cross section near the classical rainbow angle is given by

$$I_r(\theta) = \frac{2\pi(l_r + \tfrac{1}{2})}{k^2 \sin\theta_r} a_r^{-2/3} \, \text{Ai}^2\left[\frac{(\theta - \theta_r)}{a_r^{1/3}}\right] \tag{3.72}$$

and

$$f_r(\theta) = [I_r(\theta)]^{i\delta}$$

Thus the classical singularity at the rainbow angle is suppressed by the nature of the oscillating behavior of $\text{Ai}(y)$ on the bright side of the rainbow ($\theta < \theta_r$) and the rapid fall-off of the $\text{Ai}(y)$ function on the dark side ($\theta \geq \theta_r$).

Moreover, the principal maximum of Ai(y) occurs at $y \approx -1.019$; $I_r(\theta)$ reaches its largest (but finite) value at the angle

$$\theta = \theta_r - 1.019a_r \tag{3.73}$$

In other words, the rainbow angle in the semiclassical treatment is always smaller than the classical value and its difference strongly depends on the shape of the deflection function at the minimum. In addition to the primary rainbow, several supernumary rainbows may also appear because of subsidiary maxima in $[Ai(y)]^2$ at $y = -3.248, -4.820$, and so on. Hence in the semiclassical treatment both the width and amplitude of the rainbow pattern are functions of the curvature of the deflection function at the minimum; the width is simply proportional to $a_r^{1/3}$ whereas the amplitude decreases as a_r increases ($a_r^{-2/3}$ term in eq. 3.72). Since the magnitude of a_r itself increases as $E^*(=E/\varepsilon)$ decreases, as is illustrated in Fig. 3.5, the rainbow pattern observed at a low-energy collision appears to have a small amplitude and spreads out over a broad range of angles. In this case it is also clear from Fig. 3.5 that only a rather small range of angular momenta can contribute to the rainbow pattern, and because of the uncertainty principle the rainbow angle cannot be sharply localized.

At high energies (or large E^*) the contribution from the repulsive potential is significant and the overall scattering amplitude becomes

$$f(\theta) \approx [I_r(\theta)]^{1/2}e^{i\delta} + [I_c(\theta)]^{1/2}e^{i\alpha_c} \tag{3.74}$$

and

$$I(\theta) \approx I_r(\theta) + I_c(\theta) + 2(I_r \cdot I_c)^{1/2}\cos(\delta - \alpha_c) \tag{3.75}$$

The differential cross section oscillates between the limits of $I_{\max}(\theta)$ and $I_{\min}(\theta)$ where

$$I_{\max}(\theta) = (I_r^{1/2} + I_c^{1/2})^2$$

$$I_{\min}(\theta) = (I_r^{1/2} - I_c^{1/2})^2 \tag{3.76a}$$

and the angular separation of the oscillation is simply

$$\Delta\theta = \frac{2\pi}{(d/d\theta)(\delta - \alpha_c)} = \frac{2\pi}{l_r + l_c} \tag{3.76b}$$

The detailed oscillatory structure of the semiclassical differential cross section is shown in Fig. 3.10 in which the dashed curve is the classical cross section (eq. 3.11). The oscillations that have a large angular separation and the largest maximum amplitude from the primary rainbow whereas the supernumary rainbows, which have a small angular separation, are shown at small values of θ. The fine structure of this rainbow pattern in principle may

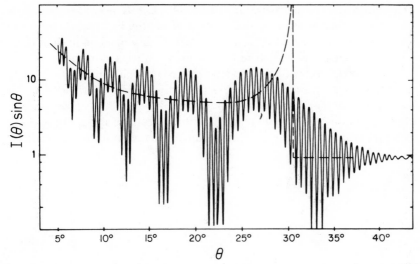

Fig. 3.10 Calculated differential cross sections, classical and semiclassical, for a Lennard–Jones (6, 12) potential.

be used to determine the well depth ε and the range parameter σ for an assumed two-parameter Lennard–Jones (6, 12) potential. The classical rainbow angle χ_r determined by E^*(eq. 3.21) is shown to be located at the point of inflection (Fig. 3.9) and lies at the outermost envelope of the primary rainbow (Fig. 3.10). The potential strength ε can be readily deduced from the observed rainbow angle with the aid of eqs. 3.21 and 3.73. To evaluate the potential size σ, the fine structure of the supernumary rainbows is required. We first determine the constant a_r from eq. 3.72 and the observed oscillation. This is then compared with χ'', which is obtained from eq. 3.6 at the same collision energy, through the expression

$$\chi = \chi_r + \chi''(b^* - b_r^*)^2 \tag{3.77}$$

where b^* is the reduced impact parameter

$$b^* = \frac{b}{\sigma} \approx \frac{l}{k\sigma} \tag{3.78}$$

Thus the magnitude of σ can be determined as

$$\sigma = \frac{1}{k}\left(\frac{\chi''}{a_r}\right)^{1/2} \tag{3.79}$$

which is obtained by comparing eq. 3.68, 3.77, and 3.78.

In molecular beam scattering experiments it is necessary to employ velocity selection to observe the rainbow structure. Even so, the resolution has to be high enough to resolve the supernumary maxima in the rainbow pattern. In fact, for most beam experiments carried out so far only the primary maximum has been clearly resolved, and in some favorable cases where a strong van der Waals force between the colliding particles is present, several supernumary maxima may be observed. Figure 3.11 shows the results observed for an Na-Hg system in which the potential well depth ε is as large as 55 meV and the number of supernumary maxima up to 7 or more has

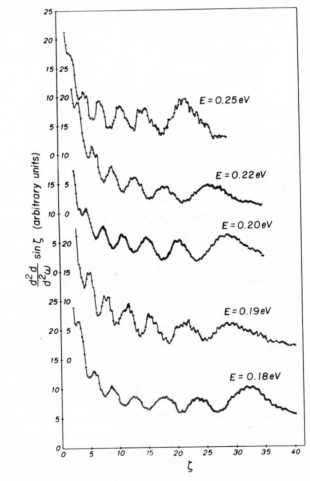

Fig. 3.11   Measured differential cross sections for Na-Hg system at five collision energies.

been observed at collision energy $E = 0.18$–0.25 eV. In addition, the rapid oscillations due to the interferences were also resolved clearly at $E = 0.25$ eV. Note that the rainbows shift to smaller angles and the higher-order super-numary rainbows disappear as the energy increases (Toennies, 1974).

3.3.3 Small-angle and Glory Scattering

For low-energy molecular collisions the small-angle scattering is an important subject since it largely reflects the long-range attractive force when collision takes place. Near the small-angle region, the scattering amplitude consists of a large contribution $f_a(\theta)$ from the l_a branch in Fig. 3.9, and a minor contribution $f_g(\theta)$ due to the confluence of the b and c branches,

$$f(\theta) = f_a(\theta) + f_g(\theta)$$
$$= [I_a(\theta)]^{1/2} e^{i\alpha_a} + [I_g(\theta)]^{1/2} e^{i\alpha_g} \qquad (3.80)$$

To evaluate $f_a(\theta)$, $P_l(\cos \theta)$ is expanded as $\theta \to 0$

$$P_l(\cos \theta) = 1 - \frac{l^2 \theta^2}{4} + \cdots$$

and eq. 3.42 can be rewritten as

$$B = \text{Im } f_a(\theta) = \frac{2}{k}(J_1 - \tfrac{1}{4}J_2\theta^2) = \frac{2J_1}{k}\left[1 - \frac{\theta^2}{4}\left(\frac{J_2}{J_1}\right)\right]$$

$$A = \text{Re } f_a(\theta) = \frac{1}{k}(J_3 - \tfrac{1}{4}J_4\theta^2) = \frac{J_3}{k}\left[1 - \frac{\theta^2}{4}\left(\frac{J_4}{J_3}\right)\right] \qquad (3.81a)$$

where

$$J_1 = \int_0^\infty l \sin^2 \eta_l \, dl \qquad J_2 = \int_0^\infty l^3 \sin^2 \eta_l \, dl$$

$$J_3 = \int_0^\infty l \sin 2\eta_l \, dl \qquad J_4 = \int_0^\infty l^3 \sin 2\eta_l \, dl \qquad (3.81b)$$

It follows that the differential cross section reduces to

$$I_a(\theta) \approx \frac{4}{k^2} J_1^2 \left[\left(1 - \frac{\theta^2}{4}\frac{J_2}{J_1}\right)^2 + \frac{1}{4}\left(\frac{J_3}{J_1}\right)^2\left(1 - \frac{\theta^2}{4}\frac{J_4}{J_3}\right)^2\right] \qquad (3.82)$$

From direct comparison between eq. 3.81b and the expression for the optical theorem, eq. 3.44, we find

$$Q = \frac{8\pi}{k^2} J_1 \qquad (3.83)$$

Making use of the phase shift derived from the Jeffreys–Born approximation (eq. 3.53) for a scattering potential $V(r) = C_n r^{-n}$:

$$\eta_l = \eta_{JB} = a_n l^{1-n}$$

$$a_n = \frac{\mu C_n K_n k^{n-2}}{\hbar^2(n-1)} \tag{3.84}$$

with the standard integrals

$$\int_0^\infty \frac{\sin^2 t \; dt}{t^{P+1}} = \frac{1}{P} \int_0^\infty \frac{\sin 2t \; dt}{t^P} = \frac{\pi 2^{P-2}}{P \sin(P\pi/2)\Gamma(P)}$$

we get

$$J_1 = \frac{a_n^{2/(n-1)}}{G(n)}$$

$$G(n) = \frac{1}{\pi} (2)^{3n-5/(n-1)} \sin\left(\frac{\pi}{n-1}\right)\Gamma\left(\frac{2}{n-1}\right) \tag{3.85}$$

and

$$I_a(\theta) = \left(\frac{kQ}{4\pi}\right)^2 H(n)\exp\left(-h(n)\frac{k^2 Q\theta^2}{8\pi}\right)$$

$$H(n) = 1 + \tan^2\left(\frac{\pi}{n-1}\right)$$

$$h(n) = \tan\frac{(2\pi/n-1)[\Gamma(2/n-1)]^2}{[2\pi\Gamma(4/n-1)]} \tag{3.86}$$

The value of $h(n)$ is unity for $n \to \infty$ and near unity for values of $n \geq 6$. Thus the important quantum correction is the leveling off of the differential cross section to a finite value as $\theta \to 0$; this yields, in turn, a finite total cross section. Since this equation is derived from a small-angle expansion of the Legendre polynomial, it becomes inaccurate at large angles. We recall the classical results of the small-angle approximation in Sec. 3.1.4, where there appears a critical angle θ_c within which eqs. 3.17 and 3.18 fail. A rough quantitative estimate of this critical angle is given by

$$\theta_c \approx \frac{\pi}{kr_0} \approx \frac{\pi\hbar}{\mu g r_0} \tag{3.87a}$$

where r_0 is the distance of closest approach. Substituting eq. 3.16 into eq. 3.87a and setting $r_0 \approx b$, we find

$$\theta_c \approx \left(\frac{\pi^2\hbar^2}{2\mu}\right)^{n/2(n-1)} (CK_n)^{1/(1-n)} E^{(2-n)/2(n-1)} \tag{3.87b}$$

and the classical differential cross section at small angles may be rewritten by comparing eqs. 3.17 and 3.18

$$I(\chi) = \frac{Q}{\pi n \chi^2} \quad \text{for} \quad \chi \geq \theta_c \tag{3.88}$$

Moreover, the condition for eq. 3.86 to be valid is given by

$$\theta < \frac{\theta_c}{(\pi^2 h(n)/2)^{1/2}} \tag{3.86a}$$

Since θ_c itself is a function of collision energy and reduced mass, the magnitude of the critical angle is found as small as 10^{-3} radian at high collision energy. For low-energy molecular scattering, it is typically less that 0.1 rad. Figure 3.12 shows a typical plot of the small-angle differential cross section from both classical and semiclassical calculations (eqs. 3.17 and 3.86) with $n = 6$; these calculations were made for K-Hg with $g = 6.35 \times 10^4$ cm/sec. The

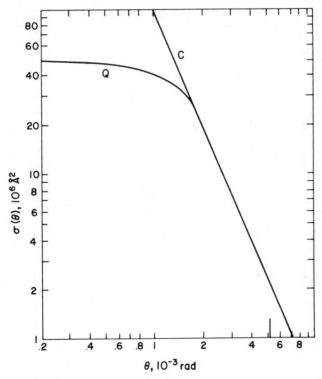

Fig. 3.12 Calculated small-angle differential cross sections for K-Hg at $g = 635$ m/sec.

arrow indicates the critical angle $\theta_c = 5.1 \times 10^{-3}$ rad (Pauly and Toennies 1968; Pauly, 1974; Greene et al., 1966).

The glory contribution $f_g(\theta)$ occurs when $\theta = 0$ (forward glory) and $\theta = -\pi, -2\pi$, and so on, (backward glory); the latter occurs only when the collision energy is very low, whereas the former always exists for a realistic, nonmonotonic potential. Here we consider only the forward glory at which the phase shift has a maximum at $l = l_g$ (Fig. 3.9). We expand

$$\eta_l = \eta_g + \tfrac{1}{2}a_g(l - l_g)^2$$

where

$$a_g = \left(\frac{\partial^2 \eta_l}{\partial l^2}\right) = \frac{1}{2}\left(\frac{\partial \chi}{\partial l}\right) < 0$$

Since the Legendre polynomials are almost independent of l in the small-angle region, the scattering amplitude of eq. 3.41 can be approximated as

$$f_g(\theta) \approx -ik^{-1}(l_g + \tfrac{1}{2})p_l(\cos\theta)e^{2i\eta_g} \int_{-\infty}^{\infty} \exp[ia_g(l - l_g)^2]d(l - l_g)$$

$$\approx k^{-1}\left(\frac{\pi}{|a_g|}\right)^{1/2}(l_g + \tfrac{1}{2})P_l(\cos\theta)\exp\left[i\left(2\eta_g - \frac{3\pi}{4}\right)\right] \qquad (3.89)$$

In analogy to eq. 3.62, eq. 3.89 can be expressed as

$$f_g(\theta) = [I_g(\theta)]^{1/2}e^{i\alpha_g}$$

where

$$I_g(\theta) = \frac{\pi(l_g + \tfrac{1}{2})^2}{k^2|a_g|}\{J_0[(l + \tfrac{1}{2})\theta]\}^2 = I_g(0)\left\{1 - \frac{[(l + \tfrac{1}{2})\theta]^2}{4} + \cdots\right\} \qquad (3.90)$$

and

$$\alpha_g = 2\eta_g - \frac{3\pi}{4}$$

Equation 3.90 yields the important results

$$\operatorname{Im} f_g(0) = [I_g(0)]^{1/2}\sin\left(2\eta_g - \frac{3\pi}{4}\right)$$

$$\operatorname{Re} f_g(0) = [I_g(0)]^{1/2}\cos\left(2\eta_g - \frac{3\pi}{4}\right) \qquad (3.91)$$

$$\frac{\operatorname{Im} f_g(0)}{\operatorname{Re} f_g(0)} = \tan\left(2\eta_g - \frac{3\pi}{4}\right)$$

Thus glory contribution alone fluctuates from zero when $\eta_g = 3\pi/4$, to infinity when $\eta_g = 5\pi/8$, to zero again when $\eta_g = 2\pi/8$. We therefore anticipate that the forward scattering ratio should show an oscillatory velocity dependence and, accordingly, the resultant total elastic cross section $Q(E)$ should also show a similar oscillatory structure.

3.3.4 Orbiting

For low-energy molecular scattering, classical orbiting may occur when the collision energy E is equal to the centrifugal barrier of the effective potential $U(r)$ and, therefore, the radial velocity of the collision pair vanishes. This implies that the particles will be trapped indefinitely. If the interaction potential is an inverse power potential $V(r) = -Cr^{-n}$, the conditions that $E = U(r)$ and $dU/dr = 0$ give the orbiting impact parameter

$$b_0 = \left(\frac{n}{n-2}\right)^{(n-2)/2n} \left(\frac{nC}{2E}\right)^{1/n} \tag{3.92}$$

We recall that for molecules that obey the Lennard–Jones (6, 12) potential, the classical orbiting occurs only when reduced orbital angular momentum $L^* = L/(2\mu\varepsilon\sigma^2)^{1/2} \le 1.569$ and $E^* = E/\varepsilon \le 0.8$ (Fig. 3.4). Since the typical value of the potential well depth is of the order of 0.01 eV, we except that the orbiting phenomenon occurs at very low collision energies.

In the region of classical orbiting there is an important quantum correction of a tunneling effect through which the collision pairs may be trapped and form a quasi-bound molecular state. Moreover, owing to the limitation of the uncertainty principle, the conditions leading to the orbiting phenomenon cannot be correctly defined in quantum mechanics. It is therefore true that in the quantum theory, our attention focuses on the possible quasi-bound states instead of the particular trapped classical trajectory. To see how the orbiting phenomenon enters in the quantum treatment, Fig. 3.13 illustrates the probability amplitudes A, B, and C that the collision particles may be captured between the classical turning points a_l and b_l. Hence we have the general expression for the wave function $\phi_l(r)$,

$$\phi_l(r) = [k_l(r)]^{-1/2} \left[A' \exp\left(i \int_{a_l}^r k_l(r)\,dr\right) + A'' \exp\left(-i \int_{a_l}^r k_l(r)\,dr\right) \right], \quad r \gg a_l$$

$$= [k_l(r)]^{-1/2} \left[B' \exp\left(i \int_{b_l}^r k_l(r)\,dr\right) + B'' \exp\left(-i \int_{b_l}^r k_l(r)\,dr\right) \right], \quad r \ll b_l$$

$$= [k_l(r)]^{-1/2} \left[C' \exp\left(i \int_{c_l}^r k_l(r)\,dr\right) + C'' \exp\left(-i \int_{c_l}^r k_l(r)\,dr\right) \right], \quad r \gg c_l$$

$$\tag{3.93}$$

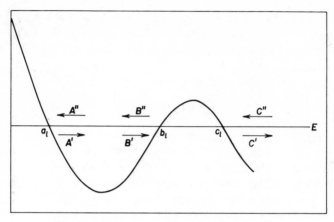

Fig. 3.13   Amplitude coefficients used in eq. 3.93.

With the aid of the semiclassical connection formulas, $\phi_l(r)$ at $r \gg c_l$ is found to be

$$\phi_l(r) \doteq [k_l(r)]^{-1/2} \sin\left(kr - \frac{(l + \frac{1}{2})\pi}{2} + \varepsilon_l\right)$$

$$\doteq e^{-ikr} - (-1)^l S_l(k) e^{ikr} \tag{3.94}$$

where $S_l(k) = \exp(2i\varepsilon_l)$ contains the relative amplitude of the incoming and outgoing parts of the wave functions.

Since the existence of a possible quasi-bound state distorts the continuum wave function of the scattering problem, the probability of finding the scattered particles at a particular distance is no longer time independent. It follows that in stationary state theory, one has to introduce the complex energy

$$E = E_r - \tfrac{1}{2}i\Gamma \tag{3.95}$$

to allow the decay of a quasi-bound state, and the probability of finding the scattered particles some distance away becomes $\phi_l^* \phi_l \exp(-\Gamma t/\hbar)$. Here E_r is the resonant energy and Γ the resonance width. Thus the scattering matrix becomes

$$S_l = \exp(2i\eta_l)\left(1 - \frac{i\Gamma}{E - E_r + \tfrac{1}{2}i\Gamma}\right) \tag{3.96}$$

and the total phase shift in the neighborhood of the resonance level is

$$\varepsilon_l = \eta_l + \arctan\left(\frac{\Gamma}{2(E_r - E)}\right) \tag{3.97}$$

$$= \eta_l + \eta_r$$

where η_l is the ordinary potential scattering phase shift, which varies slowly with the collision energy while the phase shift η_r is small except in the regions near the resonance energy. Hence the contribution from the pronounced resonance effect would be expected at some particular lth partial wave and

$$f_l(\theta) = \frac{(2l + 1)}{2ik}\, e^{2i\eta_l} P_l(\cos\theta)\left[1 - \frac{i\Gamma}{E - E_r + i\Gamma/2}\right] \tag{3.98}$$

The first term is called the potential scattering amplitude whereas the second term is called the resonance scattering amplitude, and accordingly the partial total cross section becomes

$$Q_l = \frac{4\pi}{k^2}\,(2l + 1)\left\{\sin^2\eta_l + \frac{\Gamma^2\cos 2\eta_l - 2\Gamma(E - E_r)\sin 2\eta_l}{4(E - E_r)^2 + \Gamma^2}\right\} \tag{3.99}$$

As a direct consequence the total amplitude may be expressed as

$$f_t(\theta) = f(\theta) + f_r(\theta)$$

$$f(\theta) = \frac{1}{2ik}\sum (2l + 1)(e^{2i\eta_l} - 1)P_l(\cos\theta) \tag{3.100}$$

$$f_r(\theta) = -\frac{1}{2k}\sum (2l + 1)\frac{\Gamma}{E - E_r + i\Gamma/2}\, e^{2i\eta_l} P_l(\cos\theta)$$

That is, the differential cross section contains contributions from the potential scattering, the resonance scattering, and their interference terms. Thus the classical singularity in the deflection function is replaced by the characteristic resonance behavior, and attention in quantum theory on orbiting phenomenon focuses on the possible quasi-bound states rather than on the particular trapped classical trajectory. Although these sharp quantal resonances have not yet been observed in the molecular system, they are well known in electron and high-energy nuclear scattering.

3.3.5 Total Elastic Cross Section

To evaluate the total cross section, we introduce here first the random phase (rph) approximation. For a scattering potential $V(r) = Cr^{-n}$, the phase shift must be positive and it rapidly decreases as l increases. Thus at a certain critical value L the value of $\sin^2\eta_l$ is randomly distributed over the range from zero to unity. Following this argument the rph approximation assumes that for $l < L$, $\sin^2 \eta_l$ may be replaced by the averaged value of $\frac{1}{2}$, and for $l > L$, $\sin^2 \eta_l$ is equivalent to η_l^2. The general expression for the total elastic cross section may then be divided into two parts:

$$Q = \frac{4\pi}{k^2} \sum_l (2l + 1)\sin^2 \eta_l$$

$$= \frac{8\pi}{k^2} \left[\sum_{l=0}^{L} (l + \tfrac{1}{2})\tfrac{1}{2} + \int_L^{\infty} (l + \tfrac{1}{2})\eta_l^2 \, dl \right] \tag{3.101}$$

Substituting the small phase derived from the Jeffreys–Born approximation (eq. 3.84) and replacing $l + \frac{1}{2} \approx l = bk$, we get

$$Q = \frac{2\pi L^2}{k^2} + \frac{\pi L^2}{(n - 2)k^2}$$

where the critical value can be found by setting $\eta_l = \frac{1}{2}$ in eq. 3.53. Hence we have the so-called Massey–Mohr (MM) expression for the total cross section

$$Q_{MM} = \pi \left(\frac{2n - 3}{n - 2} \right) \left(\frac{2CK_n}{(n - 1)\hbar g} \right)^{2/(n - 1)} \tag{3.102}$$

For $n = 6$ we expect that a plot of $\ln Q$ versus $\ln E$ should give a slope of $-\frac{1}{5}$ but not $-\frac{1}{3}$ as predicted from the classical mechanics. Moreover, in contrast to the classical result, the total elastic cross section is finite for $n > 2$.

If we apply this rph approximation to the simplest case of the hard sphere model

$$V(r) = \infty, \qquad r \le d$$

$$= 0, \qquad r > d$$

the radial wave function v_l (eq. 3.39) must vanish at $r \le d$ while outside the sphere it is simply a spherical Bessel function with a phase shift $\eta_l = -kd + l\pi/2$. At the low-energy limit ($\lambda \gg d$) only the s wave ($l = 0$) contributes significantly to the cross section; we have $\eta_0 = -kd$ and $I(\theta) = d^2$. The scattering is therefore isotropic but the total cross section $Q = 4\pi d^2$ is four times larger than the classical value. At the high-energy limit, on the other hand, the number of partial waves may become very large in terms of the rph approximation where $\sin^2 \eta_l$ is equivalent to $\frac{1}{2}$ for $\lambda l \ll d$ and zero at $\lambda l > d$. This yields $Q = 2\pi d^2$ or twice the classical result. The physical interpretation for the difference between quantum mechanics and the classical results can be stated as follows: At the low-energy limit, the de Broglie wavelength of a scattering particle is much larger than its size and a significant diffraction may take place, whereas at its high-energy limit the extra scattering area may be considered to be due to interference between the incident plane wave and the outgoing scattering wave behind the scattering center.

For collision energies above the classical orbiting limit, the phase shift is a smoothly varying function of l; the general expression for total elastic cross section can be obtained directly in terms of the optical theorem. Making use of eqs. 3.83, 3.85, and 3.91, we thus get

$$Q = \frac{4\pi}{k}\left[I_m f_a(0) + I_m f_g(0)\right] = P(n)\left(\frac{c_n}{\hbar g}\right)^{2/(n-1)}$$

$$+ \frac{4\pi}{k}\left[I_g(0)\right]^{1/2}\sin\left(2\eta_g - \frac{3\pi}{4}\right) \tag{3.103}$$

where

$$P(n) = \frac{\pi^2\left[2K_n/(n-1)\right]^{2/(n-1)}}{\sin[\pi/(n-1)]\Gamma(2/n-1)}$$

and

$$I_g(0) = \frac{2\pi(l_g + \frac{1}{2})}{(k^2|d_\theta/dl|)}$$

For a long-range attractive potential with $n = 6$, the first term becomes

$$Q_{\text{SLL}} = P(6)\left(\frac{C_6}{\hbar g}\right)^{2/5} = 8.083\left(\frac{C_6}{\hbar g}\right)^{2/5} \tag{3.104}$$

and is the so-called Schiff–Landau–Lifshitz approximation. Thus the total cross section is given as $Q = Q_{\text{SLL}} + \Delta Q$, in which ΔQ is the contribution from the glory scattering.

The overall results of eq. 3.103 for the Lennard–Jones (6, 12) potential is shown in Fig. 3.14; also shown is the monotonic cross section Q_{SLL} for the power law potentials. Here we plot the total cross section against the reduced relative collision velocity defined as $g^* = g/g_c$, where the characteristic velocity parameter $g_c = \varepsilon\sigma/\hbar$. The total cross section is thus divided into two regions. For collision velocities smaller than g_c, the total cross section is predominantly determined by the attractive potential, whereas for velocities larger than g_c, the total cross section is determined by the repulsion

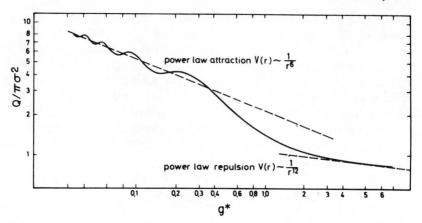

Fig. 3.14 Calculated total cross section for an assumed Lennard–Jones (6, 12) potential.

potential. The oscillation structure of the total cross section in the low-energy range is due to the variation in the forward glory scattering,

$$\Delta Q \approx \sin\left(2\eta_g - \frac{3\pi}{4}\right)$$

This means that the positions of the glory extrema occur at

$$\eta_g = \left(N - \frac{3}{8}\right)\pi \tag{3.105}$$

where the integer $N = 1, 2, \ldots$ for maxima in Q and the half integer $N = \frac{3}{2}, \frac{5}{2}, \ldots$ for minima. Thus from the observed extrema in Q, one can evaluate η_g, which is directly related to the parameters of an assumed potential form. For the Lennard–Jones (6, 12) potential, the maximum phase shift occurs at the forward glory scattering (Fig. 3.9) and can be obtained from the Jeffreys–Born approximation,

$$\eta_g \approx \frac{0.9464\varepsilon\sigma}{\hbar g} \tag{3.106}$$

and

$$N - \tfrac{3}{8} \approx 0.3012 \frac{\varepsilon\sigma}{\hbar g_N} \tag{3.107}$$

where g_N is the velocity at which the Nth maximum occurs. Thus the product of potential parameters $\varepsilon\sigma$ may be determined by a plot of $N - \frac{3}{8}$ versus g_N^{-1}. Experimentally, the total elastic cross section is measured from the attenuation of a well-collimated beam that passes through a target gas of known density n_t within a collision length s,

$$\frac{I}{I_0} = \exp(-n_t Q s) \tag{3.108}$$

Hence the variation of Q as a function of the relative collision energy may be measured.

3.3.6 Intermolecular Forces

From the foregoing theoretical description of elastic scattering, measurements of angular distribution and total cross section can be inverted to extract the detailed interaction forces along the collision paths. In principle, the absolute magnitude of measured total cross section Q may then be used to obtain information on the intermolecular potential: (1) the van der Waals constant C_n and the exponent n from eq. 3.104, (2) the product $\varepsilon\sigma$ of the assumed two-parameter potential function from the positions of extrema on the oscillatory structure in the low-energy region (eq. 3.107), and (3) the constant of the repulsive potential and its shape from the measured Q in the high-velocity region (eq. 3.102). Observation of the detailed rainbow pattern (eqs. 3.72 and 3.73), on the other hand, yields the following information: (1) the well depth ε from the angular position of the primary rainbow and (2) the size parameter σ and the shape of the potential minimum from the angular separation of supernumary rainbows. Presentation of detailed numerical inversion procedures to determine the various types of interaction potential is beyond our present scope, and readers interested in these calculations may find useful references cited at the end of this chapter (Buck, 1975; Pauly, 1974; Toennies, 1974; Bernstein, 1966, Greene et al., 1966).

Results of the elastic scattering discussed in this chapter are entirely based on the assumption that the colliding particles are structureless and the interaction force is spherically symmetric. In principle, one may argue that since the elastic scattering at low energies is generally taking place at a large impact parameter, the colliding particles may not feel the detailed structure of the interaction potential at short distances and therefore conclude that this is not a bad assumption. This conclusion is probably correct for atom-atom collision. If one of the collision particles is not an atom, however, the long-range attractive potential is in general anisotropic and results derived from the oversimplified Lennard–Jones (6, 12) potential are expected to deviate for a molecular system. In fact, even for the atom–atom collision process, the scattering potential derived from the measured elastic data is usually found more complicated than the two-parameter Lennard–Jones potential.

For the molecular collision process the concurrent inelastic or reactive channels may also occur, and they are expected to perturb the results described for elastic scattering. The origin of these possible perturbations may be summarized as follows:

1. The existence of anisotropic potential may lead to a rotational-translational coupling when one of the collision pairs is not an atomic species.
2. When the interaction potential is nonstationary, the vibrational-translational transition is also possible.
3. Curve crossing of different electronic energy states may lead to an electronic excitation at high collision energies.
4. For highly reactive species chemical reaction may take place.

For low-energy collision processes, the perturbation due to processes 2 and 3 is rather rare, whereas the effects due to 1 and 4 have been observed for many molecular systems. These effects tend to damp or quench the undulation amplitude on both total cross section and angular distribution. Moreover, if the probabilities for these processes are large, the magnitude of total elastic cross section is reduced. All these effects, in principle, can then be used to extract the information on the anisotropic scattering potential, the inelastic and reaction cross sections.

REFERENCES

Bernstein, R. B. (1966), *Adv. Chem. Phys.*, **10**, 75.

Buck, U. (1975), *Adv. Chem. Phys.*, **30**, 313.

Burhop, E. H. S. (1961), in *Quantum Theory, I. Elements*, D. R. Bates, Ed., Academic, chap. 9.

Child, M. S. (1974), *Molecular Collision Theory*, Academic.

Ford, K. W. and J. A. Wheeler (1959), *Ann. Phys.*, **7**, 259.

Goldstein, (1959), *Classical Mechanics*, Addison-Wesley.

Greene, E. F., A. L. Moursund, and J. Ross (1966), *Adv. Chem. Phys.*, **10**, 135.

Hirschfelder, J. O., C. F. Curtiss, and R. B. Bird (1954), *Molecular Theory Gases and Liquids*, Wiley.

Landau, L. D. and E. M. Lifschitz (1960), *Mechanics*, Pergamon.

Mason, E. A., J. T. Vanderslice, and C. J. G. Raw (1964), *J. Chem. Phys.*, **40**, 2153.

Mott, N. F. and H. S. W. Massey (1965), *Theory of Atomic Collisions*, Oxford University Press.

Pauly, H. (1974), in *Physical Chemistry, an Advanced Treatise*, H. Eyring, D. Henderson, and W. Jost, Eds., Academic. **6B**, 553.

Pauly, H. and J. P. Toennies (1968), in *Methods of Experimental Physics*, H. Eyring, D. Henderson, and W. Jost, Eds., Academic, **7A**, 227.

Toennies, J. P. (1974), in *Physical Chemistry, an Advanced Treatise*, H. Eyring, D. Henderson, and W. Jost, Eds., Academic, **6A**, 227.

Weston, R. E., Jr., and H. A. Schwartz (1972), *Chemical Kinetics*, Prentice-Hall.

Four

Transition
State Theory

CONTENTS

4.1   INTRODUCTION

The potential energy for a collection of atoms in their lowest state can be quite well represented as a potential hypersurface depending on as many coordinates as are required to fix the relative position of the atoms. Such a hypersurface has valleys or basins representing stable compounds. These are separated from each other by barriers with passes along which systems cross over from one vally or basin to the other. If the top of the pass is a point of no return, or nearly no return, it is a convenient place to count the rate of reaction through that pass. There is in general more than one such pass each corresponding to a different reaction mechanism. The rate of reaction is the sum of such rates. Frequently one of these passes is much lower than the others, so that reaction through the higher passes is negligible by comparison. In this case the most reactive pathway is the rate-determining mechanism. The configuration at the top of the pass is the activated complex or the transition state. The activated complex is much like any other molecule except that it has an internal translational degree of freedom that traverses the pass. If the activated state is truly a point of no return, then the number of systems passing through the pass in one direction is without influence on the number of systems passing in the reverse direction. This is true because the activated complexes for a measurable rate are far outside of each other's field of influence, so each goes its independent way oblivious of what another activated complex is doing. For this reason the rate of going from reactants to products at equilibrium persists whether or not products are present in the system. The rate at equilibrium can accordingly be used as the rate away from equilibrium if the only change in the system from equilibrium is the removal of products. For systems further disturbed one must break up the system into pools of reactants that are in equilibrium with their transition states and treat the resulting network of reactions as discussed in Chapter 1.

It is convenient to call the path leading most steeply downward in both directions from the stationary point on the hypersurface the reaction co-

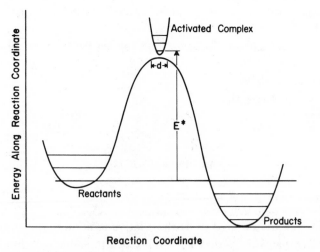

Fig. 4.1   Section through multidimensional potential surface along the reaction coordinate.

ordinate. The actual reaction trajectories through the pass in general do not follow the reaction coordinate but are statistically distributed about it.

In Fig. 4.1 we schematically represent the general situation for reactions. We have only drawn one parabola for the activated state at right angles to the reaction coordinate, but there are of course other parabolas for each vibrational degree of freedom.

4.2 CALCULATION OF THE RATE OF AN ELEMENTARY REACTION

We now calculate the rate of reaction R for this elementary process of crossing the barrier.

$$R = \kappa_{\frac{1}{2}} C_{\delta}^{\ddagger} \frac{\bar{\lambda}}{\delta} = \kappa_{\frac{1}{2}} C^{\ddagger} \frac{(2\pi m^{\ddagger} kT)^{1/2}}{h} \delta \sqrt{\frac{2kT}{\pi m^{\ddagger}}} / \delta = \kappa \frac{kT}{h} C^{\ddagger} \qquad (4.1)$$

Here κ is the transmission coefficient that is unity if the barrier top is truly a point of no return. It will differ from unity if there is leakage through the barrier or back reflection after crossing the top of the barrier. We represent the concentration of activated complexes in a length, δ, along the reaction coordinate by $C_{\delta}^{\ddagger}$. On the other hand, the concentration of activated complexes in a length δ that are in the lowest translational state at the top of the barrier we indicate by $C^{\ddagger}$. The weighted sum of states, the partition function for translation, in a one-dimensional box of length δ is $[(2\pi mkT^{\ddagger})^{1/2}/h]\delta$. This is the quantity by which $C^{\ddagger}$, the concentration in the lowest translational level, must be multiplied by in order to make the product equal to $C_{\delta}^{\ddagger}$; $\sqrt{2kT/\pi m^{\ddagger}}$ is the mean forward velocity at the top of the pass; $m^{\ddagger}$ is the effective mass associated with translation along the reaction coordinate. Like δ, $m^{\ddagger}$ cancels out of the final result. The factor $\frac{1}{2}$ in eq. 4.1 takes care of the fact that only half of the molecules $C_{\delta}^{\ddagger}$ at the top of the barrier at equilibrium are moving in the forward direction.

We now write the equation for the equilibrium constant $K^{\ddagger}$, which relates the activated complex concentration $C^{\ddagger}$ to the concentration of reactants. Accordingly,

$$K^{\ddagger} = \frac{C^{\ddagger}\gamma^{\ddagger}}{C_1\gamma_1 C_2\gamma_2\ldots} \qquad (4.2)$$

Thus

$$R = \kappa \frac{kT}{h} K^{\ddagger} \frac{C_1\gamma_1 C_2\gamma_2\cdots}{\gamma^{\ddagger}} \equiv k' \frac{\gamma_1\gamma_2\cdots}{\gamma^{\ddagger}} C_1 C_2\ldots \qquad (4.3)$$

Here the symbols γ_i represent the respective activity coefficients and C_1, C_2, and so on, represent the concentrations of substances entering into the activated complex. Where there is catalysis the appropriate concentration of the catalyst is to be included but will divide out in the equilibrium constant

$$K = \frac{k_f}{k_b} \tag{4.4}$$

where k_f and k_b are, respectively, the forward and backward rates.

It is convenient to write $\Delta G^{\ddagger} = -RT \ln K^{\ddagger}$ as for any other equilibrium. Equation 3.4 then takes the form

$$R = \kappa \frac{kT}{h} \exp[-\Delta G^{\ddagger}/RT] \frac{\gamma_1 \gamma_2 \cdots}{\gamma^{\ddagger}} C_1 C_2 \ldots \tag{4.5}$$

Here $\Delta G^{\ddagger}$ is the Gibbs free energy of activation for the standard state. For the standard state by definition $\gamma^{\ddagger} = \gamma_1 = \gamma_2 = 1$. Also

$$\Delta G^{\ddagger} = \Delta H^{\ddagger} - T \, \Delta S^{\ddagger} + \int_{p_s}^{p} \frac{\partial \, \Delta G^{\ddagger}}{\partial V} \, dp = \Delta H^{\ddagger} - T \, \Delta S^{\ddagger} + (p - p_s) \overline{\Delta V^{\ddagger}}$$

$$\tag{4.6}$$

Here $\overline{\Delta V^{\ddagger}}$ is the mean difference in volume of the activated complex, and of the molecules entering into the activated complex, and p_s is the pressure in the system in the standard state. Since values of p, of interest, are usually large compared with p_s, it follows that p_s can be neglected. If statistical mechanics is to be used to calculate $K^{\ddagger}$, it should be remembered that we have used 1 for the partition function for the reaction coordinate of the activated complex.

Equation 4.5 is applicable to both adiabatic and diabatic processes. For diabatic processes where the activated complex crosses over from one potential energy surface to another with a different electronic structure values of the transmission coefficient κ is often less than 1. In the chapter on reaction kinetics in *Quantum Chemistry* (Eyring et al., 1944) the transmission coefficient is discussed. κ is also discussed in the *Theory of Rate Processes* (Glasstone et al., 1941). When barrier leakage is important, as in the inversion of ammonia, the rate of barrier penetration must be added to the rate of passing over the barrier. Here discussions of radioactive disintegration are relevant as developed very early by Condon and Morse in their book *Quantum Mechanics* (1929). It is convenient next to develop classical mechanics briefly.

4.3 LAGRANGE'S EQUATIONS

In Chapter Two we have seen something of the utility of quantum mechanics in constructing the potential energy surfaces that play a key role in the theory of reactions. To determine the normal modes from the potential energy surface a knowledge of classical mechanics is necessary.

Newton's equations of motion are expressed by the set of equations

$$m_i \ddot{x}_i = X_i \tag{4.7}$$

There are n equations of this form, one for each of the three degrees of freedom of the $n/3$ atoms under consideration. Here m_i is the mass of an atom, $\ddot{x}_i \equiv \partial^2 x/\partial t^2$ is the atomic acceleration, and X_i is the force acting on an atom along the ith coordinate. In order to transform these n equations into an equivalent but more useful set of coordinates q, we have the n transformation equations $x_i = x_i(q_1, q_2, \ldots, q_n)$ (Glasstone et al., 1941). Now

$$dx_i = \sum_{j=1}^{n} \frac{\partial x_i}{\partial q_j} dq_j \tag{4.8}$$

so that

$$\dot{x}_i = \sum_{j=1}^{n} \frac{\partial x_i}{\partial q_j} \dot{q}_j \tag{4.9}$$

Substituting the values of the $\dot{x}_i$'s into the kinetic energy

$$T = \sum_i \tfrac{1}{2} m_i \dot{x}_i^2 \tag{4.10}$$

one gets an expression for the kinetic energy that is homogeneous and quadratic in the $\dot{q}$'s. The work, dW, in a small displacement is

$$dW = \sum_i X_i \, dx_i = \sum_j \sum_i X_i \frac{\partial x_i}{\partial q_j} dq_j = \sum_j \sum_i m_i \ddot{x}_i \frac{\partial x_i}{\partial q_j} dq_j \tag{4.11}$$

where V is the potential energy and for $X_i = -\partial V/\partial x_i$ we have

$$\sum_i X_i \frac{\partial x_i}{\partial q_j} = \sum_i -\frac{\partial V}{\partial x_i} \frac{\partial x_i}{\partial q_j} = -\frac{\partial V}{\partial q_j} \tag{4.12}$$

We can therefore write

$$dW = \sum_i X_i \, dx_i = \sum_j -\frac{\partial V}{\partial q_j} dq_j = \sum_j \sum_i \left(\frac{d}{dt} \left(m_i \dot{x}_i \frac{\partial x_i}{\partial q_j} \right) - m_i \dot{x}_i \frac{d}{dt} \frac{\partial x_i}{\partial q_j} \right) dq_j$$

$$\tag{4.13}$$

From eq. 4.9 we see that

$$\frac{\partial \dot{x}_i}{\partial \dot{q}_j} = \frac{\partial x_i}{\partial q_j} \tag{4.14}$$

We next show that

$$\frac{d}{dt} \frac{\partial x_i}{\partial q_j} = \frac{\partial \dot{x}_i}{\partial q_j} \tag{4.15}$$

Now

$$\frac{d}{dt} \frac{\partial x_i}{\partial q_j} = \sum_k \frac{\partial^2 x_i}{\partial q_k \, \partial q_j} \dot{q}_k \tag{4.16}$$

and from eq. 4.9

$$\frac{d\dot{x}_i}{\partial q_k} = \sum_j^n \frac{\partial^2 x_i}{\partial q_k \, \partial q_j} \dot{q}_j$$

or interchanging the subscripts k and j we have

$$\frac{d\dot{x}_i}{\partial q_j} = \sum_k \frac{\partial^2 x_i}{\partial q_j \, \partial q_k} \dot{q}_k \tag{4.17}$$

Since the order of taking partial derivatives is immaterial eq. 4.15 is proved. Introducing the results from eqs. 4.14 and 4.15 into eq. 4.13 gives

$$dW = \sum_j -\frac{dV}{dq_j} dq_j = \sum_j \sum_i \left(\frac{d}{dt} \left(m_i \dot{x}_i \frac{\partial \dot{x}_i}{\partial \dot{q}_j} \right) - m_i \dot{x}_i \frac{\partial x_i}{\partial q_j} \right) dq_j \tag{4.18}$$

Since this is true for arbitrary values of dq_j, it follows that

$$\frac{d}{dt} \left(m_i \dot{x}_i \frac{\partial \dot{x}_i}{\partial \dot{q}_j} \right) - m_i \dot{x}_i \frac{\partial \dot{x}_i}{\partial q_j} = -\frac{\partial V}{\partial q_j} \tag{4.19}$$

Remembering that

$$T = \sum_i \tfrac{1}{2} m_i \dot{x}_i^2 \tag{4.20}$$

and that V is independent of the $\dot{q}$'s we have

$$\frac{d}{dt} \frac{\partial L}{\partial \dot{q}_j} - \frac{\partial L}{\partial q_j} = 0 \tag{4.21}$$

where $L = T - V$. These n second-order equations of Lagrange replace the corresponding set of Newton with the advantage that a judicious selection of the q's can cause great simplification, that is, if the q's are chosen in a way to give a simple expression for the potential energy, V.

4.4 THE CANONICAL EQUATIONS OF HAMILTON

Hamilton replaced the n second-order differential equations of Newton and Lagrange with $2n$ first-order equations.

The generalized momentum p_i is defined as

$$p_i = \frac{\partial L}{\partial \dot{q}_i} \tag{4.22}$$

This is the usual definition of momentum if rectangular coordinates are used. Since for conservation systems V does not depend on the $\dot{q}_i$'s we can write

$$p_i = \frac{\partial T}{\partial \dot{q}_i} \tag{4.23}$$

Since T is quadratic and homogeneous in the $\dot{q}_i$'s we have

$$\sum_i \dot{q}_i p_i = \sum_i \dot{q}_i \frac{\partial T}{\partial \dot{q}_i} = 2T \tag{4.24}$$

It follows that

$$H = \sum_i \dot{q}_i p_i - L = T + V \tag{4.25}$$

is the total energy. Since L is simply a function of the q's $+$ $\dot{q}$'s, we have

$$dH = \sum_i \left(p_i \, d\dot{q}_i + \dot{q}_i \, dp_i \right) - \sum_i \left(\frac{\partial L}{\partial \dot{q}_i} \, d\dot{q}_i + \frac{\partial L}{\partial q_i} \, dq_i \right) \tag{4.26}$$

But also

$$\frac{d}{dt} \frac{\partial L}{\partial \dot{q}_i} - \frac{\partial L}{\partial q_i} = 0 \tag{4.27}$$

Hence

$$\frac{\partial L}{\partial q_i} = \dot{p}_i \tag{4.28}$$

Substituting eqs. (4.22) and (4.29) in (4.27) gives

$$dH = \sum_i \left(p_i \, d\dot{q}_i + \dot{q}_i \, dp_i - p_i \, d\dot{q}_i - \dot{p}_i \, dq_i \right) = \sum_i \left(\dot{q}_i \, dp_i - \dot{p}_i \, dq_i \right) \tag{4.29}$$

Since, as eq. 4.29 shows, H is only a function of the p's and q's, we must have

$$dH = \sum_i \left(\frac{\partial H}{\partial p_i} \, dp_i + \frac{\partial H}{\partial q_i} \, dq_i \right) \tag{4.30}$$

Accordingly, from eqs. 4.29 and 4.30 we get Hamilton's set of $2n$ canonical equations of motion

$$\frac{\partial H}{\partial p_i} = \dot{q}_i \quad \text{and} \quad \frac{\partial H}{\partial q_i} = -\dot{p}_i \tag{4.31}$$

Equations 4.22 and 4.25 give a straightforward way of obtaining H in terms of the p's and q's from L that was formulated in the previous section. It follows that if the p's and q's are known at any instant, the canonical equations permit the calculation of the p's and q's at any time in the future and at any time in the past for a system obeying classical mechanics. However, for very fast particles relativity theory applies and for the atomic and nuclear states wave mechanics must be used. This change replaces strict causality with very high probability in the prediction of events.

4.5 THE THEORY OF SMALL VIBRATIONS

If one expands the potential V in the neighborhood of a stationary point in a Taylor's expansion, taking the minimum or minimax as the origin, one gets

$$V = V_0 + \sum_i \left(\frac{\partial V}{\partial q_i}\right)_0 q_i + \sum_i \sum_j \left(\frac{\partial^2 V}{\partial q_i\, \partial q_j}\right)_0 q_i q_j + \sum_i \sum_j \sum_k \left(\frac{\partial^3 V}{\partial q_i\, dq_j\, \partial q_k}\right)_0 q_i q_j q_k$$

$$(4.32)$$

It is convenient to take $V_0 = 0$ and $(dV/dq_i)_0 = 0$ at a stationary point. The third derivatives, $\partial^3 V/\partial q_i\, \partial q_k\, \partial q_k$, and higher derivatives are neglected in the theory of small vibrations.

For small vibrations $\partial^2 V/\partial q_i\, dq_j$ approximates constancy over the small displacements of a stationary state so we write

$$V = \sum_i \sum_j \tfrac{1}{2} b_{ij} q_i q_j \tag{4.33}$$

Similarly, with the same type of approximations we can take for the kinetic energy

$$T = \tfrac{1}{2} \sum a_{ij} \dot{q}_i \dot{q}_j \tag{4.34}$$

Then

$$L = T - V = \tfrac{1}{2} \sum a_{ij} \dot{q}_i \dot{q}_j - \tfrac{1}{2} \sum b_{ij} q_i q_j \tag{4.35}$$

and

$$\frac{d}{dt}\frac{\partial L}{\partial \dot{q}_i} - \frac{\partial L}{\partial q_i} = \sum_j a_{ij} \ddot{q}_j + \sum_j b_{ij} q_j = 0 \tag{4.36}$$

If the system has F degrees of vibrational freedom there are F such equations. We multiply each ith equation by a constant c_i, whose value is to be determined, and add the equations. We now require that the set of c_i's give an equation for the sum of the form

$$\ddot{Q} + \lambda Q = 0 \qquad (4.37)$$

where

$$Q = \sum_j h_j q_j \qquad (4.38)$$

The equations to be satisfied in order to obtain this result are

$$\sum_i c_i a_{ij} = \frac{1}{\lambda} \sum_i c_i b_{ij} = h_j \qquad (4.39)$$

Solving the set of equations given by the equalities on the left fixes the c_i's. With the c_i's determined the remaining equalities on the right fix the value of the h's except for one arbitrary constant that measures the amplitude of the vibrations. Now the set of equations from eq. 4.39 to be solved are

$$\sum_i (\lambda a_{ij} - b_{ij}) c_i = 0 \qquad (4.40)$$

corresponding to the various values of j. They have a nontrivial solution for each of the F values of λ that satisfy the determinant:

$$|\lambda a_{ij} - b_{ij}| = 0 \qquad (4.41)$$

For each λ_i, satisfying eq. 4.41, introduced into the first $(F = 1)$ of the eq. 4.40 one obtains a set of values for $c_1/c_F \cdots c_{F-1}/c_F$; that is, the C_i's are all fixed to within an arbitrary factor c_F. As stated previously, the value of this factor C_F fixes the amplitude of the wave. With λ_2 and the corresponding c_i's determined the h's are determined and one has an explicit expression for the normal mode Q_i. The general solution of the equation

$$\ddot{Q}_i + \lambda Q_i = 0 \qquad (4.42)$$

is

$$Q_i = A_i \cos \lambda_i (t + \delta_i) \qquad (4.43)$$

We write τ_i for the period of a vibration so that $\sqrt{\lambda_i}\,\tau_i = 2\pi$ and for the frequency we have

$$v_i = \frac{1}{\tau_i} = \frac{1}{2\pi} \sqrt{\lambda_i} \qquad (4.44)$$

For an activated complex all the λ_i's are positive except the one corresponding to crossing the barrier, which is negative, giving a frequency multiplied by $\sqrt{-1}$ that singles it out.

If one is interested in how a particular q_i varies with time one can solve for it in terms of the Q_j's, obtaining

$$q_i = \sum_j q_{ij} Q_j = \sum_j q_{ij} A_j \cos(\sqrt{\lambda_j} t + \varepsilon_j) \tag{4.45}$$

As for any problem involving a determinant such as eq. 4.41 symmetry considerations may be used to greatly simplify the solutions of the determinant (Eyring, Walter, and Kimball). Those normal coordinates that form the basis for an irreducible representation vibrate with the same frequency.

To proceed further it is necessary to develop the statistical mechanical theory of equilibria between reactants and the activated complex.

4.6 STATISTICAL THERMODYNAMICS

The position of an atom is fixed by three spatial coordinates and by the corresponding three components of momentum. These six coordinates are said to specify a phase space for the atom. The phase space of n atoms is accordingly specified by $3n$ positional plus $3n$ momentum coordinates and the system is said to possess $3n$ degrees of freedom. An important problem in both equilibrium and reaction rate theory is the likelihood that any system of n atoms at equilibrium shall possess an energy, ε_i, at temperature, T, uniquely distributed in an allowed way in its $3n$ degrees of freedom. With no loss of generality we can reformulate the question what is the chance that a molecule of say, benzene (system A) in equilibrium with system B, a box composed of $s/3$ like atoms and therefore of s harmonic oscillators, will have an energy ε_i distributed in a unique allowed way in its $n = 36$ degrees of freedom when the combined system $A + B$ possesses a total energy E? We postulate that any unique allowed way of distributing the energy E is equally likely. As always in science the justification of the postulate is agreement between the model and experiment. If we represent the $n_i = (E - \varepsilon_i)/h\nu$ quanta possessed by the s oscillators by circles and the s oscillators by crosses X, then an allowed distribution is (000X)(00X)(X)(00X) where the quanta to the left of an oscillator in parenthesis are said to belong to that oscillator. Since one cross must always be on the extreme right to make sense, there are $(n_i + s - 1)!$ ways of arranging the remaining circles and crosses. However the $(s - 1)!$ ways of permuting the $(s - 1)$ interchangeable parenthesis all correspond to a single distribution. Agreement with experiment requires also that we assume the $n_i!$ permutations of the like quanta correspond to a single allowed state. We therefore have for the total allowed number of ways, N_i, of distributing n_i quanta among the s oscillators the expression

$$N_i = \frac{(n_i + s - 1)!}{(s - 1)!n_i!} \tag{4.46}$$

and for the probability p_i of this distribution occurring

$$p_i = \frac{N_i}{\sum N_i} \tag{4.47}$$

where the summation is over all possible unique ways of assigning energy, ε_i, to system A. Since the spectral distribution of A at a fixed temperature is independent of the nature, of the container B, providing it is large, we may suppose s is very large and $hv \to 0$. Then

$$N_i = \frac{1}{(s - 1)!} \prod_{i=1}^{s-1} (n_i + s - i) = \frac{1}{(s - 1)!} \prod_{i=1}^{s-1} \frac{E - \varepsilon_i + hv(s - i)}{hv}$$

$$= \frac{1}{(s - 1)!} \prod_{i=1}^{s-1} \left(\frac{E + hv(s - 1)}{hv} \left(1 - \frac{\varepsilon_i}{E + hv(s - 1)} \right) \right) \tag{4.48a}$$

If the average energy of an oscillator is γ then

$$E - \varepsilon_i = s\gamma \tag{4.48b}$$

Substituting this value of E in eq. (4.48b) and noting that $hv(s - 1)$ is negligible compared with E and also that $\varepsilon + hv(s - 1)$ is negligible compared with $s\gamma$, we have

$$N_i = \left(\frac{E}{hv} \right)^{s-1} \frac{1}{(s - 1)!} \left(1 - \frac{\varepsilon_i}{s\gamma} \right)^{s-1}$$

$$= \left(\frac{E}{hv} \right)^{s-1} \frac{1}{(s - 1)!} \left(1 - \frac{(s - 1)\varepsilon}{s\gamma} + \frac{(s - 1)(s - 2)}{2} \left(\frac{\varepsilon}{S\gamma} \right)^2 + \cdots \right)$$

$$= \left(\frac{E}{hv} \right)^{s-1} \frac{1}{(s - 1)!} \exp - \frac{\varepsilon_i}{\gamma} \tag{4.49}$$

and further

$$p_i = \frac{e^{-\varepsilon_i/\gamma}}{\sum_i e^{-\varepsilon_i/\gamma}} \tag{4.50}$$

Now for classical oscillators, Dulong and Petit established experimentally that the specific heat is R calories per degree per mole so that the average energy of our oscillator, for which $v \to 0$ and so behave classically down to absolute zero, is

$$\gamma = kT = \frac{1.98 \times 4.18 \times 10^7}{6.02 \times 10^{23}} T = 1.37 \times 10^{-16}T \qquad (4.51)$$

where $k = 1.37 \times 10^{-16}$ ergs/molecule $\cdot$ degree and eq. (4.5) becomes

$$p_i = \frac{e^{-\varepsilon_i/kT}}{\sum_i e^{-\varepsilon_i/KT}} \qquad (4.52)$$

The quantity $\sum_i e^{-\varepsilon_i/kT}$, the partition function, we indicate by f. The average energy E of a system is then

$$E = \frac{\sum \varepsilon_i e^{-\varepsilon_i/kT}}{\sum e^{-\varepsilon_i/kT}} \equiv kT^2 \frac{\partial \ln}{\partial T} \sum_i e^{-\varepsilon_i/kT} \qquad (4.53)$$

We next consider some thermodynamic relations. The first and second laws of thermodynamics tell us the heat absorbed by a system equals the increase in energy of the system plus the work done by the system and this is indicated by the equation

$$T \, dS = dE + dW \qquad (4.54)$$

where T, S, E, and W are the absolute temperature, the entropy, energy, and work done, respectively. The Helmholtz free energy A is a property defined by the equation

$$A = E - TS \qquad (4.55)$$

Thus

$$dA = dE - T \, dS - S \, dT \qquad (4.56)$$

subtracting eq. 4.54 from 4.56, we have

$$dA = -S \, dT - dW \qquad (4.57)$$

Therefore at constant temperature

$$dA = -dW \qquad (4.58)$$

so that A is a measure of the maximum work that can be gotten out of a system at constant temperature. If the work done is entirely against external pressure eq. 4.54 becomes

$$T \, dS = dE + p \, dV \qquad (4.59)$$

and

$$dA = -S \, dT - p \, dV \qquad (4.60)$$

Thus

$$\left(\frac{\partial A}{\partial T}\right)_V = -S \tag{4.61}*$$

$$\left(\frac{\partial A}{\partial V}\right)_T = -p \tag{4.62}*$$

The Gibbs free energy is defined as

$$G = A + pV = A - V\left(\frac{\partial A}{\partial V}\right)_T \tag{4.63}*$$

Thus

$$dG = dA + d(pV) = -S\,dT - dW + p\,dV + V\,dP \tag{4.64}$$

At constant temperature and pressure

$$dG = -(dW - p\,dV) \tag{4.65}$$

Accordingly, if a process is to be carried out at constant temperatures and pressure G measures the maximum available work obtainable over and above that used in pushing back the atmosphere. From eq. 4.64 we see that if the only work being done is against the atmosphere $dW = p\,dV$ and

$$G = -S\,dT + V\,dp \tag{4.66}$$

Thus

$$\left(\frac{\partial G}{\partial T}\right)_p = -S \tag{4.67}$$

and

$$\left(\frac{\partial G}{\partial p}\right)_T = V \tag{4.68}$$

Also

$$-T^2\left(\frac{\partial(A/T)}{\partial T}\right)_V = -T^2\left(\frac{\partial A}{\partial T}\right)_V \frac{1}{T} + T^2 \frac{A}{T} = TS + A = E \tag{4.69}*$$

The heat content H is defined as

$$H = E + pV = -T\left(\frac{\partial(A/T)}{\partial T}\right)_V - V\left(\frac{\partial A}{\partial V}\right)_T \tag{4.30}*$$

So that $dH = dE + p\,dV + V\,dp$ and subtracting $T\,dS = dE + p\,dV$ we have

$$dH = T\,dS + V\,dp \tag{4.71}$$

For the specific heat at constant volume C_V, we have, using eq. 4.59

$$C_V = \left(\frac{\partial E}{\partial T}\right)_V = T\left(\frac{\partial S}{\partial T}\right)_V = -T\left(\frac{\partial^2 A}{\partial T^2}\right) \qquad (4.72)*$$

The coefficient of compressibility β is defined as

$$\beta = -\frac{1}{V}\left(\frac{\partial V}{\partial P}\right)_T = \frac{1}{V(\partial P/\partial V)_T} = \frac{1}{V(\partial^2 A/\partial V^2)} \qquad (4.73)*$$

The coefficient of expansion α is defined as

$$\alpha = \frac{1}{V}\left(\frac{\partial V}{\partial T}\right)_p \qquad (4.74)$$

But

$$dp = \left(\frac{\partial p}{\partial V}\right)_T dV + \left(\frac{\partial p}{\partial T}\right)_V dT \qquad (4.75)$$

Setting $dp = 0$, we have

$$\left(\frac{\partial V}{\partial T}\right)_p = \frac{-(\partial p/\partial T)_V}{(\partial p/\partial V)_T} = \frac{-\partial^2 A/\partial T\, \partial A}{\partial^2 A/\partial V^2} \qquad (4.76)$$

and

$$\alpha = -\frac{\partial^2 A/\partial T\, \partial V}{(V(\partial^2 A/\partial V^2))} \qquad (4.77)*$$

Now the specific heat C_p at constant pressure is defined by

$$C_p = \left(\frac{\partial H}{\partial T}\right)_p = T\left(\frac{\partial S}{\partial T}\right)_p = T\left(\frac{\partial S}{\partial T}\right)_V + \left(\frac{\partial S}{\partial V}\right)_T\left(\frac{\partial V}{\partial T}\right)_p$$

$$= C_V + T\left(\frac{\partial S}{\partial V}\right)_T\left(\frac{\partial V}{\partial T}\right)_p \qquad (4.78)$$

The second equality makes use of eq. 4.59; to obtain the third equality we use $S(V, T)$; the fourth equality makes use of eq. 4.59. At this point we need Maxwell's equations. Thus if $z(x, y)$ then $dz = (\partial z/\partial x)\, dx + (\partial z/\partial y)\, dy \equiv M\, dx + N\, dy$ and since $\partial^2 z/\partial y\, \partial x = \partial^2 z/\partial x\, \partial y$, we have Euler's relation

$$\left(\frac{\partial M}{\partial y}\right)_x = \left(\frac{\partial N}{\partial X}\right)_y \qquad (4.79)$$

From the expressions for dE, dA, dG, and dH in eqs. 4.59, 4.60, 4.66 and 4.71 and using Eulers relation, eq. 4.79 we have Maxwell's relations:

$$\left(\frac{\partial T}{\partial V}\right)_S = -\left(\frac{\partial p}{\partial S}\right)_V \tag{4.80}$$

$$\left(\frac{\partial S}{\partial V}\right)_T = \left(\frac{\partial p}{\partial T}\right)_V \tag{4.81}$$

$$\left(\frac{\partial S}{\partial p}\right)_T = \left(\frac{\partial V}{\partial T}\right)_P \tag{4.82}$$

$$\left(\frac{\partial T}{\partial P}\right)_S = \left(\frac{\partial V}{\partial S}\right)_P \tag{4.83}$$

Substituting eq. 4.81 into 4.78 gives

$$C_p = C_V + T\left(\frac{\partial P}{\partial T}\right)_V\left(\frac{\partial V}{\partial T}\right)_V = C_V + TV\frac{\alpha^2}{\beta} \tag{4.84}*$$

To get the second equality in eq. 4.84 we use

$$0 = dV = \left(\frac{\partial V}{\partial T}\right)_p dT_V + \left(\frac{\partial V}{\partial p}\right)_T dp_V$$

which gives

$$\left(\frac{\partial p}{\partial T}\right)_V = -\frac{(\partial V/\partial T)_p}{(\partial V/\partial P)_T} \tag{4.85}$$

together with the definitions of α and β just derived. Since A is the maximum capability of a system to do work $(\partial A/\partial n_i)_{T,V,n_j\neq n_i}$ is the increase in this ability to do work through adding a molecule of i and is called the chemical potential μ_i. Of course, n_i is the number of i molecules in the system of interest. Thus we have

$$\mu_i = \frac{dA}{dn_i} \tag{4.86}$$

We now return to the problem of establishing the link between statistical mechanics and thermodynamics. Equating the two expressions for the energy given in eqs. 4.53 and 4.69 gives

$$-T^2\left(\frac{\partial (A/T)}{\partial T}\right) = kT^2\frac{\partial \ln \sum e^{-\varepsilon_i/kT}}{\partial T} \tag{4.87}$$

Integrating, we have

$$\frac{A}{T} = -k \ln \sum e^{-\varepsilon_i/kT} - k \ln \omega' = k \ln \sum_i \omega_i e^{-\varepsilon_i/kT} - k \ln \omega'$$

$$= -k \ln \sum_{i=1}^{\infty} \omega_0 e^{-\varepsilon_0/kT}\left(1 + \frac{\omega_i''}{\omega_0} e^{-(\varepsilon_i - \varepsilon_0)/kT} + \cdots\right) - k \ln \omega' \quad (4.88)$$

Here the integration constant $-k \ln \omega'$ is independent of the temperature. Now according to the third-law experiments that show that the entropy is zero at absolute zero we must have

$$\frac{E}{T} = -k \ln \omega_0 - k \ln \omega' + \frac{\varepsilon_0}{T} \quad (4.89)$$

The third law experiments show that $\omega_0 = 1$ and $k \ln \omega'$ is unchanged in all relevant experiments. It is therefore convenient to set $k \ln \omega' = 0$. Thus we have

$$A = -kT \ln \sum \omega_i e^{-\varepsilon_i/kT} \quad (4.90)$$

Since quantum mechanics as well as spectroscopy should, in principle, suffice to fix ω_i and ε_i, we now have two additional avenues for arriving at all thermodynamic properties.

4.7 PARTITION FUNCTIONS

With quantum-mechanical methods available for calculating approximate, and in the case of the activated complex $(H—H—H)^{\ddagger}$ nearly exact potential energy surfaces, we can frequently distinguish between possible mechanisms. Given a usable potential surface in the neighborhood of the activated complex and the requisite information of the reactants, it is possible to estimate usefully the quantities entering into the rate expression.

Classical statistical mechanics is first used to calculate the partition function for a diatomic molecule. The kinetic energy T for such a molecule is $T = \frac{1}{2}m_1(\dot{x}_1^2 + \dot{y}_1^2 + \dot{z}_1^2) + \frac{1}{2}m_2(\dot{x}_2^2 + \dot{y}_2^2 + \dot{z}_2^2)$.

In Fig. 4.2 we have the figure that indicates how we can transform from rectangular coordinates to a set of rectangular coordinates at the center of gravity of the molecule and polar coordinates for motion of the atoms about the center of gravity. The transformation of coordinates is given below and the corresponding velocities are substituted into eq. 4.91.

We take r as the distance between the two atoms and θ and ϕ as the polar coordinates of atom 1: The transformation of coordinates is

$$x_1 = x + \frac{m_2 r}{m_1 + m_2} \sin \theta \cos \phi \qquad (4.92)$$

$$y_1 = y + \frac{m_2 r}{m_1 + m_2} \sin \theta \sin \phi \qquad (4.93)$$

$$z_1 = z_1 + \frac{m_2}{m_1 + m_2} r \cos \theta \qquad (4.94)$$

$$x_2 = x - \frac{m_1 r}{m_1 + m_2} \sin \theta \cos \phi \qquad (4.95)$$

$$y_2 = y - \frac{m_1 r}{m_1 + m_2} \sin \theta \sin \phi \qquad (4.96)$$

$$z_2 = z - \frac{m_1 r}{m_1 + m_2} \cos \theta \qquad (4.97)$$

$$T = \tfrac{1}{2} m_1 \left(\dot{x} + \frac{m_2}{m_1 + m_2} \right) (\dot{r} \sin \theta \cos \phi + r \cos \theta \dot{\theta} \cos \phi - r \sin \theta \sin \phi \dot{\phi})^2$$

$$+ \tfrac{1}{2} m_1 \left(\dot{y} + \frac{m^2}{m_1 + m_2} \right) (\dot{r} \sin \theta \sin \phi + r \cos \theta \dot{\theta} \sin \phi + r \sin \theta \cos \phi \dot{\phi})^2$$

$$+ \tfrac{1}{2} m_1 \left(\dot{z} + \frac{m_2}{m_1 + m_2} \right) (\dot{r} \cos \theta - r \sin \theta \dot{\theta})^2$$

$$+ \tfrac{1}{2} m_2 \left(\dot{x} - \frac{m_2}{m_1 + m_2} \right) (\dot{r} \sin \theta \cos \phi + r \cos \theta \dot{\theta} \cos \phi - r \sin \theta \sin \phi \dot{\phi})^2$$

$$+ \tfrac{1}{2} m_2 \left(\dot{y} - \frac{m_2}{m_1 + m_2} \right) (\dot{r} \sin \theta \sin \phi + r \cos \theta \dot{\theta} \sin \phi + r \sin \theta \cos \phi \dot{\phi})^2$$

$$+ \tfrac{1}{2} m_2 \left(\dot{z} - \frac{m_2}{m_1 + m_2} \right) (\dot{r} \cos \theta - r \sin \theta \dot{\theta})^2 \qquad (4.97a)$$

If one squares the various terms one finds that all cross terms disappear. For the sum of the square terms one obtains the result

$$T = \tfrac{1}{2} M (\dot{x}^2 + \dot{y}^2 + \dot{z}^2) + \tfrac{1}{2} \mu (\dot{r}^2 + r^2 \dot{\theta}^2 + r^2 \sin^2 \theta \dot{\phi}^2) \qquad (4.98)$$

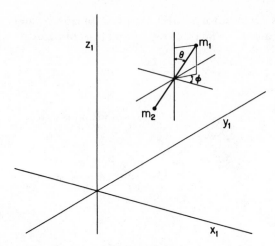

Fig. 4.2 The center of gravity x, y, z of the diatomic molecule consisting of masses m_1 and m_2, a distance r apart, rotating about the molecules center of gravity located at the origin of a moving set of coordinates whose axes remain parallel to the stationary set x_1, y_1, and z_1.

Here $M = m_1 + m_2$ and $\mu = m_1 m_2/(m_1 + m_2)$ and we take $r_1 = (r - r_0)$ where r_0 is the equilibrium distance between the two atoms. Then, remembering that $p_i = \partial T/\partial \dot{q}_i$ and $V = 0$ except for the potential energy $\frac{1}{2}gr_1^2$ between the two atoms, we have

$$H = \frac{1}{2M}(p_x^2 + p_y^2 + p_z^2) + \frac{pr_1^2}{2\mu} + \tfrac{1}{2}gr_1^2 + \frac{p_\theta^2}{2\mu r_0^2} + \frac{p_\phi^2}{2\mu r_0^2 \sin^2 \theta} \quad (4.99)$$

and the Helmholtz free energy A is

$$A = -kT \ln f_x f_y f_z f_{r_1} f_{\theta, \phi} \quad (4.100)$$

Here

$$f_x = -kT \ln \int_{x=0}^{l_x} \int_{p_x=-\infty}^{\infty} \frac{e^{-p_x^2/2mkT} \, dp_x \, dx}{h} \quad (4.101)$$

Remembering

$$\int_{-\infty}^{\infty} e^{-ax^2} \, dx = \sqrt{\frac{\Pi}{a}} \quad (4.102)$$

We have

$$f_x = \frac{(2\pi m kT)^{1/2}}{h} l_x \quad (4.103)$$

and

$$f_x f_y f_z = \frac{(2\pi mkT)^{3/2}}{h^3} l_x l_y l_z = \frac{(2\pi mkT)^{3/2}}{h^3} V \qquad (4.104)$$

$$f_{r_1} = \int_{-\infty}^{\infty} \frac{e^{-p_{r1}^2/2\mu kT}}{h} dr_1 \int_{-\infty}^{\infty} e^{-gr_1^2/2\mu T} dr_1$$

$$= \frac{\sqrt{2\pi\mu kT}}{h} \sqrt{\frac{2\pi kT}{g}} = \frac{2\pi}{h} \sqrt{\frac{\mu}{g}} kT \qquad (4.105)$$

$$f_{\theta\phi} = \frac{1}{h^2} \int_{\phi=0}^{2\pi} \int_{p_\theta=-\infty}^{\infty} \int_{\theta=0}^{\pi} \int_{p_\phi=-\infty}^{\infty} e^{-p_\theta^2/2\mu r_0^2 kT} e^{-p_\phi^2/2\mu r_0^2 \sin^2\theta kT} dp_\theta\, d\theta\, dp_\theta\, d\phi$$

$$(4.106)$$

For the moment of inertia I we write $\mu r_0^2 = I$. Hence

$$f_{\theta\phi} = \frac{1}{h^2} \int_{\phi=0}^{2\pi} \int_{p_\theta=-\infty}^{\infty} \int_{\theta=0}^{\pi} e^{-p^2/2IkT} \sqrt{2\pi IkT} \sin\theta\, d\theta\, dp_\theta\, d\phi$$

$$(4.107)$$

$$f_{\theta\phi} = \frac{1}{h^2} \int_{\phi=0}^{2\pi} \int_{p_\theta=-\infty}^{\infty} e^{-p_\theta^2/2IkT} 2\sqrt{2\pi IkT}\, dp_\theta\, d\theta = \frac{2\pi \cdot 2 \cdot \pi 2 IkT}{h^2}$$

$$= \frac{8\pi^2 IkT}{h^2} \qquad (4.108)$$

It is of interest to compare these classical statistical mechanical results with the values obtained using quantum statistical mechanics. The wave equation yields the energy for a one-dimensional translator $\varepsilon_n = n_x^2 h/8 l_x^2 m$. The Wilson–Sommerfeld quantization rule of old quantum theory yields the same result. Thus

$$nh = \oint p\, dx = \oint m\dot{x}\, dx = 2m\dot{x}l_x \text{ and } \varepsilon_n = \tfrac{1}{2}m\dot{x}^2 = \frac{1}{2}\frac{mn^2 h^2}{4m^2 l_x^2} = \frac{n_x^2 h^2}{8 l_x^2 m}.$$

The circle on the integral sign indicates the integral extends over one period. The partition function is then

$$f_x = \sum_{n_x} e^{-n_x^2 h^2/8 l_x^2 mkT} \approx \int_{n=0}^{\infty} e^{-n^2 h^2/8 l_x^2 m} kT\, dn = \frac{(2\pi mkT)^{1/2}}{h} l_x$$

in agreement with the classical result. The quantum mechanical calculation for a harmonic oscillator yields the energy $\varepsilon_n = (n + \tfrac{1}{2})h\nu$. Whence

$$f = \sum_{n=0}^{\infty} e^{-h\nu/2kT}(1 + e^{-h\nu/kT} + \cdots) = \frac{e^{-h\nu/2kT}}{1 - e^{-h\nu/kT}}$$

Or if one takes the zero of energy at $\frac{1}{2}hv$ the partition function becomes

$$f = \frac{1}{1 - e^{-hv/kT}} \underset{T \to \infty}{=} \frac{kT}{hv} \tag{4.109}$$

The high-temperature values of f, eq. 4.109, are the classical result. This may be readily seen as follows using Lagrange's equation $(d/dt)(\partial L/\partial \dot{x}) - (\partial L/\partial x) = 0$ where $L = \frac{1}{2}\mu \dot{r}_1^2 - \frac{1}{2}g r_1^2$ yields $\mu \ddot{r}_1 + g r_1 = 0$. The solution of this equation is

$$r_1 = A \sin\left(\sqrt{\frac{g}{\mu}} t + \delta\right) \tag{4.110}$$

For a period τ we have

$$2\pi = \sqrt{\frac{g}{\mu}} \tau \quad \text{or} \quad v = \frac{1}{\tau} = \frac{1}{2\pi} \sqrt{\frac{g}{\mu}} \tag{4.111a}$$

Substituting eq. 4.111a in 4.105 gives

$$f_{r_1} = \frac{kT}{hv} \tag{4.111b}$$

Both quantum mechanics and spectroscopy give for the energy of rotation

$$E_j = \frac{J(J + 1)h^2}{8\pi^2 I} \tag{4.112}$$

and for the degeneracy, $2J + 1$. Thus

$$f_{\text{rot}} = \sum_{J=0}^{\infty} (2J + 1)e^{-J(J+1)h^2/8\pi^2 IkT} \approx \int_0^{\infty} (2J + 1)e^{-J(J+1)h^2/8\pi^2 IkT}$$

Substituting $J(J + 1) = x$ and $dx = (2J + 1)dj$ gives

$$f_{\text{rot}} = \int_0^{\infty} e^{-xh^2/8\pi^2 IkT} \, dx = \frac{8\pi^2 IkT}{h^2} \tag{4.113}$$

This is the same result obtained using classical mechanics. Since only half the levels are filled for a linear molecule that is symmetrical with respect to reflection in a plane perpendicular to the molecular axis passing through the center of gravity, we write

$$f_{\text{rot}} = \frac{8\pi^2 IkT}{\sigma h^2} \tag{4.114}$$

Here σ is 2 for a symmetrical molecule and 1 otherwise.

If classical mechanics is applied to a three-dimensional rotator one obtains (Eyring et al., 1964)

$$f_{3\text{rot}} = \frac{8\pi^2(8\pi^3 ABC)^{1/2}(kT)^{3/2}}{\sigma h^3} \tag{4.115}$$

The same result is obtained by integrating the corresponding quantum-mechanical infinite sum for rotation. The symmetry number σ is equal to the number of indistinguishable positions the rigid molecule can be rotated into. For free rotation of a molecule like $CH_3C\,Cl_3$ about the carbon-carbon bond the partition function is

$$f_{1\text{rot}} = \frac{(2\pi IkT)^{1/2}2\pi}{\sigma h} \tag{4.116}$$

where

$$I = \frac{I_1 I_2}{I_1 + I_2} \tag{4.117}$$

where the two I's are the moments of inertia of the groups CH_3 and CF_3 with respect to the molecular axis. For example, the partition function for a benzene molecule is

$$f_1 = \frac{(2\pi mkT)^{3/2}}{h^3} V \frac{8\pi^2(8\pi^3 ABC)^{1/2}(kT)^{3/2}}{12h^3} \prod_{i=1}^{30} \frac{1}{1 - e^{-\theta i/T}} \tag{4.118}$$

Here $\theta_i = h\nu_i/k$ where h, k, and ν_i are Planck's constant, Boltzmann's constant, and the ith oscillator frequency, respectively, and A, B, and C are the three moments of inertia. For N such molecules we have for the partition function

$$f_N = \frac{f_1^N}{N!} = \left(f_1 \frac{e}{N}\right)^N \tag{4.119}$$

$N!$ is the symmetry number corresponding to the number of indistinguishable interchanges of N identical objects and corresponds to the fact that all such interchanges of positions correspond to a single quantum state for the system. The chemical potential for a system of N such gas molecules is

$$\mu = \frac{\partial A}{\partial N} = \frac{\partial}{\partial N}\left(-NkT\ln\left(f_1 \frac{e}{N}\right)\right) \equiv -kT\ln\left(f_1 \frac{e}{N}\right) + kT = kT\ln\frac{N}{f_1} \tag{4.120}$$

$$\mu = kT\ln\left(\frac{N/V}{f_1/V}\right) \equiv kT\ln\left(\frac{c}{F_1}\right) \equiv kT\ln\lambda \tag{4.121}$$

Here λ is defined as the absolute activity, c is the concentration in molecules per cm^3, and $F_1 = f_1/V$ is the partition function of a molecule per unit volume. We have $p = -\partial A/\partial V = (\partial/\partial V)(NkT \ln(F_1 Ve/N)) = NkT/V$. For a Van der Waals gas we have

$$A = -NkT \ln\left(F_1(V - b)e^{a/VRT}\frac{e}{N}\right) \tag{4.122}$$

An obvious check is

$$-\frac{\partial A}{\partial V} = p = \frac{NkT}{V - b} - \frac{a}{V^2} \tag{4.123}$$

All of the thermodynamic properties are readily calculated from $A(V, T)$ by making use of the starred equations in the preceding sections.

4.8   VISUALIZATION OF THE BEHAVIOR OF THREE ATOMS COLLIDING ON A LINE

Consider the atoms in a line as in the accompanying Fig. 4.3. The kinetic energy of the oscillator made up of atoms 1 and 2 is $\frac{1}{2}(m_1 m_2/m_1 + m_2)\dot{r}_1^2$ and of 3 moving with respect to the center of gravity of the oscillator made up of the atoms 1 and 2 is

$$\frac{1}{2}\frac{(m_1 + m_2)m_3}{m_1 + m_2 + m_3}\left(\frac{m_1 r_1}{m_1 + m_2} + \dot{r}_2\right)^2$$

The total kinetic energy T is then

$$\begin{aligned}
T = &\frac{1}{2}\left(\frac{m_1 m_2}{m_1 + m_2} + \frac{m_3 m_1^2}{(m_1 + m_2 + m_3)(m_1 + m_2)}\right)\dot{r}_1^2 \\
&+ \frac{1}{2}\frac{((m_1 + m_2)m_3)}{m_1 + m_2 + m_3}\dot{r}_2^2 \\
&+ \frac{(m_1 + m_2)m_3}{(m_1 + m_2 + m_3)}\frac{m_1}{(m_1 + m_2)}\dot{r}_1\dot{r}_2 \tag{4.124}
\end{aligned}$$

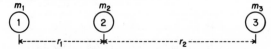

Fig. 4.3   System of three particles lying on a straight line.

that is,

$$T = \frac{1}{2M}(m_1(m_2 + m_3)\dot{r}_1^2 + 2m_1 m_2 \dot{r}_1 \dot{r}_2 + m_3(m_1 + m_2)\dot{r}_2^2) \quad (4.125)$$

If we transform eq. 4.125 so that T becomes

$$T = \frac{m_1(m_2 + m_3)}{2M}(\dot{x}^2 + \dot{y}^2) \quad (4.126)$$

by decreasing the angle between r_2 and r_1 from 90° to $(90 - \theta)°$ and by dividing r_2 by an appropriate quantity C, we find the system consisting of three atoms can be treated as a mass $m_1(m_2 + m_3)/M$ moving in the potential for which the distances r_2 are divided by C and the distances measured in the r_1 direction are unchanged. In order that the coefficient of r_2^2 be equal to that of $\dot{r}_1^2$ the quantity C must satisfy the equation

$$m_1(m_2 + m_3) = \frac{m_3(m_1 + m_2)}{C^2} \quad \text{or} \quad C = \frac{(m_3(m_1 + m_2))^{1/2}}{m_1(m_2 + m_3)} \quad (4.127)$$

To find the angle θ through which r_2/C must be rotated to eliminate the cross terms that would otherwise occur in eq. 4.126, we make the following transformations:

$$\dot{r}_1 = \dot{x} - \dot{y}\tan\theta \quad (4.128)$$

$$\dot{r}_2 = C\dot{y}\sec\theta \quad (4.129)$$

This gives

$$T = \frac{1}{2M}(m_1(m_2 + m_3)(\dot{x}^2 - 2\dot{x}\dot{y}\tan\theta + \dot{y}^2\tan^2\theta) + 2m_1 m_3(\dot{x} - \dot{y}\tan\theta)$$

$$\times (c\dot{y}\sec\theta) + m_3(m_1 + m_2)c^2\dot{y}^2\sec^2\theta) \quad (4.130)$$

For the coefficient of $\dot{x}\dot{y}$ to vanish we require that $-m_1(m_2 + m_3) 2\tan\theta + 2m_1 m_3 C\sec\theta = 0$ so that

$$\sin\theta = \frac{Cm_3}{m_2 + m_3} = \left(\frac{m_1 m_3}{(m_1 + m_2)(m_2 + m_3)}\right)^{1/2} \quad (4.131)$$

Thus we see that if we plot the potential energy surface on coordinates where the angle between r_2 and r_1 is decreased from 90° to $(90 - \theta)°$ and if in the r_2 direction there is a shortening of distances from r_2 to r_2/c, then a mass $m_1(m_2 + m_3)/M$ moving in the potential field would obey eq. 4.126. This facilitates obtaining an intuitive feeling of the behavior of the three

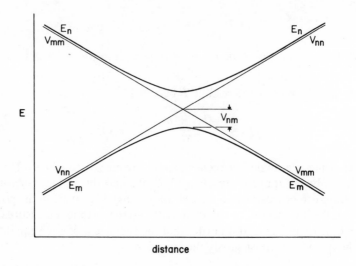

Fig. 4.4 The system in passing from left to right usually remains on the surface with energy E_m. However, for small values of the perturbation V_{nm} and low velocities V, there is the possibility of jumping to the surface whose energy is E_n. The term R_c is the distance at which the two asymptotic lines cross.

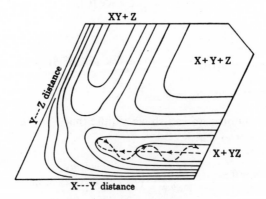

Fig. 4.5 Interconversion of relative translational and vibrational energy. Reprinted by permission from S. Glasstone, K. Laidler and H. Eyring, *The Theory of Rate Processes*, McGraw-Hill.

146

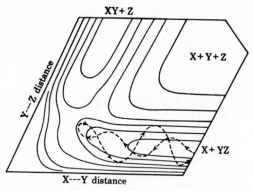

Fig. 4.6 Interconversion of energy; system possessing some vibrational energy in addition to relative translational energy. Reprinted by permission from S. Glasstone, K. Laidler, and H. Eyring, *The Theory of Rate Processes*, McGraw-Hill.

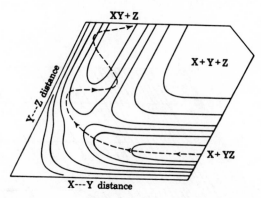

Fig. 4.7 Excess translational energy of reactants converted into vibrational energy of resultants. Reprinted by permission from McGraw-Hill.

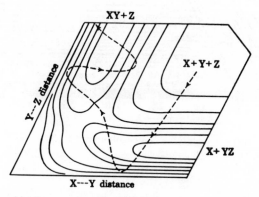

Fig. 4.8 Atom-combination reaction; Y and Z are capable of interacting and so facilitate the combination of *X* and *Y*. Reprinted by permission from McGraw-Hill.

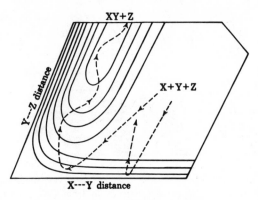

Fig. 4.9 Atom-combination reaction; Y and Z do not interact to any appreciable extent. Reaction occurs only if the system enters the curved ("non-ruled") region of the surface. Reprinted by permission from McGraw-Hill.

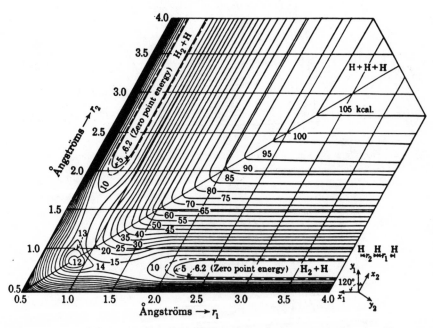

Fig. 4.10 Potential-energy surface for the system of three hydrogen atoms based on 14 percent coulombic energy. (Eyring, Gershinowitz, and Sun.)

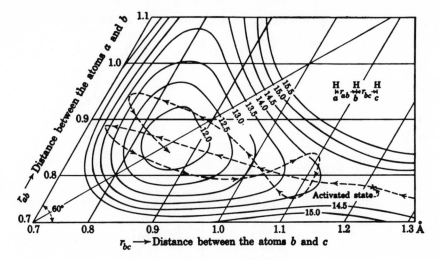

Fig. 4.11   —Path of H—H—H system in the basin at the top of the potential-energy barrier. (Hirschfelder, Eyring, and Jopley.)

atoms. The next five figures illustrate the insight resulting from the transformation. [Figures 4.10 and 4.11 are interesting in that they exhibit the first calculated trajectory made by Hirschfelder in his thesis using the then new London–Eyring–Polanyi (LEP) surface.]

4.9   THE ZENER-LANDAU RELATION FOR CROSSING BETWEEN POTENTIAL SURFACES

We consider the solution of the time-dependent Schrödinger equation

$$i\hbar \frac{\partial \Psi}{\partial t} = [\hat{H}_0 + V(t)]\Psi \tag{4.132}$$

where because of the classical relative motion of molecules (say, $R = vt$ where v is the relative velocity), the mutual interaction $V(t) = V(R)$ is time dependent. Notice that

$$\hat{H}_0 \phi_n = E_n \phi_n \tag{4.133}$$

To solve eq. 4.132, we let

$$\Psi(t) = \sum_n c_n(t)\phi_n \exp\left[-\frac{i}{\hbar} \left(E_n t + \int^t V_{nn}(t') \, dt' \right) \right] \tag{4.134}$$

by the expansion theorem. See Fig. 4.4.

It follows that by substituting eq. 4.134 into eq. 4.132, we find

$$ih\frac{dc_n}{dt} = \sum_m^{m \neq n} c_m(t)V_{nm}(t)\left[\exp -\frac{i}{\hbar}W_{mn}(t)\right] \qquad (4.135)$$

where

$$V_{nm}(t) = \langle\phi_n|V(t)|\phi_m\rangle \qquad (4.136)$$

and

$$W_{mn}(t) = (E_m - E_n)t + \int^t dt'[V_{mm}(t') - V_{nn}(t')] \qquad (4.137)$$

If the system is at ϕ_m initially, then by the perturbation method, we obtain

$$ih\frac{dc_n}{dt} = V_{nm}(t)\exp\left[-\frac{i}{\hbar}W_{mn}(t)\right] \qquad (4.138)$$

$$i\hbar c_n(t) = \int_{-\infty}^t dt\, V_{nm}(t')\exp\left[-\frac{i}{\hbar}W_{mn}(t')\right] \qquad (4.139)$$

and

$$i\hbar c_n(\infty) = \int_{-\infty}^{\infty} dt\, V_{nm}(t)\exp\left[-\frac{i}{\hbar}W_{mn}(t)\right] \qquad (4.140)$$

Notice that

$$\frac{\partial W_{mn}}{\partial t} = E_m - E_n + V_{mm}(t) - V_{nn}(t) \qquad (4.141)$$

and that setting $\partial W_{mn}/\partial t = 0$ at $R = R_c$ yields

$$E_m + V_{mm}(R_c) = E_n + V_{nn}(R_c) \qquad (4.142)$$

Equation 4.142 indicates that at $R = R_c$, the two potential curves cross.

Next we apply the saddle point method or the stationary phase approximation to calculate the integral involved in eq. 4.140. For this purpose we expand $W_{mn}(t)$ as follows:

$$W_{mn}(t) = W_{mn}(t^*) + \left(\frac{\partial W_{mn}}{\partial t}\right)_{t^*}(t - t^*) + \frac{1}{2}\left(\frac{\partial^2 W_{mn}}{\partial t^2}\right)_{t^*}(t - t^*)^2 + \cdots \qquad (4.143)$$

and choose t^* from

$$\left(\frac{\partial W_{mn}}{\partial t}\right)_{t^*} = 0 \qquad (4.144)$$

Substituting eqs. 4.143 and 4.144 into eq. 4.140 yields

$$i\hbar c_n(\infty) = V_{nm}(t^*)e^{-(i/\hbar)W_{mn}(t^*)} \int_{-\infty}^{\infty} dt \, \exp\left[-\frac{i}{2\hbar}\left(\frac{\partial^2 W_{mn}}{\partial t^2}\right)_{t^*}(t - t^*)^2\right] \quad (4.145)$$

Here we have assumed that $V_{nm}(t)$ varies slowly around t^* so that it can be approximated by $V_{nm}(t^*)$.

The integration in eq. 4.145 follows, since $\int_{-\infty}^{\infty} e^{-ax^2} \, dx = \sqrt{\pi/a}$ so that

$$i\hbar c_n(\infty) = V_{nm}(t^*)e^{-(i/\hbar)W_{mn}(t^*)}\sqrt{\frac{\pi}{(i/2\hbar)(\partial^2 W_{mn}/\partial t^2)_{t^*}}} \quad (4.146)$$

It follows that

$$|C_n(\infty)|^2 = \frac{2\pi |V_{nm}(R_c)|^2}{\hbar |(\partial^2 W_{mn}/\partial t^2)_{t^*}|} \quad (4.147)$$

Since $R = vt$ and

$$\frac{\partial^2 W_{mn}}{\partial t^2} = \frac{\partial}{\partial t}[V_{mn}(t) - V_{nn}(t)] = v\frac{\partial}{\partial R}[V_{mm}(R) - V_{nn}(R)] \quad (4.148)$$

Using eq. 4.148, eq. 4.147 becomes

$$|C_n(\infty)|^2 = \frac{2\pi |V_{nm}(R_c)|^2}{\hbar v |(\partial/\partial R)(V_{mm} - V_{nn})_{R_C}|} \quad (4.149)$$

Notice that

$$|C_m(\infty)|^2 = 1 - |C_n(\infty)|^2 = -\frac{2\pi |V_{nm}(R_c)|^2}{\hbar v |(\partial/\partial R)(V_{mm} - V_{nn})_{R_c}|} \quad (4.150)$$

Thus if $2\pi |V_{nm}(R_c)|^2/\hbar v |(\partial/\partial R)(V_{mm} - V_{nn})_{R_c}| \ll 1$, then eq. 4.150 can be rewritten as

$$|c_m(\infty)|^2 = \exp\left[-\frac{2\pi |V_{nm}(R_c)|^2}{\hbar v |(\partial/\partial R)(V_{mm} - V_{nn})_{R_c}|}\right] \quad (4.151)$$

Equations 4.149 and 4.150 are obtained by using the lowest order of the time-dependent perturbation method. Actually, when the higher-order perturbations are considered, eq. 4.151 can be derived.

4.10 BARRIER LEAKAGE

A fundamental problem in nuclear decomposition is barrier leakage. Its conspicuous importance lies in the fact that the fusion of two deuterium nuclei by collision, for example, requires temperatures approaching 100 million degrees Kelvin to form He by collision, the common mechanism for a bimolecular reaction. Ambient temperatures are thus effectively absolute zero for nuclear reactions. At absolute zero chemical reactions are also reduced to barrier leakage. Ammonia inversion is an interesting exceptional case of fast barrier leakage at ordinary temperatures. The very fast specific rate constant for barrier leakage of ammonia is $2.39 \times 10^{10}/\text{sec}$.

We now consider the theory for barrier leakage. We consider two normalized, orthogonal eigenfunctions ψ_1 and ψ_2 separated by a potential barrier but having a matching energy level E. The situation at any time can then be represented by the expression $c_1\psi_1 + c_2\psi_2$ where initially $c_1 = 1$ and $c_2 = 0$. Quantum mechanics then provides the relationship

$$\frac{dc_2}{dt} = \frac{i}{\hbar} c_1 \int \psi_2^* H' \psi_1 \, d\tau \tag{4.152}$$

In the region where the system is traversing the barrier initially with an energy E and a momentum p, it is convenient to represent ψ_1 and ψ_2^* by the equations

$$\psi_1 = e^{i/\hbar \int p \, dx} e^{-iEt/\hbar} \tag{4.153}$$

$$\psi_2^* = e^{-i/\hbar \int p \, dx} e^{iEt/\hbar} \tag{4.154}$$

and the perturbing potential

$$H' = p\dot{x} \tag{4.155}$$

Then we have for c_2 small

$$\frac{dc_2}{1 - c_2} = -\frac{i}{\hbar} \int_{x_1}^{x_2} p\dot{x} \, dt = -\frac{i}{\hbar} \int_{x=x_1}^{x_2} p \, dx \tag{4.156}$$

Here x_1 to x_2 spans the barriers width at the level of penetration.

$$\ln c_n = \frac{i}{\hbar} \int_{x=x_1}^{x_2} p \, dx = \frac{i}{\hbar} \int_{x=x_1}^{x_2} \sqrt{2m(E - V)} \, dx \tag{4.157}$$

Because $\sqrt{2m(E - V)}$ is imaginary while traversing the barrier the $\ln c_2$ is real and negative and

$$|c_2|^2 = \exp\left[-\frac{2}{\hbar} \left| \int_{x_1}^{x_2} \sqrt{2m(E - V)} \, dx \right| \right] \tag{4.158}$$

Thus for each vibration along the inversion coordinate of ammonia the probability of traversing the barrier is $|C_2|^2$ and the specific rate of crossing is

$$k_1 = \frac{e^{-E_i/kT}}{f} v_i |c_2|^2 = \frac{e^{-E_i/kT}}{f} v_i \exp\left[-\left| \frac{4\pi}{h} \int_{x_1}^{x_2} \sqrt{2m(E - V)} \, dx \right| \right] \tag{4.159}$$

where v_i is the frequency of vibration, E_i is the energy of this vibrational coordinate, and f is the partition function for the vibrational coordinate. This is an exact result where there is matching of energies in the initial and final state except for interaction with other coordinates. Otherwise it is the classical approximation that ignores the effect of discrete energy levels. The k_i values for all levels should be added to the specific rate of crossing the barrier to get the overall specific rate.

In estimating the isotope effect on the rate k_1 for barrier penetration it is obviously an oversimplification to consider only the effect of mass on $|c_2|^2$, since it also affects E_i, v_i, and f. However, if the system has widely spaced levels and is in the ground state, the factor $e^{-E_i/kT}/f$ is close to unity.

4.11 STARVATION KINETICS

An interesting place to look for fast reactions is in detonations. Shock waves accompanying detonations in solids move through the explosive at some 40 times the 700 miles an hour of sound in air. The shock initiates the decomposition of the explosive but the rate of finishing the decomposition in solids and liquids is slow, considering the temperature of the burning explosive. This is sometimes because layers have to be pealed off sequentially from the burning particle's surface so that the measured rate is the rate of reaction of a molecule in the surface layer divided by the number of molecular layers lying between the surface and the center of the particle. In a liquid, since the reaction starts from the surface of hot gas bubbles and again progresses layer after layer, the rate per molecule must again be divided by the number of layers to be burned through to obtain the measured rate.

Since this same slowness of reaction is reported in shock waves in the gas phase, another mechanism would have to be active in such cases and if so it would also be expected to appear in some solid explosives. Unimolecular reactions become bimolecular when there are not enough activating collisions to keep decomposition at its high pressure rate (Marcus and Rice, 1951). This is conveniently thought of as one form of starvation kinetics. The bond that is breaking is no longer fed fast enough to be in equilibrium with the translational degrees of freedom around it but must draw its activation energy from a starving vibrational reservoir with which it equilibrates. However, there are other ways of starving the reservoir besides simply reducing the pressure of molecules colliding with it. One such way is to introduce inefficient transmissions of energy to the reservoir. This inefficiency varies with the type of coupling made in collision. An obvious starvation process is to use a shock wave. In this case the time and/or intensity of the shock can produce a starving reservoir in equilibrium with the hidden breaking bond and lead to a slow reaction rate. That this starvation happens must be obvious to every student of initiation or dying out of detonations in both gases and solids. In fact, starvation kinetics is the name of the game. This process of starvation of shock waves was discussed at some length in the Priestley lecture of one of us (Eyring, 1975).

Starved reaction reservoirs are conveniently made by dumping measured amounts of energy into the reservoir by a variety of means. The formation of various degrees of starved reservoir reactions in mass spectra can be devised by regulating the voltage of the ionizing electrons and by the temperature of the molecules to be ionized or by having them absorb a photon of appropriate energy. Wahrhaftig, Wallenstein, Rosenstock, and Eyring developed the theory for this type of unimolecular decomposition of starved positive ions (Rosenstock et al., 1952). Molecular beam experiments provide evidence of such starved reservoir decompositions as random scattering of reaction products. Other examples come to mind but we now discuss further such starved reaction kinetics in shock waves.

The rate of the reaction in a solid cylinder may be estimated as follows. Since the pressure driving the shock wave comes from momentum transfer and since the escape of this momentum occurs over the whole unconfined surface area of the explosive, lying between the shock wave front and the Chapman Jouquet surface a distance a, we expect the velocity of the shock to fall off with the extent of this unconfined escape area. Thus we write for the relative velocity of the shock wave for a cylindrical explosive of diameter d the expression

$$\frac{D}{D^*} = \frac{\pi\, d^2/4}{\pi\, d^2/4 + \pi\, da\alpha} = \frac{1}{1 + 4a\alpha/d} \approx 1 - \frac{4a\alpha}{d}$$

We use D to indicate the velocity of the shock wave in a cylinder of diameter d, and D^* is the velocity of the shock wave in a cylinder of very large cross section where the area of the cross section is the site of the only important loss of momentum. The reason for the factor α is important. The escape rate out the side is slower per unit area than at the back since this rate must start at zero at the shock front. Thus α is the average relative rate of escape per unit area out the side of the explosive to the rate out the back where the driving pressure has reached its maximum value at the Chapman Jouguet plane. A rough estimate of α is $\frac{1}{4}$, giving us the result

$$\frac{D}{D^*} = 1 - \frac{a}{d} \tag{4.160}$$

This is the result Eyring et al. (1949) came to in a detailed investigation long ago. The observed velocity k is then estimated as approximately

$$k = \frac{D^*}{a} \tag{4.161}$$

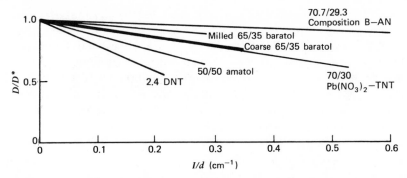

Fig. 4.12 The theoretical $\log k$ where $k = k_2 k_3/(k_2 + k_3)$ is plotted against $1/T$ (broken lines) for the decomposition of cyclopropane and cyclobutane. The circles are the experimental values (1). Here $k_2 = kT/he^{-\Delta H\ddagger/RT}e^{\Delta S\ddagger/R}$ and $k_3 = kT/he^{-D/RT}e^{-s}\sum_{i=0}^{s-1} 1/i!(D/RT)^i$. Also $\Delta H\ddagger = D$ the dissociation energy and $s = 20$ is the number of vibrational degrees of freedom feeding the reaction coordinate.

D/D^* is observed to be linearly related to $1/d$ for not too small values of d as required by our explanation. Figure 4.12 exhibits this expected linearity.

The observed rates of k in solid explosives estimated in this way are in the neighborhood of 10^5 to 10^6, far below the value to be expected if the hidden bonds to be broken were in equilibrium with the hot translational degree of freedom in the shock wave and controlled by the same mechanism. The starved state of the reservoir feeding energy into the hidden bond is understandable. A moment before it was struck by the shock wave it was part of the cold solid explosive. The molecules of explosive are suddenly volatized relatively cold and if not decomposed, decompose with the hidden bond, which breaks while equilibrating with the heat-starved reservoir, which has barely reached an energy in all its degrees of freedom equal to that required to break the fissile bond unimolecularly.

An appropriate approximate theory is readily formulated. The specific rate constant k for two unimolecular rates in series with rate constants k_2 and k_3 is

$$\frac{1}{k} = \frac{1}{k_2} + \frac{1}{k_3} = \frac{k_2 + k_3}{k_2 k_3} \qquad (4.162)$$

or

$$k = \frac{k_2 k_3}{k_2 + k_3} \qquad (4.163)$$

Now for the specific unimolecular rate constant of a reaction in equilibrium with translational degrees of freedom at temperature T we have

$$k_2 = \kappa \frac{kT}{h} e^{-\Delta H\ddagger/RT} e^{\Delta S\ddagger/R} \qquad (4.164)$$

Here the transmission constant κ is ordinarily close to unity and $\Delta H^{\ddagger}$ and $\Delta S^{\ddagger}$ are the heats and entropy of activation that can be determined by studying the rate of decomposition at low temperatures where the bond to be broken stays in equilibrium with the translational degrees of freedom.

A bond whose dissociation energy is D in equilibrium with a reservoir of s oscillators has an expression for the specific rate of decomposition, which we will now calculate. The probability p_s that such a reservoir will have an energy D or more in s degrees of freedom is

$$p_s = \sum_{i=0}^{s-1} \frac{1}{i!} \left(\frac{D}{RT} \right)^i e^{-D/RT} \tag{4.165}$$

The probability p_1 that a degree of freedom in equilibrium with this heat reservoir shall have an energy of D or more is accordingly

$$p_1 = e^{-D/\gamma} \tag{4.166}$$

where γ is the average energy of an oscillator in the reservoir. Accordingly the rate of decomposition of a bond whose energy of decomposition is D is approximately

$$k_3 = \frac{kT}{h} p_1 p_s = \frac{kT}{h} e^{-D/\gamma} \sum_{i=0}^{s-1} \frac{1}{i!} \left(\frac{D}{RT} \right)^i e^{-D/RT} \tag{4.167}$$

Obviously the decomposition of a bond is impossible until the energy in the $s + 1$ degrees of freedom of reservoir and reacting bond exceeds D and should be proceeding at a reasonable starved rate when the energy in the reservoir itself has reached D, in which case $\gamma = D/s$ and k_3 becomes

$$k_3 = \frac{kT}{h} e^{-s} \sum_{i=0}^{s-1} \frac{1}{i!} \left(\frac{D}{RT} \right)^i e^{-D/RT} \tag{4.168}$$

Equation 4.163 was applied by Eyring and Leu (1975) to the experimental results of Barnard et al (1969, 1971, 1974) and Brandley and Frend (1971) for the shock tube detonations of cyclopropane and cyclobutane that we interpreted as starvation kinetics. The results are shown in Fig. 4.13.

It is interesting that the value $s = 20$ for the number of oscillators in the reservoir is the expected number for propane, whereas in cyclobutane one might expect as many as nine more if they were all in good communication with the breaking bond. This lack of communication of part of the bonds is also observed in starvation reactions at low pressures. It is noteworthy that the extrapolated high-temperature limit for k is seen to be 10^5 to 10^6. This is far below expectations if the breaking bond is in equilibrium with the translational degrees of freedom corresponding to k_2 for which $\Delta S^{\ddagger}$ and D are

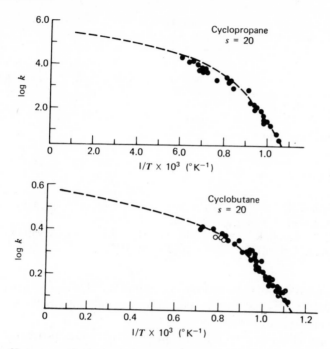

Fig. 4.13 Here we see that a reservoir with s oscillators where $s = 20$ in the equation $k = k_2 k_3/(k_2 + k_3)$ gives the observed specific rate between 10^5 and 10^6. Here the low temperature observed rate of Robertson is used.

given in Table 4.1. This same low rate for k at high translational temperatures is found for many solid explosives, as may be seen in Table 4.2 and Fig. 4.14.

This theory of starvation kinetics is suggestive of how detonations may be modulated advantageously. Thus a shaped charge focuses shock waves and, on reflection, the shock wave pass over unreacted material and in so doing speed up decomposition in the hollowed-out space, with an accompanying enhanced spewing out of decomposition products. This reflection from the sides of the hollowed-out shaped charge can bring up the pressure and therefore the velocity near the axis of the shock wave to near that for a charge of infinite diameter and yet the intensified shock wave has only to push out a plug of small cross section in the material to be penetrated. The result is the familiar intense penetrating power obtained with the Munro effect. A shock wave is highly temperature sensitive at temperature below that at which starvation kinetics takes over. As a result, the shock wave will behave very erratically if there are random holes or random changes in density in the explosive. A shock wave will ignite an explosive or die out, depending on whether it is losing energy to vibrational degrees of freedom slower or

Table 4.1
The Data for the Specific Reaction Rate k_2

	$\Delta S^{\ddagger}$(cal/mol · deg)	D(kcal/mol)	Rate of temperatures (°K)
Ammonium nitrate[a]	−5.33	37.3	490–540
EDNA[a]	−10.45	30.0	417–437
Hydrazine mononitrate[a]	−5.53	37.2	412–493
Tetryl[b]	8.88	37.4	484–533
TNT[a]	−5.86	42.3	510–550
TNT[c]	−9.63	33.3	548–583

Source: Eyring, H. *Chemical Engineering News* (1975), **53**, Number 15, by permission.
[a] Cook, M. A. and M. T. Abegg, (1956), *Ind. Eng. Chem.*, **48**, 1090.
[b] Rideal, E. K., A. J. B. Robertson, (1948), *Proc. Roy. Soc. (Lond.)* **A195**,135.
[c] Robertson, A. J. (1948), *Trans. Faraday Soc.*, **44**, 977.

faster than it is gaining energy from decomposition. To handle such problems quantitatively it is essential to take into account shifts between ordinary kinetics and starvation kinetics.

Here again the observed rates are all calculated by the procedure used for cyclopropane and cyclobutane with various values of the number of effective degrees of freedom s in the reservoir. Best concordance between the calculated k values and the k values in Table 4.2 suggest the reservoir has about 20 degrees of freedom.

The theory that the observed slow rate of burning in a detonation results from sequential burning of successive surface layers of a solid or liquid and

Table 4.2
Values of Shock Velocities, D,* and Calculated Effective Zone, Lengths, a, and Specific Reaction Rates k

Explosive	a (cm)	D^* (6)	k (sec^{-1})
70.9/29.3 Composition B-AN	0.25	7.5×10^5	3.04×10^6
70/30 Pb(NO$_3$)$_2$-TNT	0.70	6.09×10^5	8.70×10^5
Milled 65/35 Baratol	0.40	5.64×10^5	1.41×10^6
2.4-DNT	2.12	4.20×10^5	1.98×10^5
Course 65/35 Baratol	0.75	5.63×10^5	7.51×10^5
50/50 Amatol	1.25	6.93×10^5	5.54×10^5

Source: Eyring, *Chemical and Engineering News* (1975), **53**, Number 15, by permission.

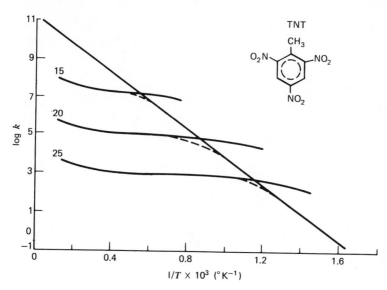

Fig. 4.14 The slopes of the lines give the value of the reaction zone, length a in accord with the equation $a = [(D^* - D)/D^*]d$. The term a is in the neighborhood of 1 cm for all the explosives considered here. Accordingly, the rate constant $k = D^*/a$ is always in the neighborhood of 10^5 or 10^6 for the observed temperatures, near $3000°K$.

the theory developed here that slow reaction arises from disequilibrium between translational and vibrational degrees of freedom of the reacting molecule both invoke the same rate-determining step, that is, slow heat transport from translational to vibrational degrees of freedom. Slow diffusion of reactants to the site of reaction is another kind of starvation kinetics frequently encountered. We can only deal appropriately with such delayed reaction rates as we recognize and sort out the exact cause. Clearly a change in the temperature dependence of reaction provides symptomatic although not conclusive evidence of the nature of the change in mechanism. Experiments designed to establish the exact cause of changes in mechanism are both needed and particularly valuable.

When material transport becomes rate limiting the slow diffusion starves ordinary thermal rate processes. For example, starvation occurs with external or internal resistance in batteries or in corrosion or with diffusion of matter through a membrane or evaporation of water from the body or from a lake into still air. For these processes involving a sequence of rate-determining steps the overall velocity v obeys the equation

$$\frac{1}{v} = \sum_i \frac{1}{v_i}$$

Here transition state theory can be applied to each of the elementary steps proceeding with the velocity v_i. Delay in oxygen transport may damage the brain, bringing on a type of senility, or it may even be fatal in aggravated cases, another example of starvation kinetics. If v_i involves parallel steps v_{ij}, we have $v_i = \sum_j v_{ij}$. Chain reactions often involve complicated interacting networks not unlike complicated electronic circuits. These not infrequently involve periodicities such as the heartbeat (Eyring and Henderson, 1978).

REFERENCES

Barnard, J. A. and R. P. Seebohn (1969), *Symp. Gas Kinetics*, Szeged, Hungary (1969).

Barnard, J. A., A. T. Cocks, and R. K. Y. Lee (1974), *J. Chem. Soc. Faraday Trans.*, **1**, 1782.

Bradley, J. N. and M. A. Freud (1971), *Trans. Faraday Soc.*, **67**, 72.

Condon, E. V. and P. M. Morse (1929), *Quantum Mechanics*, McGraw-Hill, pp. 228–231.

Eyring, H., J. Walter, and G. E. Kimball (1944), *Quantum Chemistry*, Wiley, pp. 282–298.

Eyring, H., R. E. Powell, G. E. Duffey, and R. B. Parlim (1949). *Chem. Rev.*, **45**, 69.

Eyring, H. (1975), *Chem. Eng. News*, **53**, No. 15, 27.

Eyring, H. and An-Lu Leu (1975), *Proc. Natl. Acad. Sci. U.S.*, **72**, 1717.

Eyring, H., D. Henderson, B. J. Stover, and E. M. Eyring (1964). *Statistical Mechanics and Dynamics*, 49–59, Wiley.

Glasstone, S., K. J. Laidler, and H. Eyring (1941), *The Theory of Rate Processes*, McGraw-Hill, pp. 13, 27, 146–150, 190.

Marcus, R. A. and O. K. Rice (1951), *J. Phys. Colloid Chem.*, **55**, 894.

Rosenstock, H. M., M. B. Wallenstein, A. L. Wahrhaftig, and Henry Eyring (1952), *Proc. Natl. Acad. Sci. U.S.*, **38**, 667.

Eyring, H. and D. Henderson, Eds., *Theoretical Chemistry Periodicities in Chemistry and Biology Volume 4*, Academic.

Five

Unimolecular Reactions

CONTENTS

5.1 QUASI-EQUILIBRIUM THEORY

The absolute reaction rate theory in the original form (Eyring, 1935) applies to reactions occurring at constant temperature and produces a reaction rate constant averaged over a thermal distribution of energies for the reactants. The limitation to thermal distribution is, of course, not essential. It is of great interest to consider the theory of reactions in which the energy distribution of the reactants varies radically from thermal. Various "hot atom" and "hot radical" reactions of photochemistry, nuclear and radiation chemistry, and such processes as the decomposition of ions in a mass spectrometer furnish examples in which a more flexible form of the theory is needed. In a natural extension of the absolute reaction rate theory one can calculate the decomposition rate of an isolated molecule that has fixed energy E by calculating the rate at which systems of a microcanonical ensemble pass through potential energy saddle points in configuration space (Rosenstock et al., 1952; Marcus and Rice, 1951).

It is well known that the behavior of an isolated system can be properly described in terms of a microcanonical ensemble (Eyring et al., 1964; Tolman, 1938) with systems uniformly distributed over all states having energies in the region E to $E + dE$. It is assumed here that the rate of transition of a system among the states is sufficiently rapid that the reaction yielding a different configuration has only a negligible effect on the distribution of the unreacted systems among the states. In general, it is possible to select a reaction coordinate and an activated complex defined in terms of a saddle point on a potential surface such that the change in the reaction coordinate describing the motion of the system through the saddle point can be taken to be translational motion.

Let $dW(E)$ represent the number of states of the system having energy between E and $E + dE$ with the system in the initial configuration. If the number of states is large, this number can be represented by a density function $\rho(E)$ as $dW(E) = \rho(E)\, dE$. The activated state is defined by a total energy between E and $E + dE$ with the activation energy $E_0^{\ddagger}$ and kinetic energy between ε_t and $\varepsilon_t + d\varepsilon_t$ in the reaction coordinate. The corresponding number of activated states is given by $\rho^{\ddagger}(E - E_0^{\ddagger} - \varepsilon_t)\, dE \rho_t(\varepsilon_t)\, d\varepsilon_t$. Using the usual expression for translational energy, $\varepsilon_t = n^2 h^2/8\mu^{\ddagger} l^2$, where $\mu^{\ddagger}$ is the effective "reduced mass" for the translation and l, the length in the reaction coordinate, the density of translational states in the reaction coordinate $\rho_t(\varepsilon_t)$ is given by

$$\rho_t(\varepsilon_t) = \frac{dn}{d\varepsilon_t} = \frac{l}{h}\sqrt{\frac{2\mu^{\ddagger}}{\varepsilon_t}} \tag{5.1}$$

The frequency of crossing the barrier $\gamma^{\ddagger}$ with v_t the velocity along the reaction coordinate ($\varepsilon_t = \mu^{\ddagger} v_t^2 / 2$) is given by

$$\gamma^{\ddagger} = \frac{v_t}{l} = \frac{1}{l} \sqrt{\frac{2\varepsilon_t}{\mu^{\ddagger}}} \tag{5.2}$$

Then the unimolecular rate constant $k(E)$ is given by one-half the ratio of activated complexes to initial normal molecules, multiplied by the frequency of crossing the barrier and integrated over all possible values of the translational energy in the reaction coordinate

$$
\begin{aligned}
k(E) &= \int_0^{E - E_0^{\ddagger}} \frac{\gamma^{\ddagger}}{2} \cdot \frac{\rho^{\ddagger}(E - E_0^{\ddagger} - \varepsilon_t)\, dE \rho_t(\varepsilon_t)\, d\varepsilon_t}{\rho(E)\, dE} \\
&= \frac{1}{h} \cdot \frac{1}{\rho(E)} \int_0^{E - E_0^{\ddagger}} d\varepsilon_t \, \rho^{\ddagger}(E - E_0^{\ddagger} - \varepsilon_t)
\end{aligned} \tag{5.3}
$$

The factor one-half arises from the equal probability of translation in opposite directions at equilibrium. Using the relation between $W(E)$, the total number of states, and $\rho(E)$, the density of states, $\rho(E) = dW(E)/dE$, eq. 5.3 can be written

$$k(E) = \frac{1}{h} \cdot \frac{W^{\ddagger}(E - E_0^{\ddagger})}{\rho(E)} \tag{5.4}$$

where $W^{\ddagger}(E - E_0^{\ddagger})$ is the total number of states of the activated complex for the system for all energies up to $E - E_0^{\ddagger}$.

Equations 5.3 or 5.4 can be derived by an alternate approach (Marcus, 1965). The statistical equilibrium probability of finding the activated complex in an interval $(E, E + dE)$ of energy and in an interval $(q, q + dq)$ of the reaction coordinate having a translational momentum in the range $(p, p + dp)$ is given by the ratio of quantum states of the activated complex and of the initial normal molecule or ion:

$$\frac{\rho^{\ddagger}(E - E_0^{\ddagger} - \varepsilon_t)\, dE}{\rho(E)\, dE} \cdot \frac{dp\, dq}{h} \tag{5.5}$$

It should be noted that $(dp\, dq)/h$ is the number of translational quantum states in $dp\, dq$. The corresponding probability per unit interval along q is obtained by dividing eq. 5.5 by dq. The specific unimolecular reaction rate constant $k(E)$ is obtained by multiplying the resulting ratio by the velocity $\dot{q}$ (the coordinate q is taken to be Cartesian, so that $\dot{q} = p/\mu^{\ddagger}$), and integrating over all possible values of p

$$k(E) = \int_0^p \frac{\rho^{\ddagger}(E - E_0^{\ddagger} - \varepsilon_t)}{\rho(E)} \cdot \frac{\dot{q}\, dp}{h} \tag{5.6}$$

Since $d\varepsilon_t = \dot{q}\, dp$, eq. 5.6 reduces to eq. 5.3.

Now we show that $k(E)$ averaged over the thermal energy distribution yields the ordinary reaction rate expression of the absolute reaction rate theory (Magee, 1952). The rate constant for temperature T is given by the average:

$$k(\beta) = \frac{\int_0^\infty k(E)\rho(E)\exp(-\beta E)\,dE}{\int_0^\infty \rho(E)\exp(-\beta E)\,dE} \tag{5.7}$$

The denominator in eq. 5.7 simply denotes the partition function $Q(\beta)$ (Tolman, 1938):

$$Q(\beta) = \int_0^\infty \rho(E)\exp(-\beta E)\,dE \tag{5.8}$$

where $\beta = 1/kT$.

Substituting eqs. 5.4 and 5.8 into eq. 5.7 yields

$$k(\beta) = \frac{1}{h}\cdot\frac{1}{Q(\beta)}\int_0^\infty dE\,\exp(-\beta E)W^{\ddagger}(E - E_0^{\ddagger}) \tag{5.9}$$

Introducing the new variable $E' = E - E_0^{\ddagger}$, eq. 5.9 becomes

$$k(\beta) = \frac{\exp(-\beta E_0^{\ddagger})}{hQ(\beta)}\int_0^\infty dE'\,\exp(-\beta E')W^{\ddagger}(E') \tag{5.10}$$

or

$$k(\beta) = \frac{\exp(-\beta E_0^{\ddagger})}{hQ(\beta)}\int_0^\infty dE'\,\exp(-\beta E')\int_0^{E'}\rho^{\ddagger}(\varepsilon)\,d\varepsilon \tag{5.11}$$

Interchanging the order of integrations, we obtain

$$k(\beta) = \frac{\exp(-\beta E_0^{\ddagger})}{hQ(\beta)}\int_0^\infty \rho^{\ddagger}(\varepsilon)\,d\varepsilon \int_\varepsilon^\infty dE'\,\exp(-\beta E')$$

$$= \frac{\exp(-\beta E_0^{\ddagger})}{\beta hQ(\beta)}\int_0^\infty \rho^{\ddagger}(\varepsilon)\exp(-\beta\varepsilon)\,d\varepsilon = \frac{kT}{h}\cdot\frac{Q^{\ddagger}(\beta)}{Q(\beta)}\exp(-\beta E_0^{\ddagger}) \tag{5.12}$$

which is, of course, the ordinary absolute reaction rate formula. Throughout the discussion it has been assumed that there is no reflection of reacting systems; that is, the transmission coefficient has been assumed to be unity. When the transmission coefficient is not equal to unity, the expression for $k(E)$ is given by

$$k(E) = \frac{1}{h}\cdot\frac{1}{p(E)}\sum_i \kappa_i H(E - E_0^{\ddagger} - E_i) \tag{5.13}$$

where κ_i represents the transmission coefficient and $H(x)$ the Heaviside function $(H(x) = 1, x > 0, H(x) = 0, x < 0)$. The summation in eq. 5.13 covers the quantum states of the activated complex.

As mentioned earlier, for $k(E)$ to give the actual rate constant even approximately there must be a transfer of energy among all the various degrees of freedom that is rapid compared with the calculated decomposition rate. For the cases in which the rate of energy transfer is not rapid compared with the decomposition rate, one can use the stochastic model of reaction kinetics to discuss the effect of the rate of energy relaxation on the reaction rate (Zwolinski and Eyring, 1947; Montroll and Shuler, 1958; Widom, 1963; Snider, 1965; Rankin and Light, 1967; Widom, 1971; Gelbart et al., 1971; Lin, 1972).

5.2 CALCULATION OF $W(E)$ AND $p(E)$

In the previous section we have shown that the unimolecular rate constant $k(E)$ of an isolated system can be expressed in terms $W^{\ddagger}(E - E_0^{\ddagger})$ and $\rho(E)$. In this section we discuss the calculation of the total number of states $W(E)$ and the density of states $\rho(E)$. Various methods for calculating $W(E)$ and $\rho(E)$ have been proposed (for a review of these methods, see Tou, 1967, and Forst and Prášil, 1969). Here we discuss only two theoretically interesting methods: the inversion of partition function method (Hoare and Ruijgrok, 1970; Hoare, 1970; Forst and Prášil, 1969; Forst and Prášil, 1970; Lau and Lin, 1971a) and the Darwin–Fowler method (Lin and Eyring, 1963, 1965; Tou and Lin, 1968; Lin and Ma, 1971; Lin and Lau, 1971b).

5.2.1 The Inversion of Partition Function Method

From the definition of the Heaviside function, we can express the total number of states of a system with energy E as

$$W(E) = \sum_i H(E - E_i) \tag{5.14}$$

where the summation is over all the possible states. Multiplying eq. 5.14 by $\exp(-\beta E)$ and integrating the resulting expression with respect to E from zero to infinity, we obtain

$$\int_0^\infty \exp(-\beta E)W(E)\,dE = \sum_i \int_0^\infty dE \exp(-\beta E)H(E - E_i)$$

$$= \sum_i \int_{E_i}^\infty dE \exp(-\beta E) \tag{5.15}$$

Changing the independent variable $E' = E - E_i$, eq. 5.15 becomes

$$\int_0^\infty \exp(-\beta E)W(E)\,dE = \sum_i \exp(-\beta E_i) \int_0^\infty dE' \exp(-\beta E') = \frac{1}{\beta}\,Q(\beta) \quad (5.16)$$

where $Q(\beta)$ represents the partition function of the system, $Q(\beta) = \sum_i \exp(-\beta E_i)$. Equation 5.16 implies that the Laplace transformation of $W(E)$ yields $Q(\beta)/\beta$. In order words, to obtain $W(E)$ one simply carries out the inverse Laplace transformation of $Q(\beta)/\beta$; that is,

$$W(E) = L^{-1}\left[\frac{1}{\beta}\,Q(\beta)\right] = \frac{1}{2\pi_i} \int_{\gamma-i\infty}^{\gamma+i\infty} \frac{d\beta}{\beta} \exp(\beta E)Q(\beta) \quad (5.17)$$

where the integration path is a straight line parallel to the imaginary axis, but it can be continuously deformed as long as the integral converges. L^{-1} in eq. 5.17 represents the inverse Laplace transform operator. The discussion of the Laplace transformation is given in Appendix 1. Since the state density $\rho(E)$ is related to $W(E)$ by $\rho(E) = dW(E)/dE$, from eq. 5.17 we obtain the expression for $\rho(E)$ as

$$\rho(E) = L^{-1}[Q(\beta)] = \frac{1}{2\pi i} \int_{\gamma-i\infty}^{\gamma+i\infty} d\beta \exp(\beta E)Q(\beta) = \sum_i \delta(E - E_i) \quad (5.18)$$

This relation can also be obtained directly from eq. 5.8 by the inverse Laplace transformation. Here $\delta(x)$ represents the delta function.

To show an application of eqs. 5.17 and 5.18, we consider a system of weakly coupled classical harmonic oscillators. Since the partition function of a classical harmonic oscillator is given by $Q(\beta) = 1/\beta\varepsilon_i$, where $\varepsilon_i = h\gamma_i$. The partition function for a system of N classical harmonic oscillators can be written

$$Q = \frac{1}{\beta^N \prod_{i=1}^N \varepsilon_i} \quad (5.19)$$

Substituting eq. 5.19 into eqs. 5.17 and 5.18 yields

$$W(E) = \frac{1}{2\pi_i} \cdot \frac{1}{\prod_i^N \varepsilon_i} \int_{\gamma-i\infty}^{\gamma+i\infty} \frac{d\beta}{\beta^{N+1}} \exp(\beta E) = \frac{E^N}{\Gamma(N+1)\prod_i^N \varepsilon_i} \quad (5.20)$$

and

$$\rho(E) = \frac{1}{2\Pi_i} \cdot \frac{1}{\prod_i^N \varepsilon_i} \int_{\gamma-i\infty}^{\gamma+i\infty} \frac{d\beta}{N} \exp(\beta E) = \frac{E^{N-1}}{\Gamma_{(N)}\prod_i^N \varepsilon_i} \quad (5.21)$$

Here Cauchy's residue theorem and $\exp(x) = \sum_{n=0}^\infty (x^n/n!)$ have been used. The term $\Gamma(x)$ in eqs. 5.20 and 5.21 represents the gamma function. Of course, $\rho(E)$ given in eq. 5.21 can also be obtained directly from the relation $\rho(E) = dW(E)/dE$.

Next we evaluate the contour integrals in eqs. 5.17 and 5.18 by the method of steepest descent (see Appendix 2). In the first-order approximation of the method of steepest descent $W(E)$ and $\rho(E)$ are given by

$$W(E)_1 = \frac{\exp(\beta_1^* E)Q(\beta_1^*)}{\sqrt{2\Pi\{1 + \beta_1^{*2}[[\partial^2 \log Q(\beta)]/\partial\beta^2]_{\beta=\beta_1^*}\}}} \tag{5.22}$$

where β_1^* represents the saddle point value of β and is to be determined from

$$\frac{1}{\beta_1^*} = E + \left[\frac{\partial \log Q(\beta)}{\partial\beta}\right]_{\beta=\beta_1^*} \tag{5.23}$$

and

$$\rho(E)_1 = \frac{\exp(\beta_2^* E)Q(\beta_2^*)}{\sqrt{2\Pi[[\partial^2 \log Q(\beta)]/\partial^2\beta]_{=\beta=\beta_2^*}}} \tag{5.24}$$

where β_2^* is given by

$$E = -\left[\frac{\partial \log Q(\beta)}{\partial\beta}\right]_{\beta=\beta_2^*} \tag{5.25}$$

It should be noted that $W(E)_1$ and $\rho(E)_1$ given in eqs. 5.22 and 5.24 are obtained directly from eqs. 5.17 and 5.18 by the use of the method of steepest descent, and represent the results of the first-order approximation of the method of steepest descent. The higher-order results for $W(E)$ and $\rho(E)$ can be obtained easily (Appendix 2). Another expression for $W(E)_1$ can be obtained from eq. 5.17 by replacing β by β_2^* (Hoare and Ruijgrok, 1970):

$$W(E)_1 = \frac{1}{\beta_2^*} \cdot \frac{1}{2\pi i} \int_{\gamma-i\infty}^{\gamma+i\infty} d\beta \exp(\beta E)Q(\beta) = \frac{\rho(E)}{\beta_2^*} \tag{5.26}$$

To illustrate an application of the preceding expressions for $W(E)_1$ and $\rho(E)_1$, we consider a system of weakly coupled quantum harmonic oscillators. In this case the partition function $Q(\beta)$ is given by

$$Q = \prod_{i=1}^{N} \frac{1}{1 - \exp(-\beta\varepsilon_i)} = \frac{1}{\prod_{i=1}^{N}[1 - \exp(-\beta\varepsilon_i)]} \tag{5.27}$$

Substituting eq. 5.27 into eqs. 5.22 and 5.24 yields

$$W(E)_1 = \frac{\exp(\beta_1^* E) \prod_{i=1}^{N}[1 - \exp(-\beta_1^*\varepsilon_i)]^{-1}}{\sqrt{2\Pi\{1 + \sum_{i=1}^{N}(\beta_1^*\varepsilon_i)^2\exp(\beta_1^*\varepsilon_i)/[\exp(\beta_1^*\varepsilon_i) - 1]^2\}}} \tag{5.28}$$

and

$$\rho(E)_1 = \frac{\exp(\beta_2^* E) \prod_{i=1}^{N}[1 - \exp(-\beta_2^*\varepsilon_i)]^{-1}}{\sqrt{2\Pi \sum_{i=1}^{N}[\varepsilon_i^2 \exp(\beta_2^*\varepsilon_i)]/[\exp(\beta_2^*\varepsilon_i) - 1]^2}} \tag{5.29}$$

where β_1^* and β_2^* are determined by

$$\frac{1}{\beta_1^*} = E - \sum_{i=1}^{N} \frac{\varepsilon_i}{\exp(\beta_1^* \varepsilon_i) - 1} \tag{5.30}$$

and

$$E = \sum_{i=1}^{N} \frac{\varepsilon_i}{\exp(\beta_2^* \varepsilon_i) - 1} \tag{5.31}$$

respectively. In most cases a numerical procedure is required to calculate β_1^* and β_2^* from eqs. 5.30 and 5.31. To obtain $W(E)$ and $\rho(E)$ for a system of classical harmonic oscillators, we may let $h \to 0$ in eqs. 5.22–5.31 or we may substitute eq. 5.19 into eqs. 5.22 and 5.24:

$$W(E)_1 = \frac{\exp(1 + N)}{\sqrt{2\Pi}(N + 1)^{N+(1/2)}} \cdot \frac{E^N}{\prod_i^N \varepsilon_i} \tag{5.32}$$

and

$$\rho(E) = \frac{\exp(N)}{\sqrt{2\Pi} \, N^{N-1/2}} \cdot \frac{E^{N-1}}{\prod_i^N \varepsilon_i} \tag{5.33}$$

These expressions should be compared with those given in eqs. 5.20 and 5.21. The expressions for $W(E)$ and $\rho(E)$ given in eqs. 5.20 and 5.21 reduce to $W(E)_1$ and $\sigma(E)_1$ if the Stirling's formula for $\Gamma(N)$ is introduced into eqs. 5.20 and 5.21.

To test the accuracy of the equations derived above for calculating $W(E)$, cyclopropane, water, and acetylene are chosen as test cases. A comparison of $W(E)$ for these molecules calculated by exact counting, the Darwin–Fowler method (see the next Section), and equations 5.28 and 5.26 is shown in Table 5.1. We can see that the results from the three approximation methods are in good agreement with those from the exact counting. The calculated results given in Table 5.1 are obtained by using the first-order approximation of the method of steepest descent and can be improved by going to the high-order approximations of the method of steepest descent (see Appendix 2).

Once $W(E)$ and $\rho(E)$ are obtained, the rate constant $k(E)$ of unimolecular reactions of isolated systems can be calculated by using eq. 5.4. In most cases the numerical method should be used, and the analytical expression for $k(E)$ is obtainable only for some simple systems, for example, classical oscillators. Assuming the molecule (or ion) to be a collection of N weakly coupled classical harmonic oscillators and the activated complex, a collection of $N - 1$ such oscillators, we obtain the rate constant $k(E)$ as

$$k(E) = \frac{\prod_i^N \varepsilon_i}{\prod_i^{N-1} \varepsilon_i^\ddagger} \cdot \frac{(E - E_0^\ddagger)^{N-1}}{E^{N-1}} \tag{5.34}$$

Table 5.1

Comparison of the Calculated and Exact Values of $W(E)$ for Selected Molecules

E (kcal/mole)	$W_{ex}(E)$	$W(E)^a$	$W(E)^b$	$W(E)^c$
		Cyclopropane[d]		
16	8.02×10^2	7.43×10^2	7.23×10^2	7.49×10^2
20	7.75×10^4	7.68×10^4	7.56×10^4	7.75×10^4
30	2.69×10^6	2.68×10^6	2.64×10^6	2.69×10^6
40	4.97×10^7	4.99×10^7	4.95×10^7	5.03×10^7
50	6.12×10^8	6.16×10^7	6.12×10^7	6.22×10^8
100	5.84×10^{12}	5.82×10^{12}	5.80×10^{12}	5.88×10^{12}
150	3.00×10^{15}	2.98×10^{15}	2.97×10^{15}	3.01×10^{12}
		Water[e]		
10	3	3.72	3.54	3.27
20	11	11.39	11.22	10.77
30	23	25.63	25.49	24.83
40	46	48.42	48.38	47.44
50	78	81.79	81.91	80.63
100	466	476.27	478.78	474.52
150	1405	1433.33	1442.21	1432.46
		Acetylene[f]		
2.31	5	3.84	3.58	4.34
4.61	15	14.91	14.19	16.77
9.23	94	99.26	96.07	108.86
13.84	390	401.30	392.39	433.26
18.45	1.198×10^3	1.243×10^3	1.223×10^3	1.33×10^3
23.06	3.163×10^3	3.237×10^3	3.198×10^3	3.43×10^3
27.68	7.333×10^3	7.447×10^3	7.380×10^3	7.84×10^3
36.90	2.969×10^4	3.035×10^4	3.018×10^4	3.17×10^4

[a] Results of eq. 5.28.
[b] Results of eq. 5.26.
[c] Results of eq. 5.43.
[d] Using the following rounded frequencies: 3221 (6), 1478 (3), 1118 (7), 879 (3), 750 (2), cm^{-1}.
[e] Here $v_i = 3652, 1595, 3756$ cm^{-1}.
[f] Here $v_i = 612$ (2), 729 (2), 1974 (1), 3287 (1), 3374 (1).

by substituting eqs. 5.20 and 5.21 into eq. 5.4. The corresponding expression for $k(E)$ by the use of $W(E)_1$ and $\rho(E)_1$ given in eqs. 5.32 and 5.33 is identical with eq. 5.34. This is because the errors introduced in $W^{\ddagger}(E - E_0^{\ddagger})_1$ and $\rho(E)_1$ cancel each other out.

5.2.2 The Darwin–Fowler Method

To develop the Darwin–Fowler method for calculating $W(E)$, we use the system of weakly coupled harmonic oscillators as an example. We notice that the number of states belonging to the energy value $(n_i + \frac{1}{2}g_i)hv_i$ for a g_i-dimensional isotropic oscillator (g_i is also called *degeneracy*) is equal to $(n_i + g_i - 1)!/n_i!(g_i - 1)!$ (Fowler, 1936). For a system of weakly coupled harmonic oscillators of frequencies $v_1, v_2, \ldots, v_m$, having degeneracies $g_1, g_2, \ldots, g_m$, respectively, if the total energy of the system is E, then

$$\sum_i^m n_i hv_i = E \tag{5.35}$$

where the lowest energy level has been taken as zero. For convenience, eq. 5.35 is made dimensionless by dividing both sides of eq. 5.35 by $h\langle v \rangle$. Thus

$$\sum_i n_i \frac{v_i}{\langle v \rangle} = \frac{E}{h\langle v \rangle} = \langle n \rangle \tag{5.36}$$

The quantity $\langle v \rangle$ is so chosen that the $v_i/\langle v \rangle$ values, and hence $\langle n \rangle$, are integers. The total number of states with the energy values equal to or less than E is then

$$W(E) = \sum_{\langle n \rangle = 0}^{\langle n \rangle} \sum_{n_1} \sum_{n_1} \sum_{n_m} \prod_{i=1}^m \frac{(n_i + g_i - 1)!}{n_i!(g_i - 1)!} \tag{5.37}$$

where the first summation is over the energy values and the remaining summations are taken over all values of the quantum numbers n_i's that are consistent with eq. 5.36.

Next we introduce the following generating function $G_i(z)$ for each mode of vibrations:

$$G_i(z) = \sum_{n_i=0}^{\infty} \frac{(n_i + g_i - 1)!}{n_i!(g_i - 1)!} z^{n_i v_i/\langle v \rangle} = (1 - z^{v_i/\langle v \rangle})^{-g_i} \tag{5.38}$$

In terms of these generating functions $G_i(z)$, $W(E)$ is just the summation of the coefficients of $z^{\langle n \rangle}$ in the following expansion:

$$\prod_{i=1}^m G_i(z) = \prod_{i=1}^m (1 - z^{v_i/\langle v \rangle})^{-g_i} \tag{5.39}$$

By Cauchy's residue theorem, we may express $W(E)$ by the contour integral:

$$W(E) = \frac{1}{2\Pi i} \sum_{\langle n \rangle = 0}^{\infty} \int_\gamma \frac{dz}{z^{\langle n \rangle + 1}} \sum_{i=1}^m G_i(z) = \frac{1}{2\Pi i} \int_\gamma \frac{dz}{z}$$

$$\times \frac{z^{-\langle n \rangle - 1} - 1}{z^{-1} - 1} \prod_{i=1}^m G_i(z) = \frac{1}{2\Pi i} \int_\gamma \frac{dz}{z} [\phi(z)]^{\langle n \rangle} \tag{5.40}$$

where

$$[\phi(z)]^{\langle n \rangle} = \frac{z^{-\langle n \rangle - 1} - 1}{z^{-1} - 1} \prod_{i=1}^{m} G_i(z) = \frac{z^{-\langle n \rangle - 1} - 1}{z^{-1} - 1} \prod_{i=1}^{m} (1 - z^{\nu_i/\langle \nu \rangle})^{-g_i} \quad (5.41)$$

and γ is any contour lying within the circle of convergence of these power series enclosing the origin $z = 0$. So far no approximation has been made in the derivation of $W(E)$. Next we introduce the method of steepest descent to calculate the contour integral in eq. 5.40. It is easily seen that $\phi(z)$ is an analytic function of z, which can be expanded in a series of ascending powers of z with real and positive integral coefficients; the circle of convergence is of radius unity. Consider the behavior of $\phi(z)$. Along the real axis it is continuous and tends to infinity at both $z = 0$ and $z = 1$, and so must have at least one minimum between $z = 0$ and $z = 1$. Next for the complex values of z, consider a circle of radius γ less than unity, $z = \gamma \exp(i\alpha)$; then as the modulus of a sum is never greater than the sum of moduli, it follows that $|\phi(z)|$ is always less than $\phi(\gamma)$ on this circle and decreases as it moves off the real axis. It can easily be shown that there is only one minimum between $z = 0$ and $z = 1$ (Fowler, 1936; Lin and Eyring, 1963).

The method of steepest descent (see Appendix 2) proceeds by making the contour γ pass through the saddle point of $\phi(z)$ in such a direction that the value of the integrand falls off along γ from a maximum value at the saddle point at the greatest possible rate. If θ represents the root of $\phi'(z) = 0$, this may be achieved by taking for γ the circle $|z| = 0$. If we put $z = \theta \exp(i\alpha)$, then when α is small,

$$\log \phi(z) = \log \phi(\theta) - \frac{\alpha^2 \theta^2}{2} \cdot \frac{\phi''(\theta)}{\phi(\theta)} + \cdots \quad (5.42)$$

where $\phi''(\theta)$ represents the second derivative of $\phi(z)$ with respect to z evaluated at $z = \theta$. Substituting eq. 5.42 into eq. 5.40, neglecting the higher-order terms of α in eq. 5.42, and integrating with respect to α from $-\infty$ to ∞, we obtain, to the first-order approximation of the method of steepest descent:

$$W(E)_1 = [\phi(\theta)]^{\langle n \rangle} \frac{1}{2\Pi} \int_{-\infty}^{\infty} d\alpha \exp\left[-\frac{\langle n \rangle \theta^2}{2} \cdot \frac{\phi''(\theta)}{\phi(\theta)} \alpha^2 \right]$$

$$= \frac{[\phi(\theta)]^{\langle n \rangle}}{\sqrt{2\Pi \langle n \rangle \theta^2 [\phi''(\theta)/\phi(\theta)]}} \quad (5.43a)$$

or

$$W(E) = \frac{[(\theta^{-1-\langle n \rangle} - 1)/(\theta^{-1} - 1)] \prod_{i=1}^{m} (1 - \theta^{\nu_i/\langle \nu \rangle})^{-g_i}}{\sqrt{2\Pi \langle n \rangle \theta^2 [\varphi''(\theta)/\varphi(\theta)]}} \quad (5.43b)$$

where θ is the root of $\phi'(z) = 0$ and its value in this case may be determined from the following equation:

$$\frac{\langle n \rangle + 1}{1 - \theta^{1 + \langle n \rangle}} = \frac{1}{1 - \theta} + \sum_{i=1}^{m} \frac{g_i(v_i/\langle v \rangle)}{\theta^{-v_i/\langle v \rangle} - 1} \tag{5.44}$$

For Cauchy's residue theorem to be applicable in eq. 5.40 $\langle n \rangle$ must be an integer. But for practical purposes, we may choose $\langle n \rangle$ as close to an integer as possible.

The expression for the total number of states $W(E)$ to the second-order approximation of the method of steepest descent has been obtained by Lau and Lin (1971a). To show an application we choose an artificial system of three oscillators, $v_1 = \langle v \rangle$, $v_2 = 2\langle v \rangle$, and $v_3 = 3\langle v \rangle$ with $g_1 = 3$, $g_2 = 2$, and $g_3 = 1$, respectively. In Table 5.2 are given the results of exact counting, the calculated results from the use of the first-order approximation of the method of steepest descent $W(E)_1$ and second-order results $W(E)_2$. It can be seen that the second-order approximation gives excellent agreement with the results from exact counting for the artificial model over the whole energy range. This may be due to the fact that for the artificial model both $v_i/\langle v \rangle$ and $\langle n \rangle$ are integers, which is required in the application of Cauchy's residue theorem. A numerical comparison is made of $W(E)_1$ given by eq. 5.43, and $W(E)_2$ (Lau and Lin, 1971b) with the method of the inversion of partition function $W(E)_1$ given by eq. 5.26 (Hoare and Ruijgrok, 1970), as applied to the water molecule. The results are tabulated in Table 5.3. Table 5.3 shows that $W(E)_2$ is in good agreement with exact counting and is better than $W(E)_1$. A similar comparison for other molecules is given in Table 5.1.

To compare the Darwin–Fowler method with that of the inversion of partition function, we rewrite eq. 5.40 as

$$W(E) = \frac{1}{2\Pi i} \sum_{\langle n \rangle = 0}^{\langle n \rangle} \int_\gamma \frac{dz}{z^{\langle n \rangle + 1}} \prod_{i=1}^{m} (1 - z^{v_i/\langle v \rangle})^{-g_i} \tag{5.45}$$

If the summation over $\langle n \rangle$ in eq. 5.45 is carried out, eq. 5.45 reduces to eq. 5.40. By applying the method of steepest descent (cf. eqs. 5.41–5.44), we find

$$W(E)_1 = \sum_{\langle n \rangle = 0}^{\langle n \rangle} \frac{\prod_i^m (1 - \theta^{v_i/\langle v \rangle})^{-g_i}}{\theta^{\langle n \rangle}[2\Pi \sum_i^m g_i(v_i/\langle v \rangle)^2 \theta^{v_i/\langle v \rangle}/(1 - \theta^{v_i/\langle v \rangle})^2]^{1/2}} \tag{5.46}$$

where θ is determined from

$$\langle n \rangle = \sum_i^m \frac{g_i(v_i/\langle v \rangle)}{\theta^{-v_i/\langle v \rangle} - 1} \tag{5.47}$$

Table 5.2
W(E) for Artificial Harmonic Oscillator System

$\langle n \rangle$	1	2	3	4	5	6
0	0.1862	0.2899	0.3653	0.4246	0.4716	0.7375
W_{exact}	4.00	12.00	29.0	62.0	120.00	217.00
W_1	4.4085	12.8501	30.2733	64.0720	123.7573	223.0012
W_2	3.9905	11.9917	28.9715	61.8702	120.0569	217.0916

Now let $\theta = \exp(-\beta_2^* h\langle v \rangle)$ and $\varepsilon_i = h v_i$; eqs. 5.46 and 5.47 become

$$W(E)_1 = \sum_{\langle n \rangle = 0}^{\langle n \rangle} h\langle v \rangle \frac{\exp(\beta_2^* E) \prod_i^m [1 - \exp(-\beta_2^* \varepsilon_i)]^{-g_i}}{[2\Pi \sum_i^m g_i \varepsilon_i^2 \exp(-\beta_2^* \varepsilon_i)/(1 - \exp(-\beta_2^* \varepsilon_i))^2]^{1/2}}$$

(5.48)

and

$$E = \sum_i^m \frac{g_i \varepsilon_i}{\exp(\beta_2^* \varepsilon_i) - 1}$$

(5.49)

Comparing the general term without the factor $h\langle v \rangle$ in eq. 5.48 with the expression for the density of states $\rho(E)_1$ (cf. eq. 5.29) obtained by the method of the inversion of partition function, we can see that they are identical. This is because the total number of states $W(E)$ and the density of states $\rho(E)$ are related by $(E) = \lim_{\Delta E \to 0} [W(E + \Delta E) - W(E)]/\Delta E$, and $h\langle v \rangle$

Table 5.3
Comparison of Calculated Values and the Exact Values of $W((E)$ for Water Molecule[a]

E (eV)	W_{exact}	W_1	W_2	θ	$C(\theta, \langle n \rangle)$	$W(E)^b$
0.4336	3	3.234	3.014	0.99890793	0.93147	3.54
0.8672	11	10.766	10.429	0.99929048	0.96866	11.23
1.3008	23	24.828	24.275	0.99946766	0.97773	25.52
1.7435	46	47.439	46.544	0.99957238	0.98113	48.43
2.1685	78	80.625	79.229	0.99964213	0.98268	82.00
4.3361	466	474.520	467.076	0.99980181	0.98341	479.5
6.5042	1405	1432.457	1409.996	0.99986272	0.98429	1444.0

[a] Frequencies: 3652, 2595, 3756 (cm^{-1}); $\langle v \rangle = 2 \text{ cm}^{-1}$.
[b] Results of eq. 5.26.

plays the role of ΔE in this case. Thus, as far as the calculation of the density of states is concerned, these two methods give identical results. The approaches adopted in calculating the total number of states are, however, different in these two methods. Thus one may expect that the total number of states obtained from these two methods may be slightly different (see Table 5.1).

From the preceding discussion of the calculation of $W(E)$ and $\rho(E)$, we can derive an alternate expression for calculating the quasi-equilibrium rate constant $k(E)$. For this purpose we start from eq. 5.7:

$$k(\beta)Q(\beta) = \int_0^\infty k(E)\rho(E)\exp(-\beta E)\,dE \tag{5.50}$$

Carrying out the inverse Laplace transformation yields

$$k(E)\rho(E) = \frac{1}{2\Pi i} \int_{\gamma - i\infty}^{\gamma + i\infty} d\beta \, \exp(\beta E)k(\beta)Q(\beta) \tag{5-51}$$

If $k(\beta)$ does not vary rapidly arouns the saddle point value of β given by eq. 5.25, we can replace the β value in $k(\beta)$ by β_2^* and take $k(\beta_2^*)$ out of the contour integral. We find

$$k(E)\rho(E) = k(\beta_2^*) \frac{1}{2\Pi i} \int_{\gamma - i\infty}^{\gamma + i\infty} d\beta \, \exp(\beta E)Q(\beta) = k(\beta_2^*)\rho(E) \tag{5.52}$$

It follows that

$$k(E) = k(\beta_2^*) \tag{5.53}$$

This indicates that to evaluate the quasi-equilibrium rate constant $k(E)$, we first determine β_2^* from eq. 5.25 and then substitute this β_2^* value into the corresponding thermal unimolecular rate constant (cf. eq. 5.12). In other words, β_2^* in the isolated system plays the role of β in the thermal system.

5.3 APPLICATIONS OF QUASI-EQUILIBRIUM THEORY

In Sec. 5.1, using quasi-equilibrium theory, we derived an expression for the unimolecular reaction rate constant of an isolated system in terms of the total number of states of the activated complex $W^\ddagger(E - E_0^\ddagger)$ and the density of states of the reactant $\rho(E)$. In the preceding section we discussed the approximation methods for calculating $\rho(E)$ and $W(E)$. In this section we discuss the applications of quasi-equilibrium theory to the theoretical interpretation of mass spectra and to the calculation of the translational energy of fragments of ion decomposition.

5.3.1 Translational Energy of Products of Unimolecular Decomposition

According to quasi-equilibrium theory, the translational energy of the activated complex in the reaction coordinate becomes part of the translational energy of the dissociated products. If the excess energy $E^{\ddagger}$ of the activated complex is divided into the vibrational energy ε_v and the translational energy ε_t in the reaction coordinate, then the calculation of the average translational energy is essentially the problem of calculating the distribution of vibrational energy among a collection of oscillators whose total energy is $E^{\ddagger}$ (Klots, 1964, 1971; Haney and Franklin, 1968; Lau and Lin, 1971b). The average vibrational energy of the activated complex is given by the integral

$$\bar{\varepsilon}_v = \int_0^{E^{\ddagger}} \varepsilon_v P(E^{\ddagger}. \varepsilon_v) \, d\varepsilon_v \qquad (5.54)$$

where $P(E^{\ddagger}, \varepsilon_v) \, d\varepsilon_v$ is the probability that a system with total energy $\varepsilon_v \cdot P(E^{\ddagger}, \varepsilon_v)$ can be expressed by

$$P(E^{\ddagger}, \varepsilon_v) = \frac{\rho^{\ddagger}(\varepsilon_v)}{W^{\ddagger}(E^{\ddagger})} \qquad (5.55)$$

If the average translational energy $\bar{\varepsilon}_t$ is taken to be the difference in the excess energy and the average vibrational energy, $\bar{\varepsilon}_t = E^{\ddagger} - \bar{\varepsilon}_v$, then by substituting eq. 5.55 into eq. 5.54, we obtain

$$\bar{\varepsilon}_t = E^{\ddagger} - \frac{1}{W^{\ddagger}(E^{\ddagger})} \int_0^{E^{\ddagger}} \varepsilon_v \frac{dW^{\ddagger}(\varepsilon_v)}{d\varepsilon_v} \, d\varepsilon_v \qquad (5.56)$$

Integration by part of eq. 5.52 yields

$$\bar{\varepsilon}_t = \frac{1}{W^{\ddagger}(E^{\ddagger})} \int_0^{E^{\ddagger}} W^{\ddagger}(\varepsilon_v) \, d\varepsilon_v \qquad (5.57)$$

A comparison of the calculation of the average translational energy using various methods of enumerating the total number of states has been carried out by Klots (1964), Franklin et al. (Haney and Franklin, 1968; Spatz et al., 1969), and Lau and Lin (1971a). In this section we briefly describe how to apply the method of steepest descent to calculate $\bar{\varepsilon}_t$.

Taking the Laplace transform of eq. 5.57, we obtain

$$L[\bar{\varepsilon}_t W^{\ddagger}(E^{\ddagger})] = \int_0^{\infty} dE^{\ddagger} \exp(-\beta E^{\ddagger}) \left[\int_0^{E^{\ddagger}} W^{\ddagger}(\varepsilon_v) \, d\varepsilon_v \right] \qquad (5.58)$$

where L denotes the Laplace transform operator. Integrating by parts or changing the order of integrations, we find

$$L[\bar{\varepsilon}_t W^{\ddagger}(E^{\ddagger})] = \frac{1}{\beta} \int_0^\infty dE^{\ddagger} \exp(-\beta E^{\ddagger}) W^{\ddagger}(E^{\ddagger}) = \frac{1}{\beta} L[W^{\ddagger}(E^{\ddagger})] \quad (5.59)$$

Because of eq. 4.17, that is, $W(E) = L^{-1}[(1/\beta)Q(\beta)]$, it follows that

$$L[\bar{\varepsilon}_t W^{\ddagger}(E^{\ddagger})] = \frac{Q^{\ddagger}(\beta)}{\beta^2} \quad (5.60)$$

Therefore the function $\bar{\varepsilon}_t(E^{\ddagger})W^{\ddagger}(E^{\ddagger})$ can be evaluated by taking the inverse Laplace transform of $Q^{\ddagger}(\beta)/\beta^2$:

$$\bar{\varepsilon}_t(E^{\ddagger})W^{\ddagger}(E^{\ddagger}) = L^{-1}\left[\frac{Q^{\ddagger}(\beta)}{\beta^2}\right] = \frac{1}{2\Pi i} \int_{\gamma-i\infty}^{\gamma+i\infty} d\beta \exp(\beta E^{\ddagger}) \frac{Q^{\ddagger}(\beta)}{\beta^2} \quad (5.61)$$

The contour integral in eq. 5.61 can be integrated by using the method of steepest descent. Applying the method of steepest descent (see Appendix 2), we obtain, to the first-order approximation, the following results:

$$\bar{\varepsilon}_t = \frac{1}{W^{\ddagger}(E^{\ddagger})} \cdot \frac{\exp(\beta_3^* E^{\ddagger})Q^{\ddagger}(\beta_3^*)}{\sqrt{2\Pi\beta_3^{*2}[2 + \beta_3^{*2}\{(\partial^2/\partial\beta^2)\log Q^{\ddagger}(\beta)\}_{\beta=\beta_3^*}]}} \quad (5.62)$$

where β_3^* is to be determined by

$$\frac{2}{\beta_3^*} = E^{\ddagger} + \left[\frac{\partial}{\partial\beta}\log Q^{\ddagger}(\beta)\right]_{\beta=\beta_3^*} \quad (5.63)$$

The calculation of the total number of states $W^{\ddagger}(E^{\ddagger})$ has been discussed. Thus from eq. 5.62 one can calculate $\bar{\varepsilon}_t$, the average translational energy of dissociation fragments.

An approximation can be introduced here to simplify the calculation of $\bar{\varepsilon}_t$. Suppose the factor $1/\beta$ in eq. 5.59 is replaced by the first-order value of $1/\beta_1^*$ in which β_1^* is defined in eq. 5.23. Then

$$\bar{\varepsilon}_t(E^{\ddagger})W^{\ddagger}(E^{\ddagger}) = \frac{1}{\beta_1^*} L^{-1}\left[\frac{Q^{\ddagger}(\beta)}{\beta}\right] = \frac{1}{\beta_1^*} \cdot \frac{1}{2\Pi i} \int_{\gamma-i\infty}^{\gamma+i\infty} \frac{d\beta}{\beta}$$

$$\times \exp(\beta E^{\ddagger})Q^{\ddagger}(\beta) = \frac{1}{\beta_1^*} W^{\ddagger}(E^{\ddagger}) \quad (5.64)$$

Here the relation eq. 5.17 has been used. It follows

$$\bar{\varepsilon}_t(E^{\ddagger}) = \frac{1}{\beta_1^*} \quad (5.65)$$

where β_1^* is to be determined from an equation similar to eq. 5.23 or

$$\frac{1}{\beta_1^*} = E^{\ddagger} + \left[\frac{\partial \log Q^{\ddagger}(\beta)}{\partial \beta}\right]_{\beta = \beta_1^*} \tag{5.66}$$

This expression should be compared with the result of a pressure-maintained thermal distribution, $\bar{\varepsilon}_t = kT = 1/\beta$. A slightly cruder approximation for calculating $\bar{\varepsilon}_t$ is

$$\bar{\varepsilon}_t(E^{\ddagger}) = \frac{1}{\beta_2^*} \tag{5.67}$$

Next let us apply the results of the derivation to a system of N harmonic oscillators. The partition function of the activated complex of this system is given by

$$Q^{\ddagger}(\beta) = \prod_{i=1}^{N'} [1 - \exp(-\beta \varepsilon_i^{\ddagger})]^{-1} \tag{5.68}$$

where $N' = N - 1$ for a bound activated complex and $N' = N - 3$ for a loose complex (Haney and Franklin, 1968; Spatz et al., 1969). Substituting eq. 5.68 into eq. 5.61 yields

$$\bar{\varepsilon}_t(E^{\ddagger}) = \frac{1}{W^{\ddagger}(E^{\ddagger})} \cdot \frac{\exp(\beta_3^* E^{\ddagger}) \prod_{i=1}^{N'} [1 - \exp(-\beta_3^* \varepsilon_i^{\ddagger})]^{-1}}{\sqrt{2\Pi \beta_3^{*2}[2 + \sum_{i=1}^{N'} (\beta_3^* \varepsilon_i^{\ddagger})^2 \exp(\beta_3^* \varepsilon_i^{\ddagger})/\{\exp(\beta_3^* \varepsilon_i^{\ddagger}) - 1\}^2]}} \tag{5.69}$$

where β_3^* is to be determined from

$$\frac{2}{\beta_3^*} = E^{\ddagger} - \sum_{i=1}^{N'} \frac{\varepsilon_i^{\ddagger}}{\exp(\beta_3^* \varepsilon_i^{\ddagger}) - 1} \tag{5.70}$$

When the system has an extreme high energy, β approaches zero. In this limiting case the partition function becomes classical, that is, $Q^{\ddagger}(\beta) = \prod_i^{N'} (\beta \varepsilon_i^{\ddagger})^{-1}$. The total number of states $W^{\ddagger}(E^{\ddagger})$ for the classical system has been obtained and is given in eq. 5.32. Substituting the result for $W^{\ddagger}(E^{\ddagger})$ given in eq. 5.32 into eq. 5.57, we find

$$\bar{\varepsilon}_t(E^{\ddagger}) = \frac{E^{\ddagger}}{N' + 1} \tag{5.71}$$

This classical result of $\bar{\varepsilon}_t$ can be obtained directly from eq. 5.65 with β_1^* determined by eq. 5.66 for a system of N' harmonic oscillators.

For the calculation of the average translational energy of the fragments, one may consider two possible intermediates for the activated complex, a bound ion with $N - 1$ oscillators and a loose one with $N - 3$ oscillators. The results of $\bar{\varepsilon}_t$ obtained from using eqs. 5.65 and 5.69 are given in Tables 5.4 and 5.5. The results, calculated by using eqs. 5.65 and 5.69, are in excellent

Table 5.4

Comparison of Average Translational Energy Calculated Using Different Methods of Enumerating the States for $N - 1$ Complex (in kcal/mol)

Reaction	$E^{\ddagger a}$	Observed[a]	Direct[b]	Eq. 5.63	Eq. 5.61	VWJ[b]	TW[b]
$C_2N_2^+ \rightarrow CN^+ + CN$	19	4.8	3.7	3.69	3.66	3.7	3.6
$C_2H_2^+ \rightarrow CH^+ + CH$	19	6.4	4.2	4.15	4.09	4.2	4.0
$c\text{-}C_3H_6^+ \rightarrow C_2H_2^+ + CH_4$	15	1.6	2.0	2.07	2.03	3.8	4.0
$c\text{-}C_3H_6^+ \rightarrow C_2H_3^+ + CH_3$	21	2.8	2.5	2.47	2.43	3.2	4.4
$c\text{-}C_3H_6^+ \rightarrow CH_2^+ + C_2H_4$	113	12.1	7.4	7.36	7.34	7.4	7.4

[a] Haney and Franklin, 1968.
[b] Spatz et al., 1969 who tested the methods of enumerating the states by Vestal, Wahrhaftig, and Johnston and Tou and Wahrhaftig. Indicated by initials in Table 5.4.

agreement with those obtained from the direct counting for the whole energy range. Tables 5.4 and 5.5 indicate that there are some discrepancies between the calculated results and experimental values of $\bar{\varepsilon}_t$. Franklin (Spatz et al., 1968) has attributed these discrepancies to uncertainties in the heats of formation and the experimental data as well as to a lack of knowledge of the vibrational modes existing in the ion. However, other effects, like the effect of the exclusion of disallowed states from $\rho(E)$ and $W(E)$ in unimolecular reactions (Forst and Prášil, 1970), might also be important in the calculation of $\bar{\varepsilon}_t$.

5.3.2 Application of Quasi-Equilibrium Theory of Mass Spectra

One of the important applications of quasi-equilibrium theory is to interpret the mass spectra; that is, quasi-equilibrium theory allows one to predict or or to explain the formation of various ions and the relative quantities of the ions formed at different electron energies. Let us first consider how the various ions are formed. It is believed that many of these ions originate as the result of the unimolecular decomposition of an activated complex of the parent-molecule ion, although some ions may also be formed from the unimolecular decomposition of other product ions. In other words, the application of quasi-equilibrium theory requires first the selection of a detailed "breakdown scheme" for the parent ion. This must be designed to represent the main reaction paths leading to each ion, but cannot seek to include all possible paths for reasons of complexity. There are frequently ambiguities that make the selection of a scheme somewhat arbitrary.

Table 5.5
Average Translational Energy Calculated From eqs. 5.61 and 5.63 from Various Sets of Excluded Modes (in kcal/mol)

Reaction	$E^{\ddagger a}$	$\bar{\varepsilon}_{exp}^{a}$	$\bar{\varepsilon}_{exp}^{b}$	$\bar{\varepsilon}_t$ (eq. 5.63)	$\bar{\varepsilon}_t$ (eq. 5.61)	Modes and fundamentals excluded (cm^{-1})
$c\text{-}C_3H_6^+ \rightarrow C_2H_2^+ + CH_4$	15	1.6	2.0	2.07	2.03	Ring deformation^a, 868
			2.1	2.16	2.12	Ring deformation^b, 868 CH stretch, 3024
$c\text{-}C_3H_6^+ \rightarrow C_2H_3^+ + CH_3$	21	2.8	2.5	2.47	2.43	Ring deformation^a, 868
			2.5	2.59	2.55	Ring deformation^b, 868 CH stretch, 3024
$C_2N_2^+ \rightarrow CN^+ + CN$	19	4.8	3.7	3.69	3.66	CC stretch^a, 848
			5.2	5.13	5.09	CC stretch^b, 848 (2) CCN bend, 226
$CH_2Cl_2^+ \rightarrow CH_2^+ + Cl_2$	36	8.3	5.7	5.68	5.65	CCl$_2$ stretch^a, 283
			7.1	7.08	7.04	CCl$_2$ stretch^b, 283 CCl stretch, 702 CH$_3$ rock, 899
$C_2H_2^+ \rightarrow CH^+ + CH$	19	6.4	4.2	4.15	4.09	CC stretch^a, 870
			5.5	5.67	5.58	CC stretch^b, 870 (2) CH bend, 1312
$CH_3CN^+ \rightarrow CH_2^+ + HCN$	17	2.8	3.0	2.95	2.89	CC stretch^a, 918
			3.2	3.35	3.29	CC stretch^b, 918 CH stretch 2942 C CN bend 380

Haney and Franklin, 1968.
Spatz et al., 1969

Now, for example, if we have the parent ion A and the parent ion A is decomposed competitively into the product ions P_i,

$$A \xrightarrow{k_1} P_1, A \xrightarrow{k_2} P_2, \ldots, A \xrightarrow{k_N} P_N \qquad (5.72)$$

then the amounts of reactant left and product produced at time t are given by the expressions

$$[A] = [A]_0 \exp\left(-t \sum_{i=1}^{N} k_i\right) \qquad (5.73)$$

and

$$[P_i] = \frac{k_i A_0}{\sum_{i=1}^{N} k_i} \left[1 - \exp\left(-t \sum_{i=1}^{N} k_i\right)\right] \qquad (5.74)$$

where $i = 1, 2, 3, \ldots, N$ and $[A]_0$ represents the concentration of the reactant ion present at zero time. Here we have ignored the consecutive reactions, that is, we have for simplicity ignored the decomposition of the primary product ions P_i. More complex kinetic schemes may be treated conveniently by using the Laplace transform method or the eigenvalue matrix method. From eqs. 5.73 and 5.74 we can see that to calculate the quantities of each of the ionic species, we heave to know k_i's as a function of the energy of the reactant ion and the time t. The time t is usually taken to be about 10^{-5}–10^{-6} sec. since this is the approximate time of residence of ions in the ion source, and $k_i(E)$ can be calculated by using eqs. 5.4 or 5.53, discussed in the previous section. Once $k_i(E)$ and t are determined, one can calculate the relative abundance of the ionic species as a function of the energy E of the reactant ion. In other words, one can obtain the breakdown curves of the reactant ion from the kinetic considerations by determining the amounts of each ionic species present at a given value of E. However, the internal energy E of the ion that is decomposing is not to be confused with the energy of the bombarding electrons. It is well known (Kiser, 1965) that the mass spectrum of a compound does not generally change significantly above about 30 eV electron energy, and changes only very slightly above 50 eV, up to, say, 100 eV. This implies that only some fraction of the energy of the electron is transferred to the molecule at higher electron energies and this fraction varies with the electron energy. In other words, there must be a distribution of the energies transferred to the molecules upon impact.

To illustrate how to apply quasi-equilibrium theory to mass spectra, we consider the decomposition of the propane ion $C_3H_8^+$ (Vestal, 1969; Vestal and Futrell, 1970). Briefly the procedure consists of choosing a set of reactions, assigning activation energies and activated complex configurations for each reaction, and calculating the absolute rate constants for each reaction using eq. 5.4 or eq. 5.53. From the reaction rates (cf. eq. 5.72–5.74), the relative abundance of each ion may be calculated as a function of the internal energy of the reactant ion and the time after its formation. Finally, these abundances evaluated at the time t appropriate to a particular experimental arrangement are averaged over the internal energy distributions to give a mass spectrum that may be compared with the experimental results.

Reaction Mechanism. A reasonably complete set of reactions for the propane ion including only the low-energy competing processes yielding C_2 and C_3 fragments has been proposed and is given in Fig. 5.1. Other reactions are possible, but can probably be excluded because of competition with more energetically favored reactions.

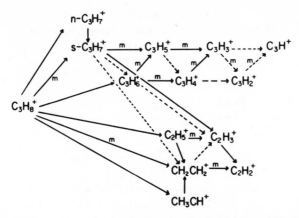

Fig. 5.1 Postulated reaction mechanism for the unimolecular decomposition of propane ions. Decompositions designated by "m" indicate that the corresponding metastable has been observed.

Activation Energies. The activation energy for a reaction

$$AB^+ \longrightarrow A^+ + B$$

may be defined operationally for an idealized experiment as:

$$E_0^{\ddagger} = AP(A^+) - I(AB^+)$$

where $I(AB^+)$ is the adiabatic ionization potential and $AP(A^+)$ is the minimum energy that must be added to the species AB in its ground state to produce the fragment ion A^+ with a nonzero rate. Difficulties inherent in the measurements of $AP(A^+)$ and $I(AB^+)$ have been discussed in considerable detail and are not given here (Chupka and Berkowitz, 1967; Vestal, 1969). For propane the adiabatic ionization potential has been assumed to be 10.64 eV, which is obtained somewhat arbitrarily by subtracting 0.43 eV from the appearance potential of $C_3H_8^+$ reported by Watanabe et al. and Inghram et al. (Watanabe et al., 1960; Inghram et al., 1961). The values of activation energies for the decomposition of the propane ion are given in Tables 5.6 and 5.7.

Activated Complexes. In most calculations using quasi-equilibrium theory, both the normal configuration and the activated complex configuration have been represented by harmonic oscillators and free rotors. The normal configuration of the ion has been assumed to be very similar to the configuration of the neutral molecule and for the most part the approximate

Table 5.6
Activation Energies for Primary Fragmentations of Propane and Deuteropropanes

No.	Reaction	$E_0^\ddagger$ (eV)			
		d_0	d_2	d_6	d_8
1	$C_3X_8^+ \to s\text{-}C_3X_7^+ + X$	0.94	0.999	0.907	0.966
2	$\to CX_2CX_2^+ + CX_4$	1.00	0.987	1.028	1.015
3	$\to CX_3CX^+ + CX_4$	1.20	1.23	1.21	1.24
4	$\to C_3X_6^+ + X_2$	1.27	1.32	1.29	1.35
5	$\to C_2X_5^+ + CH_3$	1.33	1.33	1.36	1.36
6	$\to n\text{-}C_3X_7^+ + X$	1.66	1.66	1.73	1.73

oscillator frequencies of the molecule have been used. The torsional motion of methyl groups is usually taken as a free rotation in the ion, and in many cases some of the skeletal frequencies are assumed to be somewhat lower in the ion than in the molecule. The approximate molecular frequencies are used to represent the normal configuration of the molecule ion with the CH_3 torsional motions taken as free rotations. The frequencies and moments of

Table 5.7
Activation Energies for Subsequent Reactions of the Primary Fragment Ions

Reaction No.	Reaction	$E_0^\ddagger$ (eV)
S1	$C_3H_7^+ \to C_3H_5^+ + H_2$	2.1
S2	$\to C_2H_3^+ + CH_4$	2.3
S3	$\to C_2H_4^+ + CH_3$	4.1
	$\to CH_3^+ + C_2H_4$	3.4
	$\to C_3H_6^+ + H$	3.9
S4	$C_3H_6^+ \to C_3H_5^+ + H$	2.1
S5	$\to C_3H_4^+ + H_2$	2.2
	$\to C_2H_3^+ + CH_3$	3.1
S6	$C_3H_5^+ \to C_3H_3^+ + H_2$	3.0
S7	$C_2H_5^+ \to C_2H_3^+ + H_2$	2.4
S8	$\to C_2H_4^+ + H$	3.8
S9	$C_2H_4^+ \to C_2H_2^+ + H_2$	2.7
	$\to C_2H_3^+ + H$	3.9
S10	$C_2H_3^+ \to C_2H_2^+ + H$	4.3

Table 5.8

Representation of the Normal Configuration for the
Propane Molecule-Ion

Designation	Frequency (cm$^-$)	No. of oscillators
C—H stretch	2900	8
C—H bending	1400	8
CH$_3$ deformation	1200	4
CH$_2$ deformation	800	2
C—C stretch	900	2
C—C—C bending	400	1
CH$_3$ torsion	Free rotation[a]	2

[a] The moment of inertia for the free rotor is taken as 4.6 10^{-40} g cm^2.

inertia for the normal configurations are given in Table 5.8. The parameters for the activated complex configurations are derived from the normal configurations as described later for the individual primary reactions and illustrated in Fig. 5.8 (Vestal, 1969).

For the reaction

$$C_3H_8^+ \longrightarrow S - C_3H_7^+ + H \qquad (R_1)$$

the activated complex for the formation of secondary propyl ion is taken as a tight ring structure. Both free rotations are assumed stopped and the CH$_3$ torsional frequency is taken as 300 cm^{-1}. The two CH$_3$ deformation frequencies and the C—C—C chain bending frequency are increased by a factor of 1.5. The reaction coordinate is taken as the C—H stretching of the secondary hydrogen and the corresponding C—H bending frequency is reduced by a factor of 2. The symmetry factor is 2.

For the reaction

$$C_3H_8^+ \longrightarrow CH_2CH_2^+ + CH_4 \qquad (R_2)$$

the 1,3 elimination of methane, the activated complex is taken as a ring structure similar to that used for eq. R$_1$, but slightly looser. The C—C stretching is taken as the reaction coordinate and the C—H stretching and bending frequencies for the hydrogen lost are both reduced by a factor of 2. Both free rotations are assumed stopped and are taken as torsional vibrations with frequencies of 200 cm^{-1}. The CH$_2$ deformation frequencies are increased by a factor of 2. The symmetry factor is 6.

For the reaction

$$C_3H_8^+ \longrightarrow C_3H_6^+ + H_2 \qquad (R_3)$$

the hydrogens lost are assumed to come from adjacent carbon atoms. The reaction coordinate is taken as a C—H stretching and one free rotation is stopped and taken as a torsional vibration with a frequency of 200 cm^{-1}. The symmetry factor is 4.

For the reaction

$$C_3H_8^+ \longrightarrow C_2H_5^+ + CH_3 \qquad (R_4)$$

the C—C stretching frequency is taken as the reaction coordinate; the CH$_3$ deformation frequencies for the corresponding methyl group are reduced by a factor of 4 and the C—C—C chain bending frequency is also reduced by a factor of 4. Both free rotations are active. The symmetry factor is 2.

For the reaction

$$C_3H_8^+ \longrightarrow n - C_3H_7^+ + H \qquad (R_5)$$

the reaction coordinate is taken as the C—H stretch of a primary hydrogen and the corresponding C—H bending frequency is reduced by a factor of 2, see Fig. 5.2. The two CH$_3$ deformation frequencies for the methyl group involved and C—C—C chain bending frequency are also reduced by a factor of 2. Both free rotations are assumed active. The symmetry factor is 6.

Relative Abundances. The most fundamental quantity that can be calculated using the quasi-equilibrium theory is the breakdown graph, which gives relative ion abundances at a particular time after ionization of the molecule as a function of the internal energy E of the molecular ion. The abundance of the jth product ion at a time t after formation of the reactant ion can be obtained by using an approach similar to that in eqs. 5.72–5.74 and is given by

$$A_j(E, t) = \frac{k_j(E)}{\sum_i k_i(E)} \left\{ 1 - \exp\left[1 - \sum_i k_i(E)t \right] \right\} [1 - P_j(E, t)] \quad (5.75)$$

where the summation is carried out over the set of competing reactions and $P_j(E, t)$ is the probability of farther fragmentation of the primary product ion. The relative abundance of the parent ion is given by

$$A_p(E, t) = \exp\left[-\sum_j k_j(E)t \right] \qquad (5.76)$$

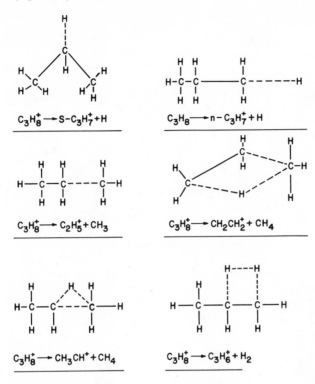

$C_3H_8^+ \longrightarrow S\text{-}C_3H_7^+ + H$

$C_3H_8 \longrightarrow n\text{-}C_3H_7^+ + H$

$C_3H_8^+ \longrightarrow C_2H_5^+ + CH_3$

$C_3H_8^+ \longrightarrow CH_2CH_2^+ + CH_4$

$C_3H_8^+ \longrightarrow CH_3CH^+ + CH_4$

$C_3H_8^+ \longrightarrow C_3H_6^+ + H_2$

Fig. 5.2 Schematic diagrams of the activated complex configuration used in the QET calculation on propane. The dashed lines indicate the bonds that are broken and the new bonds that are formed.

Hence the mass spectrum may be considered a snapshot of the extent of fragmentation at a particular time that will vary from one instrument to another. The breakdown curves for the propane ion are shown in Fig. 5.3.

Mass Spectrum. Mass spectrometric measurements normally give the relative abundances of the various ions as functions of the energy of the impacting electron and temperature, whereas the theory gives the relative abundances as a function of the internal energy of the reactant ion. Therefore, to compare theory with experiment, the internal energy distribution appropriate to given experimental parameters must be known. Recently, with the advent of photoelectron spectroscopy and photoionization derivative techniques, significant data on the internal energy distribution has become available. The internal energy distribution used in the calculations for the propane ion is derived from the measurements of Chupka and Kaminsky (1961) and is

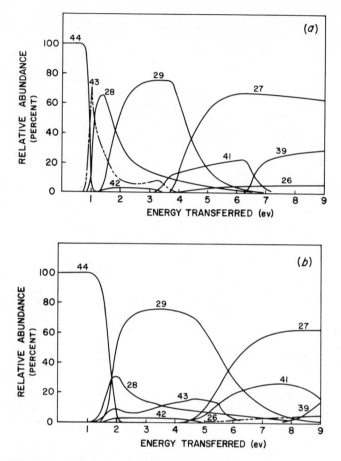

Fig. 5.3 Calculated breakdown graphs for propane at (a) 10^{-6} sec and (b) 10^{-10} sec. The curves are identified by the mass number for the fragment ion.

shown in Fig. 5.4. For the calculation of the mass spectrum at any electron energy the energy transfer diagram must be brought into the calculation. For the chosen value of electron energy, the "elementary" mass spectra must be multiplied by the weighting factor appropriate to their value of internal energy. That is, this calculation must involve multiplying the breakdown graphs of each molecule by an energy distribution function $Y(E)$ (Vestal, 1969). As mentioned earlier, the experimentally determined $Y(E)$ of Chupka and Kaminsky is used. A summation is then carried out over all relevant values of internal energy. The result of such a calculation is shown in Table 5.9. Generally, good agreement between calculation and experiment is noted.

Table 5.9
Comparison of Calculated and Experimental Mass Spectra for 70 eV electrons at 250°C; Ion Intensity as Percent of Total Ionization

C_3H_8			$CH_3CD_2CH_3$			$CD_3CH_2CD_3$			C_3D_8		
m/q	Exp	Calc	m/q	Exp	Calc	m/q	Exp	Calc	m/q	Exp	Calc
26	1.6	1.3	26	0.6	0.4	26	0.2	0.2	28	2.0	2.1
27	9.2	16.7	27	3.2	2.6	27	1.0	1.1	30	9.7	15.9
28	19.0	20.6	28	8.0	9.3	28	2.7	5.3	32	22.2	26.1
29	31.3	31.5	29	8.9	7.9	29	6.0	9.7	34	36.6	29.1
			30	19.1	22.2	30	15.9	19.7			
39	5.0	4.3	31	32.4	33.0	31	5.7	4.2	42	3.2	3.8
40	0.7	0.2				32	32.1	28.7	44	0.4	0.2
41	5.0	3.2	39	2.3	1.4				46	2.5	2.0
42	1.8	2.3	40	2.8	2.7	40	0.9	0.4	48	3.8	2.0
43	12.2	9.2	41	1.6	1.0	41	1.8	2.0	50	5.8	5.7
44	13.7	10.7	42	2.2	1.7	42	2.3	1.1	52	13.5	12.0
			43	2.3	2.1	43	0.4	0.2			
			44	3.2	2.9	44	0.7	0.9			
			45	1.9	1.2	45	2.3	1.4			
			46	10.8	11.5	46	1.7	0.2			
						47	1.3	1.2			
						48	1.3	2.6			
						49	12.0	11.0			
						50	11.6	10.5			

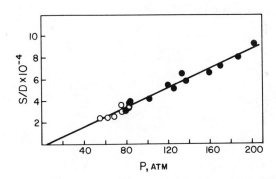

Fig. 5.4 Internal energy distribution function $Y(E)$ assumed in the present calculations. The lower energy portion from 0 to 4 eV is taken from the experiments of Chupka and Kaminsky (1961), *J. Chem. Phys.*, **35**, 1991.

187

5.4 THE RRKM THEORY OF UNIMOLECULAR REACTIONS

So far we have been concerned with the unimolecular decomposition of isolated molecules. Now we consider the unimolecular reactions of reactant molecules in the presence of collision partners. All modern theories of unimolecular reactions are based on the Lindemann scheme (Laidler, 1969),

$$A + M \underset{k_{-2}}{\overset{k_2}{\rightleftharpoons}} A^* + M \tag{5.77}$$

$$A^* \overset{k_1}{\longrightarrow} P \tag{5.78}$$

where A represents a normal reactant molecule, A^*, an energized molecule, M, a collision partner, and P, a product molecule. In eq. 5.77 the forward reaction denotes the process of energization by collision, and the backward reaction, the process of deenergization by collision. Equation 5.78 represents the reaction of the activated molecule. From eqs. 5.77 and 5.78 the rate of disappearance of A^* is given by

$$\frac{d[A^*]}{dt} = k_2[A][M] - k_{-2}[A^*][M] - k_1[A^*] \tag{5.79}$$

and the rate of formation of products is

$$\frac{d[P]}{dt} = v = k_1[A^*] \tag{5.80}$$

Application of the steady-state assumption to A^* yields

$$[A^*] = \frac{k_2[A][M]}{k_1 + k_{-2}[M]} \tag{5.81}$$

Substituting eq. 5.81 into eq. 5.80, we obtain

$$v = \frac{k_1 k_2[A][M]}{k_1 + k_{-2}[M]} \tag{5.82}$$

If $A = M$, that is, no other species but A is present in the reacting system, then eq. 5.82 reduces to

$$v = \frac{k_1 k_2[A]^2}{k_1 + k_{-2}[A]} \tag{5.83}$$

At sufficiently high pressures, $k_{-2}[A] \gg k_1$ and the rate of formation of products is given by

$$v = \frac{k_1 k_2}{k_{-2}}[A] = k_\infty[A] \tag{5.84}$$

where $k_\infty = k_1 k_2 / k_{-2}$, the first-order rate constant at high pressure. On the other hand, at low pressures $k_1 \gg k_{-2}[A]$ and eq. 5.83 reduces to

$$v = k_2[A]^2 \tag{5.85}$$

Equations 5.84 and 5.85 indicate that the kinetics of unimolecular reactions at high pressures are of the first order and at low pressures of the second order. This change from first-order to second-order kinetics has been observed experimentally for a number of reactions. Physically, eqs. 5.77 and 5.78 can be interpreted as follows. When a collision occurs between two molecules, one of them may occasionally acquire a sufficient amount of energy to enable the molecule to become a product molecule without the necessity of obtaining any additional energy; the molecule is then said to be *energized*. If the conversion of energized molecules into product molecules is slow compared with the rate of deenergization and with the rate of energization, an equilibrium is rapidly established between energization and deenergization and the equilibrium concentration of energized molecules is proportional to the concentration of normal molecules. The rate of reaction that is proportional to the concentration of energized molecules is therefore proportional to the concentration of normal molecules and the reaction kinetics is of the first order. At sufficiently low pressures the collisions cannot maintain an equilibrium supply of energized molecules and the rate of reaction depends on the rate of energization and hence is proportional to the square of the concentration of reactant molecules.

From eq. 5.82 we can define a first-order rate constant by

$$v = k^{(1)}[A] \tag{5.86}$$

where $k^{(1)}$ is given by

$$k^{(1)} = \frac{k_1 k_2 [M]}{k_1 + k_{-2}[M]} = \frac{k_\infty}{1 + k_1/k_{-2}[M]} \tag{5.87}$$

A plot of $k^{(1)}$ versus $[M]$ gives rise to a curve, shown in Figure 5.5, in which $k^{(1)}$ is constant in the high-pressure range and falls to zero at low pressures. Equation 5.87 indicates that $k^{(1)}$ becomes equal to one-half of k_∞ when $k_{-2}[M]$ is equal to k_1, that is,

$$k_{-2}[M]_{1/2} = k_1, \qquad [M]_{1/2} = \frac{k_\infty}{k_2} \tag{5.88}$$

The value of k_∞ can be obtained from experiment, and according to the simple collision theory k_2 should be equal to $Z_2 \exp(-(E^\dagger/RT))$, where Z_2 represents the bimolecular collision frequency. However, this procedure always leads to the prediction that the first-order rate constant $k^{(1)}$ should fall off at much higher pressures than observed experimentally. Since k_∞ is

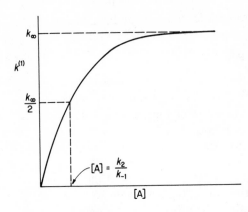

Fig. 5.5 A schematic plot of $k^{(1)}$ against $[A]$.

an experimental quantity, the error must be in the estimation of k_2. It is therefore necessary to modify the collision theory in such a way as to give larger values for k_2. A second difficulty with the Lindemann scheme becomes apparent when we write eq. 5.87 as

$$\frac{1}{k^{(1)}} = \frac{k_{-2}}{k_2 k_1} + \frac{1}{k_2[M]} = \frac{1}{k_\infty} + \frac{1}{k_2[M]} \tag{5.89}$$

Equation 5.89 implies that a plot of $1/k^{(1)}$ against the reciprocal of the concentration of M should give a straight line. However, deviations from linearity have been found experimentally (see Fig. 5.6).

Many attempts have been carried out by Hinshelwood, Kassel, Rice, Ramsperger, Marcus, and others (Laidler, 1969; Bunker, 1966) to overcome

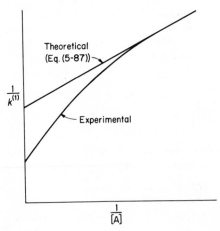

Fig. 5.6 Schematic plots of $1/k^{(1)}$ against $1/[A]$.

the difficulties associated with the Lindemann scheme. The most successful treatment so far is due to Rice and Marcus (Marcus and Rice, 1951; Marcus, 1952, 1965; Wieder and Marcus, 1962); this is now called the RKKM theory. Rather than going through the historical development of the theory of unimolecular reactions, in this section we only present the RRKM formulation.

In the RRKM theory the Lindemann scheme is written as

$$A + M \underset{k_{-2}}{\overset{k_2}{\rightleftharpoons}} A^* + M, \qquad A^* \xrightarrow{\ k_1\ } A^\dagger \xrightarrow{\hspace{1cm}} P \quad (5.90)$$

where A^* represents an energized molecule and $A^\dagger$, an activated complex. Energy is assumed to flow freely between normal modes of $A^\dagger$. The rate constants k_2 and k_1 are considered to be functions of the energy of the energized molecules. The first-order rate constant that applies to a small energy range is given by (Laidler, 1969):

$$dk^{(1)} = \frac{k_1[M] \, dk_2}{k_{-2}[M] + k_1} \quad (5.91)$$

The true first-order rate constant $k^{(1)}$ is then obtained by integrating eq. 5.91 over all possible energies:

$$k^{(1)} = \int \frac{k_1 \, dk_2 / k_{-2}}{1 + (k_1/k_{-2}[M])} \quad (5.92)$$

If the rates of energization and deenergization by collision are rapid compared with that of the conversion of energized molecules into products, the ratio dk_2/k_{-2} is equal to the fraction of molecules having energy in the range of E^* to $E^* + dE^*$ (equilibrium approximation), where E^* must be equal to or greater than the activation energy $E_0^\dagger$ at $T = 0$,

$$\frac{dk_2}{k_{-2}} = \frac{\rho_A(E^*)\exp(-\beta E^*) \, dE^*}{\int_0^\infty \rho(E^*)\exp(-\beta E^*) \, dE^*} = \frac{\rho_A(E^*)\exp(-\beta E^*) \, dE^*}{Q_A} \quad (5.93)$$

where $\rho_A(E^*)$ represents the density of states of A.

Substituting eq. 5.93 into eq. 5.92 yields

$$dk^{(1)} = \frac{k_1(E^*)\rho_A(E^*)\exp(-\beta E^*) \, dE^*}{Q_A[1 + (k_1(E^*)/k_{-2}[M])]} \quad (5.94)$$

Now from eq. 5.4, we have

$$k_1(E^*) = \frac{1}{h} \cdot \frac{W^\ddagger(E^* - E_0^\dagger)}{\rho_A(E^*)} \quad (5.95)$$

Combining eq. 5.95 with eq. 5.94, we obtain

$$dk^{(1)} = \frac{1}{hQ_A} \cdot \frac{W^{\ddagger}(E^* - E_0^{\ddagger})\exp(-\beta E^*)\, dE^*}{1 + (k_1(E^*)/k_{-2}[M])} \tag{5.96}$$

or

$$k^{(1)} = \frac{1}{hQ_A} \int_{E_0^{\ddagger}}^{\infty} \frac{W^{\ddagger}(E^* - E_0^{\ddagger})\exp(-\beta E^*)\, dE^*}{1 + (k_1(E^*)/k_{-2}[M])} \tag{5.97}$$

If we let $\varepsilon^* = E^* - E_0^{\ddagger}$, then eq. 5.97 becomes

$$k^{(1)} = \frac{\exp(-\beta E_0^{\ddagger})}{hQ_A} \int_0^{\infty} \frac{W^{\ddagger}(\varepsilon^*)\exp(-\beta\varepsilon^*)\, d\varepsilon^*}{1 + (k_1(E^*)/k_{-2}[M])} \tag{5.98}$$

When $k_{-2}[M] \gg k_1$, from eq. 5.98 we have

$$k_{\infty} = \frac{\exp(-\beta E_0^{\ddagger})}{hQ_A} \int_0^{\infty} W^{\ddagger}(\varepsilon^*)\exp(-\beta\varepsilon^*)\, d\varepsilon^* \tag{5.99}$$

or

$$k_{\infty} = \frac{\exp(-\beta E_0^{\ddagger})}{hQ_A} \int_0^{\infty} \exp(-\beta\varepsilon^*)\, d\varepsilon^* \int_0^{\varepsilon^*} \rho^{\ddagger}(\varepsilon^{\ddagger})\, d\varepsilon^{\ddagger} \tag{5.100}$$

By reversing the order of integrations in eq. 5.100, we obtain

$$k_{\infty} = \frac{\exp(-\beta E_0^{\ddagger})}{hQ_A} \int_0^{\infty} \rho^{\ddagger}(\varepsilon^{\ddagger})\, d\varepsilon^{\ddagger} \int_{\varepsilon^{\ddagger}}^{\infty} \exp(-\beta\varepsilon^*)\, d\varepsilon^* = \frac{kT}{h} \cdot \frac{\exp(-\beta E_0^{\ddagger})}{Q_A}$$

$$\times \int_0^{\infty} \rho^{\ddagger}(\varepsilon^{\ddagger})\exp(-\beta\varepsilon^{\ddagger})\, d\varepsilon^{\ddagger} = \frac{kT}{h} \cdot \frac{Q^{\ddagger}}{Q_A} \exp(-\beta E_0^{\ddagger}) \tag{5.101}$$

which is the expression obtained from the transition state theory (Glasstone et al., 1940).

If we use the expression $\rho(E) = \sum_i \delta(E - E_i)$ given in eq. 5.18, and integrate eq. 5.94 with respect to E^*, we obtain

$$k^{(1)} = \sum_i \frac{k_1(E_i)\exp(-\beta E_i)}{Q_A[1 + (k_1(E_i)/k_{-2}[M])]}$$

$$= \sum_{E_i > E_0^{\ddagger}}^{\infty} \frac{W^{\ddagger}(E_i - E_0^{\ddagger})\exp(-\beta E_i)}{hQ_A\rho_A(E_i)[1 + (k_1(E_i)/k_{-2}[M])]}$$

$$= \frac{\exp(-\beta E_0^{\ddagger})}{hQ_A} \sum_{\varepsilon_i^{\ddagger}=0}^{\infty} \frac{W^{\ddagger}(\varepsilon_i^{\ddagger})\exp(-\beta\varepsilon_i^{\ddagger})}{\rho_A(E_0^{\ddagger} + \varepsilon_i^{\ddagger})[1 + (k_1(\varepsilon_i^{\ddagger} + E_0^{\ddagger})/k_{-2}[M])]} \tag{5.102}$$

An alternate derivation of the RRKM theory based on the stochastic model is presented in Appendix Three.

5.5 APPLICATION OF THE RRKM THEORY

Many excellent reviews on the application of the RRKM theory are available (Laidler, 1969; Rabinovitch and Setser, 1964; Waage and Rabinovitch, 1970; Weston and Schwarz, 1972). Here we demonstrate how to apply the RRKM theory by examples.

5.5.1 Isomerization of Methyl Isocyanide

The thermal unimolecular isomerization of methyl isocyanide CH_3NC to acetonitrite CH_3CN,

$$CH_3NC \longrightarrow CH_3CN$$

has been investigated by Schneider and Rabinovitch (1962). The pressure dependence of unimolecular rate constants $k^{(1)}$ for CH_3NC have been studied at 199.4°C, 230.4°C, and 259.8°C. The curves of $\log k^{(1)}$ versus $\log P$ are given in Fig. 5.7. The extrapolated values of k_∞ are obtained as 7.50×10^{-5} sec^{-1} at 199.4°C and 92.5×10^{-5} sec^{-1} at 230.4°C. They were not able to obtain the k_∞ value at 259.8°C; however, they estimated it to be $k_\infty = 76.7 \times 10^{-4}$ sec^{-1} at 259.8°C.

To interpret their data they have written $k^{(1)}$ of the RRKM theory in the summation form, eq. 5.102,

$$k^{(1)} = \frac{I_r \exp(-E_0^{\ddagger}/RT)}{Q_v h} \sum_{\varepsilon_i^{\ddagger}=0}^{\infty}$$

$$\times \frac{W^{\ddagger}(\varepsilon_i^{\ddagger})\exp(-\beta\varepsilon_i^{\ddagger})}{\rho_A(E_0^{\ddagger} + \varepsilon_i^{\ddagger})[1 + I_r/\omega h \cdot W^{\ddagger}(\varepsilon_i^{\ddagger})/[\rho_A(E_0^{\ddagger} + \varepsilon_i^{\ddagger})]]} \quad (5.103)$$

where $\omega = k_{-2}[M]$. The total number of states involved in eq. 5.103 has been evaluated accurately. For the evaluation of eq. 5.103 the vibrational analysis and frequencies for CH_3NC (see Table 5.10) are known (Pillai and Cleveland, 1960), but a model for the transition state must be obtained. The activated complex may be represented as a cyclic species of C_1 or C symmetry having resonance structures

$$
\begin{array}{ccc}
H_3C-N & H_3C\cdots\ddot{N} & \overset{\oplus}{N} \\
\diagdown C & \parallel\vdots & H_3C \parallel\ominus \\
 & \underset{\cdot}{C} & C \\
(a) & (b) & (c)
\end{array}
$$

There are several ways of arriving at the structure and frequencies of the complex (Johnston, 1961). Some plausible bond orders and corresponding

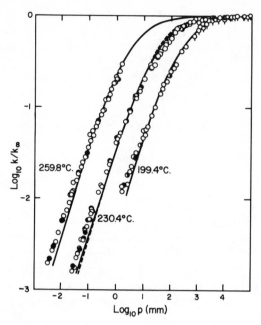

Fig. 5.7 Pressure dependence of unimolecular rate constants for CH_3NC: $\log k/k_\infty$ versus $\log p$ at 199.4°, 230.4°, 259.8°. For clarity the 260° curve is arbitrarily displaced by one $\log k$ unit to the left in the figure while the 200° curve is displaced the same distance to the right; actually, both of these curves would almost coincide with the 230° curve. Vertical marks have been placed under the 200° high-pressure points to assist in distinguishing these from the 230° data. The solid curves represent the calculated results for the 300 harmonic model, adjusted on the pressure axis to coincide with the experimental points at $\log k/k_\infty = -1$. The dotted curve at 230° is for the 600 harmonic model, similarly adjusted.

bond distances are given in Table 5.11, along with other molecular parameters. The frequency assignments for stretching modes of the activated complex were made with the aide of Badger's rule (Herschbach and Laurie, 1961) and of frequencies for cyclopropene and related structures (Table 5.10); two of the ring deformations of the complex have the nature of $N\equiv C$ and $C-N$ stretches, and it was simply assumed that such a rule applies roughly. A twisting vibration cannot be similarly treated and its assignment is more arbitrary. The reaction coordinate is closely related to an antisymmetric ring deformation, for example,

Table 5.10
Vibration Frequencies for CH_3NC Molecule and Activated Complex (cm^{-1})

	Molecule		Activated complex models[a]		
nd	Frequencies	Grouping	Bond	Frequencies 300	Frequencies 600
—H stretch	2966	2998(3)		2998(3)	2998(3)
	3014(2)				
≡C stretch	2161		Ring deformation	1990	1990
I_3 deformation	1410	1443(3)		1443(3)	1443(3)
	1459(2)				
I_3 rocking	1041(2)			1041(2)	1041(2)
—N stretch	945		Ring deformation	600	600
			Ring deformation	Translation	580(2)
NC bending	270(2)		Twist	270	560

he term r_{ce} is set at 1.46 Å, as in CH_3CH, for the standard bond distance for the calculation of
tivated complex force constants by Badger's rule. "Standard" frequencies for the C—N stretch,
≡N stretch, and twisting mode, to which the reduced force constants were applied, were 945,
60, and 880 cm^{-1}, respectively.

and it is arbitrarily designated as such. More sophisticated consideration
would relate the reaction coordinate to other of the normal vibrations. The
RRKM theory is comparatively insensitive to selection of the reaction
coordinate and to moderate vibrations in structure and frequency of the
activated complex, where these give adequate correspondence with the
observed high-pressure frequency factor A_∞, that is, the activation entropy.

Table 5.11
Structure Parameters for CH_3NC Molecule and Activated Complex

Bond	C—H	C—C	C—N	N≡C	HCH
Molecule r, Å	1.094		1.427	1.167	109°46′
Complex[a,c] r, Å	1.094[b]	1.94(0.16)	1.61(0.5)	1.20(2.5)	109°46′[b]
		1.64(0.5)			

[a] From Pauling's rule (L. Pauling (1947), *J. Am. Chem. Soc.*, 69, 542, with the constant 0.6.
[b] Molecule parameters have simply been transferred, although arguments can be made
regarding small changes from the molecule.
[c] Corresponding bond order given in parentheses.

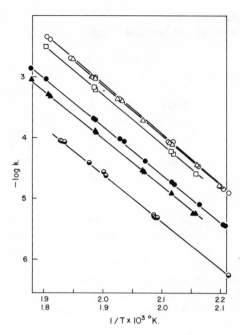

Fig. 5.8 Arrhenius plots for methyl isocyanide data: $\bigcirc$, 8 atm.; $\triangle$, 1500 mm;
$\square$, 150 mm; ●, 10 mm; ▲, 5 mm; ◗,
0.1 mm. Lines were fitted by least squares;
the lower range of abscissa values is for
0.1 mm and the higher range for all other
pressures.

The only significantly arbitrary variable above is the single-twist frequency, which has been set down in the "300 model" as coincident with the degenerate molecule bending modes. But even a variation up to 600 cm^{-1} ("600 model") for this one frequency is scarcely disastrous. All vibrational modes of the molecule and complex are here taken as active and the overall rotations as adiabatic.

Activation energy determinations over approximately 80°C ranges were made at six pressures (Fig. 5.8). Table 5.12 summarizes the Arrhenius parameters obtained by least square calculations. Activation energies were also determined as a "continuous" function of pressure between 10 and 15 mm from the two fall-off curves at 199.4°C and 230.4°C; all three curves were used down to 0.03 mm. The continuous values are in good agreement with the separate activation energy determinations (Fig. 5.9). The activation energy spread between low and high pressures is 2.0 kcal/mole. The relations between $E_0^{\ddagger}$ and the experimental high-pressure activation energy $E_a^{\ddagger}$ is given by

$$E_0^{\ddagger} = E_a^{\ddagger} - N\left[\sum_i^{11} \frac{h\nu_i^{\ddagger}}{\exp(h\nu_i^{\ddagger}/kT) - 1} - \sum_i^{12} \frac{h\nu_i}{\exp(h\nu_i/kT) - 1} + kT\right] \quad (5.104)$$

Table 5.12
Arrhenius Parameters for the Isomerization
of CH_3NC

P (mm)	$\log A \pm \sigma$	$E_a^{\ddagger} \pm \sigma$ (kcal/mole)
6075	13.61 ± 0.09^a	38.35 ± 0.21^a
1500	13.49 ± 0.10	38.11 ± 0.21
150	13.17 ± 0.14	37.69 ± 0.31
10	12.17 ± 0.08	36.58 ± 0.18
5	11.82 ± 0.08	36.26 ± 0.18
0.1	10.46 ± 0.11	36.27 ± 0.27

[a] Standard deviation of the slope and intercept.

where $v_i^{\ddagger}$ and v_i are the frequencies of the activated complex and the molecule, respectively, and N, the Avogadro constant. Then

"300 model": $E_0^{\ddagger} = E_a^{\ddagger} - 0.50 \text{ kcal} = 37.85 \text{ kcal/mole}$
"600 model": $E_0^{\ddagger} = E_a^{\ddagger} - 0.24 \text{ kcal} = 38.11 \text{ kcal/mole}$

To calculate $k^{(1)}$ given by eq. 5.103, the principal summation of eq. 5.103 has been carried out numerically, starting at $E_0^{\ddagger} = 37.85 \text{ kcal/mole}$ ("300 model") or $E_0^{\ddagger} = 38.11 \text{ kcal/mole}$ ("600 model") and being carried out up

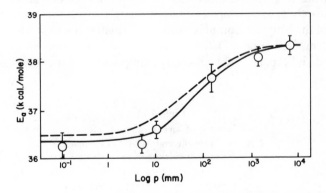

Fig. 5.9 "Continuous" Arrhenius activation energies obtained from fall-off curves versus log p, given by solid curve. Separate determinations, from the data of Fig. 5.8, are shown for comparison. The circle is shown with indication of standard deviation; calculated values from the 300 model are given by the dashed line.

to 54 kcal/mole, where further contribution becomes negligible; at 44 kcal/mole, the contribution is already about 95%. The total number of states involved in eq. 5.103 has been evaluated by the exact counting. The exact counting of the total number of states include all possible frequency combinations up to high energies and is a formidable task by other than computer methods, although easier here at the comparatively low energies involved. The problem was simplified, and the final result scarcely affected, by using six groups for the fundamental frequencies (Table 5.10), where each group is represented by a geometric mean. To evaluate the collision frequency ω, the collision parameters used are $\sigma_c(CH_3NC) = 4.5$ Å $= \sigma_c(C_2H_6)$ and $\sigma_c(N_2)$ $= 3.7$ Å. The nonideality correction to the collision frequency ω at high pressures was found to be insignificant. The deactivation probability has been equated with the collision probability. The ratio I_r is defined by

$$I_r = \frac{\sigma}{\sigma^{\ddagger}} \left(\frac{I_A^{\ddagger} I_B^{\ddagger} I_B^{\ddagger}}{I_A I_B I_C} \right)^{1/2} \tag{5.105}$$

where the symmetry number σ is 3 for the molecule and $\sigma^{\ddagger} = 1$ for an activated complex of C_1 or C_s symmetry. The moments of inertia of the molecule and the activated complex were calculated for several C—C bond orders including the "300 model" and "600 model." It is evident from Table 5.13 that the uncertainty in the C—C distance and CNC angle affects I_r by less than 15%.

The experimental fall-off curves are fairly well fitted by the "300 model." (Fig. 5.7). The calculated and observed curves overlap and the agreement of the calculated and experimental pressure values is quite striking. It is evident from eq. 5.103 that a change of I_r does not directly affect the shape of the fall-off curve but merely shifts it along the pressure axis. The "600 model" of the activated complex produces a small increase in curvature over the "300 model" (Fig. 5.7). The degree of subjective control of the fit of calculated and experimental fall-off by manipulation of the sole significantly

Table 5.13
Moments of Inertia (amu, Å²) for
CH₃NC Molecule and Activated Complex

Model	I_A	I_B	I_C	I_R
600 model	12.53	25.77	35.08	3.532
300 model	11.59	31.65	40.01	4.021
Molecule	3.23	50.29	50.29	

Table 5.14
Comparison of Experimental and Calculated High-Pressure
Frequency Factors[a] (Sec^{-1})

Model used	199.4°	230.4°	259.8°
Experimental[e]	2.42×10^{13}	2.50×10^{13}	$(2.57 \times 10^{13})^c$
300 model[b]	2.47×10^{13}	2.55×10^{13}	2.62×10^{13}
600 model	1.47×10^{13}	1.49×10^{13}	1.51×10^{13}
Anharmonicity (300) 2[d]	2.39×10^{13}	2.47×10^{13}	2.54×10^{13}
Anharmonicity (300) (in ground state and in activated complex)	2.53×10^{13}	2.62×10^{13}	2.69×10^{13}

[a] Calculated from the equation $k = Ae^{-E_0/RT}$, in which E_0 is the critical energy and not the observed activation energy.
[b] This model may easily be perturbed just a little so as to yield almost exact coincidence with the experimental values.
[c] Calculated value obtained by allowance for the discrepancy noted in footnote b.
[d] The first value refers to inclusion of anharmonicity only in the molecule, whereas the second value includes comparable anharmonicity in the activated complex.
[e] The consistency tests indicate that the experimental values are systematically low by 0.5–1.5 %.

arbitrary parameter, the twist frequency, is in fact not large. At infinite pressure the RRKM expression reduces to

$$k_\infty^{(1)} = \kappa \frac{kT}{h} \cdot \frac{I_r Q_v^{\ddagger} \exp(-E_0^{\ddagger}/RT)}{Q_v} \qquad (5.106)$$

which is identical with the conventional expression of the transition state theory. The term κ was simply set equal to unity. Good agreement between experimental and calculated high-pressure frequency factor A_∞ is particularly obtained for the "300 model" (Table 5.14). This further suggests that the frequency assignment in the "300 model" (or the "600 model") is a reasonable one, particularly since a small change within experimental error in $E_a^{\ddagger}$ produces a relatively large change in the observed frequency factor A_∞ and makes such comparison somewhat inaccurate. The efficiency of various gases in activating and deactivating collisions of this reaction has also been discussed by Rabinovitch et al. (1967).

5.5.2   Decomposition of the *sec*-Butyl Radical

A well-characterized unimolecular reaction system is the decomposition of vibrationally excited *sec*-butyl radicals (Oref et al., 1971; Placzek et al., 1966). Vibrationally excited butyl radicals were produced by the reaction of Hg-photosensitized H atoms with *cis*-2-butene:

$$H + CH_3CH = CH\,CH_3 \longrightarrow CH_3CH\,CH_2CH_3^* \quad (5.107)$$

This technique has been labeled *chemical activation* and has been extensively applied by Rabinovitch and co-workers (Rabinovitch and Flowers, 1964; Rabinovitch and Setser, 1964). The radicals undergo decomposition or collisional stabilization:

$$CH_3\dot{C}H\,CH_2CH_3^* \xrightarrow{\ k_E\ } \cdot CH_3 + CH_3CH = CH_2 \quad (D) \quad (5.108)$$

$$CH_3\dot{C}H\,CH_2CH_3^* + M \xrightarrow{\ \omega\ } CH_3\dot{C}H\,CH_2CH_3 + M \quad (S) \quad (5.109)$$

The stabilized radical can undergo various reactions; the important ones here are disproportionation and combination:

$$2\,CH_3\dot{C}H\,CH_2CH_3 \longrightarrow C_4H_8 + C_4H_{10} \quad (5.110)$$

$$2\,CH_3\dot{C}H\,CH_2CH_3 \longrightarrow C_8H_8 \quad (5.111)$$

Reaction 5.107 is exothermic by 40 kcal/mole, and this energy appears as the vibrational excitation of the *sec*-butyl radical. Because this heat of reaction is considerably larger than the activation energy for the decomposition reaction 5.108, the average energy of the reacting butyl radicals is much higher than it would be in the same decomposition conditions of thermal equilibrium. The fate of energized radicals with respect to decomposition or collisional deactivation is determined by the relative rates of reactions 5.108 and 5.109, together with secondary reactions 5.110 and 5.111 that follow these reactions. The yield of products arising from reaction 5.108 may be designated D (for decomposition) and that of products from reaction 5.109, S (for stabilization). In this case the ratio D/S is given by the expression

$$\frac{D}{S} = \frac{[C_3H_6]}{2([C_4H_{10}] + [C_8H_8])} \quad (5.112)$$

and in terms of theoretical quantities the ratio D/S takes the form

$$\frac{D}{S} = \frac{\int_{E_c}^{\infty} k_E/(\omega + k_E) f(E)\,dE}{\int_{E_c}^{\infty} \omega/(\omega + k_E) f(E)\,dE} \quad (5.113)$$

where the terms $k_E/(\omega + k_E)$ and $\omega/(\omega + k_E)$ represent the fractions of radicals with energy in a specified internal dE that undergo decomposition and collisional stabilization, respectively, and the $f(E) \, dE$ term gives the fraction of all radicals with the required energy range.

Now the average rate constant k_a is expressed in terms of the competition with the conventional specific collisional deactivation probability ω and the ratio D/S:

$$k_a = \omega \cdot \frac{D}{S} = \omega \frac{\int_{E_c}^{\infty} k_E/(k_E + \omega) f(E) \, dE}{\int_{E_c}^{\infty} \omega/(k_E + \omega) f(E) \, dE} \tag{5.114}$$

Two important forms of k_a are the high- and low-pressure limits:

$$k_{a,\,\infty} = \omega \frac{\int_{E_c}^{\infty} k_E/\omega \, f(E) \, dE}{\int_{E_c}^{\infty} f(E) \, dE} = \langle k_E \rangle \tag{5.115}$$

and

$$k_{a,\,0} = \omega \frac{\int_{E_c}^{\infty} f(E) \, dE}{\int_{E_c}^{\infty} \omega/k_E \, f(E) \, dE} = \frac{1}{\langle k_E^{-1} \rangle} \tag{5.116}$$

respectively. The ratio $k_{a,\,\infty}/k_{a,\,0}$ is a measure of the spread in energy associated with the distribution function $f(E)$. Equations 5.114–5.116 may be readily extended to include systems having two or more simultaneous decomposition reactions.

The unimolecular rate constant k_a for decomposition of chemically activated *sec*-butyl radicals was measured at pressures of 0.036–203 atm of hydrogen. The reaction enters its high-pressure region above 0.01 atm. The principal results are summarized in Table 5.15, and Fig. 5.10 shows graphically the good straight-line behavior of the S/D versus P curve, which signifies constancy of k_a over the pressures region studied.

It has been proposed on various grounds by Baetzold and Wilson (1964) and Bunker (1964) that k_a may be altered in a supra-high-pressure region because of failure of the postulate of internal randomization of the internal energy of the decomposition species. The experimental results of Rabinovitch et al. indicate that k_a is constant over the whole pressure range. The mean collisional period at the highest pressure is 2×10^{-13} sec, so that internal energy randomization or random-phase-space sampling effectively takes place on a shorter time scale. The random lifetime assumption embedded in the RRKM theory seems to hold.

The RRKM theory has been successfully applied to many unimolecular reactions (Lin and Laidler, 1968; Wieder and Marcus, 1962; Nikitin, 1966; Weston and Schwarz, 1972; Slater, 1959; Benson, 1968; Golden et al., 1971; Thomas et al., 1969), and for detailed discussion of any specific reaction, the original papers should be consulted (Forst, 1973; Robinson and Holbrook, 1972).

Table 5.15

Rates of Decomposition of Vibrationally Excited sec-Butyl Radicals at 296°K

P (meas) (atm)	P (ideal)	K_a Butane k_a Isooctane	ω 10^{10} sec^{-1}	$D/S \times 10^5$	k_a 10^7 sec^{-1}
55.0	52.8	0.57	140.6	4.22	1.78
62.1	58.5	0.58	157.5	4.14	1.96
68.7	65.1	0.58	175.2	3.98	2.09
75.4	71.3	0.56	192.8	2.78	1.61
75.7	71.6	0.62	193.7	3.35	1.95
82.5	77.3	0.55	210.5	3.05	1.93
79.6	74.7	0.58	202.6	3.22	1.96
82.6	77.4	0.56	210.6	3.07	1.94
82.6	77.4	0.55	210.7	2.60	1.65
82.9	77.7	0.67	211.5	2.98	1.89
84.4	79.0	0.51	215.3	2.53	1.63
102.0	94.0	0.50	259.8	2.38	1.86
120.2	109.5	0.60	306.8	1.84	1.70
126.0	114.9	0.52	323.3	1.96	1.90
134.0	121.6	0.55	344.1	1.54	1.60
137.3	124.7	0.55	353.9	1.71	1.82
160.1	141.7	0.67	405.0	1.52	1.85
170.5	146.5	0.63	422.2	1.39	1.76
187.8	158.6	0.65	459.6	1.25	1.72
202.5	168.7	0.65	494.2	1.08	1.60

REFERENCES

Ambortzumian, R. V. and V. S. Letokhov (1977), *Acc. Chem. Res.*, **10**, 61.

Baetzold, R. C. and D. J. Wilson (1964), *J. Phys. Chem.*, **68**, 3141.

Benson, S. W., *Thermochemical Kinetics* (1968), Wiley-Interscience.

Bunker, D. L. (1964), *J. Chem. Phys.*, **40**, 1946.

Bunker, D. L., *Theory of Elementary Gas Reactions* (1966), Pergamon.

Chupka, W. A. and M. Kaminsky (1961), *J. Chem. Phys.*, **35**, 1991.

Chupka, W. A. and J. Berkowitz (1967), *J. Chem. Phys.*, **47**, 2921.

Coggiola, M. J., P. A. Schulz, Y. T. Lee, and Y. R. Shen (1977), *Phys. Rev. Lett.*, **38**, 19.

Eyring, H. (1935), *J. Chem. Phys.*, **3**, 107.

Eyring, H., D. Henderson, B. J. Stover, and E. M. Eyring (1964), *Statistical Mechanics and Dynamics*, Wiley-Interscience.

Forst, W. and Z. Prášil (1969), *J. Chem. Phys.*, **51**, 3006.

Forst, W. and Z. Prášil (1970), *J. Chem. Phys.*, **53**, 3065.

Forst, W. (1973), *Theory of Unimolecular Reactions*, Academic.

Fowler, R. H. (1936), *Statistical Mechanics*, Cambridge University Press.

Gelbart, W. M., S. A. Rice, and K. F. Freed (1970), *J. Chem. Phys.*, **52**, 5718.

Glasstone, S., K. J. Laidler, and H. Eyring (1940), *The Theory of Rate Processes*, McGraw-Hill.

Golden, D. M., R. K. Solly, and S. W. Benson (1971), *J. Phys. Chem.*, **95**, 1333.

Haney, M. A. and J. L. Franklin (1968), *J. Chem. Phys.*, **48**, 4093.

Herschbach, D. R. and V. W. Laurie (1961), *J. Chem. Phys.*, **35**, 458.

Hoare, M. R. (1970), *J. Chem. Phys.*, **52**, 5695.

Hoare, M. R. and Th. W. Ruijgrok (1970), *J. Chem. Phys.*, **52**, 113.

Inghram, M. G., B. Steiner, and C. F. Giese (1961), *J. Chem. Phys.*, **34**, 189.

Johnston, H. S. (1961), *Adv. Chem. Phys.*, **3**, 131.

Kiser, R. W. (1965), *Introduction to Mass Spectrometry and Its Applications*, Prentice-Hall.

Klots, C. E. (1964), *J. Chem. Phys.*, **41**, 117.

Klots, C. E. (1971), *J. Phys. Chem.*, **75**, 1526.

Laidler, K. J. (1969), *Theories of Chemical Reaction Rates*, McGraw-Hill.

Lau, K. H. and S. H. Lin (1971a), *J. Phys. Chem.*, **75**, 2458.

Lau, K. H. and S. H. Lin (1971b), *J. Phys. Chem.*, **75**, 981.

Lin, S. H. and H. Eyring (1963), *J. Chem. Phys.*, **39**, 1577.

Lin, S. H. and H. Eyring (1965), *J. Chem. Phys.*, **43**, 2153.

Lin, M. C. and K. J. Laidler (1968), *Trans. Faraday Soc.*, **64**, 79, 94, 927.

Lin, S. H. and C. Y. Ma (1971), *Adv. Chem. Phys.*, **21**, 128.

Lin, S. H. (1972), *J. Chem. Phys.*, **56**, 4155.

Magee, J. L. (1952), *Proc. Natl. Acad. Sci. U.S.*, **38**, 764.

Marcus, R. A. and O. K. Rice (1951), *J. Phys. Colloid Chem.*, **55**, 894.

Marcus, R. A. (1965), *J. Chem. Phys.*, **43**, 2658.

Montroll, E. W. and K. E. Shuler (1958), *Adv. Chem. Phys.*, **1**, 361.

Nikitin, E. E., *Theory of Thermally Induced Gas Phase Reactions* (1966), Indiana University Press.

Oref, I., D. Schuetzle, and B. S. Rabinovitch (1971), *J. Chem. Phys.*, **54**, 575.

Pillai, M. E. K. and F. F. Cleveland (1960), *J. Mol. Spectrom.*, **5**, 212.

Placzek, D. W., B. S. Rabinovitch, and F. H. Dorer (1966), *J. Chem. Phys.*, **44**, 279.

Rabinovitch, B. S. and M. C. Flowers (1964), *Quart. Rev.*, **18**, 122.

Rabinovitch, B. S. and D. W. Setser (1964), *Adv. Photochem.*, **3**, 1.

Rabinovitch, B. S., S. C. Chan, and L. T. Spicer (1970), *J. Phys. Chem.*, **74**, 3160.

Rankin, C. C. and J. C. Light (1967), *J. Chem. Phys.*, **46**, 1305.

Robinson, P. J. and K. A. Halbrook (1972), *Unimolecular Reactions*, Wiley-Interscience.

Rosenstock, H. M., M. B. Wallenstein, A. L. Wahrhaftig, and H. Eyring (1952), *Proc. Natl. Acad. Sci. U.S.*, **38**, 667.

Schneider, F. W. and B. S. Rabinovitch (1962), *J. Phys. Chem.*, **84**, 4215.

Slater, N. B. (1959), *The Theory of Unimolecular Reactions*, Cornell University Press.

Snider, N. S. (1965), *J. Chem. Phys.*, **42**, 548.

Spatz, E. L., W. A. Seitz, and J. L. Franklin, *J. Chem. Phys.*, **51**, 5142.

Thomas, T. F., P. J. Conn, and D. F. Swinehart (1969), *J. Am. Chem. Soc.*, **91**, 7611.

Tolman, R. C. (1938), *The Principles of Statistical Mechanics*, Oxford University Press.

Tou, J. C. (1967), *J. Phys. Chem.*, **71**, 2721.

Tou, J. C. and S. H. Lin (1968), *J. Chem. Phys.*, **49**, 4187.

Vestal, M., *Introduction to Fundamental Processes in Radiation Chemistry* (1969), P. Ausloos, Ed., Interscience, 109.

Vestal, M. and J. Futrell (1970), *J. Chem. Phys.*, **52**, 978.

Waage, E. V. and B. S. Rabinovitch (1970), *Chem. Rev.*, **70**, 377.

Watanabe, K., T. Nakayama, and J. Mottl (1960), *J. Quant. Spectrom. Rad. Transfer*, **2**, 369.

Weston, R. and H. A. Schwarz (1972), *Chemical Kinetics*, Prentice-Hall.

Widom, B. (1963), *Adv. Chem. Phys.*, **5**, 353.

Widom, B. (1971), *J. Chem. Phys.*, **55**, 44.

Wieder, G. M. and R. A. Marcus (1962), *J. Chem. Phys.*, **37**, 1835.

Zwolinski, B. J. and H. Eyring (1947), *J. Am. Chem. Soc.*, **69**, 2702.

Six

Molecular Reaction Dynamics and Biomolecular Reactions

CONTENTS

6.1 DYNAMICS OF MOLECULAR REACTIONS

6.1.1 Introduction

In this chapter we discuss the application of collision theory to the reaction dynamics of gas phase bimolecular reactions. Unlike the elastic scattering, chemical reaction occurs at a smaller intermolecular distance where the overlap of electron cloud is appreciable and therefore makes it possible to form a new chemical bonding. Accordingly, we anticipate that for a given colliding species that is chemically active, a reactive scattering is always accompanied by an elastic scattering and the reaction cross section is, in general, smaller than the elastic cross section.

For a reactive scattering we are interested not in the conservation of particle flux but in the identity of reaction products and the energy disposal among all degrees of freedom. Although it is understood that there is very little correlation between the reaction rate and the thermodynamic laws, the latter usually provides the first-hand information about the nature of a reaction. In this regard a reaction is frequently considered as an endothermic or exothermic reaction by comparing the thermodynamic enthalpies of reactants and of products. For an endothermic reaction, additional energy— in the form of either internal energy of reactants or collision energy—has to be supplied in order to have any detectable product signal. For an exothermic reaction, on the other hand, the reaction may take place spontaneously when the reactants are encountered; however, a reaction that is exothermic may not proceed at all without the addition of external energy. Although the majority of bimolecular reactions are known to be endothermic, the studies of reaction mechanism have by now focused more on the family of exothermic reactions since they can be measured much more easily experimentally.

Since a reactive scattering usually occurs at a small intermolecular distance, the reactant internal energy states as well as the collision energy are proved to be important in governing the reaction dynamics. At a given temperature, molecules are known to have thermally equilibrated distributions on both translational energy and internal quantum states. Thus, when chemical reactions are carried out under the macroscopic condition, the measured quantities always reflect the thermally averaged values of all possible species involved. This makes it extremely difficult to obtain the detailed information governing the reaction, since most of it would be smeared out through these multiple collision processes. Increasing concern over the basic understanding of reaction dynamics has thus prompted an active pursuit of the gas phase elementary chemical reactions during recent years. The gas phase bimolecular reactions, in principle, may be studied experimentally at the microscopic level so that the parameters associated with the reaction dynamics may be monitored specifically under a given initial condition; for instance, one may be able to measure the magnitude of the reaction cross section at a specific collision energy. In addition, since the selection of the initial reactant state is feasible, the correlation between the detailed initial conditions and the corresponding measured quantities undoubtedly offers a great deal of information about the reaction dynamics.

In light of the experimental development and advanced modern computer capacity, studies of reactive scattering have excited great interest during the past two decades. Experimentally, the redefined instrumental technique makes it possible to study the major details of interesting parameters that govern the reaction while, theoretically, extensive computations have been actively carried out either to explain the experimental results or to predict the general validity of a reaction theory.

6.1.2 Collision Theory and Reaction Dynamics

This aspect of the single molecular collision process is of great interest since it offers not only the detailed physical picture of elastic collision events but also the most direct information about the chemical reactivity as well as the energy flow among all degrees of freedom of the colliding species. We have learned from the classical collision theory that every encounter between two molecules, the collision events may consist of three possible channels: elastic, inelastic, and reactive; these processes may take place simultaneously with various probabilities. For a collision process between inert gas and inert gas atoms, the probability associated with elastic scattering is almost unity at low-energy collision and the total elastic cross section is mainly attributed to the large impact parameter that leads to a small-angle deflection. This in turn implies that particles only sample the long-range attractive potential during the elastic scattering. For any two-body collision process if one of the colliding partners is not an atomic species, the presence of a large range of impact parameters suggests that in addition to elastic scattering, both reactive and inelastic collision events may occur but the probabilities are typically rather small (Laidler, 1969; Levine and Bernstein, 1974; Weston and Schwarz, 1972).

For a two-body collision process that would lead to a reactive scattering, we anticipate that molecules have to penetrate to a shorter interparticle distance so that electron exchange becomes appreciable, which makes it possible to form a chemical bonding. This means that, in contrast to elastic scattering, a small impact parameter is responsible for the reactive event and thus results in a large deflection angle. Accordingly, the distinction between elastic and reactive scatterings may be discussed in terms of the different range of impact parameters. Since the elastic process is always accompanied by reactive scattering when two reactive species approach each other, the reaction cross section may be defined as (Levine and Bernstein, 1974; Johnston, 1966)

$$Q_r(E) = 2\pi \int_0^{b_{max}} P(E, b)b \, db \qquad (6.1)$$

where $P(E, b)$ is the so-called opacity function. If inelastic scattering can be neglected, the opacity function is really the reaction probability, and in this case it is a ratio of the reaction cross section to the total collision cross section; that is, it has a value between zero and unity. To evaluate the reaction probability function $P(E, b)$, several simple analytical models have been proposed; among them the measurements of concurrent elastic angular distribution of reactive system provide a most direct and informative way to extract the $P(E, b)$ function.

The angular distribution of nonreactive scattering $I_n(E, \theta)$ can be expressed in terms of the fraction $1 - P(E, b)$ of collision that do not lead to reaction

$$I_n(E, \theta) = 2\pi b[1 - P(E, b)] \, db \qquad (6.2)$$

Since the small-angle elastic scattering is unlikely to be perturbed by the reaction, it is expected that $P(E, b)$ is practically zero at these angles and that the measured elastic scattering data at these small angles be used to extract the scattering potential; of course, this is due to the long-range attractive potential. At large angles where the small impact parameter becomes increasingly important, the nonreactive angular distribution is expected to be quenched by the appearance of the reactive collision. Because of this fact, one may be able to determine the reaction probability by comparing the differences between the measured nonreactive angular distribution and the calculated values at large angles,

$$P(E, b) = \frac{I_n^0(E, \theta) - I_n(E, \theta)}{I_n^0(E, \theta)} \qquad (6.3)$$

where $I_n^0(E, \theta) = 2\pi b\, db$ is the unquenched elastic differential cross section calculated from an assumed attractive scattering potential that gives a good fit to the measured small-angle elastic data. Thus the reaction probability may be determined if the critical angle beyond which eqs. 2.2 and 2.3 are applicable can be chosen. For low-energy collision process, this critical angle is assumed to be in the neighborhood of rainbow angle, and in spite of the fact that this is a crude, approximate method, results of this data analysis give a fairly good agreement with the measured reaction cross section of alkali atom and hydrogen halide reactions. Besides this semi-empirical approximation, we illustrate here the simplest analytical form of unit step function proposed for $P(E, b)$ that have been applied for many ion-molecule reactions.

This model postulates that all trajectories that surmount the centrifugal barrier would lead to reaction and the nonreactive scattering is mainly due to the collisions at large impact parameters outside the orbiting barrier. In other words, reaction occurs for all collisions if the impact parameter b is less than the orbiting impact parameter b_0 and no reaction for $b > b_0$,

$$
\begin{aligned}
P(E, b) &= 1, \qquad \text{for } b \leq b_0 \\
&= 0, \qquad \text{for } b > b_0
\end{aligned}
\tag{6.4}
$$

If we assume that there is no permanent activation barrier to keep the reactants apart except the centrifugal barrier associated with the large impact parameter, the condition of orbiting model for a power potential $V(r) = -C_n r^{-n}$ (eq. 1.92) leads to

$$
Q_r(E) = \pi \left[\frac{C_n(n-2)}{2E} \right]^{2/n} \left(\frac{n}{n-2} \right)
\tag{6.5}
$$

For ion-molecule reactions where $n = 4$ and $C_4 = e^2\alpha/2$, the total reaction cross section obtained from this model is

$$
Q_r(E) = \pi \left(\frac{2e^2\alpha}{E} \right)^{1/2}
\tag{6.6}
$$

where α is the polaritability of the neutral molecule. The reaction cross section calculated from this model, $\sim 100 \text{ A}^2$ for $E = 0.1$ eV and $\alpha = 3 \text{ A}^3$, is about the order observed for ion-molecule reaction at low collision energy E but gives too large value at higher E. For alkali atom and halogen molecule reactions, the magnitude of total reaction cross section predicted by the orbiting model is also in fairly good agreement with the measured results of about 100 Å^2 at thermal energy.

6.1.3   Reaction Cross Section and Rate Constant

The first empirical expression for the temperature-dependent rate constant $k(T)$ proposed by Arrhenius in 1889 was given as

$$k(T) = A \cdot \exp\left(-\frac{E_a}{RT}\right) \tag{6.7}$$

where the preexponential factor A and the activation energy E_a were considered to be constants. However, it has since been established that both parameters A and E_a are actually not constants at a wide range of temperature. Moreover, since the macroscopic reaction rate for bimolecular reaction is measured by monitoring either the concentration of reactants or temperature of the reaction system via

$$\frac{dR}{dt} = k(T)n_1 n_2 \tag{6.8}$$

the rate constant $k(T)$ obtained under this condition is subject to the restriction that the temperature is thermally equilibrated among all degrees of freedom at any given temperature. The preceding formulation of reaction rate as a function of the equilibrium thermodynamic temperature T thus greatly limits its application to real molecular system whenever the reaction rate is much faster than the rate of equilibrium. Consequently, it is recognized that the microscopic reaction cross section is a better representation than the macroscopic rate constant since the latter quantity is meaningful only when the specific condition at which it is obtained is uniquely defined.

Activated complex theory does not proceed most easily from the postulate that there be equilibration among degrees of freedom in the activated complex. Rather, you can use the rate at equilibrium away from equilibrium, providing the calculation is made for a cross section of no return, since in that case deleting the back reaction does not perturb the forward reaction. Of course even if there is equilibrium between degrees of freedom and even if you calculate for a cross section that does not fulfill the condition of no return, you will have to evaluate a transmission coefficient γ.

Consider the elementary reaction of two beams A and B

$$A(i) + B(j) \longrightarrow C(m) + D(n) \tag{6.9}$$

where i, j, m, and n refer to the internal quantum states. If the volume of intersection of the two beams is denoted by τ, the velocity distribution of the molecules in the two beams by $n_A f_A(v_A)\, dv_A$ and $n_B f_B(v_B)\, dv_B$, the relative speed by $g = |\mathbf{v}_A - \mathbf{v}_B|$, and $N_c(\Omega)$ the number of molecules C produced per unit time at a solid angle Ω, we have

$$\frac{dN_c(\Omega)}{dt} = \tau I_r\left(\frac{mn}{j; \Omega g}\right) d\Omega g(n_A f_A(v_A)\, dv_A)_i (n_B f_B(v_B)\, dv_B)_j \qquad (6.10)$$

where the differential cross section $I_r(mn/ij; \Omega g)$ is related to the reaction cross section by

$$Q_r\left(\frac{mn}{ij; g}\right) = \int d\Omega I_r\left(\frac{mn}{ij; \Omega g}\right) = 2\pi \int P\left(\frac{mn}{ij; gb}\right) b\, db \qquad (6.11)$$

The macroscopic rate of forward reaction is then obtained by summing over all possible distribution functions

$$\frac{dN_c}{dt} = -\frac{dN_A}{dt} = \sum_{ijmn} \iiint g Q_r\left(\frac{mn}{ij; g}\right)(n_A f_A(v_A)\, dv_A)_i (n_B f_B(v_B)\, dv_B)_j \qquad (6.12)$$

Substitution $n_A^i = n_A^0 x_A^i$, where x_A^i is the mole fraction of species A in quantum state i, and comparison with eq. 6.8 shows that the macroscopic rate constant

$$k(g) = \sum_{ijmn} \chi_A^i \chi_B^i \iiint g Q_r\left(\frac{mn}{ij; g}\right) f_A^i(v_A) f_B^i(v_B)\, dv_A\, dv_B \qquad (6.13)$$

depends on the distribution functions of the reactants for both translational and the internal degrees of freedom (Green et al., 1966).

At thermal equilibrium the velocity distribution functions $f_A^i(v_A)\, dv_A$ and $f_B^j(v_B)\, dv_B$ are independent of the internal energy states and may be expressed by the isotropic Maxwellian–Boltzmann equation,

$$f_A(v_A)\, dv_A = 4\pi\left(\frac{m_A}{2\pi kT}\right)^{3/2} \exp\left(-\frac{m_A v_A^2}{2kT}\right) v_A^2\, dv_A \qquad (6.14)$$

On introducing the total cross section

$$Q_r(g) \equiv \sum_{ijmn} \chi_A^i \chi_B^j Q_r\left(\frac{mn}{ij; g}\right)$$

and making a transformation from LAB to CM coordinate, the macroscopic rate constant can be reduced to a simple function of temperature,

$$k(T) = 4\pi\left(\frac{\mu}{2\pi kT}\right)^{3/2} \int_0^\infty Q_r(g) e^{-\mu g^2/2kT} g^3\, dg \qquad (6.15a)$$

or

$$k(E) = \left(\frac{2}{kT}\right)^{3/2} (\pi\mu)^{-1/2} \int_0^\infty Q_r(E) e^{-E/RT} E\, dE \qquad (6.15b)$$

Thus under the condition of thermal equilibrium, the rate constant is a product of reaction cross section, the magnitude of relative velocity, and the thermal distribution of relative collision speed

$$k(g) = \int_0^\infty Q(g)gf(g)\, dg \qquad (6.16a)$$

or

$$k(E) = \left(\frac{2}{\mu}\right)^{1/2} \int_0^\infty Q_r(E)E^{1/2}f(E)\, dE \qquad (6.16b)$$

A generalized expression for rate constant in analogy to eqs. 6.16a and 6.16b may be rewritten by replacing $f(g)\, dg$ with $p(g)\, dg$ to remove the restriction of the thermal condition. Here the function $p(g)\, dg$ can be any functional form of the actual, experimental conditions. Thus the rate constant can be readily calculated, provided that the reaction cross section and the relative collision energy function are found.

The simplest collision theory, as discussed in the previous chapter, is the hard sphere model. The total cross section, if classical mechanics is applicable, is found to be πd^2, where d is the diameter of the hard sphere. According to this model, reaction occurs if the collision energy along the line of centers exceeds a threshold value E_0. If E is the initial collision energy and E_c is the energy available along the line of centers, the impact parameters that lead to reaction are

$$b^2 = \left(1 - \frac{E_c}{E}\right)d^2 \qquad (6.17)$$

Hence for a given d and E, the maximum impact parameter occurs at $E_c = E_0$, and the reaction cross section is given

$$Q_r(E) = \pi d^2\left(1 - \frac{E_0}{E}\right), \qquad \text{for } E \geq 0$$

$$= 0, \qquad \text{for } E \leq E_0 \qquad (6.18)$$

Substituting eq. 6.18 into eq. 6.15b, the rate constant under thermal equilibrium is simplified to

$$k(T) = \pi d^2\left(\frac{8kT}{\pi\mu}\right)^{1/2} \exp\left(-\frac{E_0}{kT}\right)$$

$$= \pi d^2\langle g\rangle\exp\left(-\frac{E_0}{kT}\right) \qquad (6.19)$$

where $\langle g \rangle$ is the average speed of thermal velocity distribution at temperature T. This expression resembles the Arrhenius equation (eq. 6.7) and the two are equivalent if the preexponential factor $A = \pi d^2 (8kT/\pi\mu)^{1/2}$ and the threshold energy E_0 is equal to the empirical activation energy E_a.

At thermal energy the rate constant for ion-molecule reactions may be readily calculable from the orbiting model. Hence substitution of eq. 6.6 into eqn. 6.15b, we obtain $k(E) = 2\pi e(\alpha/\mu)^{1/2}$; that is, the rate constant for ion-molecule reaction is independent of thermal temperature of the reaction system (Henglein, 1974).

6.1.4 Energy Disposal and Energy Requirement

Given a bimolecular reaction of eq. 6.9, the law of energy conservation requires that

$$E + W + \Delta D_0 = E'_t + W' \qquad (2.20)$$

where E and W are the relative translational energy and internal energy of the reactants with E'_t and W' for the corresponding quantities of the products; the term ΔD_0 is the difference in the dissociation energies of products and reactants,

$$\Delta D_0 = D'_0 \quad (\text{products}) - D_0 \quad (\text{reactants})$$

Here reaction is called exoergic if $\Delta D_0 > 0$ and endoergic if $\Delta D_0 < 0$. For all endoergic reactions there is always an energy threshold below which no signal may be detected, whereas for exoergic reactions, there may or may not be an energy threshold. For a reaction governed by an energy threshold, there must be an energy barrier arising from the centrifugal barrier or the permanent barrier, or both, that prevents the reaction taking place. This energy barrier may be located along either the reactant path or the product retreat coordinate.

For many bimolecular reactions that are exoergic, conservation of energy flux requires that under the condition of a two-body single collision process, the energy released has to be partitioned among all possible degrees of freedom of the products. In the crossed-beam studies the recoil velocities of the reaction products can be easily measured and the fraction of reaction energy released to the product recoil energy is often expressed by the collision exothermicity ε:

$$\varepsilon = E'_t - E = \Delta D_0 + W - W' \qquad (6.21)$$

When $\varepsilon = 0$, the products carry the same kinetic energy as the incident reactants (but not necessarily the same relative velocity) and all reaction energy goes to the internal excitation; when $\varepsilon > 0$, parts of the reaction energy are released as translational energy and $\varepsilon < 0$; a fraction of reactant kinetic energy is also found in the product internal excitation. The collision exo-thermicity ε can also be indirectly obtained by monitoring the irradiation of internally excited products in the chemiluminescence measurements. In this case the measured quantity is typically the relative population of vi-brational levels of the products. If the electronic excitation is energetically inaccessible, the magnitude of the collision exothermicity can be deduced by considering the vibrational and rotational excitations. For most exoergic reactions studied at relatively low collision energy, ε is found to be positive or near zero.

For many exoergic reactions the reaction yield can be measured experi-mentally at relatively low collision energies; in accord with this experimental finding these reactions are regarded as having small or virtually zero activa-tion energy along the reaction path. There is very little correlation between the magnitude of exoergicity and that of the reaction cross section. This can be illustrated by the $Ba + Cl_2$ reaction, in which two reaction channels are energetically accessible,

$$Ba + Cl_2 \xrightarrow{\quad a \quad} BaCl \; Cl$$

$$\xrightarrow{\quad b \quad} BaCl_2 \qquad (6.22)$$

The exoergicity is 48 kcal/mole for channel a and 153 kcal/mole for channel b. Experimental results from the crossed beam studies, however, indicate that the reaction proceeds predominantly through channel a. In this particular case there is also a marked difference between the microscopic and macro-scopic conditions. If the reaction is carried out under the condition that the multiple collision process is possible, one anticipates that the ultimate product should be $BaCl_2$ rather than the radicals $BaCl$ and Cl. In view of this discrepency the significance of microscopic studies is clearly demonstrated.

For an endoergic reaction the second law of thermodynamics predicts that the reverse reaction is more favorable than the forward reaction path. To enhance the reaction rate in the forward direction, therefore, additional energy in the form of either translational or internal energy of the reactants has to be supplied to balance the energy deficit. By varying this supplementary energy, one may thus establish the energy dependence of the reaction cross section. For an endoergic reaction, since both translational energy and the internal quantum states of reactants are proved to be important in governing the reaction rate, the measurements of the endoergic reactions are of par-ticular interest for the studies of reaction dynamics (Herschbach, 1966; Miller et al., 1967).

The role of the centrifugal barrier governing the reaction has been illustrated for the ion-molecule reactions; our attention here focuses on the nature of the permanent energy barrier that is always associated with the endoergic reactions. To see the origin of the threshold energy E_0, we assume an analytical form for the reaction cross section in analogy to eq. 6.18. Hence for the expression of the rate constant $k(T)$ in eq. 6.16b, the integrand is a product of two function $Q_r(E)$ and $E^{1/2}f(E)$; that is, in order to have a large rate constant, there has to be a large overlap between these two functions. At thermal equilibrium, if there is an energy barrier, the fraction of overlap between these two quantities is then totally dependent on the magnitude of E_0 and the temperature for a given reaction. We depict a qualitative feature of these functions in Fig. 6.1.

Although the product $E^{1/2}f(E)$ rapidly decreases with E, the function $Q_r(E)$ increases with E at $E > E_0$ and is virtually equal to zero at $E \leq E_0$. For an endoergic reaction, the importance of the initial collision energy as well as the reactant internal energy states has been recognized for a number of reactions studied. The collision energy is most effective if this barrier is located in the reactant entrance channel, whereas the vibrational energy of reactants is more efficient if the barrier appears at the exit channel when products start to retreat. The significance of translational energy is obvious, since the reactants would rebound back to the entrance channel if they possess no excess energy to surmount the barrier. The importance of reactant internal energy as well as the distinction among E_0, the activation energy E_a, and the potential energy of activation $V^{\ddagger}$ is discussed in detail when we introduce the potential energy surface in Sec. 6.1.6.

6.1.5 Role of Angular Momentum and Mass Effect

In our discussion of reaction dynamics thus far, we have repeatedly emphasized the importance of the centrifugal barrier and its effect on the reaction dynamics for an exoergic reaction. Since this centrifugal potential is

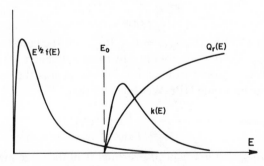

Fig. 6.1 Energy functions contributing to the rate constant in eq. 6.16b and eq. 6.18.

a result of the rotational motion of two particles centered at the relative velocity vector as they approach each other, it is conceivable that the orbital angular momentum, which is a product of reduced mass, relative velocity, and the position, also plays an important role in governing the product angular distribution and the partitioning of energy flux.

We discuss first the effect of angular momentum on the translational-rotational coupling. When the particles are far apart, there is no force acting on them and they are expected to travel in a straight-line trajectory. As interparticle distance decreases, the rotational motion takes up energy, $E_r = L^2/2\mu r^2$, and thus reduces the translational energy of the colliding particles. Accordingly, we anticipate that there is an energy flux moving back and forth between translational and rotational degrees of freedom during the entire course of the collision event. This energy flow becomes very significant if one of the reactants involved is not an atomic species. If we take the simplest three-atom system of K + HBr reaction as an example,

$$\text{K} + \text{HBr} \quad \longrightarrow \quad \text{KBr} + \text{H} \tag{6.23}$$

Conservation of angular momentum states that

$$\mathbf{J} = \mathbf{L} + \mathbf{j} = \mathbf{L}' + \mathbf{j}' \tag{6.24}$$

where $\mathbf{L}$ and $\mathbf{L}'$ denote the initial and final orbital angular momenta and $\mathbf{j}$ and $\mathbf{j}'$ denote the rotational angular momenta of HBr and KBr, respectively. Making use of the classical relation of $L = \mu g b$ and $L' = \mu' g' b'$ and the fact that $\mu' \ll \mu$ and HBr has a much larger rotational spacing than that of KBr, we expect $j \ll L$ and $L \ll j'$; that is, $J \approx L \approx j'$ for this particular system. It thus follows that the newly formed KBr product may take up a large fraction of the reaction energy available into the rotational excitation.

To estimate the product rotational excitation E_r' in connection with the angular momentum for this case, we employ here the classical mechanics, which shows that

$$E_r' = \frac{j'^2}{2I_{\text{KBr}}} \le \frac{L_{\text{max}}^2}{2I_{\text{KBr}}} \tag{6.25}$$

Here the moment of inertia $I_{\text{KBr}} = \mu_{\text{KBr}} R_{\text{KBr}}^2$ and the cutoff orbital angular momentum $L_{\text{max}} = \mu g b_{\text{max}}$ is obtained by assuming that b_{max} is the maximum impact parameter leading to reaction. Note that μ is the reduced mass of atom K and the molecule HBr. We thus find that

$$E_r' \le E \cdot \frac{\mu}{\mu_{\text{KBr}}} \left(\frac{b_{\text{max}}}{R_{\text{KBr}}} \right)^2 \tag{6.26}$$

Since $b_{\text{max}} \approx R_{\text{KBr}}$ and $\mu \approx \mu_{\text{KBr}}$, E_r' is roughly of the same magnitude as the initial collision energy E. It is therefore true that the measurements of the rotational excitation of the product KBr may be used to extract the important range of impact parameters that leads to reaction.

The second important effect of angular momentum in reaction dynamics is the characteristic of angular distribution of products observed for many bimolecular reactions. From these product angular distributions, two types of reaction mechanism may be inferred: a direct reaction and a long-lived collision complex. For a direct reaction mechanism, the reaction duration is much shorter than the rotational period of the collision complex, if it existed, so that the reactants may have the memory of their initial trajectory and thus show a strong preferential angular disposition. On the other hand, for a reaction that forms a collision complex with lifetime longer than the rotation period, $\sim 5 \times 10^{-12}$ sec, all memory of the initial trajectories would be lost after the completion of one rotation and the shape of the product angular distribution is determined by the disposal of total angular momentum $\mathbf{J}$ in eq. 6.24. Since the total angular momentum $\mathbf{J}$ and its projection M on the initial relative velocity $\mathbf{g}$ and K on the final velocity $\mathbf{g}'$ are constants of motion (conserved), the product angular distribution $I(\theta)$ is readily calculable from the relation among the angular momenta and relative velocity vectors.

To see how the product angular distribution is determined by the disposal of total angular momentum $\mathbf{J}$, we illustrate the geometrical relations among the angular momenta and relative velocity vectors for three possible cases in Fig. 6.2; in case a there is no rotational angular momenta for both reactants and products; in case b no angular momentum is present for reactants; and in case c angular momenta appear for both reactants and products. Since orbital angular momentum $\mathbf{L}$ is always perpendicular to the initial velocity vector $\mathbf{g}$ while the rotational angular momentum $\mathbf{j}$ is isotropically distributed in space, the total angular momentum $\mathbf{J}$ is cylindrically symmetric about $\mathbf{g}$. For a long-lived complex where the memory of the initial trajectory vanishes, the product must emerge with equal probability at any azimuthal angle about $\mathbf{J}$. This implies that the final relative velocity $\mathbf{g}'$ must be symmetrically distributed about $\mathbf{J}$; the overall probability distribution for the product at the scattering angle θ measured between $\mathbf{g}'$ and $\mathbf{g}$ is then given by the geometrical transformation of uniform procession $\mathbf{g}'$ about $\mathbf{J}$, which generates a cone of half-angle $\beta' = \cos^{-1}(K/J)$, followed by a uniform procession of $\mathbf{J}$ about the initial velocity $\mathbf{g}$, which generates a cone of half-angle $\beta = \cos^{-1}(M/J)$. If the magnitude of $\mathbf{J}$ and its projection M on $\mathbf{g}$ and K on $\mathbf{g}'$ are known, the relative angular distribution can be evaluated in this classical expression,

$$I(\theta) \simeq (J^2 \sin^2 \theta - M^2 - K^2 + 2MK \cos \theta)^{-1/2} \qquad (6.27)$$

Although most bimolecular reactions studied thus far proceed through the direct reaction mechanism, a few reactions appear to follow the mechanism

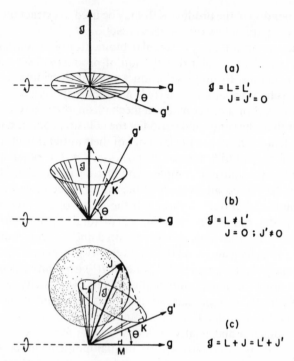

Fig. 6.2   Relationships among angular momenta and relative velocity vectors at three specified conditions. (Miller et al., 1967.)

of a long-lived collision complex. The first such experimental evidence observed in the crossed beam studies is the reaction of Cs + RbCl → CsCl + Rb; its reaction products exhibit a symmetrical angular distribution about $\theta = 90°$ with the maximum intensity at $\theta = 0$ and $180°$. Since the orbital angular momentum L is estimated to be about $400\,\hbar$ from the calculated cross section, whereas j_{mp} for RbCl at $T = 950°K$ is only about $60\hbar$, this reaction may be taken as case b and eq. 2.27 reduces to

$$I(\theta) \approx (J^2 \sin^2 \theta - K^2)^{-1/2} \qquad (6.28)$$

which suggests that the products have to be peaking at θ near 0 and $180°$ in agreement with the experimental results. Hence the consideration of angular momentum alone provides the necessary information about the product angular distribution for a reaction that proceeds through a long-lived complex. (Wheeler, 1963).

Another important factor in reaction dynamics is the mass effect. Aside from its connection with angular momentum, this effect can be easily illustrated if we consider the hydrogen atom transfer reaction of eq. 6.23. In the CM coordinate the conservation of linear momentum requires that

$$m_{KBr} v_{KBr} = m_H v_H \tag{6.29}$$

Since $m_H \ll m_{KBr}$ and $v_{KBr} \ll v_H$, the product KBr has to be confined in a small region around the velocity vector of the center of mass so that the KBr product signal has to be concentrated at a small range of the LAB angles. One may thus expect an enhanced KBr signal in the measurement of product angular distribution. To see the mass effect for a general case of

$$A + BC \longrightarrow AB + C \tag{6.30}$$

the conservation laws of both linear momentum and energy give the following relation:

$$\frac{E'_t}{E} = \cos^2 \alpha \left(\frac{v_C}{v_{BC}} \right)^2 \tag{6.31}$$

and

$$\cos^2 \alpha = \frac{m_A m_C}{(m_A + m_B)(m_B + m_C)} < 1 \tag{6.32}$$

This expression can be further simplified to $E'_t/E = \cos^2 \alpha < 1$ when we introduce the spectator stripping model that has been used to interpret results observed for many ion-molecule reactions. In its limit this model postulates that the velocity of atom C remains unchanged before and after the reaction, that is, $\mathbf{v}_C = \mathbf{v}_{BC}$. It is therefore true that the final relative translational energy E_t is always smaller than the initial collision energy E by a factor solely determined by the mass ratio; that is, we expect to have a negative collision exothermicity ε.

This behavior can be alternatively demonstrated in terms of the sudden repulsive force acting on the atoms B and C when products AB and C start to come apart. Since the force operates initially between B and C only, the center of mass of BC remains at rest. The translational energy is simply

$$E'_{BC} = \frac{P^2}{2} \left(\frac{1}{m_B} + \frac{1}{m_C} \right) \quad \text{with } P = P_B = P_C \tag{6.33}$$

As atom B recoils it strikes atom A and produces a concerted motion in AB and therefore $P_{AB} = P_B$. Hence parts of the original translational energy must go into internal excitation of AB, W', because $E'_{AB}/E'_B = m_B/m_{AB}$ and

$$W' = E'_B - E'_{AB} = \frac{P^2}{2} \left(\frac{1}{m_B} - \frac{1}{m_{AB}} \right) = \frac{E'_{BC} m_A m_C}{m_{AB} m_{BC}} = E'_{BC} \cos^2 \alpha \tag{6.34}$$

Therefore one expects that the larger the mass factor $\cos^2 \alpha$, the higher the internal excitation of the products.

6.1.6   Potential Energy Surface and Reaction Path

To give the complete description of a reaction process, the interaction potential $V(r)$ used in the central force problem is not longer valid; it has to be replaced by the potential surfaces $V(r_1, r_2, \ldots, r_n)$ where r_i denotes all interparticle distances involved in the reaction system. In doing so, the important factors in governing the reaction dynamics, such as the energy flow among all degrees of freedom of the products as well as the effect of molecular orientation may be visualized. Unfortunately, this multidimensional potential energy surface is hardly depicted physically for many-body systems except for the collinear three-atom reaction in which there are only two independent interatomic coordinates. Although a collinear three-atom potential surface is the simplest case in the domain of bimolecular reaction system, it offers an excellent physical picture that illuminates the reaction process.

Figure 6.3 shows a potential energy surface for a slightly exoergic reaction of a three-atom collinear system (eq. 6.30) and its cutoff potential energy profile along five arbitrary segments; also included in the figure are the activation energy $E_a(T)$ and the potential energy of activation $V^{\ddagger}$, defined as the difference between the absolute minima of the activated complex and of the reactants. Although the activation energy $E_a(T)$ is purely empirical, it is seen from the figure that once the potential surface for the system is constructed, it may be calculable by

$$E_a(T) = V^{\ddagger} + W^{\ddagger}(T) - W(T) \tag{6.35}$$

where $W^{\ddagger}(T)$ and $W(T)$ are the internal energies of collision of the activated complex and the reactants at the temperature T. In theory, the potential energy of activation (or simply the potential barrier) is equivalent to the threshold energy E_0 expressed in eq. 6.18. However, when the threshold energy is obtained experimentally, it usually includes a factor due to the experimental sensitivity and therefore it tends to give a higher value than is seen from the potential energy surface. To understand the effect of the potential barrier on the reaction, we have to inspect the motion of the classical trajectory along the reaction path, a line of minimum potential energy from the reactant to the product valley; this is the most probable path for the reaction to occur. If there is a potential barrier located in the reaction path the total collision energy has to be larger than the energy barrier in order to have a reactive trajectory.

It has long been recognized that the efficiency of the total energy available in the form of translational or internal energy is dependent on the location of this potential barrier; translational energy is more effective for surmount-

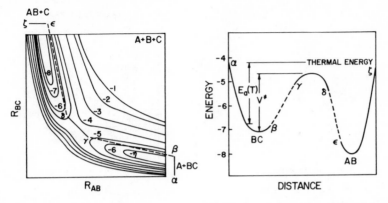

Fig. 6.3 Potential energy surface for a collinear three-atom system of eq. 6.30. (Johnston, 1966.)

ing an early barrier in the entrance valley, whereas vibrational energy is more efficient for crossing over a later barrier in the exit valley, as illustrated in Fig. 6.4. Moreover, the location of this potential barrier is also important in governing the energy disposal for both exoergic and endoergic reactions. For an early stage of the potential barrier, the reaction energy is released early along the reaction path and is converted into vibration of the newly formed product. As shown in Fig. 6.4, the early release of this energy is seen to appear initially as kinetic energy immediately after passing through the barrier; because of this initial acceleration, products do not move along the minimum energy path in the exit valley. When they are bouncing back and forth in the exit valley, the initial translational energy gradually transforms into vibrational excitation. This type of surface is often referred to as an early downhill attractive potential energy surface and is found for many exoergic reactions, notably the family of an alkali atom reacting with a halogen molecule. It is also true that a vibrational excited product can be formed on a late downhill or repulsive surface if the mass factor (eq. 6.32) is very large, that is, near unity. In this limit the exchange atom is very light (e.g., $Cl + HI \rightarrow HCl + I$ reaction) and the efficiency of conversion from exoergicity into product vibrational excitation is expected to be rather high.

This mass effect can be easily seen from the potential energy surface constructed based on the mass-weighted (or skewed) coordinates q_1 and q_2 rather than R_{AB} and R_{BC}:

$$q_1 = aR_{AB} + bR_{BC} \cos \alpha$$

$$q_2 = bR_{BC} \sin \alpha \qquad (6.36)$$

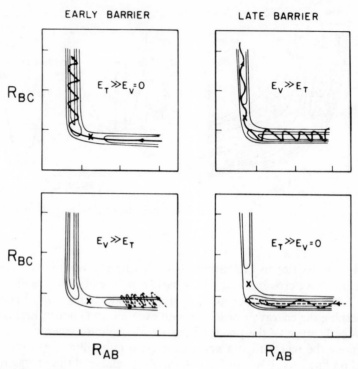

EARLY BARRIER

LATE BARRIER

R_{BC}

$E_T \gg E_V = 0$

$E_V \gg E_T$

R_{BC}

$E_V \gg E_T$

$E_T \gg E_V = 0$

R_{AB}

R_{AB}

Fig. 6.4 Influence of reactant energy for $A + BC \rightarrow AB + C$ reaction with a potential barrier (X). (Levine and Bernstein, 1974.)

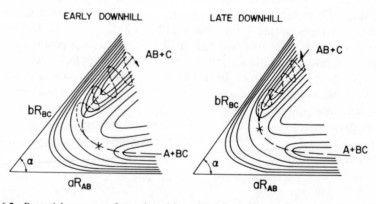

EARLY DOWNHILL

LATE DOWNHILL

bR_{BC}

AB+C

bR_{BC}

AB+C

α

A+BC

α

A+BC

aR_{AB}

aR_{AB}

Fig. 6.5 Potential energy surface, plotted in a skewed coordinates, for A + BC → AB + C reaction.

222

Here a, b, and cos α all depend on the mass of the system

$$a = \left(\frac{m_A m_{BC}}{M}\right)^{1/2}; \qquad b = \left(\frac{m_C m_{AB}}{M}\right)^{1/2}, \qquad M = m_A + m_B + m_C \quad (6.37)$$

and total kinetic energy simply becomes

$$E = \tfrac{1}{2}(\dot{q}_1^2 + \dot{q}_2^2) \quad (6.38)$$

The advantage of having the skewed coordinate representation is that the kinetic energy is specified as a point particle of unit mass moving along the coordinates (q_1, q_2); the solution of the classical equations of motion simply becomes the solution of the point particle moving on the potential surface $V(q_1, q_2)$. Figure 6.5 depicts the potential energy surfaces constructed in a skewed coordinate system. It is clearly seen that the product AB can be highly vibrationally excited if the skewed angle α is small for either an early downhill or a late downhill potential surface.

6.2 PARAMETRIC RESULTS OF A FEW WELL-STUDIED REACTIONS

6.2.1 Experimental Methods

Having discussed briefly the general, important aspects of the reaction dynamics at the microscopic level, we describe the detailed results of several important families of bimolecular reactions in this section. It is understood that the ultimate goal of an ideal experiment is to measure the details of the product energy states, product recoil velocity (both direction and magnitude), and magnitude of the reaction cross section at a given well-specified initial condition. At our present stage of instrumental technology, however, no single experiment yields all these desired physical quantities. The perception of reaction dynamics thus relies on the piecemeal information collected from various experimental techniques. In this regard we briefly discuss experimental methods used in the study of reaction dynamics (Lee et al., 1969; Polanyi and Woodall, 1972; Berry, 1973; Schafer et al., 1970).

For an exoergic reaction the chemiluminescence technique has proved to be very useful for measuring the relative population of internal excited, nascent products that would usually stabilize themselves via irradiation. A typical chemiluminescence experiment is usually carried out in a low-pressure but well-collimated flow system and the irradiation emitted from the different (v, j) states of the reaction products is measured using the infrared spectrometer. By measuring the relative intensities of emission lines of these

(v, j) states, the relative population of the internal excited, nascent products can be extracted.

Laser-induced fluorescence is another approach to determine the distribution of the product of internal energy states. In this method the nascent excited products are pumped to electronically excited states whose radiative lifetime is much shorter than those of the vibrational and rotational excited products and its irradiation can be rapidly monitored to yield the entire rotational-vibrational population distributions of the reaction products. The most productive laser application to reaction dynamics, however, is the optical pumping of the reactants to higher internal energy states. This makes it possible to measure the reaction rate as a function of reactant internal energy states and thus complements the results obtained from the measurement of velocity dependence of reaction cross section. The optical pumping of the reactants is particularly important for the reaction that is endoergic, since the reaction yield for this family of reactions is enhanced more by the reactant internal energy state than by the collision energy.

The third method to be discussed here is the crossed-beam technique. This method is an ideal approach for the study of the detailed dynamics of gaseous molecular reactions, since it provides a great deal of information that other procedures cannot provide. In principle, the crossed-beam technique can be used to determine the reaction cross section, the product translational energy as well as the internal energy states, and the product angular distribution. In addition, both reactant translational energy and internal energy states may be selected so that the initial states of reactants may be specifically defined. However, the density flux of reactants in this technique is kept very low in order to avoid any multiple collision event. The detailed selection of initial energy states of reactants is not feasible at the present stage of detection capacity. Even with this limitation, several important features observed through crossed-beam studies provide a great deal of physical insight into reaction dynamics. It is thus important for us to understand certain details about this technique. Figure 6.6 sketches a typical arrangement of the cross-beam apparatus that is used in many beam laboratories; also included in the figure is the corresponding velocity diagram in the LAB coordinate. To minimize the background scattering due to the ambient gas that would alter the recoil velocity of the reaction products and their internal energy states, both scattering chamber and detection chamber have to be kept at very low pressure. By rotating the entire housing of the detector centered at the scattering zone at which two beams intercept, the product recoil velocity, both the magnitude and the direction, can therefore be measured. These LAB distribution functions are then transformed into the CM coordinate to construct the product intensity contour map. Numerous important reaction features can then be extracted from this contour map.

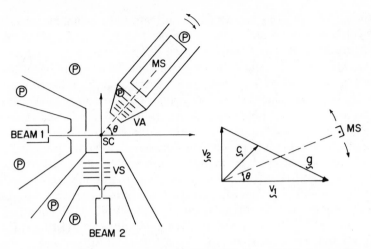

Fig. 6.6 Crossed beam studies of bimolecular reactions. SC = scattering center, P = pump, VS = velocity selector, VA = velocity analyzer, and MS = mass spectrometric detector.

6.2.2 Product Energy Distribution

For low-energy bimolecular reactions the product energy distribution generally consists of three parts—the vibrational (E'_v), rotational (E'_r), and translational (E'_t) energies—and they must be equal to the total energy available (E_{tot}), which is the summation of the initial collision energy (E), the internal energy (W) of reactants, and the reaction exoergicity (ΔD_0) of the system. Thus

$$E_{tot} = E + W + \Delta D_0 = E'_t + E'_r + E'_v \qquad (6.39a)$$

or

$$1 = \sum_i \left(\frac{E'_i}{E_{tot}}\right) = f_t + f_r + f_v \qquad (6.39b)$$

To see the energy balance of eq. 2.39, we show here results found for the well-studied system of $F + H_2$ (and D_2) reaction

$$F + H_2 \longrightarrow HF + H \qquad (6.40)$$

which is known to be highly reactive with an exoergicity $\Delta D_0 = 31$ kcal/mole. These reactions are of particular interest since they involve one of the simplest three-atom systems and thus can be used as a prototype for the study of reaction dynamics.

Experimental results obtained from both infrared chemiluminescence and chemical laser studies indicate that the product HF (or DF) is highly internally excited with the average fraction of $(f_r + f_v)$ nearly equal to 75% of the total energy available in the system; among them, one-fifth is attributed to rotational excitation. Moreover, owing to the large vibrational spacing of HF product (12.5 kcal for HF and 8.9 kcal for DF), one is also able to extract the internal vibrational population of the product HF (or DF) from the CM intensity contour map obtained in the crossed-beam studies; they are summarized in Table 6.1 (Levine and Bernstein, 1974). In spite of the discrepency in the molecular beam at $v' = 4$, it is clear that the product DF takes up a large fraction of the total reaction energy into the vibrational degree of freedom. Results of the classical trajectory calculations were also included; they were calculated from a semiempirical potential energy surface obtained from the London–Eyring–Polanyi–Sato (LEPS) method. This potential surface is shown in Fig. 6.7. It is consistent with our early discussion that an early downhill potential barrier $V^{\ddagger} = 0.9$ kcal/mole may be responsible for the highly vibrational excited HF product. For this simple system the *ab initio* calculation of the FH_2 potential energy surface is also made available; it is found very similar to the results shown in Fig. 6.7.

The application of infrared chemiluminescence to the rotational energy distribution of products is more difficult than in the measurement of vibrational distribution. Aside from the experimental limitation, this is due to the fact that the rotational-translational coupling is much stronger than the vibrational-translational transition so that the original rotational distribution function is easily distorted by the background gas. For a few favorable systems, measurements of product rotational distribution have been carried out by the infrared chemiluminescence method. In addition, the inhomogeneous electric field can also be used to obtain the averaged product rotational energy if the product is a polar molecule.

Table 6.1
Product DF Vibrational Energy Distribution

Quantum numbers	IR chemiluminescence	Chemical laser	Molecular beam	Trajectory calculations
$v' = 0$	—	0.10	0.04	0.0
1	(0.29)	0.24	0.07	0.0
2	0.67	0.56	0.18	0.45
3	1.0	1.0	1.0	1.0
4	0.66	(0.4)	3.5	0.32

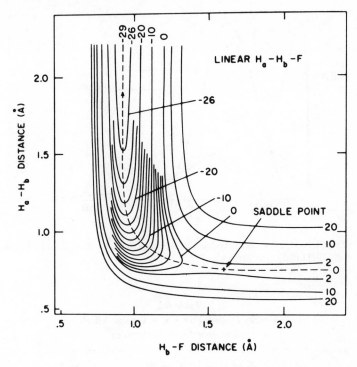

Fig. 6.7 Semiempirical potential energy surface for FH_2. (Muckerman, 1971.)

As far as the product translational energy is concerned, the crossed-beam technique is a unique, excellent approach for this measurement. During the past two decades numerous exoergic reactions have been carried out by this method; among the most extensive and careful studies are those on the family of alkali atom reactions. Figure 6.8 shows the product translational energy distributions of two well-studied exoergic reactions: $K + I_2$ and $K + CH_3I$ reactions. The exoergicity ΔD_0 for these two reactions are 40 and 24 kcal/mole, respectively. It is clearly shown that the product KI from $K + I_2$ is highly internally excited but less so than that from the $K + CH_3I$ reaction: At the peak of $P(E'_t)$ distributions the collision exothermicity ε is slightly negative for the $K + I_2$ reaction but positive for the $K + CH_3I$ reaction. This suggests that the $K + I_2$ reaction may be governed by an early downhill attractive surface whereas the $K + CH_2I$ reaction proceeds through a late downhill surface. These results, combined with other experimental findings, uniquely establish a solid background for determining the reaction dynamics. Table 6.2 lists the results found for a few alkali atom reactions.

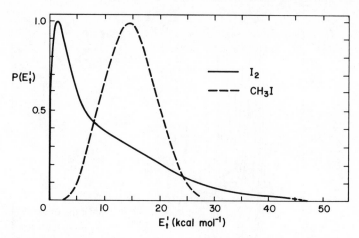

Fig. 6.8   Distribution functions of product recoil energy $P(E_{t'})$ for K $+$ I$_2$ and CH$_3$I reactions. (Gillen et al., 1971; Rulis and Bernstein, 1972.)

Table 6.2
Energy Disposal for Three Typical Alkali Atom Reactions (kcal/mole)

	K + HBr	K + Br$_2$	Cs + CH$_3$I
$\langle E \rangle$	1.5	1.23	1.5
$\langle W \rangle$	0.6	0.75	0.7
ΔD_5	4.2	45	30
f_t	0.3	0.05	0.3
f_r	0.3	0.11	0.08
f_v	0.3	0.85	—

Source: Toennies (1974). By permission.

6.2.3   Direct Reactive Collisions

Angular distributions of all these reactions except the K HBr system are found to be asymmetric in the CM coordinate; that is, the product recoil direction exhibits a strong preferential angular disposition. For the K $+$ HBr reaction the kinematic limitation discussed earlier (eq. 6.29) prevents us from giving a conclusive CM angular distribution. As we discussed earlier, these asymmetric angular distributions are characterized by a reaction time shorter than the lifetime of the collison complex and are therefore considered to proceed through a direct reaction mechanism. For a direct reaction mechanism the preferred recoil direction of products often correlates with the magnitude of the reaction cross section as well as the attenuation of the elastic scattering.

For the $K + I_2$ reaction the product KI is found to be predominantly scattering through the forward hemisphere with respect to the incident K-atom beam, as shown in Fig. 6.9 and the recoil velocity of KI is rather small. This forward scattering, in terms of classical mechanics, can be qualitatively explained by a high reaction probability $P(E, b)$ at large impact parameters (eq. 6.1) such that for every encounter at large b, the K atom picks up the I atom and carries it forward as if the other I atom were a spectator. This is an example of the stripping mechanism described in the earlier section. Thus one may anticipate that the preferred forward direction of the KI product has to be strongly correlated with a large reaction cross section; indeed, the measured cross section is rather large, $Q_r = 127 \, \text{Å}^2$. The experimental findings for the $K + I_2$ reaction may be summarized qualitatively here: (1) Most of the total reaction energy available is released as internal excitation of the KI product; (2) the product KI is predominantly confined to the forward hemisphere associated with a large reaction cross section; (3) a decrease in total reaction cross section and an increase in the forward scattering were found as the incident collision energy increased; and (4) the fall-off of elastic scattering data at wide angles is much faster than the values predicted from an approximate long-range attractive potential.

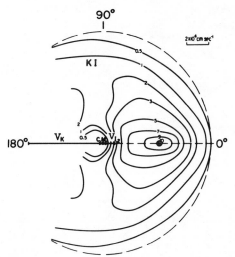

Fig. 6.9 Product KI intensity contour map for $K + I_2$ reaction. (Gillen et al., 1971.)

The KI product from the $K + CH_3I$ reaction, on the other hand, is remarkably different; it carries a very large recoil velocity into the backward hemisphere as shown in Fig. 6.10. Moreover, the reaction cross section for $K + CH_3I$ is small, $Q_r = 35 \text{ Å}^2$. This backward-scattering phenomenon is also found for the DF product from the $F + D_2$ reaction. Here again the qualitative interpretation of this rebound reaction may be stated. The reaction only takes place at a small impact parameter when the reactants are subject to the strong short-range repulsive potential driving the products KI and CH_3 in retreat. This simple interpretation is supported by the experimental evidence of elastic scattering data at wide angles. For the $K + CH_3I$ reaction, the fall-off of elastic data at wide angles is nearly parallel to those observed for a nonreactive system. Hence we have, from eq. 2.3,

$$I_r(E, \theta) = I_n^0(E, \theta) - I_n(E, \theta) = P(E, b)I_n^0(E, \theta) \qquad (6.41)$$

If $I_n^0(E, \theta)$ can be approximated by the hard sphere model, $I_n^0(E, \theta) = d^2/4$, the product angular distribution is then determined by $P(E, b)$ alone in this case.

The physical picture proposed to interpret the experimental results observed for the $K + I_2$ and CH_3I reactions is given in Fig. 6.11. The observed large cross section combined with the predominant forward scattering for the $K + I_2$ reaction is consistent with the stripping mechanism. In such a mechanism every encounter of the K atom would pick up one I atom as it moves in the forward direction without imposing much interaction on the other ones. In the extreme case, this is a spectator stripping mechanism in

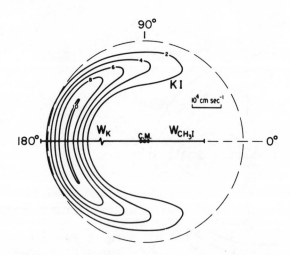

Fig. 6.10 Product KI intensity contour map for $K + CH_3I$. (Rulis and Bernstein, 1972.)

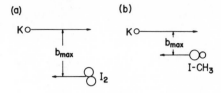

Fig. 6.11 Schematic pictures for the reaction proceeded via (*a*) stripping mechanism and (*b*) rebound mechanism.

which linear momentum of the unreacted I atom remains constant before and after the reaction. The $K + CH_3I$ reaction, is considered to occur with so small an impact parameter that the product KI has to rebound assuming the collision partners can be approximated as hard sphere; this is called a rebound mechanism.

At this point one has to be cautious, since there may be a strong coupling between the recoil speed and the direction of the newly formed products; that is, the assumption

$$I_r(E, \theta) = P(E_t')T_r(\theta) \tag{6.42}$$

is not always valid for making the data analysis. For the $K + I_2$ reaction, a weak coupling between $P(E_t')$ and $T_r(\theta)$ is actually found from various shapes of the $P(E_t')$ distribution at the different CM angle. It is because of this coupling that the results of $P(E_t')$ shown in Fig. 6.8 are in fact obtained by integrating over the measured $P(E_t')$ distributions for the important range of CM angles; for $K + I_2$, $P(E_t')$ was obtained by integrating over the forward hemisphere and for $K + CH_3I$, it was taken from $\theta \approx 40\text{--}180°$.

For a reaction that proceeds via the direct reaction mechanism, the reaction product may also exhibit a sideways peaking at the intermediate angles; for example, product KCl from the $K + CCl_4$ reaction was shown to have a maximum identity at $\theta \approx 90°$.

6.2.4 Formation of the Collision Complex and Its Decay

The intensity contour maps observed for several ion-molecule reactions and atom-molecule reactions show a symmetry about $\theta = 90°$. This symmetric feature reflects the existence of a long-lived collision complex such that before a reaction is completed, this complex may rotate several turns and lose all memory of the original direction of the colliding particles. The reaction mechanism of this product formation is governed by the statistical theory of complex formation and its decay. Figure 6.12 shows the intensity of contour maps obtained for two well-studied reactions. It is clearly seen that both reaction products have a symmetry about $\theta = 90°$; for $Cs + SF_6$, the measured product CsF peaks at $\theta = 0$ and $180°$, whereas the product C_2H_3F is most intense at $\theta = 90°$ from the $F + C_2H_4$ reaction.

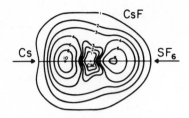

Fig. 6.12 Product intensity contour maps for reactions of $Cs + SF_6$ and $F + C_2H_4$ that proceed through an activated complex that would exhibit forward-backward symmetry in the product angular distributions. (Riley and Herschbach, 1973; Parson and Lee, 1972.)

The angular distribution for a long-lived collision complex is governed by the conservation of total angular momentum J and its projection M and K at the initial and final relative velocities g and g', respectively. If a reaction is characterized by a large reaction cross section (e.g., $Cs + SF_6$), the orbital angular momentum L may be the dominant contribution to J so that $L \approx J \approx L'$. In this case, the complex is mainly determined by its initial trajectories of approach and eq. 6.27 approaches to the limit of eq. 6.28; that is, the product distribution would show very strong forward-backward peaking. On the other hand, the reaction of the $F + C_2H_4$ system with a small reaction cross section, $Q_r \approx 2 \text{ Å}^2$, the orbital angular momentum L is likely smaller than the angular momentum j and one expects that its peaking would be rounded off and a broad spectrum of angular distribution is more likely (eq. 6.27).

Table 6.3 lists results obtained for $Cs + SF_6$ and $Cs + RbCl$ reactions; for comparison, also included in the table are the results found for the direct reaction mechanism of $K + I_2$ and $K + CH_3I$ system. The formation of a long-lived collision complex suggests that there may be a deep well along the reaction path in the potential energy surface such that the collision pairs may be held together long enough and that its decay into the products may be described in a way similar to a unimolecular reaction. Making use of the classical relation $L = \mu g b / \hbar$ and $Q_r = \pi b_{max}^2$, the minimum lifetime of this complex may be estimated by

$$\tau_{min} = 2\pi \left(\frac{I}{L_{max}} \right) \tag{6.43}$$

Table 6.3
Results Obtained from the Crossed Beam Studies of
Bimolecular Reactions[a]

Reaction	Complex intermediates		Direct reaction	
	Cs + RbCl	Cs + SF$_6$	K + I$_2$	K + CH$_3$I
$\langle E \rangle$	1.4	1.6	2.7	2.8
$\langle W \rangle$	3.5	0.5	1.1	1.1
ΔD_0	5.4	41	40.5	24
E'_t	2	1.5	1.5	16
f'_t	0.2	0.04	0.04	0.57
θ_{peak}	0, 180	0, 180	0	180
Q_r	150	110	127	35

Source: Miller *et al.* (1967).
[a] Energies in kcal/mole and cross section in Å^2.

Hence for the Cs + SF$_6$ reaction, τ_{min} is about 4×10^{-13} sec if the moment of inertia for a tight complex $I = 190$ amu Å^2 and L_{max} is $300\hbar$ whereas, for Cs + RbCl reaction, τ_{min} is about 5×10^{-12} sec.

For Cs + RbCl and Cs + SF$_6$ reactions the product velocity distributions were found to be similar to a thermal Maxwell–Boltzmann distribution. According to the approximate Rice, Ramsberger, Kassel, and Marcus (RRKM) formula of transition state theory, the product energy distributions among all degrees of freedom are randomized that T_t, T_r, and T_v are nearly equilibrated and the common temperature is given by

$$T^* = \frac{E_{tot}}{(3N - 5\frac{1}{2})R} \tag{6.44}$$

where N is the number of atoms in the complex and R is the gas constant. To obtain this simple expression, we assume that the total available energy E_{tot} is divided equally among $3N - 6$ vibrational modes and one rotation (counts as $\frac{1}{2}$) are active in the energy exchange; the other rotation modes are fixed by the centrifugal motion. Combining eq. 6.44 and E_{tot} given in Table 6.3, we get $T^* \approx 1500°K$ for Cs + RbCl reaction and $1150°K$ for the Cs + SF$_6$ reaction. These estimated values are in good agreement with the observed peak values in $P(E'_t)$ distributions if they are considered to be thermal.

The product distribution observed for the F + C$_2$H$_4$ → C$_2$H$_3$F + H reaction, on the other hand, has a broad distribution peaked at about half of the energy available, $E_{tot} \approx 14$ kcal/mole; this energy distribution is in sharp contrast to what would be expected from the RRKM theory. Moreover, this nonstatistical product translationai energy distribution is also found for a number of reactions of F atoms with olefins, dienes, and aromatic and heterocyclic molecules. The failure of RRKM theory in predicting the product energy distributions for this systems leads to the suggestion that there may be incomplete mixing of energy among all the degrees of freedom of the complex when the internal energy exchange is less rapid than the rate of the chemical reaction.

6.2.5   Energy Dependence of the Reaction Cross Section

In this section we discuss how the reaction cross section (and thus the reaction rate) varies with the initial reactant energy. This type of study is particularly important for an endoergic reaction in which the product signal is usually either small or nondetectable at all low temperatures. As we discussed earlier, this is due to the existence of a potential barrier along the reaction path that prevents the reaction from taking place. By increasing the reactant energy the probability of overcoming this potential barrier would be enhanced and thus make it possible to measure the reaction yield as a function of initial reactant energy. For an exoergic reaction, on the other hand, this potential barrier may be very small or virtually zero, so that the reaction may occur very rapidly even at very low temperatures.

The reaction cross sections predicted by the orbiting model, eq. 6.6, agree reasonably well with the observed cross sections for a number of ion-molecule reactions at low collision energy. For the $Ar^+ + D_2 \rightarrow ArD^+ + D$ reaction at $E < 1$ eV, for instance, the reaction is considered to occur whenever Ar^+ ions and D_2 molecules reach a critical distance R_c at which the ion is captured into a circular orbit around the molecule. According to this model, this critical R_c can be determined by the condition that the collision energy E is equal to the centrifugal barrier of the effective potential. Hence making use of eq. 1.7, the relation between the R_c and orbiting impact parameter can be found classically. This leads to the expression for the reaction cross section in eq. 6.6. At high energy the effective potential is repulsive everywhere and this simple orbiting model is no longer valid.

The effect of reactant energy on the total reaction cross section has been studied in great detail for a slightly endoergic reaction of $K + HCl \rightarrow KCl + H$ system. With the initial average collision energy $E = 0.9$ kcal/mole, an increase of the reaction cross section from 0.2 to 20 $Å^2$ was found in the crossed-beam measurements when the vibrational state of HCl ($E_{01} = 8.3$ kcal) is excited from $v = 0$ to $v = 1$. The enhancement of reaction yield from HCl vibrational energy is much more effective than the relative collision energy; for the same reaction with HCl at $v = 0$, a tenfold increase of reaction cross section was measured at $E = 11.5$ kcal/mole. The observed cross

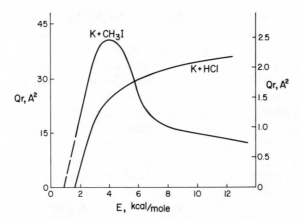

Fig. 6.13 Translational energy dependence of total reaction cross section. (Toennies, 1974; Pruett et al., 1974.)

section enhancement is similar to that observed for $H_2^+ + He \rightarrow HeH^+ + H$ reaction; in this latter case it was found that the vibrational excitation of H_2 from $v = 0$ to $v = 5$ had an even greater effect than the translational activation. These experimental results suggest that the reaction barrier is likely to be around the exit valley (Fig. 6.4) for both reactions.

For the K + HCl reaction the measured threshold energy is about 1.7 kcal; results of this translational energy dependence of reaction cross section are shown in Fig. 6.13. For this endoergic reaction there is a sharp rise in cross section at E slightly greater than E_0 but a slow increase as E further increases. Also shown in Fig. 6.13 is the exoergic reaction of the K + CH_3I system. For this system, however, the reaction cross section increases rapidly with E initially, followed by a slow decrease as E goes up.

For the K + CH_3I reaction we also note another important factor that governs the reaction, that is the orientation effect of the target molecule. It was found that this reaction occurred preferentially for K atoms colliding with the iodine end of the CH_3I molecules. Employing a six-pole electric field, the reactivity was found to be higher for the configuration of $K\text{---}I\text{---}CH_3$ than for that of $K\text{---}CH_3\text{---}I$ as the incident K atom approaches the CH_3I molecule (Odiorne et al., 1971; Brooks, 1973).

6.3 REACTION THEORY

Thus far we have repeatedly used a qualitative model interpretation to show the physical picture of the reaction dynamics for some particular systems.

In ion-molecule reactions, the orbiting model predicts a fairly good magnitude of reaction cross section for some favorable system at low collision energy. For a reaction that proceeds through a direct reaction mechanism, the stripping model accounts for the preferential forward-scattering products, whereas the rebound mechanism in association with a small impact parameter leading to reaction may be employed for the scattering process. Insofar as energy disposal and energy requirements are interesting, the description of the classical trajectory along the reaction path on the potential energy surface gives a clear, unique picture of how the energy flux is governed. In addition, the role of angular momentum plays an important role if a long-lived activated complex is formed before a reaction is completed (Herschbach, 1966; Edelstein and Pavidorits, 1971; Child, 1974).

In this section we discuss a general review of reaction theories that have been well developed to account for the various types of bimolecular reactions. Again, we make no attempt here to give the detailed numerical calculations that would be used to fit the experimental data; rather, we emphasize the validity of these reaction theories that can account for a given type of reaction.

6.3.1   Curve Crossing Model: Harpooning Mechanism

For the family of alkali atom (M) and halogen molecule (X_2) reactions, the reaction cross sections are found to be rather large, $\sim 100 \text{ Å}^2$. Since the alkali halides are known to be ionic in nature and since these alkali atoms have very low ionization potentials (3.89 eV for Cs and 5.39 eV for Li), the electron jump mechanism (harpooning) has long been invoked to explain the reaction dynamics of $M + X_2$ reactions. It postulates that an electron is transferred from M to X_2 in the entrance valley when the intersection between the covalent potential surface of $M + X_2$ and the ionic $M^+ + X_2^-$ surface occurs. In its simplest form, the radius where curve crossing occurs can be approximated as

$$R_c = \frac{e^2}{[I(M) - E(X_2)]} \tag{6.45}$$

where $I(M)$ denotes the ionization potential of the alkali atoms and $E(X_2)$ the electron affinity of the halogen molecules. The magnitude of the crossing radius thus calculated is about 4–7 Å for this family of reactions. According to this model, a reaction is completed once the electron jump has taken place. Assuming unit probability for the electron transfer at each crossing, the reaction cross section is simply equal to πR_c^2. Moreover, the electron jump model suggests that a major portion of the reaction energy released occurs when M and X_2^- are accelerated toward each other so that the product $M^+ X^-$ may be highly vibrationally excited. Thus this simple

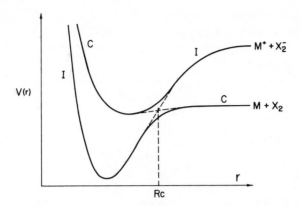

Fig. 6.14 Potential energy curves of covalent $M + X_2$ and ionic $M^+ + X_2^-$ systems.

model offers an excellent qualitative interpretation of the experimental findings of the $M + X_2$ reactions: The large reaction cross section and the highly internal excited MX product are confined mostly to the forward hemisphere.

Theoretical investigation of the detailed reaction picture indicates that the electron jump probability at the curve crossing is not always unity but a function of the electronic structure and the orientation configuration of the halogen molecules. The simple physical picture for the two-state inter-action depicted in Fig. 6.14 is very useful in the discussion of the electron jump reaction model. As the M atom and the X_2 molecule approach each other along the covalent surface (C) near the crossing point R_c, three events may occur. We denote by p the transition probability when the system jumps at every crossing, that is, the particles move along the diabatic potential surface without making an electron jump (the dashed line around R_c). An elastic scattering occurs with the total probability p^2 as the system jumps on the approach and on the retreat. On the other hand, reaction occurs with the probability $(1 - p)$ if an electron jump takes place as the particles approach and the collision pairs follow the adiabatic curve (lower CI curve). The third possibility is the system jump on the approach and an electron jump on the retreat; this leads to a probability of $p(1 - p)$ attributed to the inelastic scattering on the upper IC curve. For low-energy collision processes, the particles cannot separate as ions and may be subject to a subsequent crossing that would also contribute to the reaction channel. Accordingly, for low-energy collision processes, the probability leading to an inelastic scattering is negligible and the overall reaction probability is simply equal to $(1 - p^2)$ (Olson et al., 1970).

To evaluate the probability p that the system remains on the diabatic curve for every crossing, we employ the simplest form of the Landau–Zener formula here. According to the Landau–Zener formula, the transition probability is given as

$$p = \exp\left(-\frac{g^*}{g}\right) \tag{6.46}$$

with

$$g^* = \frac{2\pi\Delta H_{12}^2}{\hbar|F_1 - F_2|} \tag{6.47}$$

Here g denotes the radial velocity at the crossing point and

$$|F_1 - F_2| = \frac{d}{dr}|V_1 - V_2| \approx \frac{e^2}{R_c^2} \tag{6.48}$$

if the weak van der Waals forces are neglected for both $M + X_2$ and $M^+ + X_2^-$ systems. The interaction matrix ΔH_{12} may also be approximated from the semiempirical expression

$$H_{12} = \tfrac{1}{2}V_{12} = R_c^* \exp(-0.86R_c^*)$$

where

$$R_c^* = R_c[(2I)^{1/2} + (2E)^{1/2}]/2 \tag{6.49}$$

Here ΔV_{12} is the energy splitting between two potential curves at the crossing radius. In spite of its approximate nature and several limitations on this simple formula, this model has proved to be semiquantitatively correct for making estimates of transition probabilities at the intermediate energies. For the K + I$_2$ reaction at $E = 2.7$ kcal/mole, we find $R_c \approx 5.5$ Å and $\Delta V_{12} \approx 1.0$ eV. These lead to a negligibly small p or unit probability of the electron jump for every crossing. For the $M + X_2$ reaction at low collision energy, one thus finds an electron jump mechanism can be used to correlate the observed large reaction cross section.

6.3.2 Optical Model Analysis

For a collision process involving the reactive species, the elastic angular distribution at wide angles may be significantly quenched when the reactive scattering channel is accessible; the larger the difference between the observed elastic scattering data at wide angles and those calculated from the reference potential, the bigger the reaction cross section. For the alkali atom and the methyl iodide reactions, it is indeed found that both measured

$$A + BC \rightarrow A--B--C \rightarrow AB + C$$

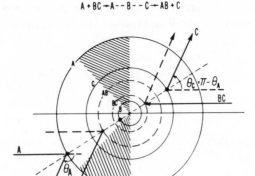

Fig. 6.15 Relations of elastic and reactive scatterings in the optical model for a rebound reaction via a collinear configuration. The unshaded area on the left-hand side shows a range of impact parameters in which the product AB is confined. (Herschbach, 1966.)

reaction cross section and fall-off of elastically scattered atom intensity at wide angles increase with the size of the alkali atom: $K \rightarrow Rb \rightarrow Cs$. For these rebound reactions, the correlation among the attenuation of elastic scattering data at wide angles, the magnitude of reaction cross section and the shape of product angular distribution can be readily illustrated by the optical model.

For simplicity we adopt here the three-body collinear hard-sphere model of $A + BC \rightarrow A--B--C \rightarrow AB + C$ reaction. Figure 6.15 shows a specular reflection approximation for collinear trajectories; sets of concentric circles indicate hard-sphere potentials from which atoms and molecules BC or AB centroids are scattered. The radii of these concentric spheres are determined by the geometrical configuration of $A--B--C$ at the onset of reaction. Thus if we assume hard-sphere interaction for these reactions where the reactive scattering mainly occurs at small impact parameters, the deflection angle of the product AB along the outgoing trajectories with exit impact parameter b' and relative translational energy E' appears to be the same as that of atom A deflected along the incoming trajectory with impact parameter b and collision energy E. For hard-sphere potentials with radii R for the reactants and R' for the products, we have

$$\theta_i = \theta_f$$

$$\frac{b}{R} = \frac{b'}{R'} \tag{6.50}$$

Hence for a backward scattering predominantly confined at $\theta_i = \pi - \theta_f \leq 3\pi/4$ as observed for alkali and methyl iodide reactions,

$$\theta_i = \frac{\pi}{4} \quad \text{and} \quad b \leq \frac{R}{(2)^{1/2}}$$

and we have

$$I_n^0(\theta) = \frac{R^2}{4} \tag{6.51}$$

The reaction probability given in eq. 2.41 is then no longer dependent on energy and

$$I_r(\theta) = I_n^0(\theta) - I_n(\theta) = \frac{R^2 P(b)}{4} \tag{6.52}$$

that is, the product angular distribution resulting from the missing part of the elastic scattering is closely related to the reaction probability. It becomes straightforward now that the reaction probability (eq. 6.5) and the total reaction cross section (eq. 6.1) can be readily calculated by evaluating the difference between the measured data $I_n^0(\theta)$ from an analogous nonreactive system and the measured $I_n(\theta)$ of the system interested. Alternatively, the $I_n^0(\theta)$ at wide angles can also be calculated from a reference potential with which the measured small angle elastic data are reproduced (Harris and Wilson, 1971; Kinsey et al., 1976).

With the foregoing discussion on the data analysis procedure, we now return to discuss the foundation of the quantum-mechanical derivation. The formulation of the optical model starts by postulating a complex interaction potential,

$$V(r) = V_0(r) + iV_1(r) \tag{6.53}$$

with the imaginary part responsible for the reaction. The associated radial wave function $\phi(r)$ is then given as

$$\phi(r) = A(r) \cdot \exp(ikr) \tag{6.54}$$

where $k^2 = 2\mu(E - V_0 - iV_1)/\hbar^2$. The complex expression k can be separated into a real and an imaginary part, $k = k_0 + ik_1$ and

$$\phi(r) = A(r)\exp(-k_1 r)\exp(ik_0 r) \tag{6.55}$$

that is, the amplitude $A(r)\exp(-k_1 r)$ decreases exponentially as a function of the radial coordinate r because of the imaginary potential $V_1(r)$. Hence the S matrix, $S_l = \exp(2i\eta_l)$, contained in the general expression of scattering amplitude

$$f(\theta) = \frac{1}{2ik} \sum (2l + 1)(S_l - 1)P_l(\cos \theta)$$

is compounded with a complex phase shift $\eta_l = a_l + ib_l$. It follows that the total elastic cross section $Q_e(E)$ and reactive cross section $Q_r(E)$ can be expressed as

$$Q_e(E) = \frac{\pi}{k^2} \sum (2l + 1)/1 - S_l/2 \qquad (6.56)$$

and

$$Q_r(E) = \frac{\pi}{k^2} \sum (2l + 1)(1 - |S_l|^2) = \frac{\pi}{k^2} \sum (2l + 1)q_l \qquad (6.57)$$

Hence in the optical model the loss of flux scattered into reactive channel is due to a phenomenological opacity function q_l that is simply equivalent to

$$q_l = 1 - \exp(-4b_l) \qquad (6.58)$$

The total collision cross section $Q_t(E)$ is now a sum of $Q_e(E)$ and $Q_r(E)$

$$Q_t(E) = Q_e(E) + Q_r(E) = \frac{2\pi}{k^2} \sum (2l + 1)(1 - ReS_l) \qquad (6.59)$$

Here again, for low-energy molecular collisions, the number of partial waves that contribute to the scattering events may be very large and we adopt a black-sphere model (used in nuclear reactions) for our illustration. According to the black-sphere model, there is unit reaction probability for all $l \leq L \approx kR$ and zero probability for $l > L$; that is,

$$\begin{aligned} S_l &= 0, && \text{for } l > L \\ &= 1, && \text{for } l \leq L \end{aligned} \qquad (6.60)$$

Hence we get

$$Q_l = Q_r = \tfrac{1}{2}Q_t = \frac{\pi L^2}{k^2} \approx \pi R^2 \qquad (6.61)$$

and the postulation of a complex potential can lead to a large reactive scattering, comparable to or even larger than the elastic scattering. Moreover, this reactive scattering seems to be confined to small l which would then lead to a large angle deflection. Thus in the optical model treatment, one can simulate fairly well the observed correlation among the fall-off of intensity of elastically scattered alkali atoms, the reaction cross section, and the shape of the alkali iodide angular distributions for this family of reactions (Emmerich, 1963).

6.3.3 Statistical Theory of Complex Formation and Its Decay

The formation of a long-lived collision complex from the crossed-beam studies lies in the fact that the product angular distribution is found to be symmetric about $\theta = 90°$. This symmetric feature can only be explained if the lifetime of the collision complex is longer than its rotation period so that the complex may rotate several turns before the products start to retreat; the estimated minimum lifetime is of the order of 10^{-12} sec. The formation of this collision complex thus suggests that there is a deep well along the reaction coordinate of the potential energy surface and the reactants may be held long enough that the breakup of this complex may be governed by the statistical densities of active rotational and vibrational states at the critical configuration associated with the transition state theory.

Figure 6.16 illustrates the energy profile along the reaction path for an exoergic reaction. Here V_a and V_a' denote the centrifugal barrier at the entrance and exit valleys, respectively; in doing so, the collision trajectory leading to a reaction is considered to follow the solid curve. It is clearly seen that the reaction proceeds first to form a collision complex held at the deep well, followed by the subsequent decay of this complex into the products on the retreat coordinate. It follows that this type of reaction may be best explained by two independent processes; the formation of the complex along the approach coordinate and the breakup of the complex on the retreat coordinate. Hence the state-to-state reaction cross section can be expressed as

$$Q_{r \to p} = Q_{ri} P_{ip} \tag{6.62}$$

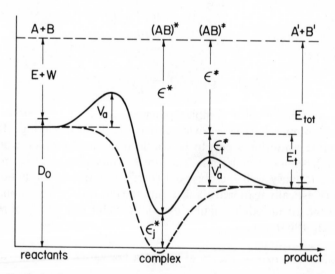

Fig. 6.16 Energy profile along the reaction path for an exoergic reaction with an intermediate collision complex. Energy diagram for the RRKM treatment. $E_{tot} = E + W + \Delta D_0 = E_t' + W'$. $\epsilon_t^{\ddagger}$ = translational energy along the reaction coordinate. $\epsilon^{\ddagger}$ = active vibrational and rotational energies.

where Q_{ri} is the cross section for forming a complex in the state i and P_{ip} is the probability for this complex to decay to the final products. To evaluate Q_{ri}, one may employ the dynamic model of the two-body collision process; for instance, the orbiting model may be employed for a system carried out under the low-energy condition. To evaluate P_{ip} we adopt here the statistical theory that has been traditionally used for unimolecular reactions.

In the transition state theory the detailed reaction mechanism may be postulated as

$$A + B \longrightarrow (AB)^*\text{---}(AB)^\ddagger \longrightarrow \text{products} \qquad (6.63)$$

In this reaction scheme the activated molecule $(AB)^\ddagger$ directly passes to final products whereas the energized complex $(AB)^*$ only has sufficient energy to become an activated molecule $(AB)^\ddagger$. For a potential profile, shown in Fig. 6.16, it is natural to consider that the molecule $(AB)^*$ is located around the deep well whereas the critical configuration of $(AB)^\ddagger$ is somewhere around the hump of the energy barrier at the exit channel.

According to the treatment of statistical theory, the breakup of the activated complex $(AB)^\ddagger$ is governed by an equal probability of passing through all open exit channels that are available under the constraints of the conserved energy and angular momentum. It is important to remark, however, that the success of this treatment can only be ensured if (1) the breakup rate of $(AB)^\ddagger$ is slower than the rate of energy randomization from the complex $(AB)^*$ to the critical stage $(AB)^\ddagger$ and (2) the rotational-translational coupling associated with the orbital motion where two degrees of rotational freedom are tied up for a tight nonlinear complex, no free energy flow among all other degrees of freedom is allowed as products start to separate from the critical region. Hence for a tight nonlinear complex $(AB)^\ddagger$ of N atoms a complete randomization among (3N−6) active vibration mode and one remaining rotation mode has to be assured in order to have a valid statistical treatment. In the statistical treatment the probability P_{ip} in eq. 6.62 for products emerging from the critical region with energy $\varepsilon^\ddagger$ is given by

$$P_{ip} = \frac{k_i(\varepsilon^\ddagger)}{\sum k_i(\varepsilon^\ddagger)} \qquad (6.64)$$

where k_i is the rate constant for decomposition of the complex at state i. This expression can be evaluated in a way similar to that for a unimolecular reaction, as discussed in Chapter 5.

For a reaction that proceeds through a long-lived collision complex, the product recoil energy distribution can also be evaluated by the statistical theory. According to the transition state approximation

$$P(E_t') \approx A(E_t')N^\ddagger(E_{tot} - E_t') \qquad (6.65)$$

where $A(E_t')$ is related to the centrifugal energy associated with separation of the products and $N^{\ddagger}$ is the energy density of active vibrations and rotations at the critical region of the activated complex. If this critical region is defined by the centrifugal barrier in the exit channel and lies sufficiently far out to be governed by the van der Waals attraction,

$$A(E_t') \approx \left(\frac{E_t'}{V_a'}\right)^{2/3}, \qquad \text{for } E_t' \leq V_a' \qquad (6.66)$$

$$\approx 1, \qquad\qquad \text{otherwise}$$

where V_a' is the maximum exit centrifugal barrier and the factor 2/3 is due to two out of three degrees of rotational freedom being coupled into the translational mode. In its classical limit the energy states can be approximated as

$$N(E_{\text{tot}} - E_t') \approx (E_{\text{tot}} - E_t')^{s + r/2 - 2} \qquad (6.67)$$

Here s and r are the number of active vibrational and rotational modes, respectively. The evaluation of s and r is dependent on whether the transition state is loose or tight; for instance, for a three-atom nonlinear complex, $s = 3$ and $r = 1$ if it is tight, and $s = 2$ and $r = 2$ if it is loose. The product energy agreement with the observed data for the $Cs + SF_6$ reaction but yields a too narrow product energy distribution for reactions between, for example, $F + C_nH_{2n}$ (Waage and Rabinovitch, 1970; Safron et al., 1972).

6.4 TRANSITION STATE THEORY

The purpose of this section is to explore in some detail the relation between the macroscopic rate constant of the absolute reaction rate theory and the corresponding microscopic reaction cross section. A number of investigations have been carried out in this area (Marcus, 1970; Morokuma et al., 1969; Light et al., 1969; Lin and Eyring, Lau 1971). The approach we adopt is as follows. We first desire the detailed specific rate constant of the absolute reaction rate theory for a bimolecular reaction and then obtain the detailed reaction cross section by using the equation given by the collision theory for the relation between rate constants and reaction cross sections (see Sec. 6.1.3; Eliason and Hirschfelder, 1959; Ross and Mazur, 1961). We also discuss the effect of disallowed states of the reactants, activated complex, and products on the rate constant and reaction cross section. It is shown that when the disallowed states are excluded, the reaction cross section obtained from the transition state theory does not diverge as the energy approaches infinity.

6.4.1 Detailed Specific Rate Constants

For a bimolecular reaction of the general type in which neither the reactant nor the product molecules are point masses but have quantized internal degrees of freedom,

$$A + B \longrightarrow C + D \tag{6.68}$$

we consider the detailed reaction

$$A_i + B_j \longrightarrow C_k + D_l, \tag{6.69}$$

where each subscript in eq. 6.69 represents the set of quantum numbers required to specify the internal state of the molecule. If C_{ai} and C_{bj} represent the concentrations of A_i and B_j, and k_{ij}^{kl} represents the detailed specific rate constant, then the detailed rate equation can be expressed as

$$-\left(\frac{dC_{ai}}{dt}\right)_{j,\,kl} = k_{ij}^{kl} C_{ai} C_{bj} \tag{6.70}$$

According to the absolute reaction rate theory, we should write eq. 6.69 as

$$A_i + B_j \longrightarrow (AB)_{mn}^{\ddagger} \longrightarrow C_k + D_l, \tag{6.71}$$

where $(AB)^{\ddagger}$ is the activated complex. In the equilibrium assumption we have

$$C_{ai} = \exp(-\alpha_a - \beta\varepsilon_{ai})Q_{a\,\mathrm{tr}}$$

$$C_{bj} = \exp(-\alpha_b - \beta\varepsilon_{bj})Q_{b\,\mathrm{tr}}$$

$$C_{mn}^{\ddagger} = \exp(-\alpha_a - \alpha_b - \beta\varepsilon_{mn}^{\ddagger\prime})Q_{\mathrm{tr}}^{\ddagger}Q_{\mathrm{rc}}^{\ddagger} \tag{6.72}$$

where $Q_{ab\,\mathrm{tr}} = Q_{a\,\mathrm{tr}}Q_{b\,\mathrm{tr}}$. If $(x_m)_{\mathrm{av}}$ represents the average velocity of the systems $Q_{\mathrm{rc}}^{\ddagger}$ the partition function of the degree of freedom corresponding to the reaction coordinate. Elimination of α_a and α_b gives

$$C_{mn}^{\ddagger} = C_{ai} C_{bj}\left(\frac{Q_{\mathrm{tr}}^{\ddagger}}{Q_{ab\,\mathrm{tr}}}\right)Q_{\mathrm{rc}}^{\ddagger}\exp[-\beta(\varepsilon_{mn}^{\ddagger\prime} - \varepsilon_{ai} - \varepsilon_{bj})] \tag{6.73}$$

where $Q_{ab\,\mathrm{tr}} = Q_{a\,\mathrm{tr}}Q_{b\,\mathrm{tr}}$. If $(\dot{x}_m)_{\mathrm{av}}$ represents the average velocity of the systems to cross over the potential barrier, then the rate of detailed reaction is given by

$$-\left(\frac{dC_{ai}}{dt}\right)_{j,\,kl} = \sum_m k_{mn}^{kl} C_{mn}^{\ddagger}\left(\frac{\langle\dot{x}_m\rangle_{\mathrm{av}}}{\delta_m^{\ddagger}}\right) \tag{6.74}$$

or

$$-\left(\frac{dC_{ai}}{dt}\right)_{j,\,kl} = \sum_m k_{mn}^{kl}\langle\dot{x}_m\rangle_{\mathrm{av}} C_{ai} C_{bj}\left(\frac{Q_{\mathrm{tr}}^{\ddagger}}{Q_{ab\,\mathrm{tr}}}\right)\left(\frac{Q_{\mathrm{rc}}^{\ddagger}}{\delta_m^{\ddagger}}\right)\exp[-\beta(\varepsilon_{mn}^{\ddagger\prime} - \varepsilon_{ai} - \varepsilon_{bj})] \tag{6.75}$$

Here the summation in eqs. 6.74 and 6.75 is over the remaining two degrees of freedom originating from the six translational degrees of reactants. Comparing eq. 6.75 and 6.70, we obtain the expression for the detailed specific rate constant as

$$k_{ij}^{kl} = \sum_m k_{mn}^{kl} \langle \dot{x}_m \rangle_{\text{av}} \left(\frac{Q_{\text{tr}}^{\ddagger}}{Q_{ab\,\text{tr}}} \right) \left(\frac{Q_{\text{rc}}^{\ddagger}}{\delta_m^{\ddagger}} \right) \exp[-\beta(\varepsilon_{mn}^{\ddagger\prime} - \varepsilon_{ai} - \varepsilon_{bj})] \qquad (6.76)$$

By the use of the relations $\langle \dot{x}_m \rangle_{\text{av}} = (2\pi m^{\ddagger} \beta)^{-1/2}$ and $Q_{\text{rc}}^{\ddagger} = (2\pi m^{\ddagger}/\beta h^2)^{1/2} \delta_m^{\ddagger}$, eq. 6.76 can also be written

$$k_{ij}^{kl} = \sum_m \left(\frac{k_{mn}^{kl}}{\beta h} \right) \left(\frac{Q_{\text{tr}}^{\ddagger}}{Q_{ab\,\text{tr}}} \right) \exp[-\beta(\varepsilon_{mn}^{\ddagger\prime} - \varepsilon_{ai} - \varepsilon_{bj})] \qquad (6.77)$$

k_{mn}^{kl} in eqs. 6.74–6.77 represents the transmission coefficient. Equation 6.77 can also be written,

$$k_{ij}^{kl} = \left(\frac{1}{\beta h} \right) \left(\frac{2\pi \mu}{\beta h^2} \right)^{-3/2} \sum_m k_{mn}^{kl} \exp[-\beta(\varepsilon_{mn}^{\ddagger\prime} - \varepsilon_{ai} - \varepsilon_{bj})] \qquad (6.78)$$

The overall rate constant is related to the detailed rate constant k_{ij}^{kl} by the relation

$$k(\beta) = \sum_{ij} \sum_{kl} \left(\frac{1}{Q_{ab}^{\text{int}}} \right) \exp[-\beta(\varepsilon_{ai} + \varepsilon_{bj})] k_{ij}^{kl} \qquad (6.79)$$

where Q_{ab}^{int} is the product of the internal partition functions of reactants. If the adiabaticity condition holds, eq. 6.79 reduces to

$$k(\beta) = \sum_{mn} \left(\frac{k_{mn}}{\beta h} \right) \left(\frac{Q_{\text{tr}}^{\ddagger}}{Q_{ab}} \right) \exp(-\beta \varepsilon_{mn}^{\ddagger\prime})$$

$$= (\beta h)^{-1} \left(\frac{Q_{\text{tr}}^{\ddagger}}{Q_{ab}} \right) \sum_{mn} k_{mn} \exp(-\beta \varepsilon_{mn}^{\ddagger\prime}) \qquad (6.80)$$

where $k_{mn} = \sum_{kl} k_{mn}^{kl}$. Introducing the average value for the transmission coefficients k and correcting the zero-point energy, we obtain

$$k(\beta) = \left(\frac{k}{\beta h} \right) \left(\frac{Q^{\ddagger}}{Q_{ab}} \right) \exp(-\beta \varepsilon_0^{\ddagger}) \qquad (6.81)$$

where $Q^{\ddagger} = Q_{\text{tr}}^{\ddagger} \sum_{mn} \exp(-\beta \varepsilon_{mn}^{\ddagger})$.

Now it should be noticed that the detailed specific rate constant k_{ij}^{kl} given earlier corresponds to the reaction from a particular internal state of reactants to an internal state of products. Generally the full details about the initial and final internal states are not obtainable. In this case the rate constants for partially resolved states of reactants and products $k_{(ij)}^{(kl)}$ are given by

$$k_{(ij)}^{(kl)} = (\beta h)^{-1}\left(\frac{Q_{tr}^{\ddagger}}{Q_{ab\,tr}}\right)\left(\frac{Q_{(ij)}^{\ddagger}}{Q_{ab(ij)}}\right)\sum_m k_{mn}^{(kl)} \exp[-\beta(\varepsilon_{mn}^{\ddagger\prime} - \varepsilon_{ai} - \varepsilon_{bj})] \quad (6.82)$$

where $Q_{(ij)}$ represents the partition function of the unspecified degrees of freedom.

An alternative way to arrive at the detailed specific rate constant k_{ij}^{kl} is as follows. In the equilibrium assumption the number of systems in an interval along the reaction coordinate between x and $x + dx$ with momentum lying between p and $p + dp$ is given by

$$dN = \left(\frac{C_a C_b}{Q_a Q_b}\right)\exp(-\beta E)\left(\frac{dx\,dp}{h}\right) \quad (6.83)$$

where E is the total energy and Q is the partition function per unit volume. Since we are concerned only with the rate of detailed reaction from the internal state (ij) of reactants to the internal state (kl) of products, the degrees of freedom of the motion of the center of mass of the system should be integrated over as eq. 6.83. It follows that

$$dC_m = \left(\frac{C_a C_b}{Q_a Q_b}\right)Q_{tr}^{\ddagger}\,\exp\left\{-\beta\left[\varepsilon_{ai} + \varepsilon_{bj} + \varepsilon_m + \left(\frac{p^2}{2\mu}\right)\right]\right\}\left(\frac{dx\,dp}{h}\right) \quad (6.84)$$

or

$$dC_m = \left(\frac{C_{ai} C_{bj}}{Q_{a\,tr} Q_{b\,tr}}\right)Q_{tr}^{\ddagger}\,\exp\left\{-\beta\left[\varepsilon_m + \left(\frac{p^2}{2\mu}\right)\right]\right\}\left(\frac{dx\,dp}{h}\right) \quad (6.85)$$

where ε_m represents the energy of the remaining two degrees of freedom mentioned. The rate of reaction from the initial state (ij) to the final state (kl) is obtained by multiplying the flux of systems past a point in the interval under discussion by the probability that those systems will react and taking into account the following possible paths:

$$-\left(\frac{dC_{ai}}{dt}\right)_{j,\,kl} = C_{ai} C_{bj}\frac{Q_{tr}^{\ddagger}}{Q_{ab\,tr}}\sum_m \exp(-\beta\varepsilon_m)\int_0^{\infty} k^{kl}(m, p)\exp\left(-\frac{\beta p^2}{2\mu}\right)\left(\frac{p\,dp}{\mu h}\right) \quad (6.86)$$

From eq. 6.86 we obtain K_{ij}^{kl} as

$$k_{ij}^{kl} = \frac{Q_{tr}^{\ddagger}}{Q_{ab\,tr}}\sum_m \exp(-\beta\varepsilon_m)\int_0^{\infty} k^{kl}(m, p)\exp\left(-\frac{\beta p^2}{2\mu}\right)\left(\frac{p\,dp}{\mu h}\right) \quad (6.87)$$

If $k^{kl}(m, p) = 0$ for $p < p_0$ and $k^{kl}(m, p) = k_m^{kl}$ for $p \geq p_0$, eq. 6.87 becomes

$$k_{ij}^{kl} = \left(\frac{Q_{tr}^{\ddagger}}{Q_{ab\,tr}}\right)(\beta h)^{-1} \sum_m k_m^{kl} \exp\left\{-\beta\left[\varepsilon_m + \left(\frac{p_0^2}{2\mu}\right)\right]\right\} \tag{6.88}$$

Equation 6.88 reduces to eq. 6.77 by the introduction of the adiabaticity condition.

6.4.2   Reaction Cross Sections and Rate Constants

It has been shown that for systems in which the distribution of the translational motion of the reaction molecules is of the Maxwell–Boltzmann type, the relation between the detailed species rate constant k_{ij}^{kl} and the reaction cross section δ_{ij} is given by

$$k_{ij}^{kl}(\beta) = \left(\frac{8\beta^3}{\pi\mu}\right)^{1/2} \int_0^{\infty} \sigma_{ij}^{kl}(E_t)E_t \exp(-\beta E_t)\, dE_t, \tag{6.89}$$

where E_t represents the relative translational energy. Equation 6.89 indicates that the detailed rate constant is related to the detailed reaction cross section by the Laplace transformation. By inverting the Laplace transformation in eq. 6.89, we obtain

$$E_t \sigma_{ij}^{kl}(E_t) = (2\pi i)^{-1} \int_{\gamma-i\infty}^{\gamma+i\infty} d\beta \, \exp(\beta E_t)\left(\frac{\pi\mu}{8\beta^3}\right)^{1/2} k_{ij}^{kl}(\beta) \tag{6.90}$$

The method of steepest descent can be applied to the integral involved in the inverse Laplace transformation in eq. 6.90.

To obtain the reaction cross section from the detailed rate constant of the absolute reaction rate theory, we substitute eq. 6.77 into eq. 6.90 to obtain

$$E_t \sigma_{ij}^{kl}(E_t) = \frac{h^2}{8\pi\mu}(2\pi i)^{-1} \int_{\gamma-i\infty}^{\gamma+i\infty} \frac{d\beta}{\beta} \sum_m k_{mn}^{kl} \exp[\beta(E_t + \varepsilon_{ai} + \varepsilon_{bj} - \varepsilon_{mn}^{\ddagger\prime})] \tag{6.91}$$

With the introduction of the integral representation of the Heaviside function $H(x)$, eq. 6.91 becomes

$$E_t \sigma_{ij}^{kl}(E_t) = \left(\frac{h^2}{8\pi\mu}\right) \sum_m k_{mn}^{kl} H(E_t + \varepsilon_{ai} + \varepsilon_{bj} - \varepsilon_{mn}^{\ddagger}) \tag{6.92}$$

If we replace k_{mn}^{kl} by an average value k_n^{kl} or if k_{mn}^{kl} is dependent on m, eq. 6.92 reduces to

$$E_t \sigma_{ij}^{kl}(E_t) = \left(\frac{h^2}{8\pi\mu}\right) k_n^{kl} W^{\ddagger}(E_t + \varepsilon_{ai} + \varepsilon_{bj} - \varepsilon_n^{\ddagger\prime}) \tag{6.93}$$

where

$$W^{\ddagger}(E_t + \varepsilon_{ai} + \varepsilon_{lj} - \varepsilon_n^{\ddagger\prime}) = \sum_m H(E_t + \varepsilon_{ai} + \varepsilon_{bj} - \varepsilon_{mn}^{\ddagger\prime})$$

the total number of states of the activated complexes with energy $E_t + \varepsilon_{ai} + \varepsilon_{bj} - \varepsilon_n^{\ddagger\prime}$ in the two degrees of freedom defined in the previous section. The total number of states of a system $W(E)$ is related to the partition function of the system (cf. Chapter 5).

$$W(E) = (2\pi i)^{-1} \int_{\gamma - i\infty}^{\gamma + i\infty} \frac{d\beta}{\beta} e^{\beta E} Q(\beta) \tag{6.94}$$

The calculation of $W(E)$ by the method of steepest descent is discussed elsewhere.

Now we turn to the relation between the over all rate constant and the reaction cross section. From eqs. 6.89 and 6.79 we have

$$k(\beta) = \left(\frac{8\beta^3}{\pi\mu}\right)^{1/2} (Q_{ab}^{int})^{-1} \sum_{ij} \exp[-\beta(\varepsilon_{ai} + \varepsilon_{bj})]$$

$$\times \int_0^{\infty} \sigma_{ij}(E_t) E_t \exp(-\beta E_t) \, dE_t, \tag{6.95}$$

If, as with Marcus, we assume that when $\sigma_{ij}(E_t) = \sum_{kl} \sigma_{ij}^{kl}(E_t)$. We let $E = E_t + \varepsilon_{ai} + \varepsilon_{bj}$ and $\varepsilon_{ij} = \varepsilon_{ai} + \varepsilon_{bj}$, then

$$\sigma_{ij}(E_t) E_t = S(E - \varepsilon_{ij}) \tag{6.96}$$

and substitution of eq. 6.96 into eq. 6.95 gives

$$k(\beta) = \left(\frac{8\beta^3}{\pi\mu}\right)^{1/2} \int_0^{\infty} S(E_t) \exp(-\beta E_t) \, dE_t \tag{6.97}$$

Carrying out the inverse Laplace transformation of eq. 6.97, we obtain

$$\sigma_{ij}(E_t) E_t = S(E - \varepsilon_{ij}) = (2\pi i)^{-1} \int_{\gamma - i\infty}^{\gamma + i\infty} dB \exp[\beta(E - \varepsilon_{ij})] \left(\frac{\pi\mu}{8\beta^3}\right)^{1/2} k(\beta) \tag{6.98}$$

If $k(\beta)$ is given by the absolute reaction rate theory,

$$(E - \varepsilon_{ij})\sigma_{ij}(E - \varepsilon_{ij}) = \frac{h^2}{8\pi\mu} (2\pi i)^{-1} \int_{\gamma - i\infty}^{\gamma + i\infty} \frac{d\beta}{\beta}$$

$$\times \exp[\beta(E - \varepsilon_{ij} - \varepsilon_0^{\ddagger})] \frac{Q_{\ddagger}^{int}(\beta)}{Q_{ab}^{int}(\beta)} k(\beta) \tag{6.99}$$

The same equation has been obtained by Marcus using another approach. This expression for the reaction cross section should be compared with that of the kinetic theory

$$\sigma(E_t)E_t = \frac{h^2}{8\pi\mu}(2\pi i)^{-1}\int_{\gamma-i\infty}^{\gamma+i\infty}\frac{d\beta}{\beta}\exp(\beta E_t)\frac{Q_{\ddagger}^{\text{int}}(\beta)}{Q_{ab}^{\text{int}}(\beta)}k(\beta) \quad (6.100)$$

Equation 6.100 holds only when $\delta_{ij}^{kl}(E_t)$ is independent of (ij) or

$$\sigma(E_t) = \left(\frac{1}{Q_{ab}^{\text{int}}}\right)\sum_{ij}\sum_{kl}\sigma_{ij}^{kl}(E_t)\exp(-\beta\varepsilon_{ij}) \quad (6.101)$$

is dependent on β. The contour integral in eq. 6.96 or 6.100 can be carried out by using the method of steepest descent. It follows that to the first-order approximation of the method of steepest descent, we have, for eq. 6.99,

$$\sigma_{ij}(E-\varepsilon_{ij})(E-\varepsilon_{ij}) = \left(\frac{h^2}{8\pi\mu}\right)\exp[\beta^*(E-\varepsilon_{ij}-\varepsilon_0^{\ddagger})]\left[\frac{Q_{\ddagger}^{\text{int}}\beta^*}{Q_{ab}^{\text{int}}(\beta^*)}\right]k(\beta^*)F(\beta^*) \quad (6.102)$$

where β^* represents the saddle point value of β determined from

$$\beta^{*-1} = E - \varepsilon_{ij} - \varepsilon_0^{\ddagger} + \left(\frac{\partial}{\partial\beta^*}\right)\left\{\log\left[\frac{Q_{\ddagger}^{\text{int}}(\beta^*)}{Q_{ab}^{\text{int}}(\beta^*)}\right]k(\beta^*)\right\} \quad (6.103)$$

and $F(\beta^*)$ is defined by

$$F(\beta^*) = \left[2\pi(1+\beta^{*2}\left(\frac{\partial^2}{\partial\beta^{*2}}\right)\left\{\log\left[\frac{Q_{\ddagger}^{\text{int}}(\beta^*)}{Q_{ab}^{\text{int}}(\beta^*)}\right]k(\beta^*)\right\})\right]^{-1/2} \quad (6.104)$$

Similar results can be obtained for $\delta(E_t)$.

In eq. 6.92, we sum over (ij) and (kl) and introduce the integral representation of the Heaviside function

$$\sum_{ij}\sum_{kl}(E-\varepsilon_{ij})\sigma_{ij}^{kl}(E-\varepsilon_{ij}) = \frac{h^2}{8\pi\mu}\sum_{ij}\sum_{kl}\sum_{m}k_{mn}^{kl}\frac{1}{2\pi i}\int_{\gamma-i\infty}^{\gamma+i\infty}\frac{d\beta}{\beta}\exp[\beta(E-\varepsilon_{mn}^{\ddagger\prime})] \quad (6.105)$$

or

$$\sum_{ij}(E-\varepsilon_{ij})\sigma_{ij}(E-\varepsilon_{ij}) = \frac{h^2}{8\pi\mu}(2\pi i)^{-1}\int_{\gamma-1\infty}^{\gamma+i\infty}\frac{d\beta}{\beta}e^{\beta E}\left[\sum_{m}\sum_{ij}k_{mn}\exp(-\beta\varepsilon_{mn}^{\ddagger\prime})\right] \quad (6.106)$$

where

$$\sigma_{ij} = \sum_{kl}\sigma_{ij}^{kl} \quad\text{and}\quad k_{mn} = \sum_{kl}k_{mn}^{kl}$$

If the adiabaticity condition holds, eq. 6.106 becomes

$$\sum_{ij} (E - \varepsilon_{ij})\sigma_{ij}(E - \varepsilon_{ij}) = \left(\frac{h^2}{8\pi\mu}\right) \sum_{mn} k_{mn} H(E - \varepsilon_{mn}^{\ddagger'}) \qquad (6.107)$$

Replacing k_{mn} by the average value k and introducing the definition of the weighted average cross section $\delta(E)$, we obtain

$$\bar{\sigma}(E) = \left(\frac{kh^2}{8\pi\mu}\right)\left[\frac{W^{\ddagger}(E - \varepsilon_0^{\ddagger})}{W_E(E)}\right], \qquad (6.108)$$

where $W_E(E)$ is the energy density function

$$W_E(E) = \sum_{ij} (E - \varepsilon_{ij})H(E - \varepsilon_{ij})$$

$$= (2\pi i)^{-1} \int_{\gamma - i\infty}^{\gamma + i\infty} \frac{d\beta}{\beta^2} e^{\beta E} Q_{ab}^{\text{int}}(\beta)$$

The contour integral in eq. 6.109 can again be evaluated by using the method of steepest descent.

Let us turn next to the collinear case, where the collision efficiency is expressed by the reaction probability $P_{ij}^{kl}(E_t)$ rather than the reaction cross section $\delta_{ij}^{kl}(E_t)$, and the relation between the detailed rate constant and the reaction probability is given by

$$k_{ik}^{kl}(\beta)_1 = \left(\frac{\beta}{2\pi\mu}\right)^{1/2} \int_0^\infty P_{ij}^{kl}(E_t)\exp(-\beta E_t)\, dE_t \qquad (6.110)$$

The absolute reaction rate constant for the collinear reaction is given by

$$k_{ij}^{kl}(\beta)_1 = (\beta h)^{-1}\left(\frac{\beta h^2}{2\pi\mu}\right)^{1/2} k_n^{kl} \exp[-\beta(\varepsilon_n^{\ddagger'} - \varepsilon_{ij})] \qquad (6.111)$$

Carrying out the inverse Laplace transformation of eq. 6.110 and substituting eq. 6.111 into the resulting equation, we obtain

$$P_{ij}^{kl}(E_t) = (2\pi i)^{-1} \int_{\gamma - i\infty}^{\gamma + i\infty} d\beta \exp(\beta E_t)\left(\frac{2\pi\mu}{\beta}\right)^{1/2} k_{ij}^{kl}(\beta)_1$$

$$= k_n^{kl}(2\pi i)^{-1} \int_{\gamma - i\infty}^{\gamma + i\infty} \frac{d\beta}{\beta} \exp[\beta(E_t + \varepsilon_{ij} - \varepsilon_n^{\ddagger'})] = k_n^{kl}, \qquad (6.112)$$

as is to be expected. Thus the reaction probability will never exceed unity. Summing eq. 6.112 over (ij) and (kl) yields

$$\sum_{kl} \sum_{ij} P_{ij}^{kl}(E - \varepsilon_{ij})$$

$$= \sum_{ij} P_{ij}(E - \varepsilon_{ij})$$

$$= (2\pi i)^{-1} \int_{\gamma - i\infty}^{\gamma + i\infty} \frac{d\beta}{\beta} e^{\beta E} \sum_{n} k_n \exp(-\beta \varepsilon_n^{\ddagger\prime})$$

$$= \sum_{n} k_n H(E - \varepsilon_n^{\ddagger\prime}) \qquad (6.113)$$

Here the adiabaticity condition has been used, and with the introduction of the average reaction probability $P(E)$, eq. 6.113 becomes

$$\bar{P}(E) = \sum_{n} \frac{k_n H(E - \varepsilon_n^{\ddagger\prime})}{W(E)} \qquad (6.114)$$

which reduces to

$$\bar{P}(E) = \frac{k[W^{\ddagger}(E - \varepsilon_0^{\ddagger})}{W(E)]} \qquad (6.115)$$

when k_n is replaced by an average quantity or when k_n is independent of n.

The experimentally determined cross sections are often the thermally averaged cross sections or rotationally averaged cross rather than the detailed cross sections $\delta_{ij}^{kl}(E_t)$ or the weighted-average cross sections $\bar{\delta}(E)$. In this case, the thermally averaged cross sections $\bar{\delta}(E_t)_{th}$ or rotationally averaged cross sections can be obtained from $\delta_{ij}^{kl}(E_t)$ given in eqs. 6.92 by the averaging relation; for example,

$$\sigma(E_t)_{th} = \sum_{ij} \sum_{kl} \left[\frac{\exp(-\beta \varepsilon_{ij})}{Q_{ab}^{int}} \right] \sigma_{ij}^{kl} \qquad (6.116)$$

and are, in general, temperature dependent.

6.4.3 Effect of Disallowed States

The purpose of this section is to show the limitation of the use of the rigid rotator–harmonic oscillator approximation in the calibration of equilibrium constants, absolute reaction rate constants, and reaction cross sections. For this purpose we first consider the dissociation of an ideal gas of the diatomic molecules AB,

$$AB \underset{}{\overset{K}{\rightleftharpoons}} A + B \qquad (6.117)$$

The equilibrium constant K can be expressed in terms of partition functions per unit volume by

$$K = \left(\frac{Q_A Q_B}{Q_{AB}}\right)\exp(-\beta D_0) \tag{6.118}$$

where D_0 is the dissociation energy of the AB diatomic molecule. In the rigid rotator–harmonic oscillator approximation, eq. 6.118 becomes

$$K = \frac{(2\pi\mu/\beta h^2)^{3/2}}{(8\pi^2 I/\sigma\beta h^2)[1/(1 - e^{-\beta h\nu})]} \frac{g_A g_B}{g_{AB}} \exp(-\beta D_0), \tag{6.119}$$

where the g factors represent the degeneracy factors of electronic states. Thus as $T\to\infty$ (or $\beta\to 0$), $K\to 0$. The reason for this contradiction of reality is that in calculating the partition function the rigid rotator–harmonic oscillator model fails to take into account the fact that the molecule dissociates with an energy beyond the dissociation energy. In other words, the effect of disallowed states is not properly taken into consideration in the calculation of the molecular partition function. To illustrate this point, for simplicity, let us consider the truncated rotator–oscillator model of molecules. For the AB molecule the internal partition function Q_{rv} according to this model is given by

$$Q_{rv}(\beta) = \left(\frac{1}{\sigma}\right)\sum_J\sum_v (2J + 1) \exp\{-\beta[BJ(J + 1) + v h\nu]\} \tag{6.120}$$

where $B = h^2/8\pi^2 I$. The summations over J and v should be carried out in such a way that the sum of rotational and vibrational energies should not exceed the dissociation energy. It is very difficult to obtain the exact analytical expression of Q_{rv} of the truncated rotator–oscillator model. To a very good approximation, Q_{rv} can be expressed as

$$Q_{rv}(\beta) = Q_v(\beta)\{Q_r(\beta) - Q_r(0)\exp[-\beta(D_0 + h\nu)]\} \tag{6.121}$$

where

$$Q_v(\beta) = (1 - e^{-\beta h\nu})^{-1}$$

$$Q_r(\beta) = \sigma^{-1} \sum_{J=0}^{Jm} (2J + 1)\exp[-\beta BJ(J + 1)]$$

$$Q_r(0) = \frac{(J_m + 1)^2}{\sigma}$$

and J_m represents the maximum rotational quantum number determined by the dissociation energy. The equilibrium constant K in this case is given by

$$K = \frac{(2\pi\mu/\beta h^2)^{3/2}}{Q_v(\beta)\{Q_r(\beta) - Q_r(0)\exp[-\beta(D_0 + h\nu)]\}} \frac{g_A g_B}{g_{AB}} \exp(-\beta D_0) \tag{6.122}$$

When $T \to \infty$, $Q_{rv}(\beta)$ and K become

$$Q_{rv}(\beta) = (\sigma h v)^{-1}[(J_m + 1)^2(D_0 + hv) - \tfrac{1}{2}BJ_m(J_m + 1)^2(J_m + 2)] \quad (6.123)$$

and

$$K = \frac{\sigma h v (2\pi\mu/\beta h^2)^{3/2}}{(J_m + 1)^2(D_0 + hv) - \tfrac{1}{2}BJ_m(J_m + 1)^2(J_m + 2)} \frac{g_A g_B}{g_{AB}} \exp(-\beta D_0) \to \infty$$

$$(6.124)$$

as is to be expected.

We turn next to the discussion of the absolute reaction rate constant of a bimolecular reaction. For simplicity we consider a bimolecular reaction of the form

$$A + BC \xrightarrow{\quad k \quad} AB + C \quad (6.125)$$

The absolute reaction rate constant in the rigid rotator–harmonic oscillator approximation is given by

$$k_{RH}(\beta) = \frac{k}{\beta h}$$

$$\frac{(8\pi^2 I_{\ddagger}/\sigma_{\ddagger}\beta h^2)[1 - \exp(-\beta h v_s)]^{-1}[1 - \exp(-\beta h v_b)]^{-1}[1 - \exp(-\beta h v_b')]^{-1}}{(2\pi\mu/\beta h^2)^{3/2}(8\pi^2 I/\sigma\beta h^2)(1 < e^{-\beta h v})^{-1}}$$

$$\times \frac{g_{\ddagger}}{g_A g_{BC}} \exp(-\beta \varepsilon_0^{\ddagger}) \quad (6.126)$$

assuming that the activated complex is linear, while the absolute reaction rate constant in the truncated rotator–oscillator approximation is given by

$$k_{TRH}(\beta) = \frac{k}{\beta h}$$

$$\frac{Q_{rv}^{\ddagger}(\beta)_s\{1 - \exp[-\beta(hv_b + v_{mb}hv_b)]\}/\{1 - \exp(-\beta h v_b)\}}{\{1 - \exp[-\beta(hv_b' + v_{mb}'hv_b')]\}/\{1 - \exp(-\beta h v_b')\}}$$

$$\times \frac{g_{\ddagger}}{g_A g_{BC}} \exp(-\beta \varepsilon_0^{\ddagger}) \quad (6.127)$$

assuming that, in the activated state, the rotation is coupled only with the stretching vibration. Thus, when $T \to \infty$, $k_{RH}(\beta) \to \infty$, and $k_{TRH}(\beta) \to 0$. The temperature dependence of the Arrhenius activation energy $\varepsilon_a^{\ddagger}$ is different for the rigid rotator–harmonic oscillator model and the truncated rotator–oscillator model. As $T \to \infty$, $\varepsilon_a^{\ddagger}(TRH) \to \varepsilon_0^{\ddagger} - \tfrac{1}{2}kT$, and $\varepsilon_a^{\ddagger}(RH) \to$

$\varepsilon_0^{\ddagger} + \frac{3}{2}kT$. This discrepancy is also reflected in the reaction cross section. To demonstrate this point we consider the weighted average reaction cross section $\bar{\sigma}(E)$. For this purpose we have to evaluate $W^{\ddagger}(E)$ and $W_E(E)$ (cf. eqs. 6.94 and 6.109). When $E \to \infty$, $Q_{\ddagger}^{int}(\beta)$ and $Q_{ab}^{int}(\beta)$ have the limiting forms in the truncated rotator–oscillator approximation,

$$Q_{\ddagger}^{int}(\beta) = \left(\frac{g_{\ddagger}}{\sigma_{\ddagger}hv_s}\right)[(J_m^{\ddagger} + 1)^2(D_{0s} + hv_s)$$
$$- \tfrac{1}{2}B_{\ddagger}J_m^{\ddagger}(J_m^{\ddagger} + 1)^2(J_m^{\ddagger} + 2)](1 + v_{mb})(1 + v_{mb}')$$

and

$$Q_{ab}^{int}(\beta) = \left(\frac{g_A g_{BC}}{\sigma hv}\right)[(J_m + 1)^2(D_0 + hv) - \tfrac{1}{2}BJ_m(J_m + 1)^2(J_m + 2)]$$

The substitution of eqs. 6.128 and 6.129 into eqs. 6.94 and 6.109, respectively, yields

$$W^{\ddagger}(E) = \left(\frac{g_{\ddagger}}{\sigma_{\ddagger}hv_s}\right)[J_m^{\ddagger} + 1)^2(D_{0s} + hv_s)$$
$$- \tfrac{1}{2}B_{\ddagger}J_m^{\ddagger}(J_m^{\ddagger} + 1)^2(J_m^{\ddagger} + 2)](1 + v_{mb})(1 + v_{mb}')$$

and

$$W_E(E) = \left(\frac{g_A g_{BC}}{\sigma hv}\right)[(J_m + 1)^2(D_0 + hv) - \tfrac{1}{2}BJ_m(J_m + 1)^2(J_m + 2)]E$$

Therefore the weighted-average cross section in the truncated rotator–oscillator approximation is given by

$$\bar{\sigma}(E)_{TRH} = \frac{kh^2}{8\pi\mu}$$

$$\frac{(g_{\ddagger}/\sigma_{\ddagger}v_s)[J_m^{\ddagger} + 1)^2(D_{0s} + hv_s) - \tfrac{1}{2}B_{\ddagger}J_m^{\ddagger}(J_m^{\ddagger} + 1)^2(J_m^{\ddagger} + 2)]}{(g_A g_{BC}/\sigma v)[J_m + 1)^2(D_0 + hv) - \tfrac{1}{2}BJ_m(J_m + 1)^2(J_m + 2)]E}(1 + v_{mb})(1 + v_{mb}')$$

as $E \to \infty$. Thus $\sigma(E)_{TRH} \to 0$, as $E \to \infty$. In the rigid rotator–harmonic oscillator approximation, Morokuma et al., (1969) have shown that $\sigma(E) \to \infty$ as $E \to \infty$. Similarly, we can show that as $E_t \to \infty$, the kinetic theory reaction cross section $\sigma(E_t)$ (cf. eq. 6.100) in the truncated rotator–oscillator approximation is of the same form as eq. 6.132 with E replaced by E_t. This energy dependence of the reaction cross section has been verified for the reactions $T + D_2$ and $T + H_2$ by using the results of Karplus et al (1966) (cf. Fig. 6.17). The small circles in Fig. 6.17 represent the data taken from the results of Karplus et al. As is to be expected from the above discussion, Fig. 6.17 indicates that, in the high-energy range, the reaction cross section varies linearly with E^{-1} (Karplus et al., 1966).

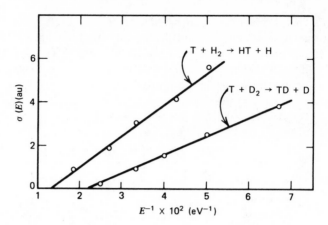

Fig. 6.17 The plot of $\sigma(E)$ versus E^{-1} in high energy range.

The numerical calculation of the weighted average cross section $\bar{\delta}(E)$ has been carried out by Lau et al. (1973) for the reaction $H + H_2$ with a linear transition complex in the RH and TRH models with H, H_2 parameters obtained from a semiempirical surface (see Chapter 2). Their calibrated results are shown in Fig. 6.18.

Figure 6.18 shows the comparison of the RH and TRH models; in the low-energy range the reaction cross sections obtained from these two models

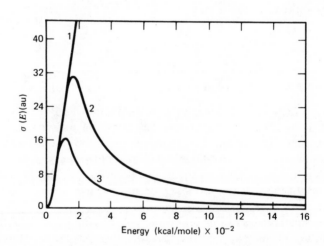

Fig. 6.18 The plot of $\sigma(E)$ versus E; curve 1-rigid rotator-harmonic oscillator model; of $D_{ob}^{\ddagger}$ 64 and $D_{ob}^{\ddagger}$ 35 kcal/mole, respectively.

agree with each other, and in the high-energy range, although the cross sections of the RH model increase with energy linearly, the cross sections of the TRH model converge as the energy increases to infinity in the manner of E^{-1}. This indicates that the reaction cross sections of the absolute reaction rate theory in the RH approximation are applicable only in the low-energy range and the absolute reaction rate constants in the RH approximation are valid only in the low-temperature range. In calculating the weighted average cross section $\bar{\delta}(E)$ of the TRH model, it is assumed that the rotation of the activated complex is coupled only with the stretching vibration of the activated complex. Figure 6.18 also shows the dependence of the cross sections on the dissociation energies of the activated complex. Curves 2 and 3 are obtained from the choices of $D_{ob}^{\ddagger} = 65$ and $D_{ob}^{\ddagger} = 35$ kcal/mole, respectively. The results in Fig. 6.18 suggest that in the high-energy range the reaction cross sections are rather sensitive to anharmonicities of normal vibrations of the reactant and activated complex.

REFERENCES

Berry, M. J. (1973), *J. Chem. Phys.*, **59**, 6229.

Brooks, P. R. (1973), *Discuss. Faraday Soc.*, **55**, 299.

Child, M. S. (1974), *Molecular Collision Theory*, Academic, p. 161.

Edelstein, S. A. and P. Davidovits (1971), *J. Chem. Phys.*, **55**, 5164.

Eliason, M. A. and J. O. Hirschfelder (1959), *J. Chem. Phys.*, **30**, 1426.

Emmerich, W. S. (1963), in *Fast Neutron Physics*, Part II, J. B. Marion, and J. L. Fowler, Eds., Wiley, p. 1057.

Gerschbach, D. R. (1966), *Adv. Chem. Phys.*, **10**, 319.

Gillen, K. T., A. M. Rulis, and R. B. Bernstein (1971), *J. Chem. Phys.*, **54**, 2831.

Green, E. F., A. L. Moursund, and J. Ross (1966), *Adv. Chem. Phys.*, **10**, 135.

Harris, R. M. and J. F. Wilson (1971), *J. Chem. Phys.*, **54**, 2088.

Henglein, A. (1974), *Physical Chemistry, Adv. Treatise*, **6B**, 509, Academic.

Johnston, H. S. (1966), *Gas-phase reaction rate theory*, Ronald.

Karplus, M., R. N. Porter, and R. D. Sharma (1966), *J. Chem. Phys.*, **45**, 3871.

Kinsey, J. L., G. H. Kwei, and D. R. Herschbach (1976), *J. Chem. Phys.*, **64**, 1914.

Laidler, K. J. (1969), *Theories of Chemical Reaction Rates*, McGraw-Hill.

Lau, K. H., S. H. Lin, and J. Eyring (1973), *J. Chem. Phys.*, **58**, 1261.

Lee, Y. T., J. D. Mcdonald, P. R. LeBreton, and D. R. Herschbach (1969), *Rev. Sci. Instrum.*, **40**, 1402.

Levine, R. D. and R. B. Bernstein (1964), *Molecular Reaction Dynamics*, Oxford University Press.

Lin, S. H. and H. Eyring (1971), *Proc. Natl. Acad. Sci. U.S.*, **68**, 402.

Marcus, R. A. (1970), *J. Chem. Phys.*, **53**, 604; and the references given therein.

Miller, W. B., S. A. Safron, and D. R. Herschbach (1969), *Discuss. Faraday Soc.*, **44**, 108.

Morokuma, K., B. C. Eu, and M. Karplus (1969), in *Kinetic Process in Gas and Plasmas*, A. R. Hochstion, Ed., Academic.

Muckerman, J. T. (1971), *J. Chem. Phys.*, **54**, 115.

Odiorne, T. J., P. R. Broods, and J. V. V. Kasper (1971), *J. Chem. Phys.*, **55**, 1980.

Olson, R. E., F. T. Smith, and E. Bauer (1970), *Appl. Opt.*, **10**, 1848.

Parson, J. M. and Y. T. Lee (1972), *J. Chem. Phys.*, **56**, 4658.

Polanyi, J. C. and K. B. Woodall (1972), *J. Chem. Phys.*, **57**, 1574.

Pruett, J. G., F. R. Grabiner, and P. R. Broods (1974), *J. Chem. Phys.*, **60**, 3335.

Riley, S. J. and D. R. Herschbach (1973), *J. Chem. Phys.*, **58**, 27.

Ross, J. and P. Mazur (1961), *J. Chem. Phys.*, **35**, 19.

Rulis, A. M. and R. B. Berstein (1972), *J. Chem. Phys.*, **57**, 5497.

Safron, S. A., N. D. Weinstein, D. R. Herschbach, and J. C. Tully (1972), *Chem. Phys. Lett.*, **12**, 564.

Schafer, T. P., P. E. Siska, J. M. Parson, F. P. Tully, Y. C. Wong, and Y. T. Lee (1970), *J. Chem. Phys.*, **53**, 3385.

Sheeler, J. A. (1963), in *Fast Neutron Physics*, Part II, J. B. Marion and J. L. Fowler, Eds., Wiley.

Toennies, J. P. (1974), *Phys. Chemistry, Adv. Treatise*, **6A**, 227, Academic.

Waage, E. V. and B. S. Rabinovitch (1970), *Chem. Rev.*, **70**, 377.

Weston, R. E. and H. A. Schwarz (1972), *Chemical Kinetics*, Prentice-Hall.

Seven

Elementary Processes in Photochemistry (I)

CONTENTS

7.1   INTRODUCTION

Let us consider a molecule D that has only two electronic states within the energy range that is of interest to us. After absorbing the incident radiation, the molecule D is excited

$$D \xrightarrow[k_r]{k_a} D^* \qquad \text{(absorption)} \tag{7.1}$$

the excited D^* will emit light and/or return to the ground state by the radiation less transition

$$D^* \xrightarrow{k_r} D + h\nu \qquad \text{(emission)} \tag{7.2}$$

$$D^* \xrightarrow{k_{nr}} D \qquad \text{(nonradiative process)} \tag{7.3}$$

These three processes are referred to as elementary (or primary) processes in photochemistry.

From eqs. 7.1–7.3, the rate expression for (D^*) can be found as

$$\frac{d(D^*)}{dt} = k_a(D) - (k_r + k_{nr})(D^*) \tag{7.4}$$

If the steady state is reached, then

$$(D^*)_s = \frac{k_a}{k_r + k_{nr})}(D) \tag{7.5}$$

Here it has been assumed that $(D^*)_s$ is negligibly small compared with (D). Removing the excitation source, eq. 7.4 then becomes

$$\frac{d(D^*)}{dt} = -(k_r + k_{nr})(D^*) \tag{7.6}$$

which can be integrated

$$(D^*) = (D^*)_s \exp[-(k_r + k_{nr})t] = (D^*)_s \exp[-(t/\tau)]. \tag{7.7}$$

Here the time origin is chosen at the steady state. Notice that

$$\frac{1}{\tau} = k_r + k_{nr} \tag{7.8}$$

The term τ is the lifetime of D^* and $\tau_r = 1/k_r$ is usually referred to as the natural (or radiative) lifetime of D^*. If we let I represent the emission intensity at time t, the eq. 7.7 indicates that the plot of log I versus t is a straight line with the slope $-(1/\tau)$, which gives us one way to determine the lifetime τ.

The quantum yield ϕ is defined as the ratio of the number of photons emitted to the number of photons absorbed. In this simple case, from eq. 7.5 we can see that the total number of photons emitted will be proportional to $k_r(D^*)_s$ and the total number of photons absorbed will be proportional to $k_a(D)$. It follows that

$$\phi = \frac{k_r}{k_r + k_{nr}} \tag{7.9}$$

From the lifetime and quantum yield measurements, one can determine both k_r and k_{nr}.

It has been found that under normal conditions the lifetime τ is independent of the wavelength of the incident excitation radiation. This is due to the fact that the vibrational relaxation that represents another elementary process is much faster than the radiative and nonradiative processes; therefore the vibrational equilibrium is established before the emission takes place (Lin, 1972a; Freed, 1976). Other cases of molecular luminescence where there is more than one electronic excited state within the energy range of interest can be discussed similarly (Turro, 1965). Fluorescence is usually referred to as the emission process between electronic states of like multiplicity, whereas phosphorescence corresponds to the emission process between electronics of different multiplicity.

Now suppose that in addition to D another molecular species A exists that can react or interact with D^*:

$$D^* + A \xrightarrow{k} B + C \tag{7.10}$$

If B and C represent molecular species that are different from A and D, then eq. 7.10 represents a photochemical reaction; and if $B = D$ and $C = A^*$ (or A), then eq. 7.10 represents the electronic energy transfer (or quenching), another elementary process.

The simplest case discussed shows us how the elementary processes are involved in photochemistry. In this and the next chapter we discuss the nature and mechanism of these elementary processes.

7.2 ABSORPTION AND EMISSION

The elementary processes absorption and emission are quite well known, and their derivations can be found in most textbooks on quantum chemistry (see, for example, Eyring et al., 1944). In this section we only sketch their derivations.

We start with the time-dependent Schrödinger equation

$$\hat{H}\Psi = i\hbar \frac{\partial \Psi}{\partial t} \tag{7.11}$$

where $\hat{H} = \hat{H}_0 + \hat{H}'$. If $\Psi_n^0(q, t)$ represents the unperturbed eigenfunctions of $\hat{H}_0$, $\Psi(q, t)$ can then be expanded in the complete set of $\Psi_n^0(q, t)$:

$$\Psi(q, t) = \sum_n (C_n H)\Psi_n^0(q, t) \tag{7.12}$$

Substituting eq. 7.12 into eq. 7.11 yields

$$\frac{dC_m}{dt} = \frac{1}{i\hbar} \sum_n C_n \langle \Psi_m^0 | \hat{H}' | \Psi_n^0 \rangle \tag{7.13}$$

This is an exact relation. Thus in any particular problem we have a set of simultaneous differential equations that can be solved for $(C_n^1 s)$. For this purpose the perturbation method can often be used (Schiff, 1968; Heitler, 1953); for example, in the first-order approximation, if the system is at the kth state initially, then *eq.* 7.13 becomes

$$\frac{dC_m}{dt} = \frac{1}{i\hbar} \langle \Psi_m^0 | \hat{H}' | \Psi_k^0 \rangle \tag{7.14}$$

For one-photon processes eq. 7.14 is sufficient; for multiphoton processes the high-order approximations have to be used (Lin and Eyring, 1977).

For induced absorption of radiation the semiclassical theory can be employed; in this case $\hat{H}'$ is given by

$$\hat{H}' = -\sum_j \frac{e_j}{mjc} \mathbf{A}_j \cdot \mathbf{p}_j \tag{7.15}$$

where $\mathbf{p}_j$ is the linear momentum of the jth particle and $\mathbf{A}_j$ represents the potential of radiation acting on the jth particle. Substituting eq. 7.15 into eq. 7.14 and ignoring the coordinate dependence of $\mathbf{A}_j$, we obtain

$$\frac{dC_m}{dt} = -\frac{1}{c\hbar} (\mathbf{A} \cdot \mathbf{R}_{mk}) \omega_{mk} \exp(it\omega_{mk}) \tag{7.16}$$

where R_{mk} represents the matrix element of electric dipole moment

$$\mathbf{R}_{mk} = \left\langle \psi_m \left| e \sum_i \mathbf{r}_i \right| \psi_k \right\rangle, \quad \psi_n^0(q, t) = \psi_n(q)\exp\left(-\frac{it}{\hbar} E_n \right)$$

and

$$\omega_{mk} = \frac{(E_m - E_k)}{\hbar}$$

If the light has the frequency ω, the time dependence of $\mathbf{A}$ may be expressed as

$$\mathbf{A} = \tfrac{1}{2}\mathbf{A}^0[\exp(it\omega) + \exp(-it\omega)] \tag{7.17}$$

Using this relation, eq. 7.16 can be integrated

$$C_m = \frac{i\omega mk}{2c\hbar}(\mathbf{A}^0 \cdot \mathbf{R}_{mk}) \frac{\exp[it(\omega_{mk} - \omega)] - 1}{\omega_{mk} - \omega} + \frac{\exp[it(\omega_{mk} + \omega)] - 1}{\omega_{mk} + \omega} \tag{7.18}$$

For absorption the second term in eq. 7.18 is unimportant and the probability that the system will be in the state m at time t is the product $|c_m|^2$, that is,

$$|C_M|^2 = \frac{\omega_{mk}^2}{c^2\hbar^2}|\mathbf{A}^0 \cdot \mathbf{R}_{mk}|^2 \frac{\sin^2[(\omega_{mk} - \omega)t/2]}{(\omega_{mk} - \omega)^2} \tag{7.19}$$

Using the representation for the delta function

$$\delta(X) = \frac{1}{\pi}\lim_{K \to \infty}\frac{1}{K} \cdot \frac{1 - \cos K\chi}{\chi^2} \tag{7.20}$$

in the long time limit we obtain the transition probability of absorption as

$$S_{mk} = \frac{|c_m|^2}{t} = \frac{\pi\omega_{mk}^2}{2c^2\hbar^2}|\mathbf{A}^0 \cdot \mathbf{R}_{mk}|^2\,\delta(\omega_{mk} - \omega) \tag{7.21}$$

If the effect of radiation damping has been included, then the delta function is replaced by the Lorentzian (Heitler, 1953; Lin and Eyring, 1977; Lin, 1966b) and S_{mk} is expressed as

$$S_{mk} = \frac{\pi\omega_{mk}^2}{2c^2\hbar^2}|\mathbf{A}^0 \cdot \mathbf{R}_{mk}|^2 \frac{1}{\pi} \cdot \frac{\Gamma_{mk}}{(\omega_{mk} - \omega)^2 + \Gamma_{mk}^2} \tag{7.22}$$

where Γ_{mk} represents the damping constant.

So far we have considered only a single frequency. If the radiation source is not monochromatic, to obtain the correct S_{mk} we have to integrate over a range of frequencies. It follows that

$$S_{mk} = \frac{\pi\omega_{mk}^2}{2c^2\hbar^2}|\mathbf{A}^0(\omega_{mk})\mathbf{R}_{mk}|^2 \tag{7.23}$$

For the isotropic system the spatial average of eq. 7.23 has to be carried out; S_{mk} becomes

$$S_{mk} = \frac{\pi\omega_{mk}^2}{6c^2\hbar^2}|\mathbf{A}^0(\omega_{mk})|^2|\mathbf{R}_{mk}|^2 \tag{7.24}$$

We may express $|A^0(\omega_{mn})|^2$ in terms of the radiation density $\rho(\omega_{mn})$

$$\rho(\omega_{mn}) = \frac{1}{4\pi} \langle |E(\omega_{mn})|^2 \rangle = \frac{\omega_{mn}^2}{8\pi c^2} |A^0(\omega_{mn})|^2 \qquad (7.25)$$

Eliminating $|A^0(\omega_{mn})|^2$ from eq. 7.24 yields

$$S_{mk} = \frac{4\pi^2}{3h^2} |R_{mk}|^2 \delta(\omega_{mk}) = \frac{2\pi}{3h^2} |R_{mk}|^2 \rho(\gamma_{mk}) = B_{k \to m} \rho(\gamma_{mk}) \quad (7.26)$$

where $\rho(\omega_{mk})$ represents the radiation density per unit ω, whereas $\rho(\gamma_{mk})$ represents the radiation density per unit v and $B_{k \to m}$ is the Einstein B coefficient. Equation 7.26 shows that the absorption rate constant S_{mk} is proportional to the light intensity (see eq. 7.1).

If the system is originally in the excited state m, then the same treatment shows that the probability of transition to the state k resulting in the emission due to the perturbing effect of the electromagnetic field is

$$B_{m \to k} \rho(v_{mk}) = B_{k \to m} \rho(\gamma_{mk}) \qquad (7.27)$$

Since a system in an excited state can emit radiation even in the absence of an electromagnetic field, the completion of the theory of radiation requires the calculation of the transition probability $A_{m \to k}$ for spontaneous emission (or the radiative rate constant). This has been accomplished by Einstein by a consideration of the equilibrium between two states of different energy. If the number of systems in the state with energy E_m is N_m and the number in the state with energy E_k is N_k, then at equilibrium we have

$$N_m[A_{m \to k} + B_{m \to k} \rho(v_{mk})] = N_k B_{k \to m} \rho(v_{mk}) \qquad (7.28)$$

or

$$\frac{Nm}{Na} = \exp\left(-\frac{h\gamma_{mk}}{kT}\right) = \frac{B_{k \to m} \rho(v_{mk})}{A_{m \to k} + B_{m \to k} \rho(v_{mk})} \qquad (7.29)$$

Solving for $\rho(v_{mk})$ and using the Planck radiation distribution for $\rho(v_{mk})$, we obtain

$$\rho(v_{mk}) = \frac{A_{m \to k}/B_{m \to k}}{\exp(hv_{mk}/kT) - 1} = \frac{8\pi l v_{mk}^3}{c^3} \cdot \frac{1}{\exp(hv_{mk}/kT) - 1} \qquad (7.30)$$

and

$$A_{m \to k} = \frac{8\pi h v_{mk}^3}{c^3} B_{m \to k} = \frac{32\pi^3 v_{mk}^3}{3c^3 h} |R_{mk}|^2 \qquad (7.31)$$

For molecular systems the adiabatic approximation can be used (see Chapter 2) and if the vibrational relaxation is fast compared with the radiative process, then the vibrational equilibrium is established within the excited electronic state before the emission takes place. In this case the radiative rate constant is given by

$$A_{b \to a} = \frac{32\pi^3}{3c^3 h} \sum_{v'} \sum_{v''} P_{bv'} |R_{bv', \, av''}|^2 v_{bv', av''}^3 \qquad (7.32)$$

where $P_{bv'}$ represents the Boltzmann factor, (a, b) refers to electronic states and (v', v'') refers to quantum states of nuclear motion. Notice that

$$R_{bv', \, av''} = \langle \Phi_b \theta_{bv'} | R | \Phi_a \theta_{av''} \rangle = \langle \theta_{bv'} | R_{ba} | \theta_{av''} \rangle \qquad (7.33)$$

where $R_{ba} = \langle \Phi_n | R | \Phi_a \rangle$ is the electronic transition moment for $b \to a$. The term R_{ba} is in general a function of nuclear coordinates. For an allowed electronic transition R_{ba} can approximately be regarded as constant, that is, $R_{bv', \, av''} \doteq R_{ba} \langle \theta_{vb'} | \theta_{av''} \rangle$. In this case, if the factor $v_{bv', \, av''}^3$ is replaced by an average value $\langle v^3 \rangle_{ba}$, then eq. 7.32 reduces to

$$A_{b \to a} = \frac{32\pi^3}{3c^3 h} |R_{ba}|^2 \langle v^3 \rangle_{ba} \qquad (7.33a)$$

For symmetry-forbidden transitions the nuclear coordinate dependence of R_{ba} becomes important and various methods have been developed for this purpose (Lee et al., 1977; Ziegler and Albrecht, 1974; Orlandi and Siebrand, 1973). Equation 7.32 for $A_{b \to a}$ should be modified if one is interested in the radiative rate constant of a molecule dissolved in a medium; the medium effect should include the correction of the effective field and light velocity.

For the purpose of our later discussion it is useful to have expressions for the absorption coefficient and intensity distribution function of emission spectra. Using eqs. 7.21, 7.25, and the Beer–Lambert law, we obtain the expression for the molecular absorption coefficient of the electronic transition $a \to b$ for an isotropic system (Lin, 1968; Hill and Lin, 1971):

$$k_{ab}(\omega) = \frac{4\pi^2 \omega}{3\hbar \alpha_a c} \sum_{v'} \sum_{v''} P_{av'} |\langle \theta_{av'} | R_{ab} | \theta_{bv''} \rangle|^2 \, \delta(\omega_{bv'', \, av'} - \omega) \qquad (7.34a)$$

where α_a is a function of the refractive index introduced to correct the medium effect. In deriving eq. 7.34a the adiabatic approximation has been used. The relation between the molecular absorption coefficient $k_{ab}(\omega)$ and the conventional molar extinction coefficient $\varepsilon(\omega)$ is given by

$$\varepsilon(\omega) = (10^{-3} N_A \log_{10} e) k_{ab}(\omega) \qquad (7.34b)$$

where N_A is Avogadro's number. Similarly, the normalized intensity distribution function of emission spectra for the electronic $a \to b$ in the adiabatic approximation can be derived (Lin et al., 1971) and is given by

$$I_{ab}(\omega)_N = \frac{4\alpha_e\omega^3}{3c^3A_{ab}} \sum_{v'} \sum_{v''} P_{av'} |\langle\theta_{av'}|\mathbf{R}_{ab}|\theta_{bv''}\rangle|^2 \, \delta(E_{av',bv''} - \hbar\omega) \quad (7.35)$$

where A_{ab} represents the radiative rate constant and α_e is the correction factor for the medium effect.

For an allowed electronic transition the radiative rate constant $A_{b\to a}$ can be determined from the absorption and emission spectra of the same electronic transition:

$$A_{b\to a} = \frac{\alpha_a\alpha_e}{\pi^2 c^2} \cdot \frac{\int k_{ab}(\omega)/\omega \, d\omega}{\int I_{ba}(\omega)_N/\omega^3 \, d\omega} \quad (7.36)$$

This is usually called the Strickler–Berg relation (Strickler and Berg, 1962). In other words, in this case the average value $\langle\omega^3\rangle_{ba}$ in eq. 7.33a is given by

$$\langle\omega^3\rangle_{ba} = \frac{1}{\int I_{ba}(\omega)_N/\omega^3 \, d\omega} \quad (7.37)$$

To show the application of eq. 7.33a for the calculation of the radiative lifetimes, in Table 7.1 are shown the calculated radiative lifetimes of the

Table 7.1
The Calculated Lifetimes of the Lowest Triplet State

| | | Calculated radiative lifetimes | | |
| | | Unscaled transition moments (sec) | Transition moments scaled according to Mulliken (sec) | Experimental radiative lifetimes (sec) |
Molecule	$E(T_0) - E(S_0)$ (cm^{-1})			
Benzene	29,000	23.2	77	—
Phenanthrene	21,700	1.5	5	26
Naphthalene	21,000	3.3	11	76
Phenalenylium	15,800	2.7	9	—
Anthracene	13,800	25.4	85	182
Azukene	11,800	35.3	118	—

$\pi \rightarrow \pi^*$ phosphorescent state of a number of aromatic hydrocarbons (Veeman and van der Waals, 1970). The experimental data (Langelaar, 1969) are also presented for comparison. The calculation of the radiative lifetimes has been carried out for π electrons using the (LCAO MO) theory. Veeman and van der Waals have attributed the uncertainty in their calculations to the neglect of interaction between σ electrons in different CC and CH bonds. To improve the results they scale their transition moments by the factor $(0.30)^{1/2}$ as suggested by Mulliken and Rieke (1941). Even after this scaling the calculated values of the radiative lifetimes are lower than those derived from a combination of phosphorescence decay and quantum yield experiments (Langelaar, 1969), but their general trend seems to be correct.

7.3 RADIATIONLESS TRANSITIONS

7.3.1 General Considerations

In molecular electronic spectroscopy *internal conversion* is defined as the rapid radiationless transitions of electronic states of like multiplicity, whereas *intersystem crossing* is defined as the spin-orbit coupling-dependent radiationless transitions (i.e., transitions between the electronic states of different multiplicity) in molecules (Kasha and McGlynn, 1957). At 4.2°K, the phosphorescent lifetimes of C_6D_6 in CH_4, Ar, Kr, and Xe were found to be 22, 25.8, 1.1, and 0.08 sec, respectively, which is thought to be due to the effect of a heavy-atom environment on multiplicity-forbidden transitions. This is usually referred to as the *external heavy-atom effect*. Similarly, it was found that these phosphorescent-lifetime measurements, which were made at low temperatures in low-molecular-weight solvents free from oxygen, give C_6H_6 (15.8 sec), C_6D_6 (25.8 sec), $C_{10}H_8$ (2.1 sec), and $C_{10}D_8$ (16.9 sec). This is referred to as the *deuterium effect* (or *isotope effect*). The lifetimes of electronic states also show the temperature effect; that is, they decrease with increasing temperature.

The theory of radiationless transitions described here is based on the breakdown of the Born–Oppenheimer adiabatic approximation (Lin, 1966a, Freed, 1976; Lin, 1978). It is assumed that the vibrational relaxation time is much shorter than that of the electronic relaxation so that the transitions always originate from a Boltzmann distribution of vibrational levels. In the cases where the vibrational relaxation time is not much shorter than that of the electronic relaxation, the thermal average over the vibrational levels cannot be taken and the vibrational relaxation has to be taken into account explicitly (Lin, 1972; Lin and Eyring, 1974).

We consider the system to be composed of a single molecule embedded in the heat bath. For convenience we assume that the condensed phase is considered here; the final results can also be used for the gaseous system. Thus the Hamiltonian for the system can be written as follows:

$$\hat{H} = \hat{T} + \hat{h}_s + \sum_a \hat{h}_\alpha + \sum_{\alpha > \beta} V_{\alpha\beta} + \sum_a V_{s\alpha} \tag{7.38}$$

where $\hat{T}$ is the kinetic energy operator of all nuclear motion, including both intra- and intermolecular vibration of the solute and solvent molecules, $\hat{h}s$ and $\sum_\alpha \hat{h}_\alpha$ are the electronic energy operators for the internal state of the solute and solvent molecules, respectively, and $\sum_{\alpha\to\beta} v_{\alpha\beta}$ and $\sum_\alpha v_{s\alpha}$, respectively, represent the potential energy of the solvent–solvent and the solute–solvent interactions. Using the adiabatic approximation, the state of the system is described by

$$\psi_{av} = \theta_{av}\Phi_a \tag{7.39}$$

where Φ_a is the electronic wave function of the system and θ_{av} designates the wave function of both intra- and intermolecular vibrations of the system.

In the adiabatic approximation an electron does not make transitions from one state to others; instead, an electronic state itself is deformed progressively by the nuclear displacements. In other words, Φ_a and θ_{av} are the solutions of the following Schrödinger equations (see Chapter Two)

$$\left(\hat{h}_s + \sum_\alpha \hat{h}_\alpha + \sum_{\alpha > \beta} U_{\alpha\beta} + \sum_a V_{s\alpha}\right)\Phi_a = U_a\Phi_a \tag{7.40}$$

and

$$[\hat{T} + U_a(R)]\theta_{av} = E_{av}\theta_{av} \tag{7.41}$$

where $U_a(R)$ is the adiabatic potential of the ath electronic state and v denotes the overall vibrational state of the nuclei. From the physical point of view, although the eigenstates described by eq. 7.39 are to be regarded as good, they are, however, not stationary in the exact sense, and the whole system can make the transition from one electronic state to another accompanied by a transition in the quantum states of nuclear motion. It is clear that in order to consider the radiationless transitions, we must look for the perturbation causing the transition between different electronic states in accordance with the approximate nature of the wave functions given by eq. 7.39. The perturbation on the system for such a process can be shown to be given by (Lin, 1966; Bixon and Jortner, 1968)

$$\hat{H}'_{BO}\psi_{av} = T\Phi_a\theta_{av} - \Phi_a\hat{T}\theta_{av} \tag{7.42}$$

If the kinetic energy operator of nuclear motion $\hat{T}$ is expressed in terms of normal coordinates, then eq. 7.42 can be written as

$$\hat{H}'_{BO}\psi_{au} = -\hbar^2 \sum_i \frac{\partial \Phi_a}{\partial Q_i} \cdot \frac{\partial \theta_{au}}{\partial Q_i} - \frac{\hbar^2}{2} \sum_i \theta_{au} \frac{\partial^2 \Phi_a}{\partial Q_i^2} \qquad (7.43)$$

where the Q_i's are normal coordinates including both intra- and inter-molecular vibrations.

From the time-dependent perturbation theory the transition probability from the state (bv') to a final state (av'') is given by the golden rule

$$W(bv' \to av'') = \frac{2\pi}{\hbar} |\langle av'' | \hat{H}'_{BO} | bv' \rangle|^2 \delta(E_{av'} - E_{bv''}) \qquad (7.44)$$

where $\delta(E_{av'} - E_{bv''})$ is the delta function. Since we assume that the vibrational relaxation is much faster than the electronic relaxation, a general result requires a Boltzmann average of the transition probability over all thermally available vibrational states of the solute and solvent molecules. Hence the total transition probability is given by summing eq. 7.44 over all initial vibrational states v' weighted by their Boltzmann factor and then summing over all final vibrational states v'' consistent with the energy conservation:

$$W(b \to a) = \frac{2\pi}{\hbar} \sum_{v'} \sum_{v''} P_{v'} |\langle av'' | H'_{BO} | bv' \rangle|^2 \, \delta(E_{av''} - E_{bv'}) \qquad (7.45)$$

where $P_{v'}$ is the Boltzmann factor. To evaluate the rate of radiationless transitions given by eq. 7.45, we assume that vibrations are harmonic. It follows that

$$P_{v'} = \prod_i \left(2 \sinh \frac{\hbar \omega'_i}{2kT} \right) \exp\left[-(v'_i + \tfrac{1}{2}) \frac{\hbar \omega'_i}{kT} \right] \qquad (7.46)$$

and

$$\theta_{av''} = \prod_i X_{av''_i}(Q''_i), \qquad \theta_{bv'} = \prod_i X_{bv'_i}(Q'_i) \qquad (7.47)$$

where $X_{av''_i} X_{bv'_i}$, and so on, are the wave functions of harmonic oscillators:

$$X_{v_i}(Q_i) = \left(\frac{\beta_i}{\sqrt{\pi} 2^{v_i} v_i!} \right)^{1/2} H_{v_i}(\beta_i Q_i) \exp(-\tfrac{1}{2}\beta_i^2 Q_i^2) \qquad (7.48)$$

where $\beta_i = (\omega_i/\hbar)^{1/2}$ and H_{v_i} is the Hermite polynomial. As is shown later, only those normal modes that have changes (or displacements) in either normal coordinates or frequencies between the electronic states concerned can contribute to the radiationless transition probability. To take into account the displacements of both normal coordinates and frequencies between the electronic states, we can express them in general as (Markham, 1959; Lin, 1966)

$$Q_i' = Q_i - d_i', \qquad Q_i'' = Q_i - d_i'' \qquad (7.49)$$

and

$$\omega_i'' = \omega_i'(1 - \rho_i) \qquad (7.50)$$

if the changes d_i', d_i'', and ρ_i are small.

In eqs. 7.49 and 7.50 it is implicitly assumed that the same classification for normal coordinates can be applied to the two different electronic states. The anharmonicity of vibrations is neglected here. In general, anharmonicities have two roles; the one is to scramble the various normal modes and the other is to cause changes in vibrational quantum numbers of a normal mode. The first effect is important in vibrational relaxation and the second effect may cause some changes in the Franck–Condon factors and hence affect the rate of radiationless transitions.

To simplify eq. 7.45 it is convenient to introduce the integral expression for the delta function

$$\delta(E_{av''} - E_{bv'}) = \frac{1}{2\pi\hbar} \int_{-\infty}^{\infty} dt \, \exp\left[\frac{it}{\hbar}(E_{av''} - E_{bv'})\right]$$

$$= \frac{1}{2\pi\hbar} \int_{-\infty}^{\infty} dt \, \exp(it\omega_{ab}) \prod_{i=1}^{N}$$

$$\times \exp\{it[v_i'' + \tfrac{1}{2})\omega_i'' - (v_i' + \tfrac{1}{2})\omega_i']\} \qquad (7.51)$$

where $\hbar\omega_{ab}$ is the difference in the electronic energies between the two electronic states; for the case of the relaxation of the electronic energy to the vibrational energy, it is negative. From eq. 7.43 we have the matrix elements for the perturbation factor due to the ith promoting mode (Lin and Bersohn, 1968) as

$$\langle av'' | \hat{H}_{BO}(i) | bv' \rangle = -\hbar^2 \langle \Phi_a \theta_{av''} \left| \frac{\partial \Phi_b}{\partial Q_i} \cdot \frac{\partial \theta_{bv'}}{\partial Q_i} \right. - \frac{\hbar^2}{2}$$

$$\times \langle \Phi_a \times \theta_{av''} | \theta_{bv'} (\partial^2 \Phi_b / \partial Q_i^2) \qquad (7.52)$$

If the derivatives of the initial electronic state vary slowly with the vibrational coordinate Q_i, we can apply the Condon approximation (Lax, 1952). Under this condition the second term on the righthand side of eq. 7.52 may be ignored and eq. 7.52 becomes

$$\langle av'' | \hat{H}_{BO}(i) | bv' \rangle = \cdot R_i(ab) \left\langle \theta_{av''} \left| \frac{\partial}{\partial Q_i} \right| \theta_{bv'} \right\rangle \qquad (7.53)$$

where

$$R_i(ab) = -\hbar^2 \left\langle \Phi_a \left| \frac{\partial}{\partial Q_i} \right| \Phi_b \right\rangle \tag{7.54}$$

Substituting eqs. 7.53, 7.51, 7.46, and 7.47 into eq. 7.45, we find

$$W(b \to a) = \frac{1}{\hbar^2} \sum_i \int_{-\infty}^{\infty} dt \, \exp(it\omega_{ab}) |R_i(ab)|^2 K_i(t) \prod_j{}' G_j(t) \tag{7.55}$$

where

$$G_j(t) = \left(2 \sinh \frac{\hbar\omega'_j}{2kT}\right) \sum_{v'_j v''_j} \exp[-(v''_j + \tfrac{1}{2})\mu''_j - (v'_j + \tfrac{1}{2})\mu'_j]. \tag{7.56}$$

and

$$\cdot |\langle X_{av''_j} | X_{bv''_j} \rangle|^2$$

$$K_i(t) = \left(2 \sinh \frac{\hbar\omega'_i}{2kT}\right) \sum_{v'_i v''_i} \exp[-(v''_i + \tfrac{1}{2})\mu''_i - (v'_i + \tfrac{1}{2})\mu'_i]$$

$$\times \left| \left\langle X_{av''_i} \left| \frac{\partial}{\partial Q_i} \right| X_{bv'_i} \right\rangle \right|^2 \tag{7.57}$$

with $\mu''_j = -it\omega''_j$ and $\lambda'_j = it\omega'_j + \hbar\omega'_j/kT$. In eq. 7.55 it has been assumed that the displacements of normal coordinates and frequencies of the promoting modes Q_i are negligible and the Condon approximation has been used. The validity and limitation of the use of the Condon approximation has been studied by Nitzan and Jortner (1972b) and by Freed and Lin (1975). Freed and Lin have proposed the Q-centroid method for treating the non-Condon effect (i.e., to take into account the nuclear coordinate dependence of the electronic matrix elements $R_i(ab)$).

To evaluate $G_j(t)$ and $K_i(t)$, we may use the Slater sum or Mehler's formula (Markham, 1959):

$$\sum_{n=0}^{\infty} \frac{1}{\sqrt{\pi} 2^n n!} H_n(x) H_n(x') \exp[-(n + \tfrac{1}{2})t - \tfrac{1}{2}(x^2 + x'^2)] = (2\pi \sinh t)^{-1/2}$$

$$\times \exp\left[-\frac{1}{4}(x + x')^2 \tanh \frac{t}{2} - \frac{1}{4}(x - x')^2 \coth \frac{t}{2} \right] \tag{7.58}$$

and $G_j(t)$ and $K_i(t)$ are given by (Lin, 1966a):

$$G_j(t) = 2\beta'_j \beta''_j \sinh \frac{\hbar\omega'_j}{2kT} [\sinh \lambda'_j \sinh \mu''_j (\beta''^2_j \coth \tfrac{1}{2}\mu''_j)$$

$$+ (\beta'^2_j \coth \tfrac{1}{2}\lambda'_j) \times (\beta''^2_j \tanh \tfrac{1}{2}\mu''_j + \beta'^2_j \tanh \tfrac{1}{2}\lambda j')]^{-1/2}$$

$$\times \exp\left[-\frac{\beta'^2_j \beta''^2_j (d''_j - d'_j)^2}{\beta'^2_j \coth \tfrac{1}{2}\lambda'_j + \beta''^2_j \coth \tfrac{1}{2}\mu''_j} \right] \tag{7.59}$$

and

$$K_i(t) = G_i(t)K_i^0(t) \tag{7.60}$$

where

$$K_i^0(t) = \frac{\omega_i'}{4\hbar}\left(1 + \coth\frac{\hbar\omega_i'}{2kT}\right)\exp(it\omega_i') + \frac{\omega_i'}{4\hbar}\left(\coth\frac{\hbar\omega_i'}{2kT} - 1\right)\exp(-it\omega_i') \tag{7.61}$$

The expression for $K_i^0(t)$ given by eq. 7.61 is valid for the promoting mode without the displacements in normal coordinate and frequency. The general expression for $K_i^0(t)$ has been obtained (Lin, 1966) and is given by

$$K_i^0(t) = \frac{1}{2}\left[\frac{\beta_i'^2\beta_i''^2}{\beta_i''^2\tanh(\lambda_{i'}/2) + \beta_i'^2\tanh(\mu_i''/2)}\right.$$
$$- \frac{\beta_i'^2\beta_i''^2}{\beta_i''\coth(\lambda_i'/2) + \beta_i'^2\coth(\mu_i''/2)}$$
$$\left. + \frac{2\beta_i'^4\beta_i''^4(d_i'' - d_i')^2}{(\beta_i''^2\coth(\lambda i'/2) + \beta_i'^2\coth(\mu_i''/2)^2}\right] \tag{7.62}$$

Substituting eq. 7.60 into eq. 7.55 yields

$$W(b \to a) = \frac{1}{\hbar}|R_i(ab)|^2\int_{-\infty}^{\infty}dt\,\exp(it\omega_{ab})K_i^0(t)\prod_j G_j(t) \tag{7.63}$$

The normal modes other than the promoting modes are usually referred to as accepting modes. Other methods have also been developed for obtaining eq. 7.63 for $W(b \to a)$ (Fischer, 1970; Freed and Jortner, 1970). A particularly interesting case occurs when both ρ_j and $(d_j'' - d_j')^2$ are small. In this case $G_j(t)$ becomes

$$G_j(t) = \left\{\exp(-\tfrac{1}{2}it\rho_j\omega_j')\coth\left(\frac{\hbar\omega_i'}{2kT} - \tfrac{1}{2}\beta_j'^2\right)(d_j'' - d_j')^2\right\}$$
$$\times \left\{\coth\frac{\hbar\omega_j'}{2kT} - \operatorname{csch}\frac{\hbar\omega_j'}{2kT} \times \operatorname{csch}\left(it\omega_j' + \frac{\hbar\omega_j'}{2kT}\right)\right\} \tag{7.64}$$

and $W(b \to a)$ is given by

$$\omega(b \to a) = \frac{1}{\hbar^2}\sum_i|R_i(ab)|^2\int_{-\infty}^{\infty}dtK_i^0(t)\exp\left[it\omega_{ab}' - \xi + \frac{1}{2}\sum_j\Delta_j^2\operatorname{csch}\frac{\hbar\omega_j'}{2kT}\right.$$
$$\left. \times \cosh\left(it\omega_j' + \frac{\hbar\omega_j'}{2kT}\right)\right] \tag{7.65}$$

where $\quad \Delta_j^2 = \beta_j'^2(d_j'' - d_j')^2, \quad \zeta = \sum_j \frac{1}{2}\Delta_j^2 \coth(\hbar\omega_j'/2kT) \quad$ and $\quad \omega_{ab}' = \omega_{ab} - \frac{1}{2}\sum \rho_j\omega_j' \coth(\hbar\omega_j'/2kT)$. The term ζ is usually called the *coupling constant* (or *strength*). The integral in eq. 7.65 can be carried out exactly (Lin and Bersohn, 1968) or approximately by using the saddle point method (Fischer, 1970; Freed and Jortner, 1970). To show the result of $W(b \to a)$ obtained by the use of the saddle point method, we consider $W(b \to a)$ at $T = 0$:

$$W(b \to a) = \frac{1}{\hbar^2} \sum_i |R_i(ab)|^2 e^{-\zeta_0} \frac{\omega_i'}{2\hbar} \int_{-\infty}^{\infty} dt \exp it\left[(\omega_{ab}' + \omega_i') + \frac{1}{2}\sum_j \Delta_j \right. $$
$$\left. \times \exp(it\omega_j') \right] \tag{7.66}$$

Applying the saddle point method to eq. 7.66 yields

$$W(b \to a) = \frac{1}{\hbar^2} \sum_i \frac{\omega_i'}{2\hbar} |R_i(ab)|^2 e^{-\zeta_0} \sqrt{\frac{4\pi}{\sum_j \omega_j'\Delta_j^2 \exp(it\omega_j')}}$$
$$\times \exp\left[it^*(\omega_{ab}' + \omega_i') + \frac{1}{2}\sum_j \Delta_j^2 \exp(it^*\omega_j')\right] \tag{7.67}$$

where t^* represents the saddle point value of t and is to be determined from

$$\frac{1}{2}\sum_j \Delta_j^2\omega_j' \exp(it^*\omega_j') = -(\omega_{ab}' + Q_i') \tag{7.68}$$

So far we have been concerned with the electronic relaxation. For the reverse process, that is, the conversion of vibrational energy into electronic energy, which is involved in delayed thermal fluorescence, the treatment should parallel the preceding one, but the temperature dependence would be different from the result for electronic relaxation because of the positive value of ω_{ab} (Lin, 1971). Furthermore, the preceding result for electronic relaxation can also be used in the gaseous phase provided the participation of molecular rotation in radiationless transitions is negligible.

7.3.2 Temperature Effect

For practical applications it is desirable to have a simple expression for $W(b \to a)$. For this purpose we may introduce an average frequency $\bar{\omega}'$ for ω_j'; in this case the t integral in $W(b \to a)$ can easily be evaluated and $W(b \to a)$ is given by (Knittel et al., 1977)

$$W(b \to a) = W_{ba}(T) = \frac{1}{\hbar^2} \sum_i |R_i(ab)|^2 \frac{\omega_i'}{4\hbar}\left[\left(\coth\frac{\hbar\omega_i'}{2kT} + 1\right)f_i(\omega_{ab} + \omega_i') \right.$$
$$\left. + \left(\coth\frac{\hbar\omega_i'}{2kT} - 1\right)f_i(\omega_{ab} - \omega_i')\right] \tag{7.69}$$

where

$$f_i(\omega_{ab} \pm \omega_i') = \frac{2\Pi}{\overline{\omega}'} e^{-\zeta} \left(\zeta_0 \bar{n} \exp\left(\frac{\Pi\overline{\omega}'}{kT} \right) \right)^{(\omega_{ba}' \pm \omega_i'/\overline{\omega}')} \sum_{v'=0}^{\infty}$$

$$\times \frac{(\zeta_0 \bar{n} \exp(\Pi\overline{\omega}'/2kT))^{2v'}}{v'!(v' - \omega_{ab}' \pm \omega_i'/\overline{\omega}')!} \tag{7.70}$$

and $\bar{n} = [\exp(\hbar\overline{\omega}'/kT) - 1]^{-1}$. When temperature is not high (i.e., $kT < \hbar\overline{\omega}'$), we may put eq. 7.70 in the following form (Lin, 1972; Knittel et al., 1977) by expanding eq. 7.70 in power series of $\exp(-\hbar\overline{\omega}'/kT)$

$$W_{ba}^*(T^*) = \frac{W_{ba}(T)}{W_{ba}(0)}$$

$$= 1 + \exp\left(-\frac{1}{T^*} \right)\left[\omega_{ba}'^* - 2\zeta_0 + \frac{\zeta_0^2}{\omega_{ba}'^*} + \frac{\zeta_0^2}{\omega_{ba}'^*(1 + \omega_{ba}'^*)} \right] \tag{7.71}$$

where $T^* = kT/\hbar\overline{\omega}'$ and $\omega_{ba}'^* = \omega_{ba}'/\overline{\omega}'$. The term ζ_0 represents the ζ value at $T = 0$. From eq. 7.71 we may expect the temperature dependence of the nonradiative rate constant to show onset behavior; that is, below a certain temperature limit, the temperature effect is small. This threshold temperature T_c^* can be found by setting the second term of eq. 7.71 equal to unity:

$$\frac{1}{T_c^*} = \log\left[\omega_{ba}'^* - 2\zeta_0 + \frac{\zeta_0^2}{\omega_{ba}'^*} + \frac{\zeta_0^2}{\omega_{ba}'^*(1 + \omega_{ba}'^*)} \right] \tag{7.72}$$

In other words, as long as T^* is below T_c^*, we may expect to have a very small temperature dependence. In Table 7.2 we present T_c^* for various combinations of $\omega_{ba}'^*$ and ζ_0. For a given $\omega_{ba}'^*$, T_c^* increases with the coupling strength ζ_0.

Qualitatively for a given energy gap $\omega_{ba}'^*$ the temperature effect on $\omega_{ba}^*(T^*)$ increases with decreasing ζ_0, and for a given coupling strength ζ_0 the tem-

Table 7.2
Calculation of T_c^*

$\omega_{ba}'^*/S_0$	0.1	1	5	10	15
10	0.4381	0.4778	0.9967		
15	0.3711	0.3890	0.5228	1.3625	
20	0.3349	0.3456	0.4123	0.6039	1.7247

Table 7.3
Determination of $|R_i(ab)|$

Compound	Transitions	ω^*	S_0	$W_{ba}^*(0)/\text{sec}^{-1}$	$|R_i(ab)|/\text{erg}^{1/2}$ sec.
$LaF_3:Ho^{3+}$	$^5F_3 \rightarrow {}^5F_4, {}^5S_2$	6	0.8	2.0×10^4	7.8×10^{-38}
$LaBr_3:Dy^{3+}$	$^6F_{3/2} \rightarrow {}^6F_{5/2}$	5	0.05	8.7×10^3	2.6×10^{-36}
$MnF_2:Dr^{3+}$	$^4F_{9/2} \rightarrow {}^4I_{9/2}$	7	0.1	1.2×10^3	5.97×10^{-35}
$KMgF_3:Co^{2+}$	$^4T_2 \rightarrow {}^4T_1$	20	2.0	1×10^2	2.7×10^{-35}
$LaBr_3:Pr^{3+}$	$^3P_1 \rightarrow {}^2P_0$	4	1.2	5.6×10^4	2.4×10^{-37}
$LaCl_3:Pr^{3+}$	$^3P_1 \rightarrow {}^3P_0$	3	0.3	3.3×10^5	9.6×10^{-37}

perature effect increases with increasing energy gap $\omega_{ba}'^*$ (Knittel et al., 1977), Notice that at $T = 0$ eq. 7.69 reduces to

$$W_{ba}(0) = \frac{\Pi}{\hbar^3} |R_i(ab)|^2 e^{-\zeta_0} \frac{\zeta_0^{\omega_{ba}'^* - 1}}{(\omega_{ba}'^* - 1)!} \qquad (7.73)$$

Here for convenience it has been assumed that there is only one promoting mode involved in the radiationless transitions. Equation 7.73 has been applied to some experimental results of transition metals and rare earth ion centers to determine the electronic matrix elements $|R_i(ab)|$ (see Table 7.3). The application of eq. 7.69 (the so-called single-configurational-coordinate model) can be demonstrated by fitting the temperature-dependent lifetime measurements of rare earth ions by Moos (1970, see Fig. 7.1 and 7.2), by

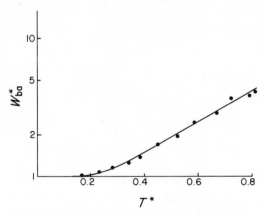

Fig. 7.1 Multiphonon transition rate against temperature of $LaF_3:Ho^{3+}$. The theoretical curve is for $^5F_3 \rightarrow {}^5F_4, {}^5S_2$ emission of six 300 cm^{-1} phonons with $S_0 = 0.8$. (Reprinted by permission, from S. H. Lin, *J. Chem. Soc. Faraday Trans.*, II, **73**, 120, 1977).

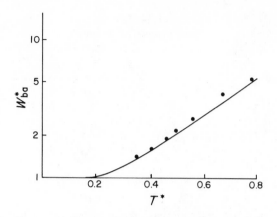

Fig. 7.2 Multiphonon transition rate against temperature of $LaBr_3:Dy^{3+}$. The theoretical fit is for $^6F_{3/2} \rightarrow {}^6F_{5/2}$ emission of five 155 cm^{-1} phonons with $S_0 = 0.05$. (Reprinted, by permission, from S. H. Lin, *J. Chem. Soc. Faraday Trans.*, II, **73**, 120, 1977).

Flaherty and DiBartolo (1973, see Fig. 7.3), and by Sturge (1973, see Fig. 7.4). The order of phonons ω^* is found by dividing $\omega_{ba} - \omega_i'$ by $\bar{\omega}$. As we can see from the preceding discussion, the expression for the nonradiative rate constant consists of two parts, one from the electronic motion (the promoting part $R_i(ab)$) and the other from the nuclear motion (the statistical part mainly determined by the Franck–Condon factor). Most recent theoretical work has been focused on the calculation of the nonradiative rate constant

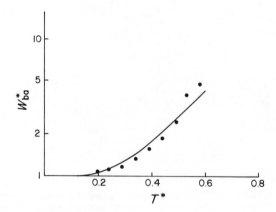

Fig. 7.3 Multiphonon transition rate against temperature of $MnF_2:Er^{3+}$. The theoretical fit is for $^4F_{9/2} \rightarrow {}^4I_{9/2}$ emission of seven 357 cm^{-1} phonons with $S_0 = 0.1$. (Reprinted, by permission, from S. H. Lin, *J. Chem. Soc. Faraday Trans.*, II, **73**, 120, 1977).

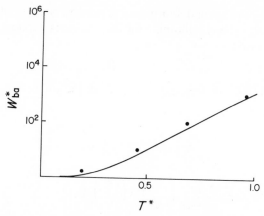

Fig. 7.4 Multiphonon transition rate against temperature of $KMgF_3:Co^{2+}$. The theoretical fit is for $^4T_2 \rightarrow {}^4T_1$ emission of twenty $345\ cm^{-1}$ phonons with $S_0 = 2.0$ (Reproduced, with permission, from S. H. Lin, *J. Chem. Soc. Faraday Trans.*, II, **73**, 120, 1977).

of one single vibronic state relative to another (see the following discussion) so that the promoting part of the rate constant is canceled if only one promoting mode is involved and hence has been concerned mainly with the Franck–Condon factor calculation (the statistical part of the rate constant).

7.3.3 Electronic Matrix Elements

Let us now briefly discuss how to calculate the electronic matrix elements involved in the promoting part of the nonradiative rate constant. Notice that

$$\langle av''\hat{H}'_{B_0}bv'\rangle = \sum_i \left[\left\langle X_{av''}|R_i(ab)|\frac{\partial X_{bv'}}{\partial Q_i} + X_{av''}|\zeta_i(ab)|X_{bv'} \right\rangle \right] \quad (7.74)$$

where $\zeta_i(ab)$ is defined by

$$\zeta_i(ab) = -\frac{\hbar^2}{2}\langle \phi_a | \frac{\partial^2 \phi_b}{\partial Q_i^2} \quad (7.75)$$

In other words, to calculate $R_i(ab)$ and $\zeta_i(ab)$, we have to find the variation of the electronic wave function with respect to nuclear motion.

We first consider the internal conversion. One way to find the normal coordinate dependence of ϕ_b is to employ the Herzberg–Teller theory by expanding the electronic Hamiltonian in power series of normal coordinates Q_i

$$\hat{h}_s = \hat{h}_s^0 + \hat{H}'_{vc} \quad (7.76)$$

and by treating the normal coordinate dependent portion of the electronic Hamiltonian $\hat{H}'_{vc}$, the Hamiltonian for the vibronic coupling, as a perturbation. It follows that

$$\hat{H}'_{vc} = \hat{H}^{(1)}_{vc} + \hat{H}^{(2)}_{vc} + \cdots \tag{7.77}$$

where $H^{(1)}_{vc}$ and $H^{(2)}_{vc}$ denote the first- and second-order vibronic couplings, respectively:

$$\hat{H}^{(1)}_{vc} = -\sum_i \sum_k \sum_\sigma e^2 \left(\frac{\partial \mathbf{r}_\sigma}{\partial Q_k}\right)_0 \cdot \frac{\mathbf{r}_{i\sigma}}{r^3_{i\sigma}} Q_k \tag{7.78}$$

and

$$\hat{H}^{(2)}_{vc} = \sum_{kl} \sum_i \sum_\sigma \frac{Z_\sigma e^2}{2r^3_{i\sigma}} \left(\frac{\partial \mathbf{r}_\sigma}{\partial Q_k}\right)_0 \cdot \left(\frac{\partial \mathbf{r}_\sigma}{\partial Q_l}\right)_0 Q_k Q_l - \sum_{kl} \sum_i \sum_\sigma \frac{3Z_\sigma e^2}{2r^5_{i\sigma}}$$
$$\times \left[\mathbf{r}_{i\sigma} \cdot \left(\frac{\partial \mathbf{r}_\sigma}{\partial Q_k}\right)_0 \mathbf{r}_{i\sigma} \cdot \left(\frac{\partial \mathbf{r}_\sigma}{\partial Q_l}\right)_0 \right] Q_k Q_l \tag{7.79}$$

Here the contribution to the vibronic coupling from the electron–electron repulsion has been ignored. In the preceding equations the summation over i refers to electrons and the summation over σ refers to nuclei.

Using the perturbation method we find

$$\Phi_b = \Phi^0_b + \sum_c{}' \frac{\langle \Phi^0_c | H'_{vc} | \Phi^0_b \rangle}{E^0_a - E^0_c} \Phi^0_c + \cdots \tag{7.80}$$

The term Φ_b given by eq. 7.43 will show the normal coordinate dependence at the electronic wave function. Other ways to find $\Phi_b(Q_i)$ have also been proposed (Lee et al., 1977). As an application we consider the internal conversion $^1A_2 \rightarrow A_1$ ($n \leftrightarrow \Pi^*$) of the carbonyl molecule. It is well known that the formaldehyde molecule has six normal modes:

$$\nu_1(a_1) = 2780 \text{ cm}^{-1}, \nu_2(a_1) = 1744 \text{ cm}^{-1}, \nu_3(a_1) = 1503 \text{ cm}^{-1},$$
$$\nu_4(b_2) = 1167 \text{ cm}^{-1}, \nu_5(b_1) = 2874 \text{ cm}^{-1}, \nu_6(b_1) = 1280 \text{ cm}^{-1}.$$

It can easily be shown that in the first-order approximation, $\langle \Phi_{^1A_2} | \partial/\partial Q_i | \Phi_{^1A_2} \rangle$ vanishes on symmetry grounds, and that in the second-order approximation $\zeta_i(ab)$ vanishes but $R_i(ab)$ exists. The calculation of $R_i(^1A_2 \rightarrow {}^1A_1)$ of formaldehyde has been carried out by Lin (1976) using simple molecular orbitals (Pople and Sidman, 1957). In his calculation of $\Phi_b(Q_i)$ using eq. 7.80, only the following electronic states are included: $A_2(n \rightarrow \Pi^*)$, $B_1(n \rightarrow \sigma^*)$, $A_1(\Pi \rightarrow \Pi^*)$, and $B_2(\sigma \rightarrow \Pi^*)$ (Henderson, 1966). The results of $R_i(^1A_2 \rightarrow {}^1A_1)$ are given by

$$R_4(^1A_2 \rightarrow {}^1A_1) = (0.0746Q_6 - 0.00351Q_5) \frac{N_A h^2}{a^2_0} \tag{7.81}$$

$$R_5(^1A_2 \to {}^1A_1) = (-0.00378Q_4)\frac{N_A\hbar^2}{a_0^2} \tag{7.82}$$

$$R_6(^1A_2 \to {}^1A_1) = (0.0950Q_4)\frac{N_A\hbar^2}{a_0^2} \tag{7.83}$$

where N_A is the Avogadro number and a_0 is the Bohr radius. Equations 7.81–7.83 indicate that Q_4 and Q_6 are major promoting modes, and for the internal conversion $^1A_2 \to {}^1A_1$ the Condon approximation cannot apply. Recently, the matrix elements $R_i(ab)$ for the internal conversion of benzene have been calculated by Lee et al., (1977).

Next we consider the intersystem crossing. In this case the question as to how the electronic matrix elements of the promoting part of the nonradiative rate constant are to be evaluated has not been settled (Lin, 1966; Bixon and Jortner, 1968; Henry and Siebrand, 1969; Lin, 1976). One way to calculate these electronic matrix elements is to invoke the vibronic coupling plus the spin-orbit coupling and/or the vibronic-spin-orbit coupling (Lin, 1966; Lin, 1967; Lin, 1976; Bixon and Jortner, 1968). In other words, $\hat{h}_s$ is assumed to include the spin-orbit coupling so that the electronic wave functions Φ_a and Φ_b are spin-mixed. To demonstrate this approach we consider the intersystem crossing $^1A_2 \to {}^3A_2$ of the carbonyl molecule. In this case we are concerned with the calculation of

$$R_i(^1A_2 \to {}^3A_2)_{M_s} = -\hbar^2\left\langle\Phi_{^1A_2}\left|\frac{\partial}{\partial Q_i}\right|\Phi_{^3A_2}\right\rangle \tag{7.84}$$

For the intersystem crossing, one has to consider the transitions involved in magnetic sublevels M_s.

The calculation of $R_i(^1A_2 \to {}^3A_2)_{M_s}$ can be accomplished by invoking the vibronic-spin-orbit coupling

$$\hat{H}'_{vso} = \frac{1}{2m^2c^2}\sum_i\sum_k Q_k\left(\frac{\partial\Gamma_i}{\partial Q_k}\right)\cdot\zeta_i \tag{7.85}$$

where $\Gamma_i = \mathbf{V}_i V x \mathbf{P}_i$ and ζ_i is the spin operator, in the first-order perturbation theory, and/or invoking the vibronic coupling $\hat{H}'_{vc}$ plus the spin-orbit coupling

$$\hat{H}'_{so} = \frac{1}{2m^2c^2}\sum_i\sigma_i\cdot\zeta_i \tag{7.86}$$

in the second-order perturbation theory. For formaldehyde it has been found that in the simple m_0 approximation, the vibronic-spin-orbit coupling between the $^1A_2$ and $^3A_2$ states vanishes (Lin, 1976) and using the second-order perturbation theory for $\hat{H}'_{so}$ and $\hat{H}'_{vc}$, $R_i(^1A_2 \rightarrow {}^3A_2)_{M_s}$ can be calculated. The results are given by

$$R_i(^1A_2 \rightarrow {}^3A_2)_y = 0 \tag{7.87}$$

$$R_4(^1A_2 \rightarrow {}^3A_2)_1 = -R_4(^1A_2 \rightarrow {}^3A_2)_{-1} = 0.072i \times 10^{-3} \frac{\hbar^2}{a_0} \sqrt{N_A} \tag{7.88}$$

$$R_6(^1A_2 \rightarrow {}^3A_2)_1 = R_6(^1A_2 \rightarrow {}^3A_2)_{-1} = -0.101 \times 10^{-3} \frac{\hbar^2}{a_0} \sqrt{N_A} \tag{7.89}$$

$$R_5(^1A_2 \rightarrow {}^3A_2)_1 = R_5(^1A_2 \rightarrow {}^3A_2)_{-1} = 0.0041 \times 10^{-3} \frac{\hbar^2}{a_0} \sqrt{N_A} \tag{7.90}$$

The calculated results indicated that there are two major promoting modes Q_4 and Q_6, the radiationless transition involving the magnetic sublevel $M_s = 0$ of $^3A_2$ is forbidden, and the radiationless transition probabilities involving the magnetic sublevels $M_s \pm 1$ are equal.

It has been observed that the electric field and high pressures affect the rate of radiationless transitions (Bogan and Fitchen, 1970; Drickamer and Frank, 1972) and theoretical models for treating their effects on radiationless transitions have been developed (Lin, 1973; Lin, 1975). Interested readers should consult the original papers.

7.3.4 Radiationless Transitions in Isolated Molecules

The radiationless transition in isolated molecules has been of recent interest both experimentally and theoretically (Lin, 1978). The isolated molecular condition in this case is realized when it is collision-free during the lifetime of the excited electronic state under consideration.

The rate constant of a single vibronic state can be observed if the bandpass of the excitation source is narrow enough that only the vibronic state under consideration is excited, and if, after the vibronic state is populated, there exist no other vibronic states of equal energy for the intramolecular vibrational relaxation (caused by the anharmonic effect) to take place. If there are other vibronic states of equal energy, then the rate constant of the single vibronic state can be measured only if the vibrational relaxation is much slower than the electronic relaxation. If there exists more than one vibronic state of equal energy and the vibrational relaxation is much faster than the electronic relaxation, then the rate constant observed is the averaged rate constant. Intermediate cases often result in nonexponential decay (Lin, 1972a).

In the following section we discuss the calculation of the radiationless transition probability of a single vibronic state and the equilibrium average (microcanonical) nonradiative decay rate constant of isolated molecules.

Only one theoretical approach is described, which is based on the use of the inverse Laplace transformation of the thermally averaged nonradiative decay rate constant (Lin, 1973); for other methods the original papers should be consulted (Nitzan and Jortner, 1971; Brailsford and Chang, 1970; Heller et al., 1972; Fischer and Schlag, 1969; Siebrand, 1971).

General Considerations. For the case in which the intermolecular vibrational relaxation is much faster than the electronic relaxation, the rate constant for an electronic relaxation (or dissipation) process can be expressed as

$$W(\beta) = \sum_i P_i W_i = [Q(\beta)]^{-1} \sum_i \exp(-\beta E_i) W_i \qquad (7.91)$$

where P_i represents the normalized Boltzmann factor, $\beta = (kT)^{-1}$, $Q(\beta)$ is the partition function of the system equals $\sum_i \exp(-\beta E_i)$, and W_i is the rate constant of a single vibronic state of an isolated molecule or the transition probability of the molecule originating from the ith vibronic state. From eq. 7.91 we can obtain W_j as

$$W_j = \lim_{E \to E_j} [\delta(E - E_j)]^{-1} (2\pi i)^{-1} \int_{\gamma - i\infty}^{\gamma + i\infty} d\beta \, \exp(\beta E) Q(\beta) W(\beta) \qquad (7.92)$$

where $\delta(E - E_j)$ represents the delta function. It has been shown that for an isolated molecule, if the vibrational relaxation due to anharmonic effect is much faster than the electronic relaxation (or dissipation), then the microcanonical average rate constant of the electronic relaxation of the isolated molecule with energy E is given by

$$W(E) = [\rho(E)]^{-1} (2\pi i)^{-1} \int_{\gamma - i\infty}^{\gamma + i\infty} d\beta \, \exp(\beta E) Q(\beta) W(\beta) \qquad (7.93)$$

where $\rho(E)$ represents the density of quantum states of the isolated molecule with energy E and is defined by

$$\rho(E) = \sum_i \delta(E - E_i) = (2\pi i)^{-1} \int_{\gamma - i\infty}^{\gamma + i\infty} d\beta \, \exp(\beta E) Q(\beta) \qquad (7.94)$$

In eq. 7.94 the summation is over the states of energy equal to E. Needless to say, eq. 7.93 reduces to eq. 7.92 when there is only one vibronic state with energy E. Equations 7.92 and 7.93 indicate that the rate constant of a single vibronic state and the average rate constant of an isolated molecule for an electronic relaxation process can be obtained from the corresponding thermally averaged rate constant by the inverse Laplace transformation. We now apply the relation given in eqs. 7.92 and 7.93 to the nonradiative decay process of isolated molecules.

Rate Constant of a Single Vibronic State.   To demonstrate the application of eq. 7.92 to the radiationless transition, we start from the expression for the thermally averaged rate constant:

$$W(b \to a) = W_{ba}(\beta) = \left[\frac{|C_i(ab)|^2}{2\hbar^2}\right]\exp\left[-\frac{1}{2}\sum_j \beta_j'^2(d_j'' - d_j')^2 \coth \tfrac{1}{2}\beta\hbar\omega_j'\right]$$

$$\times \int_{-\infty}^{\infty} dt \, \exp(i\omega_{ab}t)[(\coth \tfrac{1}{2}\beta\hbar\omega_i' + 1)\exp(i\omega_i't)$$

$$+ (\coth \tfrac{1}{2}\beta\hbar\omega_i' - 1)\exp(-i_i\omega't)]$$

$$\times \exp\left\{\frac{1}{2}\sum_j \beta_j'^2(d_j'' - d_j')^2 \operatorname{csch} \tfrac{1}{2}\beta\hbar\omega_j' \cos[\omega_j't - \tfrac{1}{2}(i\beta)\hbar\omega_j']\right\} \quad (7.95)$$

where $C_j(ab) = -\hbar^2(\omega_i'/2\hbar)^{1/2}\langle\Phi_a|\partial/\partial Q_i|\Phi_b\rangle$. In eq. 7.95 the changes in normal frequencies ν_j between the two electronic states have been ignored.

Before substituting eq. 7.95 into eq. 7.92, we notice that

$$\exp[-\tfrac{1}{2}\Delta_j^2 \coth \tfrac{1}{2}\beta\hbar\omega_j' + \tfrac{1}{2}\Delta_j^2 \operatorname{csch} \tfrac{1}{2}\beta\hbar\omega_j' \cos(\omega_j't - \tfrac{1}{2}(i\beta)\hbar\omega_j')]$$
$$= \exp\{-\tfrac{1}{2}\Delta_j^2[1 - \exp(i\omega_j't)] + \tfrac{1}{2}\Delta j^2[\exp(\tfrac{1}{2}i\omega_j't) - \exp(-\tfrac{1}{2}i\omega_j't)]^2$$
$$\times [\exp(\beta\hbar\omega_j') - 1^-]^1\}. \quad (7.96)$$

Applying the Taylor expansion to the second term in eq. 7.96 and substituting eq. 7.96 into eq. 7.95, we obtain

$$W_{bc}(\beta) = \frac{|C_i(ab)|^2}{2\hbar^2} \cdot \frac{\exp(-\tfrac{1}{2}\sum_j \Delta_j^2)}{Q(\beta)} \int_{-\infty}^{\infty} dt \, \exp\left[\frac{1}{2}\sum_j \Delta_j^2 \exp(i\omega_j't)\right]$$

$$\times \{(\coth \tfrac{1}{2}\beta\hbar\omega_j' + 1) \exp[it(\omega_{ab} + \omega_i')]$$

$$+ (\coth \tfrac{1}{2}\beta\hbar\omega_j' - 1)\exp[it(\omega_{ab} - \omega_i')]\}$$

$$\times \sum_{v_1=0}^{\infty} \sum_{n_1=0}^{v_1} \cdots \sum_{v_N=0}^{\infty} \sum_{n_N=0}^{v_N} e\, xp\left(-\beta\hbar \sum_j v_j\omega_j'\right)$$

$$\times \prod_j' \frac{v_j!\Delta_j^{2n_j}(\cos \omega_j't - 1)^{n_j}}{(v_j - n_j)!(n_j!)^2} \quad (7.97)$$

Here the relation $Q(\beta) = \prod_j [1 - \exp(-\beta\hbar\omega_j')]^{-1}$ has been used.

Substituting the equations

$$\coth \tfrac{1}{2}\beta\hbar\omega_i' + 1 = 2\sum_{n_i=0}^{\infty} \exp(-n_i\beta\hbar\omega_i'),$$

$$\coth \tfrac{1}{2}\beta\hbar\omega_i' - 1 = 2\exp(-\beta\hbar\omega_i') \sum_{n_i=0}^{\infty} \exp(-n_i\beta\hbar\omega_i') \quad (7.98)$$

and eq. 7.97 into eq. 7.92 and canceling the delta functions, the expression for the radiationless transition probability of a single vibronic state obtained is

$$
W_{bv'} = \frac{|C_i(ab)|^2}{\hbar^2} \exp\left(-\frac{1}{2}\sum_j \Delta_j^2\right) \int_{-\infty}^{\infty} dt \, \exp\left[\frac{1}{2}\sum_j \Delta_j^2 \exp(i\omega't)\right]
$$

$$
\times \left(\exp[it(\omega_{ab} + \omega_i')] \sum_{m_i=0}^{v_i'} \sum_{n_i=0}^{v_i'-m_i} \frac{(v_i' - m_i)! \Delta_i^{2n_i}}{(v_i' - m_i - n_i)!(n_i!)^2}\right.
$$

$$
\times (\cos \omega_i' t - 1)^{n_i} + \exp[it(\omega_{ab} - \omega_i')] \sum_{m_i=0}^{v_i'-1} \sum_{n_i=0}^{v_i'-1-m_i}
$$

$$
\times \frac{(v_i' - 1 - m_i)! \Delta_i^{2n_i}}{(v_i' - 1 - m_i - n_i)!(n_i!)^2} (\cos \omega_i' t - 1)^{n_i}\right)
$$

$$
\times \sum_{n_1=0}^{v_1'} \cdots \sum_{n_N=0}^{v_N'} \prod_j{}' \frac{v_j'! \Delta_j^{2n_j}}{(v_j' - n_j)!(n_j!)^2} (\cos \omega_j' t - 1)^{n_j} \tag{7.99}
$$

where v_j''s represent the vibrational quanta in the initial electronic state. In particular, when $\Delta_i = 0$, eq. 7.99 reduces to

$$
W_{bv'} = \frac{|C_i(ab)|^2}{\hbar^2} \exp\left(-\frac{1}{2}\sum_j \Delta_j^2\right) \int_{-\infty}^{\infty} \exp\left[\frac{1}{2}\sum_j \Delta_j^2 \exp(i\omega_j' t)\right]
$$

$$
\times \{(v_i' + 1)\exp[it(\omega_{ab} + \omega_i')] + v_i' \exp[it(\omega_{ab} - \omega_i')]\}
$$

$$
\times \sum_{n_1=0}^{v_1'} \cdots \sum_{n_N=0}^{v_N'} \prod_j{}' \frac{v_j'! \Delta_j^{2n_j}}{(v_j' - n_j)!(n_j!)^2} (\cos \omega_j' t - 1)^{n_j} \tag{7.100}
$$

It is not useful to carry out the integration with respect to t in eqs. 7.99 and 7.100 exactly, because the result of the integration simply gives us a delta function that keeps the energy conserved and a set of summations that cover all the possibilities of distributing the energy in the initial electronic state among the accepting modes of the final electronic state. Hence an alternate approach is adopted, that is, to carry out the integration in eqs. 7.99 and 7.100 by the method of steepest descent. For this purpose we notice that for the case $v_j' = 0$, we have

$$
W_{b0} = \frac{|C_i(ab)|^2}{\hbar^2} \exp\left(-\frac{1}{2}\sum_j \Delta_j^2\right) \int_{-\infty}^{\infty} dt
$$

$$
\times \exp\left[\frac{1}{2}\sum_j \Delta_j^2 \exp(i\omega_j' t) + it(\omega_{ab}' + \omega_i')\right] \tag{7.101}
$$

This, of course, is also equal to the nonradiative decay rate constant at $T = 0$. Equations 7.99 and 7.100 can be written in the following form:

$$W_{bv'} = \frac{|C_i(ab)|^2}{\hbar^2} \exp\left(-\frac{1}{2}\sum_j \Delta_j^2\right) \int_{-\infty}^{\infty} dt$$

$$\times \exp\left[\frac{1}{2}\sum_j \Delta_j^2 \exp(i\omega_j' t) + it(\omega_{ab} + \omega_i')\right] f_{bv'}(t) \qquad (7.102)$$

where $f_{bv'}(t)$, say for eq. 7.100, is defined by

$$f_{bv'}(t) = \left\{ \sum_{m_i=0}^{v_i'} \sum_{n_i=0}^{v_i'-m_i} \frac{(v_i' - m_i)!}{(v_i' - m_i - n_i)!(n_i!)^2} [\Delta_i^2(\cos \omega_i' t - 1)]^{n_i} \right.$$

$$+ \exp(-2it\omega_i') \sum_{m_i=0}^{v_i'-1} \sum_{n_i=0}^{v_i'-1-m_i} \frac{(v_i' - 1 - m_i)!}{(v_i' - 1 - m_i - n_i)!(n_i!)^2}$$

$$\times \left. [\Delta_i^2(\cos \omega_i' t - 1)]^{n_i} \right\} \sum_{n_1=0}^{v_1'} \cdots \sum_{n_N=0}^{v_N'} \prod_j{}' \frac{v_j'!}{(v_j' - n_j)!(n_j!)^2}$$

$$\times [\Delta_j^2(\cos \omega_j' t - 1)]^{n_j} \qquad (7.103)$$

If t^* represents the saddle point value of t of the integrand in eq. 7.101, that is, t^* is to be determined from (Appendix Two)

$$\omega_{ab} + \omega_i' + \frac{1}{2}\sum_j \Delta_j^2 \omega_j' \exp(i\omega_j' t^*) = 0 \qquad (7.104)$$

we can evaluate the integral in eq. 7.102 by the method of steepest descent as follows:

$$W_{bv'} = f_{bv'}(t^*) \frac{|C_i(ab)|^2}{\hbar^2} \exp\left(-\frac{1}{2}\sum_j \Delta_j^2\right) \int_{-\infty}^{\infty}$$

$$\times \int_{-\infty}^{\infty} dt \exp\left[\frac{1}{2}\sum_j \Delta_j^2 \exp(i\omega_j' t) + it(\omega_{ab} + \omega_i')\right]$$

$$= W_{b0} f_{bv'}(t^*) \qquad (7.105)$$

In practice, we determine t^* or it^* from eq. 7.103 first, and then substitute this t^* (or it^*) value into $f_{bv'}$ in eq. 7.105 to calculate $W_{bv'}$. It should be noticed that, in this case, W_{b0} can be identified as the rate constant at $T = 0$ or the rate constant of the zero-point state. We can also evaluate W_{b0} by using the method of steepest descent

Now let us expand eq. 9.71 in power series of $\exp(-v'_j\beta\hbar\omega'_j)$:

$$
\begin{aligned}
W_{ba}(\beta) &= \prod_j [1 - \exp(-\beta\hbar\omega'_j)] \sum_{v'} \prod_j \exp(-\beta v'_j \hbar\omega'_j) W_{bv'} \\
&= W_{b0} + \sum_j [W_{bv'}(v'_j = 1) - W_{b0}]\exp(-\beta\hbar\omega'_j) \\
&\quad + \sum_j [W_{bv'}(v'_j = 2) - W_{bv'}(v'_j = 1)]\exp(-2\beta\hbar\omega'_j) \\
&\quad + \sum_{j>k} [W_{b0} - W_{bv'}(v'_j = 1) - W_{bv'}(v'_k = 1)] \\
&\quad \times \exp[-\beta\hbar(\omega'_j + \omega'_k)] + \cdots
\end{aligned}
\tag{7.106}
$$

Here the higher-order terms beyond two phonon terms are neglected. Equation 7.106 indicates that from the measurement of the effect of temperature on radiationless transitions, one can determine the nonradiative decay rate constant of a single vibronic state. If the temperature is not high, we obtain

$$
W_{ba}(\beta) = W_{b0}\left[1 + \sum_j B_j \exp(-\beta\hbar\omega_{j'}) + \cdots \right]
\tag{7.107}
$$

Now let us estimate $W_{bv'}$ from the temperature effect measurement. Jones and Siegel have reported that for phosphorescence the B_j's for naphthalene-h_8, pyrene-d_{10}, and phenanthrene-h_{10}, and biphenyl-h_{10} are $B_1 = 3.1 \pm 0.9$, 2.6 ± 0.6, $9.7 - 2.0$, and 4.0, respectively. Using these B_j's, we obtain $[W_{bv'}(v_j = 1)/W_{b0}] = (4.1 \pm 0.9)$, (3.6 ± 0.6), $(10.7 - 3.0)$, and 5.0 for naphthalene-h_8, pyrene-d_{10}, phenanthrene-h_{10}, and biphenyl-h_{10}, respectively (Jones and Siegel, 1969).

Statistical Average Rate Constant. Next we turn to the case in which the intramolecular vibrational relaxation due to the anharmonic effect is faster than the electronic relaxation. In this case eq. 7.93 can be written

$$
W_{ba}(E) = [\rho(E)]^{-1}(2\pi i)^{-1} \int_{\gamma-i\infty}^{\gamma+i\infty} d\beta \, \exp(\beta E) Q_b(\beta) W_{ba}(\beta)
\tag{7.108}
$$

Here E is the energy of the molecule and is measured from the zero energy of the initial electronic state. The evaluation of the density of states $\rho(E)$ by the method of steepest descent is well known. If β^* represents the saddle point of the integrand in eq. 7.94, then it can easily be shown that β^* is to be determined from

$$
E = -\left[\frac{\partial}{\partial\beta} \log Q_b(\beta)\right]_{\beta=\beta^*} = \sum_j \left[\frac{\hbar\omega'_j}{\exp(\beta^*\hbar\omega'_j)} - 1\right]
\tag{7.109}
$$

In other words, β^* plays the role of temperature in the isolated system. Using the approach described in eq. 7.105 we can rewrite eq. 7.108 as

$$W_{ba}(E) = \left[\frac{W_{ba}(\beta^*)}{\rho(E)}\right] (2\pi i)^{-1} \int_{\gamma-i\infty}^{\gamma+i\infty} d\beta \, \exp(\beta E) Q_b(\beta) = W_{ba}(\beta^*) \quad (7.110)$$

Equation 7.110 indicates that the statistical average rate constant of an isolated molecule can be calculated from the corresponding rate constant of thermal systems by simply replacing the temperature β of the thermal system by the temperature β^* of the isolated molecule to be determined from energy E. Also, as far as the evaluation of $W_{ba}(E)$ is concerned, we can avoid evaluating the density of states $\rho(E)$ (also see Chapter 5).

Now to illustrate the application of eq. 7.110, we again use $W_{ba}(\beta)$ given in eq. 7.95. Of course, any other expression for $W_{ba}(\beta)$ can be used. Using eqs. 7.95 and 7.110 we have

$$W_{ba}(E) = W_{ba}(\beta^*) = \frac{|C_i(ab)|^2}{\hbar^2} \exp\left[-\frac{1}{2}\sum_j \Delta_j^2 \coth \tfrac{1}{2}(\beta^*\hbar\omega_j')\right]$$

$$\times \int_{-\infty}^{\infty} dt \, \exp(i\omega_{ab}t)\{\tfrac{1}{2}[\coth \tfrac{1}{2}(\beta^*\hbar\omega_i') + 1]$$

$$\times \exp(i\omega_i't) + \tfrac{1}{2}(\coth \tfrac{1}{2}(\beta^*\hbar\omega_i') - 1)\exp(-i\omega_i't)\}$$

$$\times \exp\left\{\frac{1}{2}\sum_j \Delta_j^2 \operatorname{csch} \tfrac{1}{2}(\beta^*\hbar\omega_j')\cos[\omega_j't - \tfrac{1}{2}(i\beta^*\hbar\omega_j')]\right\} \quad (7.111)$$

Again, we can evaluate the integral in eq. 7.111 by the method of steepest descent. It follows:

$$W_{ba}(E) = W_{bo}[1 + \bar{n}_i^* + \bar{n}_i^* \exp(-2i\omega_i't^*)]\exp\left[\sum_j \Delta_j^2\bar{n}_j^*(\cos \omega_j't^* - 1)\right]$$

$$(7.112)$$

where $\bar{n}_j^* = [\exp(\beta^*\hbar\omega_j') - 1]^{-1}$. The term t^* is to be determined from eq. 7.104. Thus, to calculate the average rate constant $W_{ba}(E)$ as a function of energy E, we determine β^* from eq. 7.109 and t^* from eq. 7.104 first, and then substitute these t^* and β^* values into eq. 7.112 to obtain $W_{ba}(E)$.

Sometimes it is convenient to have an analytical expression of $W_{ba}(E)$ to show the explicit dependence of $W_{ba}(E)$ on E. For this purpose we solve eq. 7.109 by introducing an average frequency $\langle\omega\rangle$, that is,

$$E = \frac{Nh\langle\omega\rangle}{\exp(\beta^*\hbar\langle\omega\rangle) - 1} = N\bar{n}^*\hbar\langle\omega\rangle, \qquad \bar{n}^* = \frac{E}{Nh\langle\omega\rangle} \quad (7.113)$$

where N represents the effective number of modes of the molecule. Approximating $\bar{n}_j^*$ in eq. 7.112 by $\bar{n}^*$ and substituting t^* in eq. 7.104 into eq. 7.112, we obtain

$$W_{ba}(E) = W_{b0}\left[1 + \left(\frac{E}{Nh\langle\omega\rangle}\right)\left\{1 + \left(\frac{S\langle\omega'\rangle}{\Delta\omega}\right)^{2\omega_i'/\langle\omega'\rangle}\right\}\right]$$
$$\times \exp\left[\left(\frac{E}{Nh\langle\omega\rangle}\right)\left(\sum_j A_j\right)\right] \qquad (7.114)$$

As we can see, the average rate constant $W_{ba}(E)$ depends on energy E exponentially. If we replace ω_j' by the average frequency, then eq. 7.114 becomes

$$W_{ba}(E) = W_{b0}\left\{1 + \left(\frac{E}{Nh\langle\omega\rangle}\right)\left[1 + \left(\frac{S\langle\omega\rangle}{\Delta\omega}\right)^2\right]\right\}\exp\left(\frac{E\bar{A}}{h\langle\omega\rangle}\right) \qquad (7.115)$$

If we substitute eq. 7.91 for $W_{ba}(\beta)$ into eq. 7.108, the expression for $W_{ba}(E)$ takes the form

$$W_{ba}(E) = \sum_{v'} F_{v'} W_{bv'} \qquad (7.116)$$

where $W_{bv'}$ is the decay rate constant of a single vibronic state (see eq. 7.99) and $F_{v'}$ represents the equilibrium microcanonical distribution function.

$$F_{v'} = \delta(E - E_{bv'})[\rho(E)]^{-1} = \frac{\delta(E - E_{bv'})}{[\sum_{v'} \delta(E - E_{bv'})]^{-1}} \qquad (7.117)$$

In other words, when the vibrational relaxation of isolated molecules is much faster than the electronic relaxation, the decay rate constant $W_{ba}(E)$ is simply the microcanonical average of $W_{bv'}$. Equation 7.116 can easily be recast into eq. 7.108. Also, $F_{v'}$ is expressible in the form of the Boltzmann factor with β replaced by β^* when the method of steepest descent applies.

Application. The calculation of the nonradiative lifetimes of single vibronic states ($^1B_{2u}$) of C_6H_6 and C_6D_6 has been carried out by Heller et al. (1972). In their calculation they have considered the effect of the changes of normal frequencies and coordinates between the two electronic states on the nonradiative lifetimes. Their theoretical values of the nonradiative lifetimes of single vibronic states of C_6H_6 and C_6D_6 agree well with the experimental results of Spears and Rice (1971) and Abramson et al. (1972).

First, we consider the effect of the displacements of normal coordinates on the nonradiative decay rate constant of a single vibronic state. According to Heller et al. only the totally symmetric v_1 (C—C skeletal) and v_2 (C—H stretching) modes in benzene undergo nonnegligible shifts of their equilibrium displacement in the $^1B_{2u} \rightarrow {}^3B_{1u}$ transition. The values of Δ_1^2 and Δ_2^2

Table 7.4
Effect of Normal Coordinate Changes on
Radiationless Rate Constants of Single
Vibronic States

$W_{ba}(v'_1)/W_{ba}(0)$	This work	Heller et al.
$W(1)/W(0)$	1.15	1.14
$W(2)/W(0)$	1.31	1.30
$W(3)/W(0)$	1.48	1.47

have been chosen as $\Delta_1^2 = 0.050$ and $\Delta_2^2 = 0.004$. Other quantities used are $\Delta E_{ba} = \hbar\omega_{ba} = 8200$ cm^{-1}, $\omega'_i = 1500$ cm^{-1}, $\omega'_1 = 923$ cm^{-1}, and $\omega'_2 = 3130$ cm^{-1}. If we ignore the frequency changes and consider only progressions in the v_1 mode (i.e., the v_1 mode is optically excited), then we can calculate $W_{ba}(v'_1)/W_{ba}(v'_1 = 0)$ by using eq. 7.65 and the results are given in Table 7.4.

Next we consider the effect of frequency changes on nonradiative decay. For this purpose the frequency is assumed to be due to the nontotally symmetric mode v_G ($W_{G'} = 521$ cm^{-1}), which is also chosen as the optical mode. In Table 7.5 are given the calculated results of $W_{ba}(v'_6)/W_{ba}(v'_6 = 0)$ for benzene. To compare the theory with experiment, Table 7.6 presents the calculated results of radiationless transition rate constants of the single vibronic states of C_6H_6 (Heller et al., 1972) along with the experimental data (Spears and Rice, 1971; Abramson et al., 1972); the agreement is reasonably good. However, in the calculation of the relative rate constants, it is assumed that the electronic matrix element can be canceled.

Table 7.5
Effect of Normal Frequency Changes on Radiationless
Rate Constants of Single Vibronic States

$\Delta\omega_6$ (cm^{-1})	$W(1)/W(0)$	$W(2)/W(0)$	$W(3)/W(0)$
50	1.15	1.33	1.54
100	1.40	2.15	3.34
-100	0.859	0.750	0.676
-50	0.912	0.836	0.772

Table 7.6
Radiationless Rate Constants of Single
Vibronic States

Vibronic state	Experiment	Heller et al.
$6^1 1^0$	1.19	1.15
$6^0 1^1$	1.22	1.27
$6^2 1^0$	1.27	1.32
$6^1 1^1$	1.42	1.45
$6^0 1^2$	1.73	1.62
$6^2 1^1$	1.64	1.67
$6^1 1^2$	1.94	1.84
$6^0 1^3$	2.42	2.05
$6^2 1^2$	2.67	2.12

7.4 ENERGY TRANSFER IN CONDENSED PHASE

The term *resonance energy transfer* refers to the nonradiative transfer of electronic excitation energy from an excited donor molecule D^* to an acceptor molecule A. The transfer process may take place intermolecularly and intramolecularly. It may be expected that the energy of the excited acceptor molecule A^* must be lower than that of the excited donor molecule D^* if the energy transfer process is to be efficient, and that the sensitized excitation of the acceptor molecule A by the donor molecule D must occur within the time that the donor molecule remains in the excited state. The theory of resonance energy transfer has been developed for calculating the rate of energy transfer by dipole-dipole interaction (Förster, 1948), dipole-quadrupole interaction, and exchange interaction (Dexter, 1953). This theory is presented here (Lin, 1971; 1973). The diffusion-controlled energy transfer (or quenching) process is discussed in Chapter Nine.

If the rate of vibrational relaxation is much faster than that of energy transfer, then the average energy transfer probability (or rate constant) can be written as

$$P_{ab}(T) = \frac{2\pi}{\hbar} \sum_{v'} \sum_{v''} P_{av'} |\langle \psi_{av'} | \hat{H}' | \psi_{bv''} \rangle|^2 \delta(E_{av'} - E_{bv''}) \qquad (7.118)$$

where $P_{av'}$ represents the Boltzmann factor and $\hat{H}'$, the interaction energy between the donor D and acceptor A. In eq. 7.118 $\psi_{av'}$ denotes the state in which the donor molecule is electronically excited whereas the acceptor molecule is electronically unexcited and $\psi_{bv''}$ represents the state in which the donor molecule is in the lower electronic state and the acceptor molecule is electronically excited. When we use the adiabatic approximation eq. 7.118 becomes

$$P_{ab}(T) = \frac{2\pi}{\hbar} \sum_{v'} \sum_{v''} P_{av'} |\langle \theta_{av'} | H'_{ab} | \theta_{bv''} \rangle|^2 \delta(E_{av'} - E_{bv''}) \qquad (7.119)$$

where $\theta_{av'}$ and $\theta_{bv''}$ represent the wave functions of nuclear motion of the system and $H'_{ab} = \langle \Phi_a | \hat{H}' | \Phi_b \rangle$, Φ_a and Φ_b being the electronic wavefunctions of the system.

Introducing the integral representation of the delta function into eq. 7.119, yields

$$P_{ab}(T) = \frac{1}{\hbar^2} \int_{-\infty}^{\infty} dt \sum_{v'} \sum_{v''} P_{av'} |\langle \theta_{av'} | H'_{ab} | \theta_{bv''} \rangle|^2 \exp\left[\frac{it}{\hbar}(E_{bv''} - E_{av'})\right]$$

$$(7.120)$$

Now we shall show how the preceding expression for the energy transfer probability can be reduced to that obtained by Förster and Dexter. For this purpose we consider the singlet-singlet (or triplet-singlet) transfer and triplet-triplet transfer separately.

7.4.1   Singlet-Singlet and Triplet-Singlet Transfers

In this case the matrix element H'_{ab} can be expressed as

$$H'_{ab} = \langle \Phi_a | \hat{H}' | \Phi_b \rangle \approx \left\langle \Phi_a \left| \sum_i \sum_j \frac{e^2}{\varepsilon r_{ij}} \right| \Phi_b \right\rangle$$

$$\approx \frac{1}{\varepsilon R^3} \left[\mathbf{R}_{D*D} \cdot \mathbf{R}_{AA*} - \frac{3}{R^2} (\mathbf{R} \cdot \mathbf{R}_{D*D})(\mathbf{R} \cdot \mathbf{R}_{AA*}) \right]$$

$$(7.121)$$

where $\mathbf{R}_{D*D}$ and $\mathbf{R}_{AA*}$ represent the electronic transition moments of the donor and acceptor, respectively; ε, the dielectric constant of the medium; and R, the intermolecular distance between the donor and acceptor. For both singlet-singlet and triplet-singlet transfers, it has been shown that when the exchange effect is ignored, the matrix element H'_{ab} can be approximated by the Coulomb integral for the electronic interaction between the donor and acceptor, which in turn can be reduced to the last equation in eq. 7.121 by using the multipole expansion and retaining only the dominating term.

Substituting eq. 7.121 into eq. 7.120 and assuming that the donor and acceptor are randomly oriented, we obtain

$$P_{ab}(T) = \frac{1}{3\hbar^2 \varepsilon^2 R^6} \int_{-\infty}^{\infty} dt F_A(t) F_D(t) \qquad (7.122)$$

where $F_A(t)$ and $F_D(t)$ are defined by

$$F_A(t) = \sum_{v'} \sum_{v''} P_{av'}^{(A)} |\langle \theta_{av'}^{(A)} | \mathbf{R}_{AA*} | \theta_{bv''}^{(A)} \rangle|^2 \exp(it\omega_{bv'', av'}^{(A)}) \qquad (7.123)$$

for the acceptor and

$$F_D(t) = \sum_{v'} \sum_{v''} P_{av'}^{(D)} |\langle \theta_{av'}^{(D)} | \mathbf{R}_{D*D} | \theta_{bv''}^{(D)} \rangle|^2 \exp(it\omega_{bv'', av'}^{(D)}) \qquad (7.124)$$

for the donor, respectively. In the adiabatic approximation the expression for the molecular absorption coefficient of the electronic transition $a \rightarrow b$ has been shown to be expressed as (Lin, 1968, 1971b, Hill and Lin, 1971)

$$k_{ab}(\omega) = \frac{4\pi^2 \omega}{3\hbar\alpha_a c} \sum_{v'} \sum_{v''} P_{av'} |\langle \theta_{av'} | \mathbf{R}_{ab} | \theta_{bv''} \rangle|^2 \delta(\omega_{bv'', av'} - \omega) \qquad (7.34)$$

where α_a is a function of refractive index introduced to correct the medium effect.

Introducing the integral expression for the delta function into eq. 7.34 and using eq. 7.123 yields

$$
\begin{aligned}
k_{ab}^{(A)}(\omega) &= \frac{2\pi\omega}{3\hbar\alpha_a c} \int_{-\infty}^{\infty} dt \, \exp(-it\omega) \sum_{v'} \sum_{v''} P_{av'}^{(A)} |\langle \theta_{av'}^{(A)} | \mathbf{R}_{AA*} | \theta_{bv''}^{(A)} \rangle|^2 \\
&\quad \times \exp(it\omega_{bv'', av'}^{(A)}) \\
&= \frac{2\pi\omega}{3\hbar\alpha_a c} \int_{-\infty}^{\infty} dt \, \exp(-it\omega) F_A(t)
\end{aligned} \qquad (7.125)
$$

Inverting the Fourier transformation, we obtain

$$F_A(t) = \frac{3\hbar\alpha_a c}{4\pi^2} \int_{-\infty}^{\infty} d\omega \, \exp(it\omega) \frac{k^{(A)}(\omega)}{\omega} \qquad (7.126)$$

If we substitute 7.126 into 7.34 the energy transfer probability is given by

$$P_{ab}(T) = \frac{\alpha_a c}{2\hbar\pi^2 \varepsilon^2 R^6} \int_{-\infty}^{\infty} dt \int_{-\infty}^{\infty} d\omega \, \exp(it\omega) \frac{k_{ab}^{(A)}(\omega)}{\omega} F_D(t) \qquad (7.127)$$

Carrying out the integration with respect to t, we find

$$P_{ab}(T) = \frac{ac}{\pi\varepsilon^2 R^6} \int_{-\infty}^{\infty} d\omega \, \frac{k_{ab}^{(A)}(\omega)}{\omega} \sum_{v'} \sum_{v''} P_{av'}^{(D)} |\langle \theta_{av'}^{(D)} | \mathbf{R}_{D*D} | \theta_{bv''}^{(D)} \rangle|^2 \delta(E_{av', bv''} - \hbar\omega)$$

$$(7.128)$$

It has been shown that the normalized intensity distribution function of emission for the electronic transition $a \to b$ in the adiabatic approximation can be written (Lin et al., 1971b)

$$I_{ab}(\omega)_N = \frac{4\alpha_e \omega^3}{3c^3 A_{ab}} \sum_{v'} \sum_{v''} P_{av'} |\langle \theta_{av'} | \mathbf{R}_{ab} | \theta_{bv''} \rangle|^2 \delta(E_{av', bv''} - \hbar\omega) \quad (7.35)$$

where A_{ab} represents the transition probability of spontaneous emission or the radiative rate constant and α_e the correction factor for medium effect. Intensity distribution function given in eq. 2.13 is in terms of number of quanta. In terms of energy, the intensity is proportional to the fourth power of radiation frequency. Combining eq. 7.35 with eq. 7.128, we obtain

$$P_{ab}(T) = \frac{3\alpha_a c^4 A_{ab}^{(D)}}{4\pi\alpha_e \varepsilon^2 R^6} \int_{-\infty}^{\infty} \frac{d\omega}{\omega^4} k_{ab}^{(A)}(\omega) I_{ab}^{(D)}(\omega)_N \quad (7.129)$$

This is exactly the same expression as obtained by Förster and Dexter. From the preceding derivation, we can see that eq. 7.129 is valid regardless of whether the electronic transitions of the donor and acceptor involved in the energy transfer are symmetry-allowed or symmetry-forbidden but vibronic-allowed.

7.4.2 Triplet-Triplet Transfers

Next we turn to the triplet-triplet transfer. In this case we use the Condon approximation to rewrite eq. 7.120 as

$$P_{ab}(T) = \frac{|H'_{ab}|^2}{\hbar^2} \int_{-\infty}^{\infty} dt f_A(t) f_D(t) \quad (7.130)$$

Again if the donor and acceptor molecules are randomly oriented, we have to replace $|H'_{ab}|^2$ by $|H'_{ab}|^2_{av}$, that is, we have to average the matrix element $|H'_{ab}|^2$ over all possible relative orientations. The functions $f_A(t)$ and $f_D(t)$ in eq. 7.130 are defined by

$$f_A(t) = \sum_{v'} \sum_{v''} P_{av'}^{(A)} |\langle \theta_{av'}^{(A)} | \theta_{bv''}^{(A)} \rangle|^2 \exp(it\omega_{bv'', av'}^{(A)}) \quad (7.131)$$

and

$$f_D(t) = \sum_{v'} \sum_{v''} P_{av'}^{(D)} |\langle \theta_{av'}^{(D)} | \theta_{bv'}^{(D)} \rangle|^2 \exp(it\omega_{bv'', av'}^{(D)}) \quad (7.132)$$

The functions $f_A(t)$ and $f_D(t)$ are closely related to $F_A(t)$ and $F_D(t)$; if we assume that the electronic transitions involved are symmetry-allowed or if the Franck–Condon approximation holds, then

$$f_D(t) = \frac{1}{|\mathbf{R}_{AA^*}|^2} F_D(t) \quad (7.133)$$

and

$$f_A(t) = \frac{1}{|\mathbf{R}_{AA^*}|^2} F_A(t) = \frac{3\hbar\alpha_a c}{4\pi^2} \frac{1}{|\mathbf{R}_{AA^*}|^2} \int_{-\infty}^{\infty} d\omega \exp(it\omega) \frac{k_{ab}^{(A)}(\omega)}{\omega} \quad (7.134)$$

For symmetry-allowed transitions, from eq. 7.34 we have

$$\int_{-\infty}^{\infty} d\omega \frac{k_{ab}^{(A)}(\omega)}{\omega} = \frac{4\pi^2}{3\hbar\alpha_a c} |\mathbf{R}_{AA^*}|^2 \quad (7.135)$$

Combining eq. 7.135 with eq. 7.134, we obtain

$$f_A(t) = \frac{\int_{-\infty}^{\infty} d\omega \exp(it\omega) k_{ab}^{(A)}(\omega)/\omega}{\int_{-\infty}^{\infty} d\omega k_{ab}^{(A)}(\omega)/\omega} = \int_{-\infty}^{\infty} d\omega \exp(it\omega) \sigma_{ab}^{(A)}(\omega) \quad (7.136)$$

where $\sigma_{ab}^{(A)}(\omega)$ is defined by

$$\sigma_{ab}^{(A)}(\omega) = \frac{k_{ab}^{(A)}(\omega)/\omega}{\int_{-\infty}^{\infty} d\omega k_{ab}^{(A)}(\omega)/\omega} \quad (7.137)$$

which represents the normalized distribution function for absorption. Substituting eq. 7.136 into eq. 7.127 yields

$$P_{ab}(T) = \frac{|H'_{ab}|^2}{\hbar^2} \int_{-\infty}^{\infty} dt \int_{-\infty}^{\infty} d\omega \exp(it\omega) \sigma_{ab}^{(A)}(\omega)$$

$$\times \sum_{v'} \sum_{v''} P_{av'}^{(D)} |\langle \theta_{av'}^{(D)} | \theta_{bv''}^{(D)} \rangle|^2 \exp(it\omega_{bv'', av'}^{(D)}) \quad (7.138)$$

When we carry out the integration with respect to t, eq. 7.138 becomes

$$P_{ab}(T) = \frac{2\pi}{\hbar^2} |H'_{ab}|^2 \int_{-\infty}^{\infty} d\omega \sigma_{ab}^{(A)}(\omega)$$

$$\times \sum_{v'} \sum_{v''} P_{av'}^{(D)} |\langle \theta_{av'}^{(D)} | \theta_{bv''}^{(D)} \rangle|^2 \delta(\omega_{av', bv''}^{(D)} - \omega)$$

$$= \frac{2\pi}{\hbar^2} |H'_{ab}|^2 \int_{-\infty}^{\infty} d\omega \sigma_{ab}^{(A)}(\omega) \eta_{ab}^{(D)}(\omega) \quad (7.139)$$

where the spectral distribution function of emission $\eta_{ab}^{(D)}(\omega)$ is defined by

$$\eta_{ab}^{(D)}(\omega) = \sum_{v'} \sum_{v''} P_{av'}^{(D)} |\langle \theta_{av'}^{(D)} | \theta_{bv''}^{(D)} \rangle|^2 \delta(\omega_{av', bv''}^{(D)} - \omega) \quad (7.140)$$

which is normalized. Equation 7.139 is similar to that derived by Dexter.

7.4.3 Temperature Effect, Energy Gap Law, and Isotope Effect

From the preceding discussion we can see that the expressions for the energy transfer probability and radiationless transition probability are closely related and the theoretical technique employed in radiationless transitions can be used in energy transfer. The method of steepest descent has been introduced in simplifying the expression for the nonradiative rate constant and the resulting expressions are simple enough that the numerical calculation of the radiationless rate constant is now possible. In view of this development in radiationless transitions, we present some theoretical results in energy transfer obtained from the use of this theoretical method. The derivation has been given in the previous section and is not repeated here. The resulting expression for $P_{ab}(T)$ in the Condon approximation is given by (Lin, 1972)

$$P_{ab}(T) = P_{ab}(0)\exp\left(\sum_j \bar{n}_j A_j\right) \tag{7.141}$$

where $\bar{n}_j = \{\exp(\hbar\omega'_j/kT) - 1\}^{-1}$ and $P_{ab}(0)$ represents the energy transfer probability at $T = 0$. The summation in eq. 7.141 covers the normal vibrations of both donor and acceptor. The A_j are defined by (Lin 1972)

$$A_j = \tfrac{1}{2}\Delta_j^2\left[\left(\frac{\Delta E_{ab}}{S\hbar\omega'}\right)^{\omega'_j/2\bar{\omega}'} - \left(\frac{\Delta E_{ab}}{S\hbar\omega'}\right)^{\omega'_j/2\bar{\omega}'}\right]^2 \tag{7.142}$$

where ΔE_{ab} represents the energy gap between the excited donor and ground state acceptor, Δ_j is related to changes in normal coordinates between the two electronic states under considerations, $S = \tfrac{1}{2}\sum_j \Delta_j^2$, and $\bar{\omega}'$ denotes a certain average frequency which usually can be approximated by the maximum normal frequency like the C—H stretching.

The rate constant of energy transfer at $T = 0$, $P_{ab}(0)$, evaluated by using the method of steepest descent, can be expressed in the energy gap law form

$$P_{ab}(0) = \frac{|H'_{ab}|^2_{av}}{\hbar^2} \exp\left(-\frac{\Delta E_{ab}}{\hbar\bar{\omega}'} \ln \frac{\Delta E_{ab}}{S\hbar\bar{\omega}'}\right)C_{ab} \tag{7.143}$$

where C_{ab} is defined by

$$C_{ab} = \sqrt{(4\pi)}e^{-S} \exp\left[\frac{1}{2}\sum_i \Delta_i^2\left(\frac{\Delta E_{ab}}{S\hbar\bar{\omega}'}\right)^{\omega'_i/\bar{\omega}'}\right]\left[\sum_i \Delta_i^2\omega'_i\left(\frac{\Delta E_{ab}}{S\hbar\bar{\omega}'}\right)^{\omega'_i/\bar{\omega}'}\right]^{-1/2} \tag{7.144}$$

When temperature is not high, eq. 7.141 can be simplified as (Lin, 1972)

$$P_{ab}(T) = P_{ab}(0)\left[1 + \sum_j A_j \exp\left(-\frac{\hbar\omega'_j}{kT}\right) + \cdots\right] \tag{7.145}$$

The quantities A_j are transferable to and from other processes like emission, absorption, radiationless transitions, and so on. Since the Condon approximation is used in eq. 7.141 and 7.143 the preceding expressions for the energy transfer probability cannot be used for the singlet-singlet transfer or triplet-singlet transfer in which the electronic transitions involved are symmetry-forbidden but vibronic-allowed. However, the derivation can be easily extended to include the case in which the electronic transitions involved in the energy transfer are symmetry-forbidden but vibronic-allowed, we do not present the derivation here.

7.4.4 Application

From eq. 7.129 we can see that $P_{ab}(T)$ for the singlet-singlet transfer can be conveniently written as

$$P_{ab}(T) = \frac{1}{\tau_D} \left(\frac{R_c}{R}\right)^6 = P_{ab}(R) \tag{7.146}$$

where τ_D is the lifetime of D^* and R_c is the critical separation of D and A for which energy transfer from D^* to A and decay from D^* are equally probable. For the triplet-triplet transfer, Dexter assumes that $|H'_{ab}|^2$ takes the exponential from $|H'_{ab}| \alpha \exp(-AR)$. In this case the energy transfer probability

$$P_{ab}(T) = \frac{1}{\zeta_D} \exp[-A(R - R_c)] = P_{ab}(R) \tag{7.147}$$

where A is constant.

In the presence of acceptors A the decay function I/I_0 (see eq. 7.7) of D^* is in general not exponential. To take into account the acceptor concentration effect on the donor emission intensity or decay function $\phi(t)$, one has to consider the spatial distribution of A around D^*. Following Siegel and Eisenthal (1964) and Inokuti and Hirayama (1965), we can express the decay function of D^* for flash excitation as

$$\phi(t) = \exp\left(-\frac{t}{\tau_D}\right) \lim_{\substack{N \to \infty \\ V \to \infty}} \left[\int_V \exp\{-tP_{ab}(R)\} W(R) \, dV\right]^N \tag{7.148}$$

where N represents the number of A molecules in the volume V. In eq. 7.147 the limit $N \to \infty$, $V \to \infty$ is so chosen that N/V, the concentration of A, is finite, and $W_{(R)}$ represents the distribution function of A molecules around D^*.

To show the calculation of $\phi(t)$ using eq. 7.148, we consider the singlet-singlet transfer. Introducing a new integration variable in eq. 7.148,

$$y = \frac{t}{\tau_D} \left(\frac{R_c}{R}\right)^6 \tag{7.149}$$

Table 7.7
Singlet-Singlet Transfer by the
Resonance Mechanism

Donor	Acceptor	R_c
1-Chloroanthracene	Perylene	41
1-Chloroanthracene	Rubrene	38
9-Cyanoanthracene	Rubrene	84

and assuming a random spatial distribution of A around D^*, that is, $W(R)\,dV = 4\Pi R^2\,dR/V$, we obtain

$$\phi(t) = \exp\left(-\frac{t}{\tau_D}\right) \lim_{N \to \infty} \left[\tfrac{1}{2}y^{1/2} \int_{y_v}^{\infty} y_v^{-3/2} \exp(-y)\,dy\right]^N \quad (7.150)$$

where $y_v = (t/\tau_D)(R_c/R_v)^6$ and $v = (4\Pi/3)R_v^3$. For $y_v \to 0$ we can use the approximation

$$\int_{y_v}^{\infty} y^{-3/2} \exp(-y)\,dy = 2y_v^{-1/2} \exp(-y_v) - 2P(\tfrac{1}{2}) + 0[y_v^{1/2}] \quad (7.151)$$

Substituting eq. 7.151 into eq. 7.150 yields

$$\phi(t) = \exp\left[-\frac{t}{\tau_D} - \Gamma\left(\frac{1}{2}\right)\frac{C}{C_0}\left(\frac{t}{\tau_D}\right)^{1/2}\right] \quad (7.152)$$

where $P(\tfrac{1}{2})$ is the gamma function, C is the concentration of A, and $C_0 = \tfrac{3}{4}\Pi R_c^3$. The decay function for steady-state excitation $\phi_s(t)$ is shown to be

Table 7.8
Critical Separation Distances for Energy Transfer
from Donor Triplet to Acceptor Singlet

Donor	Acceptor	R_c
Phenanthrene	Fluorescein	35
2-Acetonaphthone	Fluorescein	43
Triphenylamine	Chlorophyll a	54
4-Phenylbenzaldehyde	Chrysoidine	33
N,N-Dimethylamine	9-Methylanthracene	24

Table 7.9
Triplet-Triplet Transfer in Rigid Media at 77°K

Donor[a]	$E_T{}^b$	$E_S{}^c$	R_c	Acceptor[a]	$E_T{}^b$	$E_{S_1}{}^c$
Benzophenone	69	74	13	Naphthalene	61	89
Benzaldehyde	72	76	12	Naphthalene	61	89
Carbazole	70	84	15	Naphthalene	61	89
Diphenylamine	72	89	13	Biphenyl	65	97

[a] Donor and acceptor concentration about 1.0 M in ethanol and ether solvent. Excitation provided by 3660 Å light.
[b] Energy of lowest triplet level in kcal/mole.
[c] Energy of lowest singlet level in kcal/mole.

related to the decay function for flash excitation $\phi(t)$ as follows (Siegel and Eisenthal, 1964)

$$\phi_s(t) = \frac{\int_t^\infty (t^1) \, dt^1}{\int_0^\infty (t^1) \, dt^1} \tag{7.153}$$

The preceding discussion (eqs. 7.148–7.153) describes how the decay functions of D^* can be obtained and analyzed to obtain physical parameters once the energy transfer probability $P_{ab}(T)$ is obtained. In Tables 7.7–7.9 are shown some experimental results of R_c for singlet-singlet, triplet-singlet, and triplet-triplet transfers (Turro, 1965).

7.5 ENERGY TRANSFER AND QUENCHING IN GASES

When a molecule (donor D^*) is in an excited electronic state and collides with another molecule (acceptor A) the electronic energy may transfer to the acceptor

$$D^* + A \longrightarrow D + A^* \quad \text{(transfer)}$$

or the electronic energy may transform into the energy of nuclear motion of A and D through collision,

$$D^* + A \longrightarrow D + A \quad \text{(quenching)}$$

For convenience, the first case will be called the *energy transfer* and the second will be called the *quenching of electronic excitation*.

Another way for energy transfer or quenching to take place is through the formation of a long-lived collision intermediate or complex and the redistribution or transformation of the electronic energy (nonradiative process) before the donor and acceptor molecules move apart. This case involves several elementary processes and is not discussed in this chapter.

Since from the theoretical viewpoint both energy transfer and quenching can be treated similarly, only the treatment of energy transfer is presented in this chapter.

7.5.1   General Considerations

In the gas phase the luminescent intensity of the excited donor in the presence of quencher (or acceptor molecules) can be described by the well-known Stern–Volmer equation

$$\frac{I}{I_0} = \frac{1}{1 + \beta P_A} \tag{7.154}$$

where I_0 is the luminescent intensity of excited donor in the absence of quencher (or acceptor) and I, the luminescent intensity in the presence of quencher (or acceptor). The term β in eq. 7.154 is a constant and P_A, the partial pressure of acceptors. Equation 7.154 can be derived from the consideration of the following kinetic mechanisms:

$$D \xrightarrow{k_a} D^*; \qquad D^* \xrightarrow{k_f} D;$$

$$D^* + A \xrightarrow{k_c} D + A^* \quad \text{(or } A\text{)} \tag{7.155}$$

The first equation represents the excitation of donors by absorption of radiation; k_a denotes a collection of constants including the donor absorption coefficient and the incident radiation intensity. The second equation represents the unimolecular deexcitation of excited donors; the rate constant k_f includes both radiative and nonradiative processes. The last equation represents the collisional energy transfer from the donor to the acceptor.

If the steady state for D^* is reached, then the concentration of excited donors can be obtained as

$$C_{D^*} = \frac{k_a C_D}{k_f + k_c C_A} \tag{7.156}$$

and the intensity of donor luminescence is given by

$$I = \frac{k_r k_a C_D}{k_f + k_c C_A} \tag{7.157}$$

where k_r represents the radiative rate constant of donors. From eq. 7.157 we obtain the intensity of excited donor luminescence in the absence of acceptors as

$$I_0 = \frac{k_r k_a C_D}{k_f}$$

Combining this relation with eq. 7.157 yields

$$\frac{I}{I_0} = \frac{k_f}{k_f + k_c C_A} = \frac{1}{1 + \tau k_c C_A} \tag{7.158}$$

where $\tau = 1/k_f$. Equation 7.48 reduces to eq. 7.154 as C_A is proportional to P_A. From the preceding discussion we can see that the central problem in the energy transfer of electronic excitation in the gas phase lies in the evaluation of the bimolecular rate constant of collisional energy transfer k_c or equivalently the evaluation of inelastic collisional cross sections.

To treat the inelastic collision between the donor and acceptor molecules quantum mechanically, we consider the solution of the Schrödinger equation,

$$\left[\hat{H}_A + \hat{H}_D - \left(\frac{\hbar^2}{2\mu} \right) \nabla_r^2 + V(\mathbf{r}, \mathbf{r}_A, \mathbf{r}_D) \right] \Psi = (E_A + E_D + \tfrac{1}{2}\mu v^2)\Psi \tag{7.159}$$

where the donor molecule is distinguished by the suffix D and the acceptor or quencher molecule by the suffix A. The terms $-(\hbar^2/2\mu)\nabla_r^2$ and $\frac{1}{2}\mu v^2$ represent the kinetic energy operator and kinetic energy of the relative motion between A and D. $V(\mathbf{r}, \mathbf{r}_A, \mathbf{r}_D)$ is the interaction energy between A and D. In eq. 7.159 the motion of the center of mass of the complete system has been separated. In other words, we have chosen a coordinate frame in which the center of mass of the complete system is at rest. Equation 7.159 can usually be solved by introducing the expansion (Mott and Massey, 1965; Massey et al., 1971)

$$\Psi(\mathbf{r}, \mathbf{r}_A, \mathbf{r}_D) = \sum_m \psi_m(\mathbf{r}, \mathbf{r}_A) F_m(\mathbf{r}) \tag{7.160}$$

in which the $F_m(\mathbf{r})$ are to be determined from

$$\left[\frac{\hbar^2}{2\mu} \nabla_r^2 + (E_T - E_m) \right] F_m(\mathbf{r}) = \iint V(\mathbf{r}, \mathbf{r}_A, \mathbf{r}_D)\Psi\psi_m^* \, d\mathbf{r}_A \, d\mathbf{r}_D = \sum_n F_n(\mathbf{r})$$

$$\times \iint V(\mathbf{r}, \mathbf{r}_A, \mathbf{r}_D)\psi_n\psi_m^* \, d\mathbf{r}_A \, d\mathbf{r}_D = \sum_n F_n(\mathbf{r})V_{mn}(\mathbf{r}) \tag{7.161}$$

where E_T is the total energy of the system.

If the $m = 0$ state denotes the initial state of the system, then the function $F_0(\mathbf{r})$ must represent the sum of an incident and scattered wave; thus $F_0(\mathbf{r})$ must have the asymptotic form

$$F_0(\mathbf{r}) \sim e^{ik_0 z} + \frac{1}{r} e^{ik_0 r} f_0(\theta \phi) \tag{7.162}$$

where $k_0 = \mu v / \hbar$.

The function $F_m(\mathbf{r})$ must represent scattered waves only and so have the asymptotic form

$$F_m(\mathbf{r}) \sim \frac{1}{r} e^{ik_m r} f_m(\theta \phi) \tag{7.163}$$

where $k_m^2 = (2\mu/\hbar^2)(E_0 + \frac{1}{2}\mu v^2 - E_m)$. From eq. 7.163, the differential cross section of an inelastic collision can be obtained as

$$\sigma_{0, m}(\theta, \phi) = \left(\frac{k_m}{k_0}\right) |f_m(\theta \phi)|^2 \tag{7.164}$$

With the aid of Green functions, eq. 7.161 may be solved in the integral form (Mott and Massey, 1965; Massey et al., 1971)

$$F_n(\mathbf{r}) \sim -\frac{\mu}{2\pi\hbar^2} \frac{\exp(ik_n r)}{r} \iiint \exp(-i\mathbf{k}_n \cdot \mathbf{r}) \psi_n^* V \Psi \, d\mathbf{r}_A \, d\mathbf{r}_D \, d\mathbf{r} \tag{7.165}$$

Many approximation methods have been developed to solve eq. 7.161 (Mott and Massey, 1965; Massey et al., 1971). Dickens, Linnett, and Sovers (1962) in discussing the collisional energy transfer between atoms and diatomic molecules, that is,

$$A^* + BC \longrightarrow A + (BC)^*$$

have employed the distorted wave method [i.e., assuming that only $V_{00}(\mathbf{r})$ and $V_{nn}(\mathbf{r})$ are nonzero in obtaining $F_n(\mathbf{r})$] to calculate the energy transfer cross section. In their calculation they assumed simple approximate functional forms for the matrix elements V_{nn}, V_{00}, and so on, of interaction potential rather than calculating them theoretically.

In this chapter we use the Born approximation, which is sufficient for semiquantitative investigation. Within the validity of the first Born approximation, we may replace Ψ in eq. 7.165 by the product of an incoming plane wave and the wave function of the initial unperturbed molecular state; $\Psi = \exp(i\mathbf{k}_0 \cdot \mathbf{r})\psi_0(\mathbf{r}_A, \mathbf{r}_D)$. Thus the Born amplitude for the collision between A and D^* for the transition from the state $\psi_0(\mathbf{r}_A, \mathbf{r}_D)$ to the state $\psi_m(\mathbf{r}_A, \mathbf{r}_D)$ is given by

$$F_m(\mathbf{r}) = -\frac{\mu}{2\pi\hbar^2} \frac{\exp(ik_m r)}{r} \iiint \exp[i(\mathbf{k}_0 - \mathbf{k}_m) \cdot \mathbf{r}] \psi_m^* V \psi_0 \, d\mathbf{r}_A \, d\mathbf{r}_D \, d\mathbf{r} \tag{7.166}$$

and the corresponding differential cross section can be written as

$$\sigma_{0,m}(\theta\phi) = \frac{\mu^2 k_m}{4\pi^2\hbar^4 k_0}\left|\iiint V(\mathbf{r},\mathbf{r}_A,\mathbf{r}_D)\exp(i\mathbf{K}\cdot\mathbf{r})\psi_m^*\psi_0\,d\mathbf{r}_A\,d\mathbf{r}_D\,d\mathbf{r}\right|^2 \quad (7.167)$$

where $\mathbf{K} = \mathbf{k}_0 - \mathbf{k}_m$. In the adiabatic approximation eq. 7.167 can be rewritten as

$$\sigma_{av',bv''}(\theta,\phi) = \frac{\mu^2 k_{bv''}}{4\pi^2\hbar^4 k_{av'}}\left|\iiint V(\mathbf{r},\mathbf{r}_A,\mathbf{r}_D)\exp(i\mathbf{K}\cdot\mathbf{r})\psi_{bv''}^*\psi_{av'}\,d\mathbf{r}\,d\mathbf{r}_A\,d\mathbf{r}_D\right|^2$$

$$(7.168)$$

and the total cross section for the transition $av' \to bv''$ is given by

$$Q_{av',bv''} = \frac{2\pi}{k_{av'}k_{bv''}}\int_{|k_{av'}-k_{bv''}|}^{k_{av'}+k_{bv''}}\sigma_{av',bv''}(K)K\,dK \quad (7.169)$$

We next discuss the inelastic collisions for the singlet-singlet (or triplet-singlet) transfer and triplet-triplet transfer separately.

7.5.2 Singlet-Singlet and Triplet-Singlet Transfers

As has been pointed out by Förster and Dexter, the energy transfer between D^* and A involves two electrons; hence for the interaction energy $V(\mathbf{r},\mathbf{r}_A,\mathbf{r}_D)$, we may ignore the one-electron operators and consider only the electron-electron interaction between D and A. Thus if Φ_a and Φ_b represent the electronic wave functions of the initial and final states, respectively, then we have

$$\langle\Phi_b|V(\mathbf{r},\mathbf{r}_A,\mathbf{r}_D)|\Phi_a\rangle = V_{ba} = \left\langle\Phi_b\left|\sum_i\sum_j\frac{e^2}{r_{ij}}\right|\Phi_a\right\rangle = 2\left\langle X_{D^*}X_A\left|\frac{e^2}{r_{ij}}\right|X_D X_{A^*}\right\rangle$$

$$- \left\langle X_{D^*}X_A\left|\frac{e^2}{r_{ij}}\right|X_{A^*}X_D\right\rangle = V_{ba}^{(c)} + V_{ba}^{(e)} \quad (7.170)$$

where $X_{D^*}, X_D)$ and (X_{A^*}, X_A) are the molecular orbitals (or atomic orbitals) of the donor and acceptor molecules, respectively. The first term in eq. 7.170 represents the Coulomb interaction $V_{ba}^{(c)}$ and the second term, the exchange interaction $V_{ba}^{(e)}$. In discussing the energy transfer in the solid phase the exchange interaction is usually neglected (cf. eq. 7.121), because the donors and acceptors are rather far apart; in the gas phase, this may not be the case, because during the collision between A and D^*, the A and D^* molecules may actually come close together so that there is a considerable overlap of electronic charge.

Substituting eq. 7.170 into 7.168 yields

$$\sigma_{av', bv''}(\theta, \phi) = \frac{\mu^2 k_{bv''}}{4\pi^2 \hbar^4 k_{av'}} | U^{(c)}_{bv'', av'} + U^{(e)}_{bv'', av'} |^2 \qquad (7.171)$$

where

$$U^{(c)}_{bv'', av'} = \left\langle \theta_{bv''} \left| \int d\mathbf{r} \exp(i\mathbf{K} \cdot \mathbf{r}) V^{(c)}_{ba} \right| \theta_{av'} \right\rangle \qquad (7.172)$$

and

$$U^{(e)}_{bv'', av'} = \left\langle \theta_{bv''} \left| \int d\mathbf{r} \exp(i\mathbf{K} \cdot \mathbf{r}) V^{(e)}_{ba} \right| \theta_{av'} \right\rangle \qquad (7.173)$$

Tentatively, we ignore the exchange effect (i.e., the term, $U^{(e)}_{bv'', av'}$), and return to this point after we discuss the triplet-triplet transfer. To simplify $U^{(c)}_{bv'', av'}$, we notice that $r_{ij} = |(\mathbf{r}_{Di} - \mathbf{r}_{Aj})|$. Making use of this relation and the Bethe integral (Bethe, 1930; Mott and Massey, 1965),

$$\int \frac{\exp(i\mathbf{K} \cdot \mathbf{r}_i)}{|\mathbf{r}_i - \mathbf{r}_j|} d\mathbf{r}_i = \frac{4\pi}{K^2} \exp(i\mathbf{K} \cdot \mathbf{r}_j) \qquad (7.174)$$

the integration over $d\mathbf{r}$ can be readily performed to yield

$$U^{(c)}_{bv'', av'} = \frac{4\pi}{K^2} \langle \theta_{bv''} | [\sqrt{2} \langle X_{D*} | e \exp(i\mathbf{K} \cdot \mathbf{r}_{Di}) | X_D \rangle]$$
$$\times [\sqrt{2} \langle X_A | e \exp(-i\mathbf{K} \cdot \mathbf{r}_{Aj}) | X_{A*} \rangle] | \theta_{av'} \rangle \qquad (7.175)$$

and the differential cross section in this case is given by

$$\sigma^{(c)}_{av', bv''}(\theta, \phi) = \frac{4\mu^2 k_{bv''}}{K^4 \hbar^4 k_{av'}} | \langle \theta_{bv''} | [\sqrt{2} \langle X_{D*} | e \exp(i\mathbf{K} \cdot \mathbf{r}_{Di}) | X_D \rangle]$$
$$\times [\sqrt{2} \langle X_A | e \exp(-i\mathbf{K} \cdot \mathbf{r}_{Aj}) | X_{A*} \rangle] | \theta_{av'} \rangle |^2 \qquad (7.176)$$

To obtain some qualitative features of eq. 7.176, we use the relation

$$\exp(i\mathbf{K} \cdot \mathbf{r}_i) = 1 + (i\mathbf{K} \cdot \mathbf{r}_i) + \cdots$$

To the dipole approximation eq. 7.176 becomes

$$\sigma^{(c)}_{av', bv''}(\theta\phi) = \frac{4\mu^2 k_{bv''}}{\hbar^4 k_{av'}} | \langle \theta_{bv''} | (\hat{\kappa} \cdot \mathbf{R}_{D*D})(\hat{\kappa} \cdot \mathbf{R}_{AA*}) | \theta_{av'} \rangle |^2 \qquad (7.177)$$

where $\mathbf{R}_{AA*}$ and $\mathbf{R}_{D*D}$ represent the transition moments of the acceptor and donor, respectively, and $\hat{\kappa}$ denotes the unit vector along the direction of $\mathbf{K}$.

Comparing 7.177 with the corresponding expression for the energy transfer probability of singlet-singlet (or triplet-singlet) transfers in the solid phase, we can see that they are closely related. If the Franck–Condon principle applies, eq. 7.177 can be written

$$\sigma_{av',bv''}^{(c)}(\theta, \phi) = \frac{4\mu^2 k_{bv''}}{\hbar^4 k_{av'}} |\mathbf{R}_{D*D}|^2 |\mathbf{R}_{AA*}|^2 |\langle \theta_{bv''}|(\hat{\kappa} \cdot \hat{r}_{D*D})(\hat{\kappa} \cdot \hat{r}_{AA*})|\theta_{av'}\rangle|^2$$

$$(7.178)$$

where $\hat{r}_{D*D}$ and $\hat{r}_{AA*}$ are unit vectors,

$$\hat{r}_{D*D} = \frac{\mathbf{R}_{D*D}}{|\mathbf{R}_{D*D}|} \quad \text{and} \quad \hat{r}_{AA*} = \frac{\mathbf{R}_{AA*}}{|\mathbf{R}_{AA*}|}$$

The wave functions $\theta_{av'}$ and $\theta_{bv''}$ of nuclear motion include the wave functions of both molecular rotation and vibration, and because of the existence of the factors $(\hat{\kappa} \cdot \hat{r}_{D*D})$ and $(\hat{\kappa} \cdot \hat{r}_{AA*})$ in the rigid rotator approximation the rotational motion of A and D can either absorb or deliver one rotational quantum in the energy transfer. In discussing the role of molecular rotation in the energy transfer, we can take the axis of quantization to lie along $\mathbf{K}$. Substituting eq. 7.178 into eq. 7.169, we obtain the approximate expression for the total cross section of $av' \rightarrow bv''$ as

$$Q_{av',bv''}^{(c)} = \frac{16\pi\mu^2 k_{bv''}}{\hbar^4 k_{av'}} |\mathbf{R}_{D*D}|^2 |\mathbf{R}_{AA*}|^2 |\langle \theta_{bv''}|(\hat{\kappa} \cdot \hat{r}_{D*D})(\hat{\kappa} \cdot \hat{r}_{AA*})|\theta_{av'}\rangle|^2 \quad (7.179)$$

This indicates that the energy transfer cross sections vary linearly with respect to the squares of the transition moments of A and D. Chiu and Gordon (Gordon and Chiu, 1971; Chiu, 1971), in discussing the electronic energy transfer of NO molecules, have also used the first Born approximation to calculate the energy transfer cross sections. In their calculation they have used for the matrix element of the electron-electron interaction between the donor and acceptor molecules

$$V_{ba} = \langle \Phi_b | V(\mathbf{r}, \mathbf{r}_A, \mathbf{r}_D) | \Phi_a \rangle = \frac{1}{r^3} \left[(\mathbf{R}_{D*D} \cdot \mathbf{R}_{AA*}) - \frac{3}{r^2} (\mathbf{r} \cdot \mathbf{R}_{D*D})(\mathbf{r} \cdot \mathbf{R}_{AA*}) \right]$$

$$(7.180)$$

that is, the dipole-dipole interaction between A and D as in the solid phase and ignored the exchange effect completely. In the Franck–Condon approximation the energy transfer (or quenching) cross sections obtained by using the electronic dipole-dipole interaction equation 7.180 also vary linearly with $|\mathbf{R}_{D*D}|^2 |\mathbf{R}_{AA*}|^2$.

7.5.3 Triplet-Triplet Transfers

In this case it has been shown that the electronic interaction between A and D (i.e., $\langle \Phi_b | V(\mathbf{r}, \mathbf{r}_A, \mathbf{r}_D) | \Phi_a \rangle$) is of the exchange type and the differential cross section of triplet-triplet transfers can thus be written

$$\sigma_{av', bv''}(\theta, \phi) = \frac{\mu^2 k_{bv''}}{4\pi^2 \hbar^4 k_{av'}} | U_{bv'', av'}^{(e)} |^2 = \frac{\mu^2 k_{bv''}}{4\pi^2 \hbar^4 k_{av'}} | \langle \theta_{bv''} |$$

$$\times \int d\mathbf{r} \, \exp(i\mathbf{K} \cdot \mathbf{r}) V_{ba}^{(e)} | \theta_{av'} \rangle |^2 \tag{7.181}$$

A general analytical expression for the exchange integral $V_{ba}^{(e)}$ is not available. However, in discussing the triplet-triplet transfer in the solid phase, Dexter assumed that the exchange integral $V_{ba}^{(e)}$ is proportional to the overlap integral, that is,

$$V_{ba}^{(e)} = \frac{\omega e^2}{R_0} \exp\left(-\frac{r}{L}\right) \tag{7-182}$$

where R_0 and L are constant. The ω factor in eq. 7.182 should in reality be a function of the relative orientation between D^* and A but for our purpose of semiquantitative discussion, we for convenience regard ω to be constant. Under these conditions we have

$$\int V_{ba}^{(e)} \exp(i\mathbf{K} \cdot \mathbf{r}) \, d\mathbf{r} = \frac{8\pi\omega e^2}{LR_0(K^2 + 1/L^2)^2} \tag{7.183}$$

$$\sigma_{av', bv''}(\theta, \phi) = \frac{16\mu^2 k_{bv''}}{L^2 R_0^2 (K^2 + 1/L^2)^4} | \langle \theta_{bv''} | \theta_{av'} \rangle |^2 \tag{7.184}$$

and

$$Q_{av', bv''} = \frac{16\mu\pi^2}{3\hbar^4 k_{av'}^2} \cdot \frac{\omega^2 e^4}{L^2 R_0^2} \left[\frac{1}{(K_{min}^2 + L^{-2})^3} - \frac{1}{(K_{max}^2 + L^{-2})^3} \right] | \langle \theta_{bv''} | \theta_{av'} \rangle |^2 \tag{7.185}$$

where $K_{min} = |k_{av'} - k_{bv''}|$ and $K_{max} = k_{av'} + k_{bv''}$.

If we approximate the exchange integral $V_{ba}^{(e)}$ by the product of the overlap integral and Coulomb interaction as

$$V_{ba}^{(e)} = \frac{\omega e^2}{r} \exp\left(-\frac{r}{L}\right) \tag{7.186}$$

we have

$$\int V_{ba}^{(e)} \exp(i\mathbf{K} \cdot \mathbf{r}) \, d\mathbf{r} = \frac{4\pi\omega e^2}{K^2 + L^{-2}} \tag{7.187}$$

$$\sigma_{av', bv''}(\theta, \phi) = \frac{4\mu^2 k_{bv''}}{\hbar^4 k_{av'}} \frac{\omega^2 e^4}{(K^2 + L^{-2})^2} | \langle \theta_{bv''} | \theta_{av'} \rangle |^2 \tag{7.188}$$

and

$$Q_{av', bv''} = \frac{16\pi\mu^2 k_{bv''}}{h^4 k_{av'}} \frac{\omega^2 e^4}{(K_{min}^2 + L^{-2})(K_{max}^2 + L^{-2})} |\langle \theta_{bv''} | \theta_{av'} \rangle|^2 \quad (7.189)$$

Thus we can see that as the rate of triplet-triplet transfers in solid phase, the cross section of triplet-triplet transfers for each transition is proportional to the corresponding Franck–Condon factor.

7.5.4 Relation Between Rate Constants and Cross Sections

The relation between the chemical reaction cross section and the bimolecular reaction rate constant has been thoroughly discussed by Hirschfelder and Eliason (Eliason and Hirschfelder 1959; Lin et al., 1971a) and Light, Ross, and Shuler (1969). Their results can be directly applied to the energy transfer. For a given energy transfer cross section $Q_{av', bv''}$ for the transition $av' \rightarrow bv''$, its corresponding detailed quenching constant $k_{av', bv''}$ is given by

$$k_{av', bv''}(T) = \int_0^\infty Q_{av', bv''} f(v) v \, dv \quad (7.190)$$

where $f(v)$ represents the Maxwell-Boltzmann velocity distribution of relative motion between A and D. Here

$$f(v) = 4\pi \left(\frac{\mu}{2kT} \right)^{3/2} \exp\left(\frac{-\mu v^2}{2kT} \right) v^2.$$

If the rate of internal energy equilibration is much faster than that of energy transfer, then the overall quenching constant can be obtained from eq. 7.190 by carrying out the Boltzmann average over the internal states and summing over all the possible ways of energy transfer, that is,

$$k_{a,b}(T) = \sum_{v'} \sum_{v''} P_{av'} k_{av', bv''}(T) = \sum_{v'} \sum_{v''} P_{av'} \int_0^\infty Q_{av', bv''} f(v) v \, dv \quad (7.191)$$

where $P_{av'}$ represents the Boltzmann factor of internal states. The experimentally observed cross section $Q_{a,b}$ is related to $k_{a,b}(T)$ by

$$\bar{Q}_{a,b} = \frac{k_{a,b}(T)}{\bar{v}} = \sqrt{\left(\frac{\pi\mu}{8kT} \right)} k_{a,b}(T) \quad (7.192)$$

where $\bar{v}$ is the average relative speed, $\sqrt{(8kT/\pi\mu)}$.

Let us first discuss the rate constants of singlet-singlet and triplet-singlet transfers. From eq. 7.191 we can see that the rate constant can be obtained once the cross section is calculated quantum mechanically. For example, if we use eq. 7.190 as the energy transfer cross section, then the corresponding approximate rate constant will be given by

$$k_{a,b}^{(c)}(T) = \frac{16\pi\mu^2}{\hbar^4} |R_{D*D}|^2 (R_{AA*})^2 \sum_{v'} \sum_{v''} P_{av'} |\langle\theta_{bv''}|(\hat{\kappa}\cdot\hat{r}_{D*D})(\hat{\kappa}\cdot\hat{r}_{AA*})|\theta_{av'}\rangle|^2$$

$$\times \int_0^\infty f(v)\frac{k_{bv''}}{k_{av'}} v\,dv$$

$$= \frac{16\pi\mu^2}{\hbar^4} |R_{D*D}|^2 |R_{AA*}|^2 \sum_{v'} \sum_{v''} P_{av'} |\langle\theta_{bv''}|(\hat{\kappa}\cdot\hat{r}_{D*D})(\hat{\kappa}\cdot\hat{r}_{AA*})$$

$$\times |\theta_{av'}\rangle|^2 D(E_{av'} - E_{bv''}) \tag{7.193}$$

where $D(E_{av'} - E_{bv''})$ is defined by

$$D(E_{av'} - E_{bv''}) = \int_0^\infty f(v)\frac{k_{bv''}}{k_{av'}} v\,dv$$

$$= \int_0^\infty f(v)\sqrt{\left\{1 + \frac{2}{\mu v^2}(E_{av'} - E_{bv''})\right\}}v\,dv \tag{7.194}$$

The energy transfer rate constant given in eq. 7.193 should be compared with the energy transfer probability in the solid phase (cf. eq. 7.119)

$$P_{ab}(T) = \frac{2\pi}{\hbar} |H'_{ab}|^2_{av} \sum_{v'} \sum_{v''} P_{av'} |\langle\theta_{bv''}^{(v)}|\theta_{av'}^{(v)}\rangle|^2 \delta(E_{av'} - E_{bv''}) \tag{7.195}$$

where $\theta_{bv''}^{(v)}$ and $\theta_{av'}^{(v)}$ are the vibrational wave functions. Here for the purpose of comparison, the Condon approximation has been used. The similarity between eq. 7.195 and eq. 7.193 is striking; note that

$$|H'_{ab}|^2_{av} = \frac{2}{3\varepsilon^2 R^6} |R_{D*D}|^2 |R_{AA*}|^2$$

and the operators $(\hat{\kappa}\cdot r_{D*D})$ and $(\hat{\kappa}\cdot r_{AA*})$ affect only the rotational wave functions. Although the energy conservation in the solid-state energy transfer is controlled by the delta function $\delta(E_{av'} - E_{bv''})$, the energy conservation in the collisional energy transfer is regulated by the translational energy of the relative translational motion through the factor $D(E_{av'} - E_{bv''})$.

Next we consider the triplet-triplet transfer. Substituting eq. 7.185 into eq. 7.191, we obtain the rate constant for the triplet-triplet transfer as

$$k_{a,b}(T) = \frac{16\pi\mu^2\omega^2 e^4}{3\hbar^4 L^2 R_0^2} \sum_{v'} \sum_{v''} P_{av'} |\langle\theta_{bv''}|\theta_{av'}\rangle|^2 D_t(E_{av'} - E_{bv''})_1 \tag{7.196}$$

where

$$D_t(E_{av'} - E_{bv''})_1 = \int_0^\infty f(v) \left[\frac{1}{(K_{min}^2 + L^{-2})^3} - \frac{1}{(K_{max}^2 + L^{-2})^3} \right] \frac{v}{k_{av'}^2} \, dv$$

(7.197)

If eq. 7.189 is used as the cross section, then the rate constant is given by

$$k_{a,b}(T) = \frac{16\pi\mu^2\omega^2 e^4}{\hbar^4} \sum_{v'} \sum_{v''} P_{av'} |\langle \theta_{bv''} | \theta_{av'} \rangle|^2 D_t(E_{av'} - E_{bv''})_2$$

(7.198)

where

$$D_t(E_{av'} - E_{bv''})_2 = \int_0^\infty f(v) \frac{k_{bv''}}{k_{av'}(K_{min}^2 + L^{-2})(K_{max}^2 + L^{-2})} v \, dv$$

(7.199)

We should again compare the triplet-triplet transfer or eq. 3.198 with the corresponding energy transfer probability in the solid phase eq. 3.195. In this case

$$|H'_{ab}|_{av}^2 = \frac{\omega^2 e^4}{R_0^2} \exp\left(-\frac{2r}{L} \right)$$

Again the similarity between the gaseous energy transfer rate constant and the solid-state energy transfer probability should be noticed.

7.5.5 Application

As mentioned before, for singlet-singlet and triplet-singlet transfers the electron-electron interaction between A and D consists of two parts; one is the Coulomb interaction, and the other, the exchange interaction. In other words, both Coulomb and exchange interactions contribute to the cross sections of singlet-singlet and triplet-singlet transfers. In Sect. 7.5.2 we discussed only the contribution from the Coulomb interaction to the singlet-singlet (or triplet-singlet) transfer cross section and postponed the discussion of the contribution from the exchange interaction, as the theoretical treatment of the contribution from the exchange interaction to the singlet-singlet (or triplet-singlet) transfer is similar to that of the triplet-triplet transfer.

From eq. 7.171, if we neglect the cross term between $U_{bv'',av'}^{(c)}$ and $U_{bv'',av'}^{(e)}$, the differential cross section for singlet-singlet and triplet-singlet transfers can be written

$$\sigma_{av',bv''}(\theta\phi) = \sigma_{av',bv''}^{(c)}(\theta, \phi) + \sigma_{av',bv''}^{(e)}(\theta, \phi)$$

(7.200)

where

$$\sigma_{av',bv''}^{(c)} = \frac{\mu^2 k_{bv''}}{4\pi^2\hbar^4 k_{av'}} |U_{bv'',av'}^{(c)}|^2$$

(7.201)

and

$$\sigma^{(e)}_{av', bv''} = \frac{\mu^2 k_{bv''}}{4\pi^2 \hbar^3 k_{av'}} |U^{(e)}_{bv'', av'}|^2 \tag{7.202}$$

From 7.200 it follows that

$$k_{a,b}(T) = k^{(c)}_{a,b}(T) + k^{(e)}_{a,b}(T) \tag{7.203}$$

and

$$\bar{Q}_{a,b} = \bar{Q}^{(c)}_{a,b} + \bar{Q}^{(e)}_{a,b} \tag{7.204}$$

Now in the dipole approximation, (cf. 7.177–7.179) the cross sections $\sigma^{(c)}_{av', bv''}$ and $Q^{(c)}_{av', bv''}$ are proportional to the product of the squares of the transition moments R_{D*D} and R_{AA*}. When the Franck–Condon principle applies, the radiative rate constant for an electronic transition is proportional to the square of the corresponding transition moment. Thus for singlet-singlet and triplet-singlet transfers we may expect the energy transfer cross section to vary linearly with respect to the radiative rate constants of the donors for a given acceptor, and vice versa, provided that the Franck–Condon factors in 7.193 are not affected significantly. This has been verified by Breuer and Lee (1972) in their measurements of singlet-singlet energy transfer from several fluorinated and methylated benzenes to cyclopentanone. Their results also show that the exchange effect in the singlet-singlet (or triplet-singlet) transfers is significant. We may also expect that as in radiationless transitions, the deuterium substitution on donors or acceptors will slow down the rate of energy transfer, which as in radiationless transitions can be explained by the Franck–Condon factors in eqs. 7.193 or 7.196 or 7.198. This has also been experimentally demonstrated (Lee *et al.*, 1969; Schmidt and Lee, 1968, 1970). Furthermore, for singlet-singlet transfers, the contributions to the rate constants or cross sections originate from both Coulomb and exchange interactions, whereas for triplet-triplet transfers, the contribution to the quenching rate constants or cross sections can only come from the exchange interaction. This suggests that in most cases the quenching rate constants or cross sections of singlet-singlet transfers will be bigger than those of triplet-triplet transfers, which has also been borne out experimentally.

From the singlet-singlet transfer, the rate constant contributed from the Coulomb interaction $k^{(c)}_{a,b}$ has been shown to be closely related to the transition probability of the singlet-singlet transfer in the solid phase. Thus we may expect $k^{(c)}_{a,b}$ to vary linearly with the product of the radiative rate constant of the donor and the dipole-dipole spectral overlap J_{d-d} between the donor emission and acceptor absorption (see eq. 7.129). Similarly, we may expect the rate constant contributed by the exchange interaction $k^{(e)}_{a,b}$ to vary linearly with respect to the exchange spectral overlap J_{exc} between the

Table 7.10

Values of the Exchange and Dipole-Dipole Contributions to the Singlet-Singlet Electronic Energy Transfer

	Donor benzene ($^1B_{2u}$)			Donor p-difluorobenzene ($^1B_{2u}$)		
Acceptor	(Å²)	(Å²)	(Å²)	(Å²)	(Å²)	(Å²)
acetone	33 ± 2	32.12	0.53	56 ± 2	40.5	15.7
acetone-d_6	—	—	—	54 ± 2	—	—
2-butanone	33 ± 2	—	—	—	—	—
3-pentanone	41 ± 2	40.25	0.88	84 ± 3	56.1	28.2
2,4-dimethyl-3-pentanone	28 ± 2	26.47	1.38	95 ± 5	42.7	52.0
2,2,4,4-tetra-methyl-e-pentanone	9.5 ± 1.2	7.76	1.73	84 ± 3	13.7	70.1
cyclopentanone	65	63.65	1.82	156	93.6	62.4

donor emission and acceptor absorption (see eq. 7.139). The experimental results of Lee et al. indeed demonstrate the existence of the contribution from the exchange interaction in the singlet-singlet transfer. In Table 7.10 are shown some results of the exchange and dipole-dipole contributions to the singlet-singlet transfer (Loper and Lee, 1975).

As mentioned before, the quenching of electronic excitation can be treated in the manner shown for energy transfer and is not discussed here. However, it has been shown theoretically that the reason that the quenching rate constants are always smaller than the energy transfer rate constants is that in the quenching process the vibration of the acceptor scarcely participates in accepting the electronic excitation energy, whereas in the energy transfer process only part of the excited electronic energy of the donor is converted into the energy of nuclear motion; and the polar acceptor molecules are better quenchers than nonpolar acceptor molecules.

REFERENCES

Abramson, A. S., K. G. Spears, and S. A. Rice (1972), *J. Chem. Phys.*, **56**, 2291.

Bethe, H. A. (1930), *Ann. Phys.*, **5**, 325.

Bixon, M. and J. Jortner (1968), *J. Chem. Phys.*, **48**, 715.

Bogan, L. D. and D. B. Fitchen (1970), *Phys. Rev.*, **B1**, 4122.

Brailsford, A. D. and T. Y. Chang (1970), *J. Chem. Phys.*, **53**, 3108.

Breuer, G. M. and E. K. C. Lee (1972), *Chem. Phys. Lett.*, **14**, 407.

Chiu, Y.-N. (1971), *Chem. Phys. Lett.*, **14**, 407.

Dexter, D. L. (1953), *J. Chem. Phys.*, **21**, 836.

Dickens, P. G., J. W. Linnett, and O. Sovers (1962), *Trans. Faraday Soc.*, **33**, 52.

Drickamer, H. G. and C. W. Frank (1972), *Ann. Rev. Phys. Chem.*, **23**, 39.

Eisenthal, K. B. and S. Siegel (1964), *J. Chem. Phys.*, **41**, 652.

Eliason, M. A. and J. O. Hirschfelder (1959), *J. Chem. Phys.*, **30**, 1426.

Engleman, R. and J. Jortner (1970), *Mol. Phys.*, **18**, 145.

Eyring, H., J. Walter, and G. E. Kimball (1944), *Quantum Chemistry*, Wiley.

Fischer, S. and E. W. Schlag (1969), *Chem. Phys. Lett.*, **4**, 393.

Fischer, S. (1970), *J. Chem. Phys.*, **51**, 3195.

Flaherty, J. M. and B. DiBartolo (1973), *J. Lumin.*, **8**, 51.

Förster, Th. (1948), *Ann. Phys.*, **2**, 55.

Freed, K. F. and J. Jortner (1970), *J. Chem. Phys.*, **52**, 6272.

Freed, K. F. and S. H. Lin (1975), *Chem. Phys.*, **11**, 409.

Freed, K. F. (1976), Topics in Applied Physics, Springer-Verlag, pp. 36–85.

Gordon, R. G. and Y.-N. Chiu (1971), *J. Chem. Phys.*, **55**, 1469.

Heitler, W. (1953), *Quantum Theory of Radiation*, Oxford University Press.

Heller, D. F., K. F. Freed, and W. M. Gelbart (1972), *J. Chem. Phys.*, **56**, 2309.

Henderson, J. R. (1966), *J. Chem. Phys.*, **44**, 3496.

Henry, B. R. and W. Siebrand (1969), *Chem. Phys. Lett.*, **3**, 90; *J. Chem. Phys.*, **51**, 2396.

Hill, C. O. and S. H. Lin (1971), *Trans. Faraday Soc.*, **67**, 2833.

Inokuti, M. and F. Hirayama (1965), *J. Chem. Phys.*, **43**, 1978.

Jones, P. F. and S. Siegel (1969), *J. Chem. Phys.*, **50**, 1134.

Kasha, M. and S. P. McGlynn (1957), *Ann. Rev. Phys. Chem.*, **1**, 403.

Knittel, D. R., H. Raiszdadeh, H. P. Lin, and S. H. Lin (1977), *J. Chem. Soc. Faraday Trans. II*, **73**, 120.

Langelaar, J. (1969), Thesis, University of Amsterdam.

Lax, M. (1952), *J. Chem. Phys.*, **20**, 1752.

Lee, E. K. C., M. W. Schmidt, R. G. Shortridge, Jr., and G. A. Haninger, Jr. (1969), *J. Phys. Chem.*, **73**, 1805.

Lee, S. T., Y. H. Yoon, S. H. Lin, and H. Eyring (1977), *J. Chem. Phys.* **66**, 4349.

Light, J. C., J. Ross, and K. E. Shuler (1969), in *Kinetic Processes in Gases and Plasmas*, A. R. Hochstim, Ed., Academic, p. 281.

Lin, S. H. (1966), *J. Chem. Phys.*, **44**, 3759.

Lin, S. H. (1967), *J. Chem. Phys.*, **46**, 279.

Lin, S. H. (1968), *Theor. Chim. Acta*, **10**, 301.

Lin, S. H. and R. Bersohn (1968), *J. Chem. Phys.*, **48**, 2732.

Lin, S. H. (1971a), *Mol. Phys.*, **21**, 853.

Lin, S. H. (1971b), *J. Chem. Phys.*, **55**, 354.

Lin, S. H., L. Colangelo, and H. Eyring (1971a), *Proc. Natl. Acad. Sci. U.S.*, **68**, 2135.

Lin, S. H., K. H. Lau, and H. Eyring (1971b), *J. Chem. Phys.*, **55**, 5657.

Lin, S. H. (1972a), *J. Chem. Phys.*, **56**, 4155.

Lin, S. H. (1972b), *Theor. Chim. Acta*, **23**, 98.

Lin, S. H. (1972c), *J. Chem. Phys.*, **56**, 2648.

Lin, S. H. (1973a), *J. Chem. Phys.*, **59**, 4458.

Lin, S. H. (1973b), *J. Chem. Phys.*, **58**, 5760.

Lin, S. H. (1973c), *Proc. Roy. Soc. (Lond.)*, A335, 51.

Lin, S. H. and H. Eyring (1974), *Ann. Rev. Phys. Chem.*, **25**, 39.

Lin, S. H. (1975), *J. Chem. Phys.*, **62**, 4500.

Lin, S. H. and H. Eyring (1975), *Proc. Natl. Acad. Sci. U.S.*, **72**, 4205.

Lin, S. H. (1976), *Proc. Roy. Soc. (Lond.)*, **A352**, 57.

Lin, S. H. and H. Eyring (1977), *Proc. Natl. Acad. Sci. U.S.*, in press.

Lin, S. H. Ed. (1978), *Radiationless Transition*, Academic.

Loper, G. L. and E. K. C. Lee (1975), *J. Chem. Phys.*, **63**, 264.

Markham, J. J. (1959), *Rev. Mod. Phys.*, **31**, 956.

Massey, H. S. W. et al., Eds. (1971), *Electronic and Ionic Impact Phenomena*, Vol. 3, Oxford University Press.

Moos, H. W. (1970), *J. Lumin.*, **1**, 106.

Mott, N. F. and H. S. W. Massey (1965), *The Theory of Atomic Collisions*, Oxford University Press.

Mulliken, R. S. and C. A. Rieke (1941), *Rep. Prog. Phys.*, **8**, 231.

Nitzan, A. and J. Jortner (1971), *J. Chem. Phys.*, **55**, 1355.

Nitzan, A. and J. Jortner (1972a), *J. Chem. Phys.*, **56**, 2079.

Nitzan, A. and J. Jortner (1972b), *J. Chem. Phys.*, **56**, 3360.

Orlandi, C. and W. Siebrand (1973), *J. Chem. Phys.*, **58**, 4513.

Pople, J. A. and J. W. Sidman (1957), *J. Chem. Phys.*, **27**, 1270.

Schiff, L. I. (1968), *Quantum mechanics*, McGraw-Hill.

Schmidt, M. W. and E. K. C. Lee (1968), *J. Am. Chem. Soc.*, **90**, 5919.

Schmidt, M. W. and E. K. C. Lee (1970), *J. Am. Chem. Soc.*, **92**, 3579.

Siebrand, W. (1971), *J. Chem. Phys.*, **54**, 363.

Spears, K. G. and S. A. Rice (1971), *J. Chem. Phys.*, **55**, 5561.

Strickler, S. J. and R. A. Berg (1962), *J. Chem. Phys.*, **37**, 814.

Sturge, M. D. (1973), *Phys. Rev.*, **B8**, 8.

Turro, N. J. (1965), *Molecular Photochemistry*, Benjamin.

Veeman, W. S. and J. H. van der Waals (1970), *Mol. Phys.*, **18**, 63.

Ziegler, L. and A. C. Albrecht (1974), *J. Chem. Phys.*, **60**, 3558.

Eight

Elementary Processes in Photochemistry (II)

CONTENTS

8.1 MOLECULAR ENERGY TRANSFER IN GASES

8.1.1 Introduction

In this section we discuss the subject of molecular energy transfer occurring in the course of binary collision processes. Although inelastic cross sections are typically small compared with elastic cross sections, the vast majority of molecular collision processes involves energy flow between the colliding partners. This exchange of energy may occur between translational and internal degrees of freedom or among the various internal degrees of freedom.

Under the condition of thermal equilibrium, the relative population in various energy levels is a function of ambient temperature and is given by the Boltzmann distribution:

$$\frac{n_i}{N} = \frac{g_i \cdot \exp(-\varepsilon_i/kT)}{\sum_i g_i \exp(-\varepsilon_i/kT)} \tag{8.1}$$

where n_i is the number of molecules in the energy level ε_i and N is the total number of molecules. The degeneracy g_i is the number of possible quantum states of the molecule at ε_i and T is the common temperature of the vessel containing the molecules. When the Hamiltonian for the molecules concerned is separable, the energy of the ith level can be written as a sum of electronic, vibrational, rotational, and translational energies, and each of these has its own distribution function defined as in eq. 8.1. Hence in the event of inelastic collision, the exchange of energy among various degrees of freedom can lead to a transient population distribution that is no longer characterized by a given common temperature. Therefore to have a meaningful discussion of molecular energy transfer, one has to describe the equilibrium and temperature within one of the internal degrees of freedom relative to the others and relative to the vessel temperature.

Various processes can lead to a transient population distribution; in particular, the selective application of laser excitation to a particular vibrational model of molecules has received much attention during recent years. Since the probability that vibrational or rotational energy will be lost by spontaneous infrared radiation is usually very low, this vibrational excitation may bring about a very high vibrational temperature and studies on its subsequent relaxation caused by the binary collisional deexcitation are of particular interest in inelastic scattering.

Relaxation Time and Collision Number. The classical theory of gas kinetics is the starting point of discussion of collisional relaxation processes. At ordinary temperatures the time duration (Δt) between two successive collisions of a given molecule at 1 atm is about 10^{-9} sec. Thus for a molecule A_1 that is not electronically excited, the collisional deactivation is the primary process accounting for molecular relaxation (Flygare, 1968):

$$A_1 + M \underset{k_{01}}{\overset{k_{10}}{\rightleftharpoons}} A_0 + M \qquad (8.2a)$$

Here the rate constant k_{01} refers to the collisional excitation and k_{10} to collisional deactivation. In addition, if A_1 is an electronically excited species and is an electric dipole-allowed transition, the molecule may return to the ground state A_0 via spontaneous radiation:

$$A_1 \xrightarrow{k_2} A_0 + hv \qquad (8.2b)$$

where k_2 is the reciprocal of the natural lifetime, that is, $k_2 = 1/\tau_n$.

To determine the rate at which the disturbed population of a given mode relaxes towards equilibrium via the binary collision process, eq. 8.2a, we follow the conventional rate expression for bimolecular reactions,

$$-\frac{dA_1}{dt} = \frac{dA_0}{dt} = (k_{10}A_1 - k_{01}A_0)M \qquad (8.3)$$

Since at thermal equilibrium, $-dA_1/dt = dA_0/dt = 0$ and $A_1/A_0 = k_{01}/k_{10} = \exp(-E_{10}/kT)$ with $E_{10} = E_1 - E_0$, the change of molecular concentration ΔA_1 from equilibrium is then

$$-\frac{d\Delta A_1}{dt} = [k_{10}(A_{1,eq} + \Delta A_1) - k_{01}(A_{0,eq} - \Delta A_1)]M$$

$$= (k_{10} + k_{01})\Delta A_1 M$$

$$= \frac{\Delta A_1}{\tau}$$

with

$$\tau = \frac{1}{M(k_{10} + k_{01})} \qquad (8.4)$$

Here τ is called the *collisional relaxation time*, a measure of how soon the equilibrium condition can be achieved after perturbation. Since the magni-

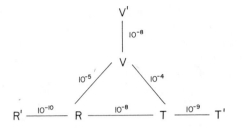

Fig. 8.1 Schematic diagram for the energy transfer processes during binary collisions. The numbers are typical values of bulk relaxation times in units of atm sec.

tude of τ itself depends on the concentration of ambient gas, the conventional expression for relaxation time is often referred to a pressure at 1 atm. Figure 8.1 shows the typical relaxation time, in units of atm · sec, for various processes of molecular energy transfer. Thus one can see clearly that vibrational-rotation (V-R) and vibrational-translation (V-T) are the least probable energy transfer processes, whereas the rotational-translation (R-T) energy exchange occurs very effectively (Flygare, 1968).

In contrast to the translational degree of freedom, both vibrational and rotational energies are quantized and, in general, are not freely interchanged with translational energy during the collision process. Thus the transition probability between either rotational or vibrational energy and translational energy is always less than unity such that a molecule may undergo a considerable number of collisions before it can gain or lose its interval energy. If p is the transition probability per collision, the average number of collisions required for a molecule to gain or lose its interval energy is then $Z_c = 1/p$. The collision number Z_c thus defined is closely related to the relaxation time τ by

$$Z_c = Z\tau \tag{8.5}$$

Here Z is the number of collisions one molecule suffers per second. Note that both quantities Z and τ are inversely proportional to the gas pressure and the collision number is dimensionless. If the efficiency of molecular energy transfer is expressed in terms of the collision number, it is in the range of 1–10 for the fast R-T transition and about 10^6 for the slow V-T transfer at room temperature.

Example: Classical Treatment of V-T Transfer. To gain physical insight into the problem of molecular energy transfer, we begin here the classical treatment of the V-T transition. Although inelastic collision processes that involve the internal energy states that are quantized can only be solved properly by means of quantum mechanics, the example given below, the forced oscillator, is extremely valuable since it, in fact, not only provides a great deal of physical insight into the V-T transfer but also leads to a rather good approximate result, provided that the translational energy available greatly exceeds the amount of energy transferred during the collision.

Consider a mass m executing a simple harmonic motion and subject to an external time-dependent force $F(t)$. Under these conditions the equation of motion is given by (Landau and Lifshitz, 1960)

$$\ddot{x} + \omega^2 x = \frac{F(t)}{m} \tag{8.6}$$

where ω is the frequency of the oscillator. The general solution of this inhomogeneous linear differential equation is $x(t) = x_0(t) + x_1(t)$ where $x_0(t)$ is a general solution of the corresponding homogeneous equation and $x_1(t)$ is a particular solution of eq. 8.6. By introducing $p = \dot{x} + i\omega x$, eq. 8.6 is now reduced to

$$\dot{\rho} - i\omega\rho = \frac{F(t)}{m} \tag{8.7}$$

The homogeneous equation corresponding to eq. 8.7 can be easily solved and is given as $\rho_0 = A \exp(i\omega t)$, where A is a constant. Thus one seeks the particular solution of eq. 8.7 with the form $\rho_1 = A(t)\exp(i\omega t)$. The coefficient $A(t)$ can be found by substituting into eq. 8.7 to give

$$\dot{A}(t) = \frac{F(t)}{m} \exp(-i\omega t) \tag{8.8}$$

Integration over time then gives $A(t)$ and since the complete solution is $\rho = \rho_0 + \rho_1$, we obtain

$$\rho = \left(\int_0^t \frac{F(t)}{m} e^{-i\omega t} \, dt + \rho_0 \right) e^{i\omega t} \tag{8.9}$$

where ρ_0 now stands for ρ at $t = 0$. The total energy of the oscillator is

$$E = \tfrac{1}{2}m(\dot{x}^2 + \omega^2 x^2) = \tfrac{1}{2}m|\rho|^2 \tag{8.10}$$

and by setting $t \to -\infty$ as the initial state where $\rho = 0$, the change of oscillator energy as a result of action of the external force is given

$$\Delta E = \tfrac{1}{2}m|\rho(\infty)|^2$$

$$= \frac{1}{2m} \left| \int_{-\infty}^{\infty} F(t) e^{-i\omega t} \, dt \right|^2 \tag{8.11a}$$

This classical expression for the V-T energy transfer is very useful in a number of physical problems. If the force is finite only for a time short compared to ω^{-1}, then $\exp(-i\omega t)$ will be essentially constant within this period of time and

$$\Delta E = \frac{1}{2m} \left| \int_{-\infty}^{\infty} F(t) \, dt \right|^2 \tag{8.11b}$$

That is, the amount of energy transferred is expected to be proportional to the square of the impulse delivered in the collision. Thus, if the force is impulsive, the instant contact between the oscillator and its collision partner may lead to a highly excited oscillator. This is particularly true if the oscillator is so loose that one of its ends may be regarded as a spectator and the change in relative motion is totally converted into vibrational motion of the oscillator.

On the other hand, if $F(t)$ varies very slowly and acts during many periods of oscillation, then because of the oscillations in the integrand of eq. 8.11a, the amount of energy transferred is very small and the oscillator experiences little perturbation. This can be shown explicitly if we assume here that $F(t)$ is constant F_0 at $0 \le t \le T$ and zero for $t > T$. Hence eq. 8.11a becomes (Mahan, 1970; Levine and Bernstein, 1974)

$$\Delta E = \frac{2}{m} \left(\frac{F_0}{\omega}\right)^2 \sin^2\left(\frac{\omega T}{2}\right) \tag{8.12}$$

Thus one sees clearly that the amount of energy transferred is proportional to ω^{-2}; that is, it becomes harder to excite the oscillator as the vibration frequency increases. In this particular instance the amount of energy transferred reaches a maximum value as the force acts for one-half of the vibration period; that is, $T = \pi/\omega$ and $\Delta E = 2/m(F_0/\omega)^2$. We also note that classically a more effective energy transfer occurs for a light oscillator than for a heavy one if other parameters are unchanged (Bates, 1961).

Interaction Force. In our earlier discussion of elastic scattering we simplified the collision problems by assuming a spherically symmetric intermolecular potential; that is, the collision pairs are considered to be structureless. This assumption, however, becomes inadequate when we describe inelastic scattering in which the colliding molecules not only suffer a deflection but also undergo a change in their internal energy such that one or both of the collision pairs must suffer, to some extent, an internal distortion before the collision is completed. Thus, to give a full description of inelastic scattering it is necessary that the internal motions of colliding particles have to be taken into account.

For the sake of simplicity, we illustrate classically a collision process between an atom A and a diatomic molecule BC. If we assume that the molecule BC is a rigid rotor, the Hamiltonian for the system may be written as

$$H = \tfrac{1}{2}\mu\dot{R}^2 + \frac{L^2}{2\mu R^2} + V(R, \gamma) + E_r$$

$$= \tfrac{1}{2}\mu\dot{R}^2 + \frac{L^2}{2\mu R^2} + V_s(R) + V_a(R, \gamma) + E_r \tag{8.13}$$

where γ is the angle between the position vector R and the axis of the rotor. Here the first three terms are identical to the central force problem used to describe the elastic scattering. In addition, we have to include the asymmetric part of the interaction potential $V_a(R, \gamma)$ and the rotational energy $E_r = j^2\hbar^2/2I$ that participates in the collision. This orientation dependence of the anisotropic potential is in general present in all collision processes that involve molecular species and can appear in the repulsive force as well as the attractive force; for the attractive part of interaction potential, this anisotropic force is well understood theoretically, whereas for the repulsive potential, the existence of an anisotropic potential is less clear. Generally speaking, for a heteronuclear diatomic molecule where there exists a permanent dipole, $V_a(R, \gamma)$ can be comparatively large and leads to a large transition probability for rotational energy transfer. For a homonuclear diatomic molecule, the magnitude of $V_a(R, \gamma)$ decreases and the R-T transition probability is proportionately decreased. Since the R-T transfer is a rapid process and may occur when the colliding particles are still a few angstroms away, the long-range attractive force is certainly as important as the short-range repulsion.

The significance of internal rotational energy E_r may be best understood if we introduce the mass effect. For a light molecule like H_2, the rotational quanta are rather large and it is expected that the R-T transfer will be less effective. On the other hand, a large transition probability may occur for heavy molecules where rotational energy spacing is small. Because of this mass effect, one finds that at room temperature, a few collisions are sufficient to have an R-T energy transfer for most heavy molecules, whereas 200–300 collisions are required for H_2 molecules collided with inert gases.

For the less effective vibrational energy transfer, we have learned from the example given in the previous section that the interaction force on which transition is induced has to act in a short time, compared with that of the vibration period. Because the long-range attractive force varies so gradually, the internal degrees of freedom may adjust themselves adiabatically, the V-T transition is often treated by considering the exponential repulsion alone. To illustrate the V-T energy transfer, we apply the result of forced oscillation to a collinear collision process of an atom A with a harmonic oscillator BC, as depicted in Fig. 8.2, by postulating that the transition is

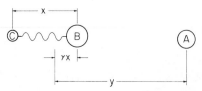

Fig. 8.2   Collinear collision between atom A and a harmonic oscillator BC.

induced by an exponential repulsion between atoms A and B (Levine and Bernstein, 1974):

$$V(R_{AB}) = \frac{Ce^{-R}AB}{L} \tag{8.14}$$

Here C is a constant, R_{AB} the distance between atoms A and B, and L the characteristic length. In terms of the distance y, measured from the center of mass of the diatomic BC to atom A, we get $R_{AB} = y - \gamma x$, where $\gamma = m_C/m_{BC}$, and the equations of motion are

$$\ddot{x} + \omega^2(x - x_e) = -\frac{\gamma C}{\mu_{BC}L} \exp\left[-\frac{(y - \gamma x)}{L}\right] \tag{8.15a}$$

$$\mu_{AB}\ddot{y} = \frac{c}{L} \exp\left[-\frac{(y - \gamma x)}{L}\right] \tag{8.15b}$$

Here x_e is the equilibrium separation of the oscillator BC. If we assume explicitly $F(t)$ to be independent of x, x on the right-hand side of eq. 8.15 may be replaced by its equilibrium distance x_e. This is approximately correct whenever the vibrational motion of BC is much slower than the time the strong force is acting between atoms A and B. Equation 8.15b can now be solved for $y(t)$ by differentiation, followed by integration to give

$$\ddot{y} + \frac{\dot{y}^2}{2L} = \frac{g^2}{2L}$$

where g is the initial velocity of the projectile. The solution of this equation is

$$y(t) = L \cdot \ln\left(\frac{2C'}{\mu_{AB}g^2} \cosh^2 \frac{gt}{2L}\right) \tag{8.16}$$

and

$$F(t) = -\frac{C'\gamma}{L} e^{-y/L} = -\frac{\gamma\mu_{AB}g^2}{2L \cosh^2(gt/2L)} \tag{8.17}$$

where $C' = C \cdot \exp(\gamma x_e/L)$. Thus the amount of energy transferred is found from eq. 8.11a:

$$\Delta E = 2\mu_{BC}\left(\frac{m_A}{m_{AB}}\right)^2 (\pi\omega L)^2 \operatorname{csch}^2\left(\frac{\pi\omega L}{g}\right) \tag{8.18}$$

If the initial relative collision energy $E_t = \frac{1}{2}\mu_{A-BC}g^2$, one finds

$$\frac{\Delta E}{E_t} = 4\frac{m_A m_B m_C m_{ABC}}{m_{AB}^2 m_{BC}^2}\left(\frac{\pi\omega L}{g}\right)^2 \operatorname{csch}^2\left(\frac{\pi\omega L}{g}\right) \tag{8.19a}$$

$$= 4\cos^2\beta \sin^2\beta\left(\frac{\pi\omega L}{g}\right)^2 \operatorname{csch}^2\left(\frac{\pi\omega L}{g}\right) \tag{8.19b}$$

Here β is the skewed angle, defined in our discussion of the potential energy surface and $\cos^2 \beta = m_A m_C / m_{AB} m_{BC} \le 1$. Thus by postulating that the vibrational transition is induced by the short-range repulsion, we find that the amount of energy transferred is mainly determined by the argument $\pi \omega L / g$ in the hyperbolic cosecant function. Since this function goes to zero as its argument approaches infinity and goes to infinity as the argument goes to zero, eq. 8.19 indicates that vibrationally inelastic collisions are favored by small ω, small L, and large g.

Using the relation $p = \Delta E / \hbar \omega$ for a harmonic oscillator and in the limit where $\pi \omega L / g \gg 1$, the transition probability for the V-T energy transfer is reduced to

$$p = 8 \mu_{BC} \left(\frac{m_A}{m_{AB}} \right)^2 \frac{(\pi \omega L)^2}{\hbar \omega} e^{-2 \pi \omega L / g} \tag{8.20}$$

If we further assume that $\mu_{AB} g^2 / 2 \gg \hbar \omega$, this transition probability may be thermally averaged to give

$$\ln p(T) = A - B T^{-(1/3)} \tag{8.21}$$

where A and B are constants. This is the famous Landau–Teller formula that $\ln p(T)$ is inversely proportional to $T^{1/3}$ when the probability itself is small (Landau and Liftschitz, 1960).

Although the attractive force does not participate directly in the V-T transfer of atom-diatomic collisions, it may become operative for polyatomic system where many vibrational modes are available and their energy spacings are somehow smaller. Thus one expects that for more complicated molecules, the fast V-V process is more important than the V-T transfer. This implies that for V-V transitions the attractive force is as important as the repulsive force.

Finally, we examine the electronic transitions for a slow molecular collision process. According to the adiabatic approximation the motions of electron and nucleus can be decoupled and the motion of colliding particles can be described by the potential energy surface constructed for a given electronic state. Provided that these two electronic states are well separated, the molecular motion of these colliding particles does not cause a transition from one surface to the other. However, when a transition between two electronic states occurs, a strong nonadiabatic interaction is operative in the region where two adiabatic potential surfaces intercept or come close together.

For diatomic molecules of the same symmetry, no curve crossing should take place. If they do cross in lower-order approximation, the effect of the neglected terms tends to repel these two surfaces around the crossing

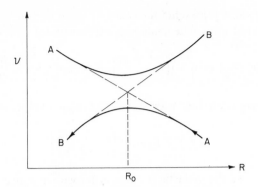

Fig. 8.3 Curve crossing model of two potential energy surfaces. Solid curves refer to adiabatic condition while the dashed curves represent the crossing diabatic levels.

point and eventually leads to an avoidance of crossing. For polyatomic molecules, since there are several internuclear distances, the curve crossing can occur even if they have the same symmetry. We depict a schematic picture of curve crossing in Fig. 8.3. Thus to determine the electronic transition one has to find out the magnitude of the transition probabilities for an electron to jump from one surface A to surface B when two particles approach each other as well as when they fly apart at or near the crossing point. If this transition probability is unity for each avoided crossing, the cross section may then simply be given as πR_0^2. This curve crossing model works reasonably well, at least semiquantitatively, to correlate the relation between the cross section and the distance at which curve crossing occurred for a number of alkali, mercury, and cadmium quenching experiments.

Elastic and Inelastic Scattering. We now return to describe a general formulation of collision theory that includes the energy exchange between translational and internal degrees of freedom of the two colliding particles. Here we follow the partial wave treatment used in our earlier discussion of elastic scattering. When the inelastic channel is open to the collision particles, the net flux density $\mathbf{j}$ between the incident and outgoing waves for the elastic channel will not vanish at large $\mathbf{r}$ and the phase shift ε_l has to be a complex variable, $\varepsilon_l = \eta_l + i\xi_l$ where η_l is responsible for the elastic scattering and ϕ_l for the inelastic channel.

For the purpose of our discussion we start from a normalized asymptotic wave function for the elastic scattering at channel i:

$$u_i^l(r) \sim A_i\left[\exp\left(-i\left(k_i r - \frac{l\pi}{2}\right)\right) - \exp\left(i\left(k_i r - \frac{l\pi}{2}\right)\right)\right] \qquad (8.22)$$

where the first term represents the incident wave and the second term the outgoing scattered wave. The analogous expression for inelastic scattering is

$$v_i^l(r) \sim A_i \exp\left[-i\left(k_i r - \frac{l\pi}{2}\right)\right] - B_i \exp\left[i\left(k_i r - \frac{l\pi}{2}\right)\right] \qquad (8.23)$$

In addition, there is another wave appearing in channel $j(j \neq i)$,

$$v_j^l(r) \sim B_j \exp\left[i\left(k_j r - \frac{l\pi}{2}\right)\right] \qquad (8.24)$$

Thus, to determine the inelastic process, what we are concerned with is the magnitude of the amplitude B_j with respect to the coefficient A_i in the incident wave. Here it is convenient to introduce the scattering matrix S_{ij}, defined by the relation

$$B_j = S_{ij} A_i \qquad (8.25)$$

Note that if the incident wave is a superposition of components in more than one channel, eq. 8.25 becomes $B_j = \sum S_{ij} A_i$. In this notation the flux of the outgoing wave in channel j is

$$\frac{4\pi}{2l+1} |B_j|^2 = \frac{4\pi}{2l+1} |S_{ij}|^2 |A_i|^2 \qquad (8.26a)$$

and the flux of the outgoing wave in channel i is

$$\frac{4\pi}{2l+1} |S_{ii}|^2 |A_i|^2 \qquad (8.26b)$$

where the factor $4\pi(2l+1)$ is due to the angular component of $P_l(\cos\theta)$ for each partial wave l. Applying the law of conservation of current density, eq. 8.26 leads to a unitary S matrix:

$$\sum_j |S_{ij}|^2 = 1 \qquad (8.27)$$

Hence we are now ready to calculate the partial cross sections for elastic and inelastic scattering. By the definition of cross section and with $|A_i|^2 = (2l+1)^2/4k_i^2$ determined from the total incident flux, we get

$$Q_{ij}^l = \frac{\pi(2l+1)}{k_i^2} |S_{ij}|^2$$

$$Q_e^l = Q_{ii}^l = \frac{\pi(2l+1)}{k_i^2} |1 - S_{ii}|^2 \qquad (8.28)$$

The total partial inelastic cross section through all possible channels j is then

$$
\begin{aligned}
Q_{in}^l &= \frac{\pi(2l + 1)}{k_i^2} \sum_{j \neq i} |S_{ij}|^2 \\
&= \frac{\pi(2l + 1)}{k_i^2} \left(\sum_j |S_{ij}|^2 - |S_{ii}|^2 \right) \\
&= \frac{\pi(2l + 1)}{k_i^2} (1 - |S_{ii}|^2)
\end{aligned}
\tag{8.29}
$$

and the total cross section with the orbital angular momentum quantum number l is then

$$
Q_t^l = Q_e^l + Q_{in}^l = \frac{\pi(2l + 1)}{k_i^2} \{|1 - S_{ii}|^2 + 1 - |S_{ii}|^2\}
$$

$$
= \frac{\pi(2l + 1)}{k_i^2} \{2 - (S_{ii} + S_{ii}^*)\}
\tag{8.30}
$$

The preceding equations can be further simplified when we introduce $S_{ii} = \exp(2i\varepsilon_l)$, where $\varepsilon_l = \eta_l + i\phi_l$. This leads to

$$
Q_{in}^l = \frac{\pi(2l + 1)}{k_i^2} (1 - e^{-4\phi_l}) = \frac{2\pi(2l + 1)}{k_i^2} e^{-2\phi_l} \sinh 2\phi_l
$$

$$
Q_e^l = \frac{2\pi(2l + 1)}{k_i^2} (\cosh 2\phi_l - \cos 2\eta_l) e^{-2\phi_l}
\tag{8.31}
$$

$$
Q_t^l = \frac{2\pi(2l + 1)}{k_i^2} (1 - e^{-2\phi_l} \cos 2\eta_l)
$$

It is seen that since Q_{in}^l cannot be less than zero, ϕ_l can never be negative and $\exp(-4\phi_l)$ can never exceed unity. When $\phi_l = 0$, the scattering is purely elastic and $Q_t^l = Q_e^l = 4\pi(2l + 1)/k_i^2$ at $\eta_l = (n + \frac{1}{2})\pi$. On the other hand, when ϕ_l is very large such that $Q_{in}^l = Q_e^l = \pi(2l + 1)/k_i^2$ and Q_t^l is $2\pi(2l + 1)/k_i^2$. In other words, the inelastic cross section can be as large as the elastic cross section at a given l (Levine and Bernstein, 1974).

The partial wave treatment is very useful when the physical problem concerned involves light particles in a slow collision process so that only a few partial waves are included in the calculations. For heavy particles, however, the partial wave treatment proves to be laborious and the inelastic cross section is often calculated from the averaged transition probability:

$$
Q_{in}(E) = 2\pi \int_0^\infty p(E, b)b \, db
\tag{8.32}
$$

Here $p(E, b)$ is the averaged transition probability over all possible inelastic channels typically observed in the bulk system.

8.1.2   Transition Probability

Formulation of Inelastic Scattering Theory.   In the quantum treatment of molecular energy transfer, it is necessary to seek solutions to the Schrödinger equation in which the Hamiltonian for the system may be given as

$$\hat{H}(R, r) = -\frac{\hbar^2}{2m} \nabla_R^2 + \hat{H}^0(r) + \hat{H}_{12}(R, r) \qquad (8.33)$$

where $\hat{H}^0(r)$ is the Hamiltonian for a noninteracting system and $\hat{H}_{12}$ is the intermolecular potential that is a function of the relative position R and the integral coordinates r. One thus looks for the wave function $\psi(R, r)$ that satisfies

$$\hat{H}\psi(R, r) = E\psi(R, r) \qquad (8.34)$$

The problem of inelastic scattering is then discussed on the basis of the wave function $\psi(R, r)$.

Before making a full expansion of $\psi(R, r)$, it is important to recognize the significance of molecular energy transfer under the high- and low-velocity conditions. For a slow collision process where the transition is faster than the relative motion, an adiabatic approximation is preferable. On the other hand, the diabatic formulation is necessary for the fast collision process in which the transition is slower than the nuclear motion. Since perturbation under these two extremes is very different, the approximate method used to solve eq. 8.34 is also different. As a consequence, we define here the adiabaticity (or Massey) parameter ρ,

$$\rho = \tau_c \omega \qquad (8.35a)$$

Here τ_c is the duration of molecular collision and ω the characteristic frequency of the state under consideration. If L is the length over which the interaction acts effectively and ΔE is the amount of energy transferred, that is, $\Delta E = \hbar\omega$, the adiabaticity parameter may be rewritten as

$$\rho = \frac{L|\Delta E|}{\hbar v} \qquad (8.35b)$$

Here v is the velocity at the instant when transition occurs. At the high-velocity limit where the collision energy is much larger than ΔE and $\rho \ll 1$, we are in the sudden limit. On the other hand, ΔE may be larger than the collision energy at low collision rates such that $\rho > 1$ and the process is called adiabatic (Child, 1974).

Importance of this adiabaticity parameter may be best understood if we consider the V-T transfer where the vibration quanta are generally large compared with kT at room temperature. In this case one often finds that $\rho > 1$ and the transition probability under the influence of such a slowly varying perturbation is rather small. On the other hand, for a high-velocity collision process, say at high temperature, the instant contact between the oscillator and its colliding partner may cause a large displacement from its equilibrium position, and thus a large transition probability is expected as shown in eq. 8.11a.

We now return to a discussion of the expansion of $\psi(R, r)$ based on these two conditions. In its diabatic formulation, the perturbation causing transitions is taken to be the interaction potential $\hat{H}_{12}$ and $\psi(R, r)$ may be expanded in terms of the unperturbed orthonormal eigenstates $\phi_m(r)$

$$\psi(R, r) = \Sigma_m F_m(R)\phi_m(r) \tag{8.36}$$

where ϕ_m has to satisfy $H^0(r)\phi_m(r) = E_m\phi_m(r)$. After substitution into eq. 8.34, multiplication by $\phi_n^*(r)$, and integration over $\mathbf{r}$, we then have

$$[\nabla_R^2 + k_n^2]F_n(R) = \Sigma_m U_{nm}(R)F_m(R) \tag{8.37}$$

where

$$k_n^2 = \frac{2\mu(E - E_n)}{\hbar^2} \quad \text{and} \quad U_{nm}(R) = \frac{2\mu}{\hbar^2}\langle\phi_n(r)|H_{12}(R, r)|\phi_m(r)\rangle$$

The diagonal terms U_{nn} are then contributed to the elastic scattering whereas the off-diagonal terms U_{nm} are responsible for inelastic processes that occur. The solution for the generalized scattering theory would thus include an asymptotic form of a plane wave together with an outgoing spherical wave

$$F_n(R) \sim e^{ik_n z} + f_{nn}(\theta, \phi)\frac{e^{ik_n R}}{R} \tag{8.38a}$$

and outgoing waves in all other channels

$$F_m(R) \sim f_{nm}(\theta, \phi)\frac{e^{ik_m R}}{R} \tag{8.38b}$$

in analogy to the expressions in eq. 8.23 and eq. 8.24. The differential cross sections $I_e(\theta, \phi)$ for elastic and $I_{in}(\theta, \phi)$ for inelastic channels are then given in terms of the associated scattering amplitudes (Jackson and Mott, 1932):

$$I_e(\theta, \phi) = |f_{nn}(\theta, \phi)|^2 \tag{8.39a}$$

and

$$I_{in}(\theta, \phi) = \frac{k_m}{k_n}|f_{nm}(\theta, \phi)|^2 \tag{8.39b}$$

where the velocity factor for the inelastic differential cross section enters because of the possible exchange of translational and internal degrees of freedom during the collision.

If the transition of the subsystems is faster than the relative motion, $\rho > 1$, the expansion of $\psi(R, r)$ in eq. 8.36 has to be replaced by the adiabatic internal states $\chi_m(R, r)$. In other words, the perturbation is taken to be the relative motion and eq. 8.36 becomes (Mott and Massey, 1965)

$$\psi(R, r) = \Sigma_m a_m(R)\chi_m(R, r) \tag{8.40}$$

where $\chi_m(R, r)$ satisfies

$$[H^0(r) + H_{12}(R, r)]\chi_m(R, r) = w_m(R)\chi_m(R, r) \tag{8.41}$$

and the eigenvalue w_m now depends on the position vector $\mathbf{R}$. Similarly, multiplication by χ_n^* and integration over r yields

$$[\nabla_R^2 + k_n^2(R)]A_n(R) = \Sigma[A(R) + B(R)]A_m(R) \tag{8.42}$$

where

$$k_n^2(R) = \frac{2m(E - w_n(R))}{\hbar^2}$$

$$A(R) = -2\langle\chi_n(R, r)|\nabla_R|\chi_m(R, r)\rangle\nabla_R$$

$$B(R) = -\langle\chi_n(R, r)|\nabla_R^2|\chi_m(R, r)\rangle$$

This is essentially the Born–Oppenheimer formula for decoupling the motions of electrons and nuclei. For the scattering problem, however, both eq. 8.37 and 8.42 are very hard to solve and approximate methods are sought.

Perturbation Theory.   In this section we describe the quantum treatment of molecular energy transfer when the diabatic formulation is valid. These approximate methods, however, are applicable only if the interaction leading to inelastic scattering is weak; in other words, this is the case when inelastic scattering occurs at the limit of large impact parameters and high energies.

We begin with the Born approximation. This approximation is obtained by replacing F_m on the right-hand side of eq. 8.37 by its zeroth-order approximation, $F_m = \exp(i\mathbf{k}_o \cdot \mathbf{R})\sigma_{om}$ where subscript o denotes the initial state. Thus we obtain a simple expression for the first-order Born approximation in the inelastic channel (also see Sec. 7.5):

$$(\nabla_R^2 + k_n^2)F_n = U_{no}e^{i\mathbf{k}_o \cdot \mathbf{R}} \tag{8.43}$$

that is, the distortions of the incident wave due to U_{oo} and of the scattered wave by U_{nn} are neglected and no intermediate states are involved in the transition. In general, the Born approximation should give good results in the limit of high energies where both relevant elastic and inelastic cross sections are small so that the assumptions made earlier may be approximately correct.

Another less drastic approximation, the distorted wave approximation (DWA), retains all matrix elements relating to initial and final states, U_{oo}, U_{on}, and U_{nn} but ignores others. This two-state approximation reduces the problem to a set of two coupled equations:

$$(\nabla_R^2 + k_o^2 - U_{oo})F_o = U_{on}F_n$$
$$(\nabla_R^2 + k_n^2 - U_{nn})F_n = U_{no}F_o \qquad (8.44)$$

This approximation, however, still presents a difficulty since they cannot be decoupled unless U_{on} is small. Thus, provided that U_{on} is small, the scattering problem can then be adequately solved by the DWA approximation. In this case the resulting F_o may then be used to solve for F_n in the second expression of eq. 8.44. This method is an improvement over the Born approximation in that the incident and outgoing waves are described not by plane waves, but by waves distorted by an approximate static potential. Under certain conditions, particularly at low energies, this leads to a major improvement over the Born approximation.

Example: The DWA Treatment of the V-T Transition. One of the important applications of the DWA approximation is the V-T transition that was derived by Jackson and Mott in 1932. For the purpose of our discussion we consider only a collision between atom A and a harmonic oscillator BC with an interaction potential

$$V(y, x) = C \cdot e^{-(y - \gamma x)/L} = a(y) \cdot e^{\gamma x/L} \qquad (8.45a)$$

identical to eq. 8.14 used in the classical treatment. Thus, the interaction matrix elements in eq. 8.44 becomes

$$U_{on}(y) = a(y)\frac{2\mu}{\hbar^2}\langle\phi_o|e^{\gamma x/L}|\phi_n\rangle$$

$$= \frac{2\mu}{\hbar^2}a(y)V_{on} \qquad (8.45b)$$

Note that μ is the reduced mass for the system $A - BC$. Now, if we consider only the transition process between $v = 1$ and $v = 0$ and provided that the probability of the V-T transition is small, the interaction matrix U_{o1}, in the first expression of eq. 8.44 may be set to zero. In addition, we also assume that the diagonal element $V_{nn} \approx 1$ (i.e., $V_{00} \approx V_{11} \approx 1$), eq. 8.44 then reduces to

$$\left(\nabla^2 + k_0^2 - \frac{2\mu}{h_2} a(y)\right) F_0(y) = 0 \tag{8.46a}$$

$$\left(\nabla^2 + k_1^2 - \frac{2\mu}{h^2} a(y)\right) F_1(y) = \frac{2\mu}{h^2} a(y) V_{10} F_0(y) \tag{8.46b}$$

The asymptotic solution for $F_0(y)$ can then be readily solved:

$$F_0(y) = 2\cos(k_0 y + \eta_0) = 2f_0(y) \tag{8.47a}$$

To determine $F_1(y)$, we seek $F_1(y) = b(y) f_1(y)$, where $f_1(y)$ satisfies

$$\left(\nabla^2 + k_1^2 - \frac{2\mu}{h^2} a(y)\right) f_1(y) = 0 \tag{8.48}$$

and similarly it can be expressed as

$$F_1(y) = b(y) \cdot \cos(k_1 y + \eta_1) \tag{8.47b}$$

Substituting eq. 8.47b into eq. 8.46b and rearranging yields

$$\frac{db}{dy} = \frac{4\mu}{h^2} V_{10} T_{10} \sec^2(k_1 y + \eta_1) \tag{8.49}$$

where

$$T_{10} = a(y)\langle f_1 | f_0 \rangle \tag{8.50}$$

Thus we have

$$b(y) = \frac{4\mu}{h^2 k_1} V_{10} T_{10}[\tan(k_1 y + \eta_1) + \text{const}] \tag{8.51}$$

and

$$F_1(y) = b(y) f_1(y) = -\frac{4i\mu V_{10} T_{10}}{h^2 k_1} e^{+i(k_1 y + \eta_1)} \tag{8.52}$$

The transition probability is then given by

$$p_{10} = \frac{k_1}{k_0} \left| \frac{4i\mu V_{10} T_{10}}{h^2 k_1} \right|^2 = \frac{16\mu^2 V_{10}^2 T_{10}^2}{h^4 k_0 k_1} \tag{8.53}$$

That is, the transition probability will be mainly determined by the magnitudes of the translational overlap T_{10} and the vibrational matrix element V_{10}. The evaluation of V_{10} for a harmonic oscillator is straightforward. From the standard normalized vibration wave functions $\phi_n(x)$ of a simple harmonic oscillator, we find

$$V_{10} = \int_{-\infty}^{\infty} e^{\gamma x/L} \phi_1(x)\phi_0(x)\, dx \qquad (8.54)$$

where

$$\phi_0(x) = \left(\frac{\alpha}{\pi}\right)^{1/4} e^{-\alpha x^2/2}$$

$$\phi_1(x) = \left(\frac{\alpha}{\pi}\right)^{1/4} (2\alpha)^{1/2} \chi e^{-\alpha x^2/2}$$

$$\alpha = \frac{2\pi\mu_{BC} v}{\hbar}$$

$$\Delta E = \hbar\omega = h\nu$$

To determine the translational overlap, one has to solve the distorted wave function $f_n(y)$ from eq. 8.48:

$$\left(\nabla^2 + k_n^2 - \frac{2C\mu}{\hbar^2} e^{-y/L}\right) f_n(y) = 0 \qquad (8.55)$$

By introducing

$$p = \frac{2L}{a^{1/2}} e^{-y/2L} \quad \text{and} \quad q = 2kL. \qquad (8.56)$$

where $a = \hbar^2/2C\mu$, eq. 8.55 becomes equivalent to

$$\left(\frac{d^2}{dp^2} + \frac{1}{p}\frac{d}{dp} + \frac{q^2}{p^2} - 1\right) f_n(p) = 0 \qquad (8.57)$$

and can be solved by means of the modified Bessel function

$$f_n(p) = \beta_n K_{nq_n}(p) = \frac{i\pi\beta_n}{2} \frac{I_n q_n(p) - I - nq_n(p)}{\sinh \pi q_n} \qquad (8.58)$$

where

$$\beta_n = \left(q_n \sinh \frac{\pi q_n}{\pi}\right)^{1/2}$$

With the aid of the integral formula (Abramowitz and Stegun, 1965)

$$I = \int_0^{\infty} K_{nq_0}(p) K_{nq_1}(p) p\, dp = \frac{\pi^2(q_0^2 - q_1^2)}{4(\cosh \pi q_0 - \cosh \pi q_1)}$$

the translational overlap can readily be solved, to give

$$\frac{T_{10}^2}{k_0 k_1} = \frac{1}{k_0 k_1} \left| \int_{-\infty}^{\infty} c e^{-y/L} f_0(y) f_1(y) \, dy \right|^2$$

$$= \frac{1}{k_0 k_1} \frac{\hbar^4}{4\mu^2} \left(\frac{\beta_0 \beta_1}{2L} \right)^2 \left| \int_0^{\infty} K_0 q_0 K_1 q_1 \, p \, dp \right|^2$$

$$= (\pi \hbar \omega L^2)^2 W(q_0, q_1) \tag{8.59}$$

where

$$W(q_0, q_1) = \frac{\sinh \pi q_0 \sinh \pi q_1}{(\cosh \pi q_0 - \cosh \pi q_1)^2} \tag{8.60}$$

Combining eq. 8.53, 8.54, and 8.59, the transition probability is given as

$$p_{10} = 8\mu^2 \frac{m_C}{m_B m_{BC}} \frac{(\pi \omega L)^2}{\hbar \omega} W(q_0, q_1) \tag{8.61}$$

For the V-T transition at room temperature, it is generally true that $1 \ll (q_0 - q_1) \ll \pi q_0, \pi q_1$, and eq. 8.59 reduces to

$$\frac{T_{10}^2}{k_0 k_1} = (\pi \hbar \omega L^2) \cdot e^{-2\pi \omega L/\bar{g}} \tag{8.62}$$

where $\bar{g} = (g_0 + g_1)/2$. Thus the transition probability reduces to

$$p_{10} = 8\mu^2 \frac{m_C}{m_B m_{BC}} \frac{(\pi \omega L)^2}{\hbar \omega} e^{-2\pi \omega L/\bar{g}} \tag{8.63}$$

That is, we arrive at the same conclusion obtained from the classical treatment that the transition probability is favored for a collision process with small ω, small L, and large g. It is this result from which the SSH theory (due to Schwartz, Slawsky, and Herzfeld) developed and was successfully applied to vibrational relaxation in gases (Schwartz, Slawsky, and Herzfeld, 1952).

It should be noted that eq. 8.63 is identical to the classical result given in eq. 8.20 if the reduced mass μ is taken in a way similar to the classical treatment by setting $\mu = m_A m_B / m_{AB}$. Moreover, if the probability for collision on both ends of the oscillator BC is included, the mass factor in eq. 8.63 changes to $\mu/2[(1/\mu_{BC}) - (2/m_{BC})]$. This means that it is much easier for a light oscillator to make a V-T transition than it is for a heavy one as found in the classical treatment (eq. 8.12).

Time-Dependent Perturbation Theory. In its adiabatic formulation the physical problem of molecular energy transfer is customarily approached using the semiclassical treatment in which the relative motion of colliding particles is considered classically while the other internal degrees of freedom are treated quantum mechanically. Since the mass difference between the electrons and nuclei is so large, this method, in general, presents no problem in a slow collision process. Thus instead of solving the wave function $\psi(R, r)$ in eq. 8.34, we are now seeking for the wave function $\phi(r, t)$ that describes the system along a classical trajectory $R(t)$. In this way the wave function $\phi(r, t)$ under the influence of the time-dependent interaction field $\hat{H}_{12}(R(t), r)$ will satisfy

$$i\hbar \frac{\partial \phi(r, t)}{\partial t} = [\hat{H}^0(r) + \hat{H}_{12}(R(t), r)]\phi(r, t) \tag{8.64}$$

which is the semiclassical analog of the full quantal expression in eq. 8.34. Although the time-dependent wave function is in general more difficult than a stationary one, eq. 8.64 is easier to solve than eq. 8.34, since the former involves fewer degrees of freedom.

It thus follows that the interaction term $\hat{H}_{12}$ may be approximated as a time-dependent perturbation $\hat{H}_{12}(R(t))$, which induces transitions between different stationary states $\phi_s(r)$ of the unperturbed Hamiltonian $\hat{H}^0$. Therefore we look for $\phi_n(r, t)$ that can be expanded as

$$\phi_n(r, t) = \Sigma_s a_{ns}(t)\phi_s(r)\exp(-i\omega_s t) \tag{8.65}$$

where $\omega_s = E_s/\hbar$. Substitution of eq. 8.65 into eq. 8.64, multiplication of $\phi_m^*(r)\exp(i\omega_m t)$, and integration over $\mathbf{r}$ gives

$$i\hbar\dot{a}_{nm} = \Sigma_s a_{ns} V_{ms} \exp(i\omega_{ms} t) \tag{8.66}$$

where

$$V_{ms} = \langle \phi_m | H_{12}(R(t)) | \phi_s \rangle$$

$$\omega_{ms} = \frac{(E_m - E_s)}{\hbar} \tag{8.67}$$

Since one term in eq. 8.66 contains no oscillating factor as ω_{mm} vanishes, it sometimes makes an enormous contribution when the equations are integrated with respect to time and tends to increase the difficulty of obtaining satisfactory solutions to these equations. This can be eliminated by using the perturbed eigenvalues instead of the unperturbed ones in the exponential terms by introducing the new coefficient

$$c_{ns} = a_{ns} \exp\left(\frac{i}{\hbar} \int_0^t V_{ss}\, dt\right) \tag{8.68}$$

which converts eq. 8.66 to

$$i\hbar\dot{c}_{nm} = \sum_{s \neq m} c_{ns} V_{ms} \exp\left(\frac{i}{\hbar} \int_0^t \omega_{ms}\, dt\right) \tag{8.69}$$

where

$$\omega_{ms} = (E_m + V_{mm}) - (E_s + V_{ss})$$

The collision is thus treated as a transient perturbation in which $\hat{H}_{12}$ approaches zero as $t \to \pm\infty$. If the system is initially prepared in state n, the probability of a collision-induced transition to the final state m is simply given by

$$p_{nm} = |a_{nm}(+\infty)|^2 = |c_{nm}(+\infty)|^2 \tag{8.70}$$

The actual evaluation of p_{nm} then involves some approximate technique. If the interaction between the collision pairs is small, we replace the set of a_{ns} in eq. 8.66 by the zeroth-order approximation, $a_{ns} = \delta_{ns}$; this yields (Nikitin, 1968; Bates, 1962)

$$p_{nm} = \frac{1}{\hbar^2} \left| \int_{-\infty}^{\infty} V_{nm}(R)e^{i\omega_{nm}t}\, dt \right|^2 \tag{8.71a}$$

These solutions are approximately valid if $\sum_{s \neq n} |a_{ns}|^2 \ll 1$ for all t. Following the treatment of a straight-line trajectory with initial velocity g, eq. 8.71a can be rewritten as

$$p_{nm} = \frac{1}{\hbar^2 g^2} \left| \int_{-\infty}^{\infty} V_{nm}((b^2 + z^2)^{1/2})e^{i\omega_{nm}z/g}\, dz \right|^2 \tag{8.71b}$$

Since at high velocities where the exponential factor is essentially unity, the probability or the inelastic cross section decreases as the square of the relative velocity. On the other hand, the exponential factor oscillates at low g and leads to cancellation within the integral. Owing to the fact that only off-diagonal interaction terms are considered, this can be regarded as the semiclassical equivalent of the Born approximation described earlier. If eq. 8.69 rather than eq. 8.66 is used for the analysis, a semiclassical version of the distorted wave approximation can also be obtained. The first-order approximation, however, may not give a satisfactory result at very slow collisions and higher-order perturbation or the so-called PSS (perturbed-stationary-state) method may be required. The formulation of the PSS method assumes the colliding particles are at rest at some distance R_0 at which the method of the Born–Oppenheimer approximation may be used to describe the interaction terms. Provided that the duration of the collision is longer than the period of molecular rotation and vibration, and that the distortion is nearly adiabatic, the PSS method is preferable to the higher-order perturbation theory.

Finally, before we close this section, it is important to consider the conditions for the validity of such a semiclassical approximation. One of the necessary conditions is that the de Broglie wavelength be much smaller than the distance characterizing the spatial dependence of the potential field so that the change in the localized de Broglie wavelength itself is also small. Another condition is that the classical trajectories should not be appreciably perturbed in the presence of the interaction $\hat{H}_{12}$. This latter condition requires that the kinetic energy associated with the relative motion be much larger than the total energy transferred. In this respect the straight-line trajectory may be a good approximation in describing the relative motion.

Two-State Approximation. As an application of the time-dependent perturbation, we examine a system restricted to only two levels. In this case the infinite set of coupled equations, eq. 8.69, reduces to

$$\dot{c}_p = -\frac{i}{\hbar} c_q V_{pq} \exp\left(\frac{i}{\hbar} \int_0^t \omega'_{pq} \, dt^0\right)$$

$$\dot{c}_q = -\frac{i}{\hbar} c_p V_{gp} \exp\left(-\frac{i}{\hbar} \int_0^t \omega'_{pq} \, dt^0\right) \qquad (8.72)$$

by dropping the subscript n and setting $m = p$, $s = q$ in eq. 8.69. We can now proceed to apply these two-level approximation results to various processes.

The first example to be carried out here is the curve crossing model in which the first-order perturbed energies cross at some radius R_0, such that $\omega'_{pq}(R_0) = 0$. The near R_0 we may postulate that $\partial V_{pq}/\partial t = 0$ and $\omega'_{pq} = \alpha t$ so that c_q may be neglected in eq. 8.72, giving

$$\ddot{c}_p - i\alpha t \cdot \dot{c}_p + \frac{V_{pq}^2}{\hbar^2} c_p = 0 \qquad (8.73)$$

Given the boundary conditions that $c_p(-\infty) = 0$, $c_q(-\infty) = 1$, we find

$$|c_p(\infty)|^2 = 1 - e^{-2\gamma} \qquad (8.74)$$

with

$$\gamma = \frac{\pi |V_{pq}(R_0)|^2}{\hbar^2 (\partial \omega'_{pq}/\partial t)|_{R=R_0}}$$

Equation 8.74 yields the transition probability for each crossing. However, to evaluate the transition probability of the actual trajectory we have to include the crossover that occurs once when colliding particles approach each other and once when they retreat. Thus the overall transition probability is given as (Nikitin, 1974)

$$p_{pq} = 2|c_p(\infty)|^2 \{1 - |c_p(\infty)|^2\}$$
$$= 2e^{-2\gamma}(1 - e^{-2\gamma}) \qquad (8.75)$$

This is the famous Landau–Zener formula; it is clear that the probability is small in two regions, where γ is very large or very small, and has a maximum possible value of $\frac{1}{2}$ when $\exp(-2\gamma)$ is equal to $\frac{1}{2}$. For the vast majority of molecular collisions, $\exp(-2\gamma)$ is less than unity; eq. 8.75 may be approximated as $p_{pq} = 2 \cdot \exp(-2\gamma)$.

In this derivation the assumption that $\omega'_{pq} = \alpha t$ enables us to evaluate γ by employing the straight-line trajectory. This gives

$$\frac{\partial \omega'_{pq}}{\partial t} = g \frac{\partial \omega'_{pq}}{\partial R} = \frac{g}{\hbar} \left(\frac{\partial V_{pp}}{\partial R} - \frac{\partial V_{qq}}{\partial R} \right) \Bigg|_{R=R_0}$$

and

$$\gamma = \frac{\pi |V_{pq}|^2}{g\hbar(\partial V_{pp}/\partial R - \partial V_{qq}/\partial R)} = \frac{\pi |V_{pq}|^2}{g\hbar |F_{pp} - F_{qq}|} \tag{8.76}$$

Thus, when g is small, γ is small and p_{pq} is small; when g is large, γ is small and p_{pq} is small again. However, the Landau–Zener formula breaks down at both very low and very high collision energies. At low energies there is no allowance for tunneling as collision energy falls below the crossing radius, and the assumption $\omega'_{pq} = \alpha t$ is no longer true. On the other hand, the assumption $V'_{pq} = $ constant would eventually fail at high energies.

If no curve crossing occurs or occurs only at very small values of R, then the diagonal matrix elements V_{qq} and V_{pp} are negligible and, setting $t = z/g$, eq. 8.72 becomes

$$\frac{\partial \dot{c}_p}{\partial z} = -\frac{i}{\hbar q} c_q V_q \exp(i\omega_{pq} t)$$

$$\frac{\partial \dot{c}_p}{\partial z} = -\frac{i}{\hbar q} c_q V_{pq} \exp(i\omega_{pq} t) \tag{8.77}$$

These two coupled equations may be solved exactly, for example, if

$$E_{ap} + E_{bp} = E_{aq} + E_{bq} \tag{8.78}$$

The transition probability, with the boundaries $c_p(-\infty) = \delta_{pq}$, has the form

$$p_{pq} = |c_p(\infty)|^2 = \frac{\Lambda^2}{\Omega^2} \sin^2\left(\frac{\Omega}{\hbar g}\right) \tag{8.79}$$

where

$$\Omega = \int_{-\infty}^{\infty} V_{pq} \, dz$$

$$\Lambda = \int_{-\infty}^{\infty} V_{pq} \, e^{i\omega_{pq} z/q} \, dz$$

This is the Zener–Rosen formula; it may be used in calculating the transition probability in a near-resonant energy transfer. In general we expect the transition probability to be small for this case if

$$|\omega_{pq}\,\delta t| = \left|\frac{\Delta E\,\delta L}{hg}\right| \gg 1$$

that is, the adiabatic condition.

Next we consider the molecular energy transfer under resonant conditions, for instance, HCl $(v = 1)$ + HCl $(v = 0) \to$ HCl $(v = 0)$ + HCl $(v = 1)$. In this case $\omega_{pq} = 0$ and $V_{pp} = V_{qq}$ as well as $V_{pq} = V_{qp}$ so that eq. 8.72 reduces to

$$\dot{c}_p = -\frac{i}{\hbar}c_q V_{pq}$$

$$\dot{c}_q = -\frac{i}{\hbar}c_p V_{pq} \tag{8.80}$$

Taking the summation and difference, with the boundary condition $c_p(-\infty) = \delta_{pq}$, this gives

$$(c_p + c_q)|_{t\to\infty} = \exp\left(-\frac{i}{\hbar}\int_{-\infty}^{\infty}V_{pq}\,dt\right)$$

$$(c_p - c_q)|_{t\to\infty} = -\exp\left(\frac{i}{\hbar}\int_{-\infty}^{\infty}V_{pq}\,dt\right) \tag{8.81}$$

Thus the transition probability is given by

$$P_{pq} = |c_p(\infty)|^2 = \sin^2\left\{\frac{1}{\hbar}\int_{-\infty}^{\infty}V_{pq}\,dt\right\} \tag{8.82}$$

To illustrate its application we assume here $V_{pq} = V_0\exp(-\alpha R)$ and a straight-line trajectory such that

$$R^2 = b^2 + g^2t^2$$

Thus the integrand of eq. 8.82 has the value

$$I = \frac{2}{g}\int_b^{\infty}\frac{V_0 e^{-\alpha R}R\,dR}{(R^2 - b^2)^{1/2}}$$

and under the condition that $\alpha b \gg 1$ this yields

$$p_{pq}(b, g) = \sin^2\left\{\frac{V_0 e^{-\alpha b}}{hg}\left(\frac{2\pi b}{\alpha}\right)^{1/2}\right\} \tag{8.83}$$

This approximate result indicates an important feature: at a fixed energy, p_{pq} is small but monotonically increasing as the impact parameter b decreases. As b further decreases, p_{pq} starts to oscillate between zero and unity. Thus for $b < b_0$ we have $p_{pq} = \frac{1}{2}$ and for $b \geq b_0$, it is appropriate to set

$$p_{pq} = \frac{2\pi b V_0^2}{\alpha \hbar^2 g^2} e^{-2\alpha b} \tag{8.84}$$

In analogy to the Massey–Mohr random phase approximations described in elastic scattering (see eq. 8.102), these approximations are very useful for calculating total inelastic cross sections for resonant energy transfer.

8.1.3 Molecular Energy Transfer

With the foregoing discussion of the general features about molecular energy transfer and the approximate methods used to calculate these transition probabilities, we present in this section the observed experimental results and their theoretical interpretation. Since the conditions governing these processes differ remarkably from each other, our discussion is confined to emphasizing a specific subject without considering the other energy transitions that may occur concurrently during the binary collision process. In addition, we are interested in processes carried out at or near room temperature, that is, in the slow collision energy region.

Generally speaking, the rotational excitation or deactivation in thermal energy collisions is by far the most efficient process whereas the V-T transition is much less effective, in this latter case, the transition is described using the adiabatic condition. For electronic energy transfer, on the other hand, the curve crossing model is often used, which is certainly beyond the adiabatic limit. Here the nonadiabatic transition of the sudden electron jump is introduced at or near the crossing point of two adiabatic potential surfaces.

Various experimental techniques can be used to study energy transfer during the binary collision process. Among them the use of laser excitation is most interesting and promising. If the laser source is sufficiently monochromatic and intense, it is possible that one can move selectively a significant fraction of the molecules of interest into the excited state and then measure the subsequent collisional relaxation of this particular state of the species under consideration. In principle, this selective laser application provides an excellent experimental approach for the study of molecular energy transfer (Nikitin, 1974; Amme, 1975).

Rotational Energy Transfer. For the majority of molecules rotational quanta are rather small and are easily transferred to translational degrees of freedom. In addition, the R-R transition also occur very rapidly, and under favorable conditions it can be much faster than the R-T transfer. Typically, the collision number for R-T transfer is less than 10, corresponding to a relaxation time of 10^{-9} sec at 1 atm pressure. Since rotational quanta are generally small, molecules are usually well populated over a wide range of rotational energy states at thermal equilibrium. Thus the abundance of available rotational states makes it possible that the multiple quantum jump may occur in the binary collision process, in contrast to what is allowed by the optical selection rule; for instance, $\Delta j = \pm 1$ for heteronuclear diatomic and $\Delta j = \pm 2$ for momonuclear molecules (Takayanagi, 1963; van den Bergh et al., 1973).

Owing to the rapid exchange of R-R and R-T transitions, it is difficult to study these transition processes experimentally. This is particularly true when an experiment is carried out in a bulk system where molecules undergo many collisions and the signal of interest may be lost somewhere else before it is received by the detector. Experimental measurements of rotational transitions have received considerable attention lately, using microwave-microwave double resonant spectroscopy, and the molecular beam technique makes it possible to study these processes on the basis of state-state selection.

Theoretical treatment of rotational transitions is also difficult. Since rotational quanta are so small that at room temperature there are numerous rotational states populated and participating in the collision, the theoretical calculations of rotational transitions are much harder than the V-T transfer. In the theoretical calculation of rotational transitions, molecules under consideration are often treated as rigid rotors so that the vibrational effect may be neglected. This is in fact a fairly good approximation for the molecules that are in the low-lying vibronic state where the vibrational motion is restricted to a small internuclear distance.

As discussed earlier, the fast rotational energy transfer occurs as a result of the angular dependence of the interaction potential; the larger the anisotropic part of the potential, the higher the probability that a rotational transition may be induced. Thus for a polar-polar molecular collision the transition probability is expected to be higher than for an atom-homonuclear diatomic system. In view of the significance of this anisotropic potential and the difference in thermal population of rotational quanta present in heavy and light molecules, it is convenient to discuss the various cases separately: light molecules, heavy molecules, polar diatomic molecules, and polyatomic molecules.

Light Molecules. In the case of light molecules we are particularly concerned with hydrogen molecules and their isotopes. Because of its light mass, the hydrogen molecule does not thermally populate many rotational states at room temperature, thereby making a full quantum calculation feasible. In fact, the hydrogen molecules are the only case for which the quantal treatment of the DWA approximation is satisfactory.

At room temperature the normal hydrogen molecules are composed of one-quarter para-H_2 (even j) and three-quarters ortho-H_2 (odd j). Since rotational quanta of hydrogen molecules are well separated, the probabilities of rotational transitions for these molecules are usually found to be rather small at room temperature; for instance, the collision number for p-H_2 in a $j = 2$–0 transition is reported to be in the range of 200–300. For these light molecules the multiple quantum jump, say $j = 4$–0 in p-H_2 is possible but negligible at room temperature.

With the refined molecular beam technique, it is feasible now to study rotational transitions for a single collision. Although the results are rather crude, 0.6 eV collisions of Li^+-H_2 yield differential cross sections for inelastic scattering about an order of magnitude smaller than for elastic scattering and the averaged energy transfer is only about 3% of the total kinetic energy available.

Heavy Molecules. Since the rotational spacings of heavy molecules are small, the rotational quantum number j as well as the change in j during the binary collisions are expected to be much larger than unity; thus a full quantum calculation proves to be difficult. Although the time-independent S matrix or the PSS method has been attempted, theoretical calculations have largely been carried out by the classical trajectory method. Because the rotational quanta of these heavy molecules are typically small, the rotational transition rate is expected to be very fast; for instance, the averaged collision number is 5.3 for N_2 and 4.1 for O_2. The efficiency of R-T transition increases proportionally to the increase of reduced mass of diatomic systems, as illustrated in Fig. 8.4. The variation of these transition probabilities can be

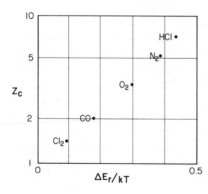

Fig. 8.4 Rotational collision numbers of the diatomic system.

alternatively correlated with the amount of mean energy transfer ΔE_r; that is, the smaller the ΔE_r, the larger the transition probability (or the smaller the collision number). Thus one expects the heavy molecules to have a faster R-T transition than the light ones. In Fig. 8.4 the collision number is plotted against the function $\Delta E_r/kT$, where T is the experimental temperature at $300°K$.

Polar Diatomic Molecules. For a collision process involving two polar molecules, the anisotropic potential arises from the strong permanent dipole interaction, which extends to a very large intermolecular distance, leading one to expect that the rotational motion would be perturbed to a large extent. Indeed, extremely large cross sections for R-R transition have been reported for polar molecular systems.

For diatomic molecules, such as NO and CO, the collision numbers for R-T transition are somewhat smaller than the homonuclear O_2 or N_2. In particular, the HCl molecule is most interesting; the averaged collision number is closer to that for spacing. This is due in part to the large permanent dipole moment of HCl, but also to the selection rule, which for N_2 and O_2 is $\Delta j = \pm 2$, improving the energy match.

Polyatomic Molecules. For polyatomic molecules the efficiency of rotational transition is expected to be very high. This is because of the abundance of rotational states available in polyatomic molecules. Since R-R transition is in general a very fast process, this would involve a significant increase in rotational relaxation for polyatomic molecules. For instance, results observed in the microwave-microwave double resonance experiment indicate that the R-R transition in OCS molecules is substantially faster than the R-T transfer.

Vibrational-Translational Energy Transfer. Since vibrational quanta are generally large compared with RT at room temperature, transition probability for V-T transfer is rather small, as in the cases where the adiabatic condition is maintained. In calculating the vibrational transition probabilities, it is customary to start from the breathing sphere model, in which the interaction potential is obtained by averaging over the molecular orientation and the fast rotational transition is thus neglected. Strictly speaking, this approach is valid only when the anisotropic part of the interaction potential is very small.

For the slow V-T transition the long-range attractive potential is expected to have little effect on the transition and an exponential function for the repulsive potential is commonly used (see eq. 8.14 and 8.45a) to calculate transition probabilities. However, although the attractive potential does not induce directly the V-T transition appreciably, it affects the V-T transition probabilities indirectly by accelerating the relative motion when two particles approach each other.

For atom-diatomic collision processes, since there is only one fundamental frequency involved and since in its low-lying vibrational states, the diatomic molecule may be regarded as a simple harmonic oscillator, theoretical treatment of vibrational relaxation for an atom-diatomic system is considerably easier than rotational relaxation or vibrational relaxation in polyatomic molecules. Moreover, for the majority of diatomic molecules at room temperature, $hv > kT$ and the optical allowed transition $\Delta v = \pm 1$ is also observed in the binary collision process.

For the atom-diatom collision process, the SSH theory proves to be very successful for most data observed thus far; it predicts reasonably well the temperature dependence of the V-T relaxation process. For diatomic molecules, Cl_2, CO, N_2, O_2, and others, the calculated relaxation time gives good agreement with experimental data over a wide temperature range, as shown in Fig. 8.5. For most atom-diatom systems involving low-lying states, the change of vibrational quanta $\Delta v = \pm 1$ is most common. However, the multiple quantum jump $\Delta v > 1$ is possible for a highly excited vibrational state but is expected to be several orders of magnitude smaller. Also, for a given diatomic molecule the probability of the V-T transition increases with decreasing reduced mass of the colliding particles; for instance, H_2 and OH are more active in causing the CO vibrational relaxation $v = 1 \rightarrow 0$ than Ne or than CO itself.

For polyatomic molecules, since more vibrational modes are available and their energy spacings are somewhat smaller, the transition probability for vibrational relaxation is expected to be higher and agreement between theory and experiment is likely to be worse than for the diatomic case. This is because the interaction potential for polyatomic molecules is more compli-

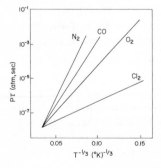

Fig. 8.5 Temperature dependence of the vibrational relaxation of diatomic system.

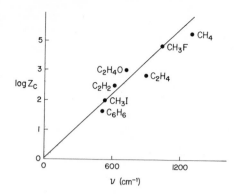

Fig. 8.6 Lambert-Salter plot of the empirical relation between the collision number and the lowest vibrational frequency.

cated than that for diatomic molecules and the theoretical calculation based on the SSH theory becomes less accurate. In addition, the effects of rotational transition as well as V-V transition become more important for polyatomic than diatomic molecules and cause greater discrepency between theory and experimental results.

For polyatomic systems Lambert and Salter found a simple empirical relation for the V-T relaxation and the magnitude of the lowest fundamental frequency: The logarithm of the collision number is proportional to the magnitude of the lowest vibrational frequency. Figure 8.6 shows the Lambert–Salter plot for a number of polyatomic molecules that contain more than one hydrogen atom. In addition, this relation also holds for halogenated molecules having different slopes (Millikan and White, 1963; Lambert and Salter, 1959).

Finally, before we close this section, it should be noted that the V-T transition observed for NO molecules is abnormal; the observed V-T relaxation rate is several times higher than is predicted by SSH theory. As pointed out by Nikitin, this is due to the very fact that when two $NO(X^2\pi)$ molecules collide, the interaction potential splits into $^1\Sigma_g^+$ and $^3\Sigma_g^-$ with a substantial translational overlap between them that would lead to a higher transition probability.

Vibrational-Vibrational Energy Transfer. For a collision process involving two polyatomic molecules, a direct transfer of vibrational-vibrational energy may occur either within the same molecule or into another molecule (Nikitin, 1960). If energy separation between these two active vibrational modes is small, we have a near-resonant transfer of vibration energy, whereas if the vibration quanta of two active modes are equal, this is the resonant situation.

A direct transfer of vibrational energy between the colliding particles is a fast process. For near resonance or resonance, the probability for V-V energy

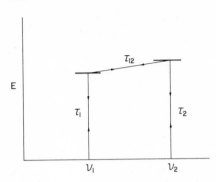

Fig. 8.7 Schematic diagram for V–V collisional energy transfer.

transfer will be either proportional to temperature if the short-range repulsion is dominant, or inversely proportional to the temperature if transition is due to the long-range attraction. If the two active vibrational quanta are well separated, that is, in the far-resonant case, the contribution from the long-range force rapidly vanishes as in a V-T transition.

The possibility for the V-V interchange between two active vibrational modes with frequency v_1 and v_2 is illustrated in Fig. 8.7. To make it happen, there are three distinct processes participating:

1. Transfer of translational and/or rotational energies with $v = 0 \to 1$ excitation of the mode v_1 with relaxation time τ_1.

2. Transfer of translation and/or rotational energies with $v = 0 \to 1$ excitation of another mode v_2 with relaxation time τ_2.

3. V-V transition between v_1 and v_2 with the aid of translational and/or rotational degrees of freedom to achieve energy balance.

If v_2 is not much larger than v_1, one finds that $\tau_2 \approx \tau_1 \gg \tau_{12}$ and the condition of steady-state equilibrium between v_1 and v_2 may be achieved. On

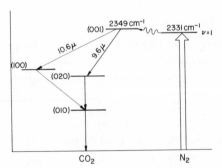

Fig. 8.8 Energy level diagram for the CO_2 laser.

the other hand, if v_1 and v_2 are well separated, the condition $\tau_2 \gg \tau_{12} > \tau_1$ applies and V-T relaxation or the v_1 mode is the dominant process.

Intermolecular vibrational energy transfer is very interesting since observation of these energy transfers through binary collisions offers a great deal of information about the mechanism of vibrational relaxation. For instance, if we consider here a mixture of HCl and DCl molecules, the possible energy transfer processes are somewhat similar to the scheme shown in Fig. 8.7 with v_1 referring to DCl and v_2 to HCl. If a small fraction of HCl molecules is originally pumped from $v_2 = 0$ to $v_2 = 1$, its relaxation can proceed via

$$HCl(1) + HCl(0) \longrightarrow HCl(0) + HCl(1) \qquad (a)$$
$$+ HCl(0) \longrightarrow 2\,HCl(0) \qquad (b)$$
$$+ DCl(0) \longrightarrow HCl(0) + DCl(1) \qquad (c)$$
$$+ DCl(0) \longrightarrow HCl(0) + DCl(0) \qquad (d)$$

where processes (a) and (c) are due to V-V transitions whereas (b) and (d) involve V-T transitions. Thus it is seen that in principle one can monitor various vibration relaxation processes under specific initial conditions and that the overall relaxation mechanism can be established.

In connection with the V-V transition, the widely used CO_2 gas laser is certainly a good example for a demonstration. For the purpose of our discussion, a simple energy scheme is shown in Fig. 8.8 for this powerful laser source in which the gas mixture is composed of CO_2, N_2, and He gases. The performance of this laser starts from the electron excitation of N_2 molecules (Lengyel, 1971),

$$N_2(v = 0) + e^- \longrightarrow N_2(v = 1) + e^-$$

followed by the near-resonant V-V transition between $N_2(v = 1)$ and $CO_2(000)$,

$$N_2(1) + CO_2(000) \rightleftharpoons N_2(0) + CO_2(001)$$

The very intense light emission, 10.6 μ and 9.6 μ, is then established, provided that the population inversion of the asymmetric stretch (001) state is constantly maintained through the complicated binary collision processes (Earl and Herm, 1974; Karl et al., 1967; Berry, 1970).

8.2 TIME-DEPENDENT BEHAVIOR OF A RELAXATION SYSTEM

From the experimental viewpoint it is desirable to have knowledge of the time-dependent behavior of the population distribution in a relaxation measurement. For this purpose we discuss the master equation approach in this section and show how to obtain and solve the master equation for a relaxation (Lin, 1974; Lin and Eyring, 1974).

8.2.1 The Liouville Equation

Suppose we have an ensemble consisting of a large number N of identical systems and let $\psi^{(k)}$ be the wave function of the kth system. If we consider the representation

$$\psi^{(k)} = \sum_n c_n^{(k)} \psi_n \tag{8.85}$$

then the density matrix is defined by (Tolman, 1938)

$$\rho_{nm} = \frac{1}{N} \sum_{k=1}^{N} c_m^{(k)} c_n^{(k)} \tag{8.86}$$

The diagonal elements of the density matrix are given by

$$\rho_{nn} = \frac{1}{N} \sum_{k=1}^{N} |c_n^{(k)}|^2 \tag{8.87}$$

which represents the probability that a system chosen at random from the ensemble will be in the state n. The density matrix is the quantum analog of the density function in classical statistical mechanics.

Some important properties of the density matrix are given in the following discussion. First we have

$$\text{Trace}(\hat{\rho}) = T_r(\hat{\rho}) = 1 \tag{8.88}$$

which shows that the probability of finding a system chosen at random from the ensemble in some state is unity. The ensemble average $\langle A \rangle$ of a physical quantity A is given by

$$\langle A \rangle = T_r(\hat{A}\hat{\rho}) = T_r(\hat{\rho}\hat{A}) \tag{8.89}$$

Next we find the equation of motion for $\hat{\rho}$. From eq. (8.85) and the time-dependent Schrödinger equation, we obtain

$$i\hbar \frac{\partial c_m^{(k)}}{\partial t} = \sum_n c_n^{(k)} H_{mn} \tag{8.90}$$

noticing that ρ_n is independent of time t. It follows that

$$v\hbar \frac{\partial \rho_{mn}}{\partial t} = \sum_j (H_{mj}\rho_{jn} - \rho_{mj}H_{jn}) \tag{8.91}$$

which can be expressed as

$$\frac{\partial \hat{\rho}}{\partial t} = -\frac{i}{\hbar}(\hat{H}\hat{\rho} - \hat{\rho}\hat{H}) = -\frac{i}{\hbar}[\hat{H}, \hat{\rho}] = -i\hat{L}\hat{\rho} \qquad (8.92)$$

where L represents the Liouville operator. Equation 8.92 is the so-called quantum-mechanical Liouville equation, and is the quantum analog of Liouville's theorem in classical statistical mechanics. The Liouville equation forms a basis for the investigation of nonequilibrium statistical mechanics.

8.2.2 The Pauli Equation

As mentioned in the previous section, the diagonal element of the density matrix gives us the probability that a system is in a particular state at a certain t. It is thus important to obtain a rate equation for the diagonal density matrix element. For this purpose we write the total Hamiltonian of the system as $\hat{H} = \hat{H}_0 + \lambda \hat{H}'$, where $\hat{H}'$ represents the perturbation Hamiltonian and λ is the perturbation parameter. The Liouville equation in this case becomes

$$\frac{\partial \hat{\rho}}{\partial t} = -i\hat{L}_c\hat{\rho} - i\hat{L}'\hat{\rho} \qquad (8.93)$$

To apply the perturbation method to eq. 8.93 we expand $\hat{\rho}$ as follows:

$$\hat{\rho} = \hat{\rho}^{(0)} + \lambda\hat{\rho}^{(1)} + \lambda^2\hat{\rho}^{(2)} + \cdots \qquad (8.94)$$

We choose the representation so that the unperturbed energy is diagonal. From eqs. 8.93 and 8.94, we find

$$\frac{\partial \rho_{nm}}{\partial t} = -\frac{i}{\hbar}(E_n - E_m)\rho_{nm}^{(0)} \qquad (8.95)$$

in the zeroth-order approximation. If we assume that $\rho_{nm} = 0$ for $n \neq m$ at $t = 0$, then eq. 8.95 can be solved to yield

$$\rho_{nm}^{(0)} = 0 \quad (n \neq m); \qquad \rho_{nn}^{(0)} = \rho_{nn}(0) \qquad (8.96)$$

where $\rho_{nn}^{(0)}$ represents $\rho_{nn}^{(t)}$ at $t = 0$. Similarly, in the first-order approximation we have

$$\frac{\partial \rho_{nm}^{(1)}}{\partial t} = -\frac{i}{\hbar}(E_n - E_m)\rho_{nm}^{(1)} - \frac{i}{\hbar}H'_{nm}(\rho_{mm}^{(0)} - \rho_{nn}^{(0)}) \qquad (8.97)$$

which can easily be solved

$$\rho_{nm}^{(1)} = \frac{H'_{nm}}{E_n - E_m}(\rho_{nn}^{(0)} - \rho_{mm}^{(0)})\left[1 - \exp\left\{-\frac{it}{\hbar}(E_n - E_m)\right\}\right] \qquad (8.98)$$

Notice that $\rho_{nm}^{(1)} = 0$.

In the second-order approximation the rate equation for $\rho_{nm}^{(2)}$ is given by

$$\frac{\partial \rho_{nn}^{(2)}}{\partial t} = -\frac{i}{\hbar} \sum_k (H'_{nk}\rho_{kn}^{(1)} - \rho_{nk}^{(1)}H'_{kn}) \tag{8.99}$$

Substituting eq. 8.98 into eq. 8.99 yields

$$\frac{\partial \rho_{nm}^{(2)}}{\partial t} = \sum_k (\rho_{nk}H)(\rho_{kk}^{(0)} - \rho_{nn}^{(0)}) \tag{8.100}$$

where

$$k_{nk}(t) = \frac{2}{\hbar} \frac{|H'_{nk}|^2}{E_n - E_k} \sin \frac{t(E_n - E_k)}{\hbar} \tag{8.101}$$

In the long time limit eq. 8.101 becomes the golden rule expression

$$k_{nk}(\infty) = k_{nk} = \frac{2\Pi}{\hbar} |H'_{nk}|^2 \delta(E_n - E_k) \tag{8.102}$$

Here the following relation has been used

$$\delta(x) = \frac{1}{\Pi} \lim_{k \to \infty} \frac{\sin kx}{x} \tag{8.103}$$

Therefore to the second-order approximation and in the long time limit, the rate expression for ρ_{nm} is given by

$$\frac{\partial \rho_{nn}}{\partial t} = \sum_k k_{nk}(\rho_{kk} - \rho_{nn}) \tag{8.104}$$

This is the so-called Pauli master equation. Notice that in equilibrium $\rho_{nm} = \rho hk$; in other words, the states of equal energy are equally probable. In the next section, we shall apply the Pauli equation to find the master equation for a relaxation phenomenon. It should be noted that the Pauli master equation given by eq. 8.104 is valid only to the second-order approximation with respect to the perturbation Hamiltonian $\hat{H}'$, and that the generalized master equation for $\rho_{nn}(t)$, which is valid to all orders in the perturbation, has been obtained (Zwanzig, 1964; Résibois, 1961; Montroll, 1960).

8.2.3 Master Equations for Relaxation Phenomena

To demonstrate the application of the Pauli master equation, we consider the vibrational relaxation of a quest molecule in a heat bath. For this purpose we require knowledge of the interaction perturbation that causes this type of vibrational relaxation to take place; the interaction perturbation involves the many-body interaction and requires a many-dimensional potential surface. To avoid this difficulty, we adopt a phenomenological approach; that is, we assume a particular type of perturbation that is physically reasonable and proceed to derive the corresponding master equation for vibrational relaxation.

We assume that the total density matrix $\hat{\rho}$ can be written as a product of the density matrices of the system and heat bath $\hat{\rho} = \hat{\rho}^{(s)}\hat{\rho}^{(b)}$, with $\hat{\rho}^{(b)}$ being expressed in the equilibrium Boltzmann distribution. It follows that the Pauli master equation of the system can be given as

$$\frac{d}{dt}\rho_{m_sm_s} = \sum_{n_s} K_{m_sn_s} \exp(\beta E_{m_s}) \left[\exp(-\beta E_{m_s})\rho_{n_sn_s} - \rho_{m_sm_s}\exp(-\beta E_{n_s})\right]$$

(8.105)

where

$$K_{m_sn_s} = \frac{2\pi}{\hbar}\sum_{n_b}\sum_{m_b}\rho_{n_bn_b}|\langle m_s m_b|\hat{H}'|n_s n_b\rangle|^2\delta(E_{m_sm_b} - E_{n_sn_b})$$
(8.106)

Notice that $\rho_{n_bn_b} = \exp(-\beta E_{n_b})/Q_b(\beta)$, where $Q_b(\beta) = \sum_{n_b} \times \exp(-\beta E_{n_b})$, the partition function of the heat bath.

First we consider the case in which the molecular oscillator Q_i interacts linearly with the medium,

$$\hat{H}' = Q_i F_{ib}.$$
(8.107)

In this case each molecular oscillator relaxes independently and from eq. 8.105 we obtain

$$\frac{d\rho_{m_i}^{(i)}}{dt} = K_i\{(m_i + 1)\rho_{m_i+1}^{(i)} - [m_i + (m_i + 1)e^{-\beta\hbar\omega_i}]\rho_{m_i}^{(i)} + m_i e^{-\beta\hbar\omega_i}\rho_{m_i-1}^{(i)}\}$$

(8.108)

Here for simplicity we use the notation $\rho_{m_sm_s} \equiv \rho_{m_s}$ and K_i is defined by

$$K_i = \frac{2\pi}{\hbar}\cdot\frac{1}{2\beta_i^2}\sum_{m_b}\sum_{n_b}\rho_{n_b}|\langle m_b|F_{ib}|n_b\rangle|^2\delta(E_{m_b} - E_{n_b} - \hbar\omega_i)$$
(8.109)

with $\beta_i^2 = \omega_i/\hbar$. Equation 8.107 indicates that F_{ib} plays the role of "force" acting on the ith molecular mode by the medium. The master equation for vibrational relaxation of molecular oscillators given in eq. 8.108 is valid only in the lowest order of approximation (the second order in this case) with respect to $\hat{H}'$.

So far we have considered the case in which the molecular oscillators can lose one quantum to the heat bath or gain one quantum from the heat bath during vibrational relaxation, and molecular oscillators relax independently.

Next we consider another possibility for which the interaction Hamiltonian is given by

$$\hat{H}' = Q_i Q_j F_{ijb} \tag{8.110}$$

In other words, during vibrational relaxation the vibrational energy of one molecular oscillator is allowed to flow to another. In this case, substituting eq. 8.110 into eqs. 8.105 and 8.106, we find the master equation to be given by

$$\frac{d\rho_{m_s}}{dt} = K_{ij}\{(m_i + 1)m_j[\rho_{m_i+1\,m_j-1} - \rho_{m_im_j}e^{\beta\hbar(\omega_j-\omega_i)}]$$

$$+ m_i(m_j + 1)[\rho_{m_i-1\,m_j+1}e^{\beta\hbar(\omega_j-\omega_i)} - \rho_{m_im_j}]\} \tag{8.111}$$

where

$$K_{ij} = \frac{2\pi}{\hbar}\frac{1}{4\beta_i^2\beta_j^2}\sum_{n_b}\sum_{m_b}\rho_{n_b}|\langle m_b|F_{ijb}|n_b\rangle|^2\,\delta(\hbar\omega_j - \hbar\omega_i + E_{m_b} - E_{n_b}) \tag{8.112}$$

The vibrational relaxation described by the perturbation Hamiltonian given by eq. 8.110 does not take place by itself; it is usually accompanied by the type of vibrational relaxation described by eq. 8.109. In this case the master equation for vibrational relaxation is given by

$$\frac{d\rho_{m_s}}{dt} = \sum_i K_i\{(m_i + 1)\rho_{m_i+1} - [m_i + (m_i + 1)e^{-\beta\hbar\omega_i}]\rho_{m_i} + m_i e^{-\beta\hbar\omega_i}\rho_{m_i-1}\}$$

$$+ K_{ij}\{(m_i + 1)m_j[\rho_{m_i+1\,m_j-1} - \rho_{m_im_j}e^{\beta\hbar(\omega_j-\omega_i)}] + m_i(m_j + 1)$$

$$\times [\rho_{m_i-1\,m_j+1}e^{\beta\hbar(\omega_j-\omega_i)} - \rho_{m_im_j}]\} \tag{8.113}$$

In other words, the vibrational excitation can directly give up its excess energy to the heat bath; it can also be redistributed among the molecular oscillators before dissipating its excess energy to the heat bath through low-frequency molecular oscillators. Notice that at $T = 0$, eqs. 8.110 and 8.113 reduce to

$$\frac{d\rho_{m_i}^{(i)}}{dt} = K_i[(m_i + 1)\rho_{m_i+1}^{(i)} - m_i\rho_{m_i}^{(i)}] \tag{8.114}$$

and

$$\frac{d\rho_{m_s}}{dt} = \sum_i K_i[(m_i + 1)\rho_{m_i+1} - m_i\rho_{m_i}]$$

$$+ K_{ij}[(m_i + 1)m_j\rho_{m_i+1\,m_j-1} - m_i(m_j + 1)\rho_{m_im_j}] \tag{8.115}$$

respectively. Here we have assumed that $\omega_i > \omega_j$.

The types of master equations for vibrational relaxation given by eqs. 8.108 and 8.113 are quite general and can be applied to describe not only the vibrational relaxation in solids and gases but also the vibrational relaxation in liquids; the difference lies in the rate constants K_i and K_{ij}, which in turn are determined by the choice of the interaction Hamiltonian. For example, let us consider the master equation given by eq. 8.108. Introducing the integral representation for the delta function in eq. 8.109 yields

$$K_i = \left(\frac{1}{\hbar^2}\right) \cdot (\tfrac{1}{2}\beta_i^2)G(\omega_i) \tag{8.116}$$

where $G(\omega_i)$ is defined by

$$G(\omega_i) = \int_{-\infty}^{\infty} dt\ e^{it\omega_i}\langle F_{ib}(0)F_{ib}(t)\rangle_{\mathrm{av}} \tag{8.117}$$

Here $\langle F_{ib}(0)F_{ib}(t)\rangle_{\mathrm{av}}$ represents the correlation function

$$\langle F_{ib}(0)F_{ib}(t)\rangle_{\mathrm{av}} = \sum_{n_b} \rho_{n_b}\langle n_b|F_{ib}(0)F_{ib}(t)|n_b\rangle \tag{8.118}$$

where

$$F_{ib}(t) = \exp\left(-\frac{it\hat{H}_0}{\hbar}\right)F_{ib}\exp\left(\frac{it\hat{H}_0}{\hbar}\right).$$

Equation 8.116 was the starting equation used by Zwanzig to criticize the isolated binary collision theory of vibrational relaxation in liquids (Zwanzig, 1961; see also Sec. 8.1).

The solution of the type of master equation given by eq. 8.108 has been accomplished by Montroll and Shuler for various initial conditions. For example, if the ith oscillator is initially populated in the u_ith state, the solution of eq. 8.108 is given by (Montroll and Shuler, 1957)

$$\rho_{m_i}^{(i)}(t) = \frac{(1 - e^{\theta_i})e^{u_i\theta_i}}{(e^{-\tau_i} - e^{\theta_i})}\left(\frac{e^{-\tau_i} - 1}{e^{-\tau_i} - e^{\theta_i}}\right)^{m_i + u_i} F(-m_i, -u_i, 1; \alpha_i^2) \tag{8.119}$$

where $F(-m_i, -u_i, 1; \alpha_i^2)$ represents the hypergeometric function, $\alpha_i = \sinh(\theta_i/2)[\sinh(\tau_i/2)]^{-1}$, $\theta_i = \hbar\omega_i/kT$, and $\tau_i = K_i t(1 - e^{-\theta_i})$. Notice that at $T = 0$ eq. 8.119 becomes (Fleming et al. 1974)

$$\rho_{v_i}(t) = \frac{v_i!}{v_i!(u_i - v_i)!}[1 - \exp[-K_i(\infty)t]]^{u_i - v_i}\exp[-v_iK_i(\infty)t] \tag{8.120}$$

This result implies that the initial state u_i decays exponentially with the rate constant $u_iK_i(\infty)$ and the lower states v_i decay exponentially only at longer times, each with a different rate constant $v_iK_i(\infty)$. Another simple case that can be considered is $U_i = 0$, that is, the heating of the ith mode by the heat bath. In this case eq. 8.119 becomes

$$\rho_{v_i}(t) = \frac{(e^{\theta_i} - 1)(1 - e^{-t_i})^{v_i}}{(e^{\theta_i} - e^{-t_i})^{v_i+1}} \tag{8.121}$$

When $t \to \infty$, we obtain the equilibrium distribution

$$\rho_{v_i}(\infty) = (e^{\theta_i} - 1)e^{-\theta_i(v_i+1)} \tag{8.122}$$

The particular results of eqs. 8.120 and 8.121 have also been obtained by Nitzan and Jortner (1973), using an entirely different approach.

In concluding the discussion of this section, it should be noted that although we have only shown how to derive the master equations for vibrational relaxation from the Pauli equation, the master equations for other relaxation phenomena can be derived similarly and are not discussed here. The master equations for vibrational relaxation given in this section can be applied to gaseous and condensed phases. The multimode vibrational relaxation of polyatomic molecules in gaseous and condensed phases can be treated by using eq. 8.113 (Moore, 1973; Weitz and Flynn, 1974).

8.3 VIBRATIONAL RELAXATION IN CONDENSED PHASE

Using a mode-locked laser with resolution on a picosecond scale, it has now become possible to directly measure the lifetime of an optical phonon and a normal mode of a polyatomic molecule dissolved in condensed media. Alfano and Shapiro report the direct measurement of the lifetime of the 1086 cm^{-1} optical phonon in calcite, which is found to be 22 ± 4 psec at $100°K$ and 8.5 ± 2 psec at $297°K$. Laubereau and Kaiser (1975) report on an investigation of the fundamental $\bar{1}0$ phonons of diamond; the relaxation times are found to be 2.9 ± 0.3 psec and 3.4 ± 0.3 psec for 295 and $77°K$, respectively. Laubereau and Kaiser also report the first direct measurements of the vibrational lifetimes of simple normal modes of polyatomic molecules in liquids; they find the lifetime of 5 ± 1 psec for the totally symmetric CH_3 vibration 2939 cm^{-1} for CH_3CCl_3 and the lifetime af 20 ± 5 psec for the CH_3 vibration 2928 cm^{-1} of C_2H_5OH at $300°K$. The dephasing time of the 459 cm^{-1} vibration of CCl_4 has been directly measured by von der Linde et al.; they find it to be 4.0 ± 0.5 psec (Lin and Eyring, 1974; Laubereau and Kaiser, 1975; Lin, 1978).

Recently several models of vibrational relaxation in condensed media have been proposed (Sun and Rice, 1965; Nitzan and Jortner, 1973; Diestler,

1974; Lin et al., 1976; Nitzan et al., 1974; Lin, 1976). Only two important ones are described in this chapter. One of them is based on the use of the short-range repulsive force to describe the coupling the medium oscillators (phonon) and the system oscillator (vibron). The other is based on the idea that the frequencies of the system oscillators (vibrons) are much larger than those of the medium oscillators (phonons) so that the adiabatic type of approximation can be used to separate these two types of vibration; the vibrational relaxation is then induced by the Born–Oppenheiner coupling in exactly the same way as the electronic relaxation (see Chapter 7).

8.3.1 Repulsive Potential Model

In eqs. 8.108–8.118 we demonstrated how to obtain the rate constant of vibrational relaxation once we know the coupling between the relaxing modes and accepting modes. We now discuss a model for calculating the rate constant of vibrational relaxation in condensed media based on the assumption that only the short-range repulsive portion of the intermolecular force is assumed to be effective in vibrational relaxation. This assumption is reasonable, as it is known from the collisional vibrational relaxation in gases that the long-range attractive force is ineffective in vibrational relaxation. Consider $K_i(\beta)$ given by eq. 8.109. In this case the "force" F_{ib} is given by

$$F_{ib} = C_i e^{-\alpha_i \gamma_i} \qquad (8.123)$$

where γ_i represents the instantaneous distance between the center of mass of the molecule and the nearest neighbor solvent atom. In other words, we consider a system of a monatomic lattice with a single substitutional molecular lattice. The presence of an impurity alters the vibrational spectrum of the solid in the locality of the impurity and the localized modes introduced by the impurity play an important role in vibrational relaxation.

Now we expand the relative displacement between the center of mass of the molecule and the nearest neighbor in terms of the normal coordinates of all the contributing normal modes of the solid:

$$F_{ib} = C_i \exp\left(-a_i \gamma_{i0} - a_i \sum_{i\mu} \gamma_{i\mu} Q_\mu\right) \qquad (8.124)$$

where γ_{i0} represents the corresponding equilibrium distance of γ_i. Substituting eq. 8.123 into eq. 8.109 yields

$$K_i = \frac{2\pi}{\hbar} \frac{C_i'^2}{2\beta_i^2} \sum_{m_b} \sum_{n_b} \rho_{n_b} \prod_\mu |\langle m_\mu | e^{-\alpha_i \gamma_{i\mu} Q_\mu} | n_\mu \rangle|^2 \delta(E_{m_b} - E_{n_b} - \hbar\omega_i) \qquad (8.125)$$

where $C_i' = C_i e^{-\alpha_i \gamma_i 0}$ and $\beta_i = (\omega_i/\hbar)^{1/2}$. Introducing ω_i, the integral representation for the delta function, we obtain

$$K_i = \frac{C_i'^2}{2\hbar^2 \beta_i^2} \int_{-\infty}^{\infty} dt \, e^{-it\omega_i} \prod_\mu G_\mu(t) \tag{8.126}$$

where

$$G_\mu(t) = \sum_{n_\mu} \sum_{m_\mu} \rho_{n_\mu} |\langle m_\mu | e^{-\alpha_i \gamma_{i\mu} Q_\mu} | n_\mu \rangle|^2 \exp[it(m_\mu - n_\mu)\omega_\mu] \tag{8.127}$$

Using the Slater sum, eq. 8.127 can easily be simplified to yield (see also Sec. 7.3)

$$G_\mu(t) = \exp\left\{ \frac{a_i^2 \gamma_{i\mu}^2}{2\beta_\mu^2} \left[\coth \frac{\hbar\omega_\mu}{2kT} + \operatorname{csch} \frac{\hbar\omega_\mu}{2kT} \cosh\left(it\omega_\mu + \frac{\hbar\omega_\mu}{2kT} \right) \right] \right\} \tag{8.128}$$

Substituting eq. 8.128 into eq. 8.126, we find

$$K_i = \frac{C_i'^2}{2\hbar\omega_i} e^S \int_{-\infty}^{\infty} dt \exp\left[-it\omega_i + \sum_\mu \frac{a_i^2 \gamma_{i\mu}^2}{2\beta_\mu^2} \operatorname{csch} \frac{\hbar\omega_\mu}{2kT} \cosh\left(it\omega_\mu + \frac{\hbar\omega_\mu}{2kT} \right) \right] \tag{8.129}$$

where

$$S = \frac{1}{2} \sum_\mu^N \frac{a_i^2 \gamma_{i\mu}^2}{\beta_\mu^2} \coth \frac{\hbar\omega_\mu}{2kT}$$

The summation over μ covers all the contributing normal modes in receiving the vibrational energy.

The simplification of eq. 8.129 can be carried out in a similar manner as was done for radiationless transitions. For example, eq. 8.129 can also be expressed as (Lin and Bersohn, 1968)

$$K_i = \frac{\pi C_i'^2}{\omega_i} e^S \sum_{v_1} \cdots \sum_{v_N} \sum_{v_1'} \cdots \sum_{v_N'} \prod_\mu \frac{\exp(v_\mu \hbar\omega_\mu/kT)}{v_\mu! v_\mu'!}$$
$$\times \left(\frac{a_i^2 \gamma_{i\mu}^2 \bar{n}_\mu}{2\beta_\mu^2} \right)^{v_\mu + v_\mu'} \delta\left[\sum_\mu \hbar\omega_\mu (v_\mu - v_\mu') - \hbar\omega_i \right] \tag{8.130}$$

In particular, at $T = 0$, eq. 8.130 reduces to

$$K_i = \frac{\pi C_i'^2}{\omega_i} e^S \sum_{v_1} \cdots \sum_{v_N} \prod_\mu \frac{1}{v_\mu!} \left(\frac{a_i^2 \gamma_{i\mu}^2}{2\beta_\mu^2} \right)^{v_\mu} \delta\left(\sum_\mu v_\mu \hbar\omega_\mu - \hbar\omega_i \right)$$
$$\tag{8.131}$$

The summations over v_μ cover all the possibilities of distributing the vibrational energy $\hbar\omega_i$ among the accepting oscillators.

Now if

$$\frac{1}{2}\sum_\mu \frac{a_i^2\gamma_{i\mu}^2}{\beta_\mu^2} \gg 1$$

(this corresponds to the strong coupling case in radiationless transitions), we may expand $\cosh[it\omega_\mu + (\hbar\omega_\mu/2kT)]$ in power series of t and carry out the integration with respect to t:

$$K_i = \frac{C_i'^2}{2\hbar\omega_i}\left(\frac{4\pi}{\sum_\mu (a_i^2\gamma_{i\mu}^2\omega_\mu^2)/\beta_\mu^2\,\coth(\hbar\omega_\mu/2kT)}\right.$$
$$\left. \times \exp\left[2S - \frac{(\omega_i - \sum_\mu (a_i^2\gamma_{i\mu}^2\omega_\mu/2\beta_\mu^2))^2}{\sum_\mu (a_i^2\gamma_{i\mu}^2\omega_\mu^2/\beta_\mu^2)\coth(\hbar\omega_\mu/2kT)}\right]\right. \tag{8.132}$$

In particular, if the temperature is high enough that $\hbar\omega_\mu/kT \ll 1$ for all the accepting normal modes, then eq. 8.132 can be simplified as

$$K_i = \frac{C_i'^2}{2\hbar\omega_i}\left(\frac{2\pi}{kTa_i^2\sum_\mu \gamma_{i\mu}^2}\right)^{1/2}$$
$$\times \exp\left[2kTa_i^2\sum_\mu \frac{\gamma_{i\mu}^2}{\omega_\mu^2} - \frac{[\omega_i - (\hbar a_i^2/2)\sum_\mu \gamma_{i\mu}^2]^2}{2kTa_i^2\sum_\mu \gamma_{i\mu}^2}\right] \tag{8.133}$$

Equation 8.133 indicates that the temperature dependence of K_i has the Arrhenius form only when $2kTa_i^2\sum_\mu (\gamma_{i\mu}^2/\omega_\mu^2)$ is negligible compared with $(\omega_i - (\hbar a_i^2/2)\sum_\mu \gamma_{i\mu}^2)^2/2kT(a_i^2\sum_\mu \gamma_{i\mu}^2)$ and when the temperature is reasonably high.

Sun and Rice (1965) have also treated the vibrational relaxation of a molecule dissolved in a monatomic lattice using the same interaction potential as that given by eq. 8.123. However, they have employed the binary collision theory of vibration deactivation for the relaxation process with the motion of collision partners determined by the local modes of the lattice and calculated the collision frequency by using an analysis similar to that of the Slater theory of unimolecular reactions. The rate constant of vibrational relaxation has been expressed in the Arrhenius form.

To estimate the order of magnitude of K_i, we notice that at $T = 0$, eq. 8.132 reduces to

$$K_i = \frac{C_i'^2}{2\hbar\omega_i}\left(\frac{4\pi}{\sum_\mu (a_i^2\gamma_{i\mu}^2\omega_\mu^2/\beta_\mu^2)}\right)^{1/2}\exp\left[2S - \frac{(\omega_i - \sum_\mu (a_i^2\gamma_{i\mu}^2\omega_\mu/2\beta_\mu^2))^2}{\sum_\mu (a_i^2\gamma_{i\mu}^2\omega_\mu^2/\beta_\mu^2)}\right]$$
$$\tag{8.134}$$

Now $\omega_i \doteq 10^2 - 10^3$ cm^{-1}, $\omega_\mu = 0 - 10^2$ cm^{-1}, $S \doteq 1 - 10$, and $C_i' \doteq 10^5 - 10^7$ erg/cm $\cdot$ g$^{1/2}$. It follows that K_i can take a wide range of values, covering from 10^{12} sec^{-1} to 1 sec^{-1} or longer.

Next we consider the weak coupling case defined by $\zeta \le 1$. In this case the integration in eq. 8.129 can be carried out by the saddle point method. The result is given by (Lin et al., 1976)

$$K(T) = \frac{C'^2}{\hbar\omega}\left(\frac{\pi}{\sum_u \Delta_u^2 \omega_u^2 \operatorname{csch}(\hbar\omega_u/2kT)\cosh(it^*\omega_u + \hbar\omega_u/2kT)}\right)^{1/2}$$
$$\times \exp\left(S - it^*\omega + \sum_u \frac{\Delta_u^2}{2}\operatorname{csch}\left(\frac{\hbar\omega_n}{2kT}\right)\cosh\left[it^*\omega_u + \left(\frac{\hbar\omega_u}{2kT}\right)\right]\right)$$

(8.135)

where for convenience the subscript i has been omitted from K_i and t^* represents the saddle point value of t and is to be determined from

$$\omega = \sum_u \frac{\Delta_u^2 \omega_u}{2}\operatorname{csch}\frac{\hbar\omega_u}{2kT}\sinh\left(it^*\omega_u + \frac{\hbar\omega_u}{2kT}\right)$$ (8.136)

To demonstrate the behavior of the temperature dependence of the vibrational relaxation rate constant $K(T)$, we introduce an average frequency ω_m for ω_u (in other words, we assume all the accepting modes to have the same frequency that corresponds to the Einstein model of solids). In this case eq. 8.129 becomes

$$K(T) = \frac{\pi C'^2}{\hbar\omega\omega_m}e^S \sum_{l=0}\frac{(S_0\bar{n}_m)^l}{l!}\frac{(S_0\bar{n}_m e^{\hbar\omega_m/kT})^{\omega/\omega_m + l}}{(l + \omega/\omega_m)!}$$ (8.137)

where $\bar{n}_m = (e^{\hbar\omega_m/kT} - 1)^{-1}$ and $S_0 = \sum_u \Delta_u^2/2$, the S value at $T = 0$. In particular, for $T = 0$, eq. 8.137 reduces to

$$K(0) = \frac{\pi C'^2}{\hbar\omega\omega_m}e^{S_0}\frac{S_0^{\omega/\omega_m}}{(\omega/\omega_m)!}$$ (8.138)

For convenience, we define the dimensionless quantities T^*, ω^*, and $K^*(T^*)$ as follows:

$$T^* = \frac{kT}{\hbar\omega_m}, \qquad \omega^* = \frac{\omega}{\omega_m}, \qquad K^*(T^*) = \frac{K(T)}{K(0)}$$ (8.139)

It follows that

$$K^*(T^*) = \exp\left[S_0\left(\coth\frac{1}{2T^*} - 1\right)\right](\bar{n}_m e^{1/T^*})^{\omega^*}\omega^*! \sum_{l=0}\frac{(S_0\bar{n}_m)^l}{l!}\frac{(S_0\bar{n}_m e^{1/T^*})^l}{(l + \omega^*)!}$$

(8.140)

Similarly, eqs 8.132 and 8.135 can be written as

$$K^*(T^*)_{sc} = \left(\tanh \frac{1}{2T^*}\right)^{1/2} \exp\left[\left(1 - \tanh \frac{1}{2T^*}\right)\right.$$

$$\left. \times \left(\frac{(\omega^* - S_0)^2}{2S_0} + 2S_0 \coth \frac{1}{2T^*}\right)\right] \tag{8.141}$$

and

$$K^*(T^*)_{wc} = \frac{\exp\{S_0[\coth(1/2T)^* - 1]\}}{[1 + S_0^2/\omega^{*2} \operatorname{csch}^2/1/2T^*)]^{1/4}}$$

$$\times \exp\left[-\omega^* \log\left(e^{-1/2T^*} \sinh \frac{1}{2T^*}\right.\right.$$

$$\left.+ \sqrt{\frac{S_0^2}{\omega^{*2}} e^{-1/T^*} + e^{-1/T^*} \sinh^2 \frac{1}{2T^*}}\right)$$

$$\left. + \sqrt{\omega^{*2} + S_0^2 \operatorname{csch}^2 \frac{1}{2T^*}} - \omega^*\right] \tag{8.142}$$

respectively.

Now we quantitatively compare the behaviors of $K^*(T^*)$, $K^*(T^*)_{sc}$, and $K^*(T^*)_{wc}$ and also quantitatively study how the coupling constant S_0 and energy gap ω^* affect the temperature dependence of the vibrational relaxation rate constant.

In Fig. 8.9 we show the plot of $K^*(T^*)$ versus T^* for various S_0 values. As we can see from this figure, the temperature effect on the rate constant increases with increasing coupling strength S_0. In the lower temperature range (below $T^* = 0.3$) the temperature effect is very small, and above $T^* = 0.3$ the rate constant increases with T^* exponentially (see also Fig. 8.10.) Recalling that $T^* = kT/\hbar\omega_m$, this indicates that for crystals with strong force constants, the temperature effect may be expected to be quite small up to reasonably high temperatures (determined by $T = 0.3\hbar\omega_m/k$). In other words, whether there will be a strong temperature dependence or not depends on the temperature range measured and on the property of crystals. For many solids

$$\theta_E\left(\theta_E = \frac{\hbar\omega_m}{kT}\right)$$

is in the range 100–300°K, although instances are known that fall above or below this range; for example, diamond has a θ_E value of 1320°K.

Thus if $\bar{\omega}_m = 100 \text{ cm}^{-1}$, then for T below 40°K, we expect to have a small temperature dependence; and if $\bar{\omega}_m = 500 \text{ cm}^{-1}$, then for T below 220°K,

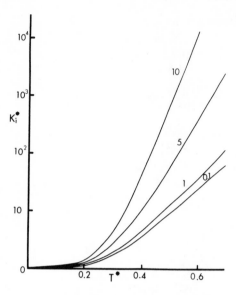

Fig. 8.9 Temperature dependence of vibrational relaxation rate constants for various S_0 values with $\omega^* = 15$. [Reprinted, with permission, from S. H. Lin, *J. Chem. Phys.*, **61**, 3810 (1974).]

we expect to have a small temperature dependence. A similar phenomenon may be expected to happen in electronic radiationless transition (see Sec. 7.3). In the low-temperature range, from eq. 8.140 we obtain

$$K^*(T^*) = 1 + e^{-1/T^*} \frac{[(\omega^* + S_0)^2 + (\omega^* + 2S_0)]}{\omega^* + 1} + \cdots \quad (8.143)$$

Here we have assumed that $(S_0 e^{-1/T^*}) \ll 1$. A similar equation has been derived for electronic relaxation.

If we let T_c^* represent the onset temperature above which the temperature effect on the vibrational relaxation sets in, then it can be determined from eq. 8.143 by setting the second term on the right-hand side of eq. 8.143 equal to unity,

$$T_c^* = \frac{1}{\log(\omega^* + 2S_0 + S_0^2/1 + \omega^*)} \quad (8.144)$$

Dubost (1976) has observed that the temperature dependence of the vibrational fluorescence decay time of CO in solid argon is very slow between 8°K and 30°K when highly purified samples are studied. Tinti and Robinson also report that the vibrational relaxation time in the $A^3\Sigma_u^+$ state of N_2 in solid rare gases lies between 0.4 and 3.3 sec and that the temperature effect on the relaxation rate is found to be small over the range 1.7–30°K. Allamandolla and Nibler (1974) have employed the optical double resonance technique to measure the lifetimes of the $v = 1$ and $v = 2$ states of matrix isolated C_2^- and obtained half-lives of 0.2 and 1.2 msec, respectively, in an argon matrix of $16°K$, and 0.3 and 1.3 msec in an N_2 matrix. They also report that no significant temperature variation of these rate constants has been found in the temperature range $14–24°K$. Recently Brus and Bondybey (1975) have measured the vibrational relaxation of some diatomic radicals and ions in rare gas lattices and emphasized the importance of the pseudorotational local mode in vibrational relaxation. They have also studied the temperature effect and found that the rates of vibrational relaxation of NH are independent of temperature for $T \le 25°K$ in Ar and $T \le 37°K$ in Kr (Legay, 1977). These observations of the temperature effect on vibrational relaxation seem to be consistent with what has been discussed earlier.

In Fig. 8.9 we present the plot of K^* versus T^* for T^* in the range between 0 and 0.8. From this figure we can see that for $S_0 > 1$, the rate constant changes very rapidly with T^*. Figure 8.11 shows the effect of the energy gap on the temperature dependence of the rate constant. The temperature effect is bigger for larger energy gaps. The same tendency is also observed in electronic relaxation. This tendency is more pronounced for weaker coupling (i.e., small S_0 values).

Table 8.1 shows the comparison of K^* and K_{wc}^*. The agreement between K^* and K_{wc}^* is excellent for all ω^*, T^*, and S_0 values. The slight deviation between K^* and K_{wc}^* only appears in the small energy gap (or ω^*) and high T^* and high S_0. Below $S_0 = 1$, for $\omega^* = 10, 15$, and 20, the agreement between

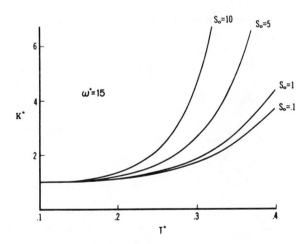

Fig. 8.10 Temperature dependence of vibrational relaxation rate constants in the low temperature range. [Reprinted, with permission, from S. H. Lin, *J. Chem. Phys.*, **61**, 3810 (1974).]

K^* and K^*_{wc} is even better than that shown in Table 8.1. It should be noted that we are comparing $K(T)/K(0)$ and $K(T)_{wc}/K(0)_{wc}$ and have not compared $K(0)$ and $K(0)_{wc}$. However, it can easily be shown that the difference between $K(0)$ and $K(0)_{wc}$ is simply the use of the Stirling formula for $\omega^*!$, and the accuracy of the Stirling formula for the factorial function is well known.

Although it has been claimed that K_{sc} holds for $\zeta \gg 1$, the limitation and validity of the strong coupling approximation have never been critically examined. To compare $K(0)_{sc}$ and $K(0)$, we show some numerical results of $K(0)_{sc}/K(0)$,

$$\frac{K(0)_{sc}}{K(0)} = \frac{1}{\sqrt{2\pi S_0}} \frac{\omega^*!}{S_0^{\omega^*}} \exp\left(S_0 - \frac{(\omega^* - S_0)^2}{2S_0}\right) \qquad (8.145)$$

in Fig. 8.12 as a function of ω^* and S_0. Figure 8.12 shows that at $T^* = 0$, for a given ω^*, increasing S_0 values do not always improve the accuracy of $K(0)_{sc}$; actually for a given ω^* value $K(0)_{sc}$ only applies for a certain range of S_0 values. Beyond this range of S_0 values, the accuracy of $K(0)_{sc}$ gets worse with increasing S_0 values.

The conditions under which the weak coupling approximation and strong coupling approximation hold have also been analytically examined by Lin et al (1976).

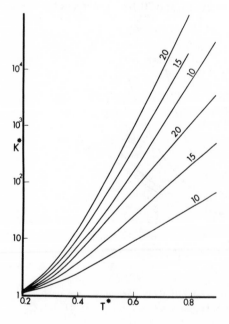

Fig. 8.11 Effect of the energy gap on the temperature dependence of vibrational relaxation rate constant for $\omega^* = 20$, 15, and 10. For the first three curves $S_0 = 5$, and for the next three curves $S_0 = 0.1$. [Reprinted, with permission, from S. H. Lin, *J. Chem. Phys.*, **61**, 3810 (1974).]

Table 8.1
Comparison of K^* and K^*_{wc}

ω^*	S_0	T^*	K^*_{wc}	K^*
20	1.0	0.7	4.592×10^2	4.593×10^2
15	1.0	0.7	1.174×10^2	1.174×10^2
10	1.0	0.7	3.019×10	3.020×10
20	1.0	1.1	1.225×10^5	1.226×10^5
15	1.0	1.1	9.456×10^3	9.459×10^3
10	1.0	1.1	7.405×10^2	7.412×10^2
20	1.0	1.5	1.652×10^7	1.652×10^7
15	1.0	1.5	4.651×10^5	4.653×10^5
10	1.0	1.5	1.346×10^4	1.348×10^4
20	5.0	0.7	9.122×10^3	9.132×10^3
15	5.0	0.7	2.684×10^3	2.690×10^3
10	5.0	0.7	8.923×10^2	8.978×10^2
20	5.0	1.1	9.458×10^7	9.482×10^7
15	5.0	1.1	1.044×10^7	1.049×10^7
10	5.0	1.1	1.531×10^6	1.546×10^6
20	5.0	1.5	7.967×10^{11}	7.996×10^{11}
15	5.0	1.5	4.217×10^{10}	4.244×10^{10}
10	5.0	1.5	3.492×10^9	3.534×10^9
20	10	0.7	8.683×10^5	8.710×10^5
15	10	0.7	3.738×10^5	3.759×10^5
10	10	0.7	2.340×10^5	2.366×10^5
20	10	1.1	2.832×10^{12}	2.846×10^{12}
15	10	1.1	7.044×10^{11}	7.099×10^{11}
10	10	1.1	3.434×10^{11}	3.476×10^{11}
20	10	1.5	1.627×10^{19}	1.636×10^{19}
15	10	1.5	2.863×10^{18}	2.886×10^{18}
10	10	1.5	1.212×10^{18}	1.227×10^{18}

8.3.2 Born–Oppenheimer Coupling Model

In this section we discuss the theoretical model for vibrational relaxation of vibrons in condensed media based on the breakdown of the adiabatic separation of vibron (high frequencies) and phonon (low frequencies) motion (Lin, 1976).

Let the total Hamiltonian of the system consisting of vibrons and phonons be written as

$$\hat{H} = \hat{T}_Q + \hat{T}_q + V(q, Q) \tag{8.146}$$

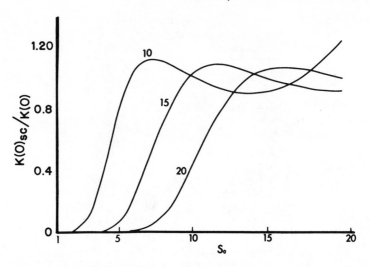

Fig. 8.12 Comparison of the rate constant in the strong coupling approximation K_{sc} and the exact rate constant for various S_0 values at $T = 0$. [Reprinted, with permission, from S. H. Lin, *J. Chem. Phys.*, **61**, 3810 (1974).]

where the q's and Q's are normal coordinates of vibrons and phonons, respectively, and $\hat{T}_Q$ and $\hat{T}_q$ are the kinetic energy operators of phonons and vibrons. If we let $\lambda = (\omega_Q/\omega_q)^{1/2}$, where ω_Q and ω_q denote any typical frequencies of phonons and vibrons, we can show that the adiabatic separability of the total Hamiltonian into the vibron motion and phonon motion is determined by the magnitude of λ. The situation here is similar to that of separating the motion of molecular systems into nuclear and electronic motion. Thus to solve the Schrödinger equation of the vibron-phonon system

$$\hat{H}\Psi = E\Psi \tag{8.147}$$

we first consider the solution of the Schrödinger equation of the vibron part:

$$\hat{h}_q \phi_v(q, Q) = U_v(Q)\phi_v(q, Q) \tag{8.148}$$

where

$$\hat{h}_q = \hat{T}_q + V(q, Q) \tag{8.149}$$

Next we set

$$\Psi = \sum_v \phi_v(q, Q) \oplus_v(Q) \tag{8.150}$$

It follows that

$$[\hat{T}_Q + U_v(Q) + \langle \phi_v | \hat{H}'_{BO} | \phi_v \rangle - E] \oplus_v = -\sum_{v'}{}' \langle \phi_v | \hat{H}'_{BO} | \phi_{v'} \rangle \oplus,$$

(8.151)

where $\hat{H}'_{BO}$ represents the Born–Oppenheimer (BO) coupling

$$\hat{H}'_{BO} \phi_v \oplus_v = -\hbar^2 \sum_\alpha \frac{\partial \phi_v}{\partial Q_\alpha} \cdot \frac{\partial \oplus_v}{\partial Q_\alpha} - \frac{\hbar^2}{2} \sum_\alpha \frac{\partial^2 \phi_v}{\partial Q_\alpha^2} \oplus_v$$

(8.152)

When $\langle \phi_v | \hat{H}'_{BO} | \phi_{v'} \rangle = 0$ for $v \neq v'$, the solution becomes adiabatic; that is,

$$\Psi_{nv}(q, Q) = \phi_v(q, Q) \oplus_v(q, Q) \oplus_{nv}(Q)$$

(8.153)

Thus, as in electronic relaxation, when the adiabatic wave functions are used as a basis set, the BO coupling can be used as the perturbation for vibrational relaxation.

Now suppose that $V(q, Q)$ takes the form (Burke and Small, 1974)

$$V(q, Q) = \frac{1}{2} \sum_i B_{ii} q_i^2 + \frac{1}{2} \sum_a b_{\alpha\alpha} Q_\alpha^2$$

$$+ \frac{1}{2} \sum_{i\alpha} B_{i\alpha} q_i Q_\alpha + \frac{1}{3!} \sum_{ij\alpha} D_{ij\alpha} q_i q_j q_\alpha$$

$$+ \frac{1}{3!} \sum_{i\alpha\beta} D_{i\alpha\beta} q_i Q_\alpha Q_\beta$$

$$+ \frac{1}{4!} \sum_{ij\alpha\beta} E_{ij\alpha\beta} q_i q_j Q_\alpha Q_\beta + \cdots$$

(8.154)

In this case we obtain

$$U_v(Q) = \sum_i (v_i + \tfrac{1}{2})\hbar\omega_i + \frac{1}{2} \sum_\alpha b_{\alpha\alpha} Q_\alpha^2 + \frac{1}{3!} \sum_{i\alpha} D_{ii\alpha} \frac{(v_i + \tfrac{1}{2})\hbar}{\omega_i} Q_\alpha$$

$$+ \frac{1}{4!} \sum_{i\alpha\beta} E_{ii\alpha\beta} \frac{(v_i + \tfrac{1}{2})\hbar}{\omega_i} Q_\alpha Q_\beta - \frac{1}{8} \sum_{\alpha\beta} \frac{B_{i\alpha} B_{i\beta}}{\omega_i^2} Q_\alpha Q_\beta + \cdots \quad (8.155)$$

We can easily show that $U_v(Q)$ given by eq. 8.155 is valid to the second-order approximation with respect to λ. But one may expect that the last two terms that will bring in the mixing of phonon modes and the change of phonon frequencies are less important than the term $\frac{1}{2}\sum_\alpha b_{\alpha\alpha} Q_\alpha^2$. Thus in the lowest-order approximation we have

$$U_v(Q) = \sum_i (v_i + \tfrac{1}{2})\hbar\omega_i + \frac{1}{2} \sum_\alpha b_{\alpha\alpha} Q_\alpha^2 + \frac{1}{3!} \sum_{i\alpha} D_{ii\alpha} \frac{(v_i + \tfrac{1}{2})\hbar}{\omega_i} Q_\alpha \quad (8.156)$$

which can be written as

$$U_v(Q) = \sum_i (v_i + \tfrac{1}{2})\hbar\omega_i + \frac{1}{2}\sum_\alpha b_{\alpha\alpha} Q_{v\alpha}^2 \qquad (8.157)$$

where

$$Q_{v\alpha} = Q_\alpha + \frac{1}{6b_{\alpha\alpha}}\sum_i D_{ii\alpha}\frac{(v_i + \tfrac{1}{2})\hbar}{\omega_i} = Q_\alpha + \Delta Q_{v\alpha} \qquad (8.158)$$

Here the term

$$-\sum_\alpha \frac{1}{18 b_{\alpha\alpha}}\left[\sum_i D_{ii\alpha}\frac{(v_i + \tfrac{1}{2})\hbar}{\omega_i}\right]^2$$

has been ignored in comparison with $\sum_i (v_i + \tfrac{1}{2})\hbar\omega_i$. Equation 8.157 indicates that in the lowest-order approximation both vibron and phonon oscillators are harmonic; the origins of phonon oscillators are shifted by $\Delta Q_{v\alpha}$ depending on the vibron states.

Thus in particular for the $V(q, Q)$ given by eq. 8.154 the zero-order basis set to be chosen for vibrational relaxation of vibrons is given by

$$\Psi_{nv}^{(0)} = \phi_v^{(0)}(q) \oplus_{nv}^{(0)}(Q_v) \qquad (8.159)$$

Both vibron and phonons are harmonic. Other effects, for example, like $\phi_n(q, Q)$, can be found by the perturbation method. Notice that the BO coupling comes into effect on the wave function in the third-order approximation of λ and on the energy in the fifth-order approximation of λ. When the anharmonic effect is important, the higher-order approximations (higher than the second order) should be included. In the following discussion of vibrational relaxation the potential function $V(q, Q)$ of the vibron-phonon is assumed to take the form given by eq. 8.155.

As pointed out before, the success of the use of the adiabatic approximation in the vibron-phonon problem depends largely on the ratio ω_Q/ω_q. Suppose that the ratio of ω_q/ω_Q is 10. Then if the anharmonic coupling of the form

$$\hat{H}'\alpha q \prod_\alpha Q_\alpha \qquad (8.160)$$

is used, it will require an order of 10 with respect to λ in $\hat{H}'$ to induce the vibrational relaxation. On the other hand, if the BO coupling is employed as a perturbation for vibrational relaxation, it will only require an order of 2 in $\hat{H}'_{BO}$ with respect to λ to induce the vibrational relaxation; actually, because of the implicit dependence of $\phi_v(q, Q)$ on Q, the order of approximation becomes slightly higher than 2.

Using the adiabatic approximation as the basis set and the BO coupling as the perturbation, the detailed rate constant for the vibron transition $v' \to v$ is given by

$$W_{v'-v} = W_{v'v}(\beta)$$

$$= \frac{2\pi}{\hbar} \sum_n \sum_m P_n |\langle mv| \dot{H}'_{BO} |nv'\rangle|^2 \delta(E_{mv} - E_{nv'}) \qquad (8.161)$$

Here we have assumed that the phonon heat bath is in thermal equilibrium, that is, the vibrational relaxation of phonons is much faster than that of vibrons. Notice that

$$\langle mv| \hat{H}'_{BO} |nv'\rangle = -\hbar^2 \sum_\alpha \left\langle \phi_{v\oplus mv} \left| \frac{\partial \phi_{v'}}{\partial Q_\alpha} \cdot \frac{\partial \oplus_{nv'}}{\partial Q_\alpha} \right. \right\rangle$$

$$- \frac{\hbar^2}{2} \sum_\alpha \left\langle \phi_{v\oplus mv} \left| \frac{\partial^2 \phi_{v'}}{\partial Q_\alpha^3} \oplus_{nv'} \right. \right\rangle \qquad (8.162)$$

The second term in the BO coupling is smaller than the first term. We come back to this point later. Since the process required to simplify eq. 8.161 is exactly the same as that involved in electronic relaxation, we simply write down the result (see Sec. 7.3):

$$W_{v'v}(\beta) = \sum_\alpha \frac{|R_\alpha(v'v)|^2}{\hbar^2} \int_{-\infty}^{\infty} dt \; e^{(it/\hbar)E_{vv'}}$$

$$\times \left[\frac{\omega_\alpha}{4\hbar} \left(\coth \frac{\hbar\omega_\alpha}{2kT} + 1 \right) e^{it\omega\alpha} + \frac{\omega_\alpha}{4\hbar} \left(\coth \frac{\hbar\omega_\alpha}{2kT} - 1 \right) e^{-it\omega\alpha} \right] \prod_\sigma G_\sigma(t)$$

$$(8.163)$$

or

$$\omega_{v'v}(\beta) = \sum_\alpha \frac{|R_\alpha v'v|^2}{\hbar^2} \left[\frac{\omega_\alpha}{4\hbar} \left(\coth \frac{\hbar\omega_\alpha}{2kT} + 1 \right) f_\alpha(\omega_{vv'} + \omega_\alpha) \right.$$

$$\left. + \frac{\omega_\alpha}{4\hbar} \left(\coth \frac{\hbar\omega_\alpha}{2kT} - 1 \right) f_\alpha(\omega_{vv'} - \omega_\alpha) \right] \qquad (8.163a)$$

where

$$R_\alpha(v'v) = -\hbar^2 \left\langle \phi_v \left| \frac{\partial \phi_{v'}}{\partial Q_\alpha} \right. \right\rangle \qquad (8.164)$$

and

$$G_\sigma(t) = \exp\left\{ -\tfrac{1}{2}\beta_\sigma^2 \Delta Q_{vv',\sigma}^2 \left[\coth \frac{\hbar\omega_\sigma}{2kT} - \operatorname{csch} \frac{\hbar\omega_\sigma}{2kT} \cos\left(\omega_\sigma t - \frac{i\hbar\omega_\sigma}{2kT} \right) \right] \right\}$$

$$(8.165)$$

with

$$\Delta Q_{vv',\sigma} = \frac{1}{6b_{\sigma\sigma}} \sum_i D_{ii\sigma}(v_i - v_i') \frac{\omega_i}{\hbar} \tag{8.166}$$

Equation 8.166 shows that the normal coordinate displacements of phonon modes depend linearly on the quantum numbers of the vibron or vibrons participating in the vibrational relaxation.

Just as in electronic relaxation, the coupling constant (or the Huang-Rhys factor) S defined by

$$S = \sum_\sigma \tfrac{1}{2}\beta_\sigma^2 \Delta Q_{vv',\sigma}^2 \coth \frac{\hbar\omega_\sigma}{2kT} \tag{8.167}$$

plays an important role in vibrational relaxation. For $S \le 1$, the weak coupling case· conventionally the integral with respect to t in eq. 8.163 is evaluated by the saddle point method, and for $S \gg 1$, the strong coupling case, conventionally the integral in eq. 8.163 is evaluated by expanding $\cos[\omega_\sigma t - (i\hbar\omega_\sigma/2kT)]$ in power series of t up to t^2.

The integral with respect to t in eq. 8.163 can be evaluated exactly (cf. eq. 8.130)

$$f_\alpha(\omega_{vv'} \pm \omega_\alpha) = 2\pi\, e^{-S} \sum_{v_1} \cdots \sum_{v_N} \sum_{v_1} \cdots \sum_{v_N} \prod_\sigma \frac{1}{v_\sigma! v_\sigma'!} \left(\frac{\beta_\sigma^2 \Delta Q_{vv',\sigma}^2 \bar{n}_\sigma}{2}\right)^{v_\sigma'}$$

$$\times \left(\frac{e^{\beta\hbar\omega_\sigma}\beta_\sigma^2 \Delta Q_{vv',\sigma}^2 \bar{n}_\sigma}{2}\right)^{v_\sigma} \delta\left[\omega_{vv'} \pm \omega_\alpha + \sum_\sigma (v_\sigma - v_\sigma')\omega_\sigma\right] \tag{8.168}$$

where $\bar{n}_\sigma = (e^{\beta\hbar\omega_\sigma} - 1)^{-1}$. If an average frequency ω_m can be introduced for phonon frequencies ω_σ, then eq. 8.168 can be simplified as

$$f_\alpha(\omega_{vv'} \pm \omega_\alpha) = \frac{2\pi}{\omega_m} e^{-S}(S_0\bar{n}_m e^{\beta\hbar\omega_m})^{(\omega_{v'v}\mp\omega_\alpha)/\omega_m} \sum_{v'=0}^{\infty}$$

$$\times \frac{(S_0\bar{n}_m e^{(1/2)\beta\hbar\omega_m)2v'}}{v'!(v' \mp (\omega_{v'v} \mp \omega_\alpha/\omega_m))!} \tag{8.169}$$

where S_0 represents the value of S at $T = 0$. If there is only one promoting mode and temperature is not high, then we have

$$\frac{W_{v'v}(\beta)}{W_{v'v}(\infty)} = 1 + e^{-\beta\hbar\omega_m}\left(\frac{\omega_{v'v}}{\omega_m} - 2S_0 + \frac{S_0^2}{\omega_{v'v}/\omega_m}\right.$$

$$\left. + \frac{S_0^2}{(\omega_{v'v}/\omega_m)[1 + (\omega_{v'v}/\omega_m)]}\right) + \cdots \tag{8.170}$$

Equation 8.170 yields the threshold temperature T_c below which the temperature effect is negligible as

$$\frac{\hbar\omega_m}{kT_c} = \log\left(\frac{\omega_{v'v}}{\omega_n} - 2S_0 + \frac{S_0^2}{(\omega_{v'v}/\omega_m)} + \frac{S_0^2}{(\omega_{v'v}/\omega_m)[1 + (\omega_{v'v}/\omega_m)]}\right) \quad (8.171)$$

Next let us consider $\langle\phi_v|\partial\phi_{v'}/\partial Q_\alpha\rangle$ in $R_a(v'v)$. For this purpose we have to find the dependence of $\phi_{v'}$ on normal coordinates of phonons. Since $V(q, Q)$ takes the form given by eq. 8.155 the terms other than $\frac{1}{2}\sum_i B_i q_i^2$ can be regarded as a perturbation to find $\phi_{v'}(q, Q)$

$$\phi_{v'}(q, Q) = \phi_{v'}^{(0)}(q) + \sum_{v''}{}' \frac{\langle\phi_{v''}^{(0)}|\frac{1}{2}\sum_{i\alpha} B_{i\alpha}q_i Q_\alpha|\phi_{v'}^{(0)}\rangle}{E_{v'}^{(0)} - E_{v''}^{(0)}} \phi_{v''}^{(0)} + \cdots \quad (8.172)$$

Here for simplicity we have retained only the dominating term. It follows

$$\left\langle\phi_v\left|\frac{\partial\phi_{v'}}{\partial Q_\alpha}\right\rangle = \frac{\langle_v^{(0)}|\frac{1}{2}\sum_i B_{i\alpha}q_i|\phi_{v'}^{(0)}\rangle}{E_{v'}^{(0)} - E_v^{(0)}} \quad (8.173)$$

In this case, because $\phi_v^{(0)}$ and $\phi_{v'}^{(0)}$ are simply a product of harmonic oscillator wavefunctions of vibrons, from eq. 8.173 we can see that $\langle\phi_v|\partial\phi_{v'}/\partial Q_\alpha\rangle$ does not vanish only when v and v' differ by ± 1 for one vibron mode. In other words, in this case only the vibrational relaxation of one vibron mode is permitted.

Now suppose the ith vibron mode is involved in vibrational relaxation. It follows that for $v' \to v' - 1$, we have

$$\left\langle\phi_{v'-1}\left|\frac{\partial\phi_{v'}}{\partial Q_\alpha}\right\rangle = \frac{B_{i\alpha}}{2\omega_i}\sqrt{\frac{v_{i'}}{2\hbar\omega_i}} \quad (8.174)$$

and

$$\Delta Q_{vv',\sigma} = -\frac{D_{ii\sigma}\omega_i}{6b_{\sigma\sigma}\hbar}, \quad (8.175)$$

and that for $v' \to v' + 1$, we have

$$\left\langle\phi_{v'+1}\left|\frac{\partial\phi_{v'}}{\partial Q_\alpha}\right\rangle = -\frac{B_{i\alpha}}{2\omega_i}\sqrt{\frac{v_i' + 1}{2\hbar\omega_i}} \quad (8.176)$$

and

$$\Delta Q_{vv',\sigma} = \frac{D_{ii\sigma}\omega_i}{6b_{\sigma\sigma}\hbar} \quad (8.177)$$

Notice that although $\langle\phi_v|\partial\phi_{v'}/\partial Q_\alpha\rangle$ depends on the quantum state of the vibron, $\Delta Q_{vv',\sigma}$ is independent of the vibron state.

Next we consider the case in which two vibron modes are involved in the vibrational relaxation. In this case we have

$$\left\langle \phi_v \left| \frac{\partial \phi_{v'}}{\partial Q_\alpha} \right\rangle = \frac{\langle \phi_v^{(0)} | (1/3!) \sum_{ij} D_{ij\alpha} q_i q_j | \phi_{v'}^{(0)} \rangle}{E_{v'}^{(0)} - E_v^{(0)}} \right. \tag{8.178}$$

Here again only the dominating term is considered. It follows that for $v_i' \to v_i' + 1$, $v_j' \to v_j' - 1$,

$$\left\langle \phi_{v_i'+1v_j'-1} \left| \frac{\partial \phi_{v_i'v_j'}}{\partial Q_\alpha} \right\rangle = \frac{D_{ij\alpha}}{6(\omega_j - \omega_i)} \sqrt{\frac{v_j'(v_i'+1)}{4\omega_i\omega_j}} \right. \tag{8.179}$$

and

$$\Delta Q_{vv',\sigma} = \frac{(D_{ii\sigma}\omega_i - D_{jj\sigma}\omega_j)}{6\hbar b_{\sigma\sigma}} \tag{8.180}$$

Other multimode relaxation processes can be treated similarly.

It is important to notice that according to the present model of vibrational relaxation, once the potential function of the system $V_{(q,Q)}$ is specified, the time-dependent behavior of the relaxing oscillator or oscillators (multimode relaxation) and the vibrational relaxation rate constants can be obtained; this model has also been applied to infrared absorption in condensed media.

8.3.3 Comparison of the Two Models

From the results presented in previous sections we can see that the type of perturbation

$$\hat{H}' = C_i q_i \prod_\alpha Q_\alpha \tag{8.181}$$

can be used for the vibrational relaxation of multiphonon processes that involve only three or less phonons. In other words, eq. 8.181 is most appropriate for describing the relaxation of one phonon into the phonon bath (phonon-phonon scattering) and the relaxation of low-frequency vibron (or low-frequency molecular vibration, that is, $\omega_q/\omega_Q \leq 3$) into the phonon bath. For the relaxation of high-frequency vibrons or molecular vibrations, the BO coupling is to be used as the perturbation for inducing the vibrational relaxation; in this case, the short-range repulsive potential can also be used to describe the multiphonon vibrational relaxation.

Now let us compare the vibrational relaxation induced by the BO coupling and that by the short-range repulsive potential.

Comparing eq. 8.144 with eq. 8.171 we can see that as long as $S_0 \leq (\omega_i/\omega_m)^2$, the T_c given by eq. 8.144 is smaller than the T_c given by eq. 8.171. In other words, the temperature effect sets in earlier for the repulsive potential model.

Let us now compare these two models quantitatively. For this purpose we assume that there is only one promoting mode. Using eq. 8.169 we find

$$
\frac{W_{v'v}(\beta)}{W_{v'v}(\infty)} = W_{v'v}^*(\beta) = e^{S_0 - S}(\bar{n}_m e^{\beta \hbar \omega_m})^{\omega_{v'v}^*}(\omega_{v'v}^* - 1)!
$$

$$
\times \left[\sum_{n=0}^{\infty} \frac{(S_0 \bar{n}_m e^{\beta \hbar \omega_m/2})^{2n}}{n!(n - 1 + \omega_{v'v}^*)!} + S_0^2 \bar{n}_m^2 e^{\beta \hbar \omega_m} \sum_{n=0}^{\infty} \frac{(S_0 \bar{n}_m e^{\beta \hbar \omega_m/2})^{2n}}{n!(n + 1 + \omega_{v'v}^*)!} \right]
$$

(8.182)

where $\omega_{v'v}^* = \omega_{v'v}/\omega_m$. Here for simplicity the average quantity of phonon frequencies has been introduced. Figure 8.13 shows the comparison of the temperature effect on vibrational relaxation predicted by the repulsive potential model

$$
K_i^*(\beta) = \frac{K_i(\beta)}{K_i(\infty)} = e^{S - S_0}(\bar{n}_m e^{\beta \hbar \omega_m})^{\omega_i^*} \omega_i^*! \sum_{n=0}^{\infty} \frac{(S_0 \bar{n}_m e^{\beta \hbar \omega_m/2})^{2n}}{n!(n + \omega_i^*)!}
$$

(8.141)

and by the BO coupling model with the vibrational relaxation rate constant given by eq. 8.182. For convenience the dimensionless quantities have been used. Here the energy gap [$\omega_{v'v}^*$ for eq. 8.182 and ω_i^* for eq. 8.140 is chosen to be 15 and the S_0 values of 0.1, 1, 5, and 10 are used to see how the coupling strength affects the temperature effect. As we can see from Fig. 8.13 the temperature effect on vibrational relaxation predicted by these two models is quite different; the temperature effect predicted by the repulsive potential model increases with the increasing coupling strength S_0, whereas the temperature effect predicted by the BO coupling model decreases with the increasing S_0 values. Thus the measurement of the temperature dependence of the vibrational relaxation rate constant can determine which model is to be preferred.

8.3.4 Effect of Molecular Rotation on Vibrational Relaxation

Recently Brus and Bondybey (1975; see also Legay, 1977) have measured the vibrational relaxation of OH and OD($A^2\Sigma^+$) and NH and ND($A^3\Pi$) in rare-gas lattices; they have found that the vibrational relaxation rate in OH and NH is higher that that in OD and ND despite the larger vibrational energy gaps in OH and NH. This inverse deuterium effect has been attributed by Brus and Bondybey to the participation of the pseudorotation in the vibrational relaxation. The experimental data of Brus and Bondybey and others are presented in Table 8.2 (Legay, 1977); these results are mainly for the vibrational relaxation of small molecules.

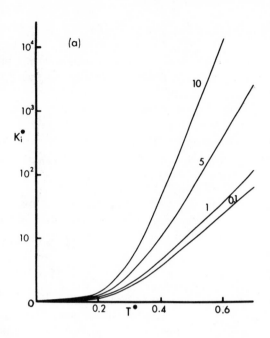

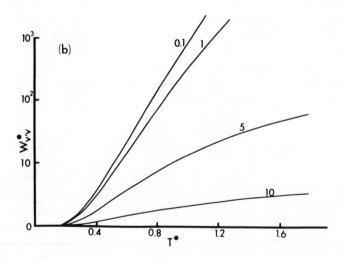

Fig. 8.13 Temperature effect on vibrational relaxation (for $\omega^* = 15$); (a) K_i^* versus T^* and (b) $W_{v'v}^*$ versus T^* for S_0 values of 0.1, 1, 5, and 10. [Reprinted, with permission, from S. H. Lin, *J. Chem. Phys.*, **61**, 3810 (1974).]

Table 8.2
Nonradiative Relaxation Rate Constants

Molecule	Matrix	Vibrational transition	Frequency (cm^{-1})	$k\ (sec^{-1})$	$T\ (^\circ K)$
N_2	N_2	$(4\to3)$–$(13\to12)$	2245–1988	4.8–15.6	4.2
$N_2(A^3\Sigma_u^+)$	Ne, Ar, Kr, Xe	$(1\to0)$–$(4\to3)$	1434–1347	0.3–2.5	1.7–30
CO	Ne, Ar	$(1\to0)$–$(8\to7)$	2138–1937	<10	8–24
	CO	$(1\to0)$– $\cdots$	2138	<50	8–68
NO	Ar	$(1\to0)$–$(7\to6)$	1841–1701	<20	8
HCl	Ar	$(1\to0)$	2871	8.3×10^2	9
		$(2\to1)$	2768	4.0×10^3	9
DCl	Ar	$(2\to1)$	2028	4.0×10^2	9
$C_2^-(X^2\Sigma_g^+)$	Ar	$(1\to0)$	$\simeq1770$	31	18
	Kr	$(1\to0)$	$\simeq1770$	32	18
	Xe	$(1\to0)$	$\simeq1770$	28	18
$OH(A^2\Sigma^+)$	Ne	$(1\to0)$	2970 (gas)	0.9×10^5	4.2
		$(2\to1)$	2784 (gas)	4.0×10^5	4.2
$OD(A^2\Sigma^+)$	Ne	$(1\to0)$	2200 (gas)	0.39×10^5	4.2
		$(2\to1)$	2099 (gas)	1.4×10^5	4.2
$NH(A^3\Pi)$	Ar	$(1\to0)$	2977	1.2×10^6	4.2–25
		$(2\to1)$	2718	6.2×10^6	4.2–25
	Kr	$(1\to0)$	2953	17.8×10^6	4.2–25
$ND(A^3\Pi)$	Ar	$(1\to0)$	2217	$\leq2\times10^4$	4.2
	Kr	$(1\to0)$	2214	1.4×10^6	4.2
NH_3	Ar	$\nu_2(1\to0)$	974	$>2\times10^6$	8
	N_2	$\nu_2(1\to0)$	970	6.6×10^4	8
CH_3F	Kr	$\nu_3(1\to0)$	1036	9.1×10^4	15
CD_3F	Kr	$\nu_3(1\to0)$	987	9.1×10^3	15
$NH(X^3\Sigma^-)$	Ar	$(1\to0)$	3131	5.3×10^3	4–30
$ND(X^3\Sigma^-)$	Ar	$(1\to0)$	2316	3.2×10	4–30

Source: Legay (1977).

Both the repulsive potential model and the BO coupling model can be applied to treat the rotational effect on vibrational relaxation. Since the approach is exactly the same for both models, we describe only the treatment due to the repulsive potential model.

For convenience, we assume that the relaxing molecule is diatomic, and the "force" in $\hat{H}'$ given by eq. 8.108 can be expressed as

$$F = F_q F_r \tag{8.183}$$

where F_q represents the part of the "force" term coupled with lattice phonons and F_r represents the part of F coupled with the "molecular rotation." Substituting eq. 8.183 into eq. 8.109 yields

$$K(T) = \frac{1}{2h\omega\mu} \int_{-\infty}^{\infty} dt \, \exp(-it\omega)k_r(t)k_q(t) \tag{8.184}$$

where μ is the reduced mass, ω is the molecular vibrational frequency, and $k_r(t)$ and $k_q(t)$ are defined by

$$k_r(t) = \sum_{v_r} \sum_{v_{r'}} P_{v_r} |\langle \theta_{v_r} | F_r | \theta_{v_{r'}} \rangle|^2 \exp\left[\frac{it}{\hbar}(E_{v_{r'}} - E_{v_r})\right] \tag{8.185}$$

and

$$k_q(t) = \sum_{v_q} \sum_{v_{q'}} P_{v_q} |\langle \theta_{v_q} | F_q | \theta_{v_{q'}} \rangle|^2 \exp\left[\frac{it}{\hbar}(E_{v_{q'}} - E_{v_q})\right] \tag{8.186}$$

Now if we define

$$K_r(T, \omega') = \frac{1}{2h\mu\omega} \int_{-\infty}^{\infty} dt \, \exp(-it\omega')k_r(t) \tag{8.187}$$

eq. 8.184 can be rewritten as

$$K(T) = \int_{-\infty}^{\infty} d\omega' K_r(T, \omega')I_q(\omega - \omega') \tag{8.188}$$

where

$$I_q(\omega - \omega') = \sum_{v_q} \sum_{v_{q'}} P_{v_q} K\theta_{vq}|F_q|\theta_{v_{q'}})|^2\delta\left[(\omega' - \omega) + \frac{1}{\hbar}(E_{v_{q'}} - E_{v_q})\right]$$

$$= \frac{1}{2\pi} \int_{-\infty}^{\infty} dt \, k_q(t)\exp[it(\omega' - \omega)] \tag{8.189}$$

Notice that if the proportional constant in F is absorbed in F_r, then the expression $(\omega/\omega')K_r(T, \omega')$ represents the vibrational relaxation constant in which the vibrational energy $\hbar\omega'$ is relaxed into the molecular rotation. In other words, $K_r(T, \omega)$ represents the vibrational relaxation rate constant for the case where all the vibrational energy is relaxed into the "rotational" energy. To calculate $K_r(T, \omega')$, we need an explicit expression for F_r; according to Freed and Metiu (1977), F_r for a planar rotator can be expressed as

$$F_r(\phi) = v_0 \exp(\alpha \cos n\phi) \tag{8.190}$$

where V_0, α, and n are constant. Using eq. 8.190, $K_r(T, \omega')$ becomes

$$K_r(T, \omega') = \frac{\pi}{\mu\omega'} \sum_m \sum_{m'} P_m |\langle m' | F(\rho) | m \rangle|^2\delta(\hbar\omega' + E_m + E_{m'}) \tag{8.191}$$

where (m, m') denote the quantum numbers of the "molecular rotation." If the rotation is completely free, we obtain

$$K_r(T, \omega') = \frac{\pi}{\mu\omega} \sum_m \sum_{m'} P_m |I_{m'-m}|^2 \delta(\hbar\omega' + E_m - E_{m'}) \qquad (8.192)$$

where $E_m = m^2 n^2 \hbar^2 / 2I = m^2 n^2 \hbar\omega_r$ and

$$I_{m'-m} = \frac{V_0}{2\pi} \int_0^2 \pi \, d\phi' \exp[\alpha \cos \phi' - i(m' - m)\phi'] \qquad (8.193)$$

In most cases $\hbar\omega' \gg E_m E_{m'}$; in this case the density of states $\rho(E_{m'})$ can be introduced into eq. 8.192:

$$K_r(T, \omega') = \frac{\pi}{\mu\omega} \sum_m P_m |I_{m'-m}|^2 \rho(E_{m'}) \qquad (8.194)$$

where $\rho(E_{m'}) = (E_{m'} n^2 \hbar\omega_r)^{-1/2} = (|m'| n^2 \hbar\omega_r)^{-1}$ and $E_{m'} = E_m + \hbar\omega'$.

To evaluate $I_{m'-m}$, the saddle point method can be used; to the first-order approximation, we find for $m'' > 0$

$$I_{m''} = \frac{V_0}{(2\alpha\pi \cos \phi^*)^{1/2}} \exp(\alpha \cos \phi^* - im''\phi^*) \qquad (8.195)$$

where ϕ^* is to be determined from

$$\exp(i\phi^*) = \frac{m''}{\alpha} + \left[1 + \left(\frac{m''}{\alpha}\right)^2\right]^{1/2} \qquad (8.196)$$

For $m''/\alpha \gg 1$, eq. 8.195 reduce to

$$I_{m''} = \frac{V_0}{(2\pi m'')^{1/2}} \exp\left(m'' - m'' \log \frac{2m''}{\alpha}\right) \qquad (8.197)$$

The exact result of $I_{m''}$ is given by (Knittel and Lin, to be published)

$$I_{m''} = V_0 \sum_{\kappa=0}^{\infty} \frac{(\alpha/2)^{m''+2k}}{k!(m''+k)!} \qquad (8.198)$$

Using eq. 8.197, the expression for $K_r(T, \omega')$ at $T = 0$ can be calculated from eq. 8.194 as

$$K_r(0, \omega') = \frac{V_0}{2h\mu\omega\omega'} \exp\left(\gamma\sqrt{\omega'/\omega_r}\right) \qquad (8.199)$$

where $\gamma = 2/n[\log(2/\alpha)\sqrt{(\omega'/\omega_r)} - 1]$; eq. 8.199 shows that when $\omega \doteq \omega'$, the plot of $\log K_r(\infty, \omega)$, versus $\sqrt{\omega/\omega_r}$ is linear approximately; this has been verified experimentally (Legay, 1977). In the low-temperature range eq. 8.194 can be simplified as (Knittel and Lin)

$$\frac{K_r(T, \omega')}{K_r(0, \omega')} = 1 + \exp\left(-\frac{n^2\hbar\omega_r}{kT}\right)\left(\sqrt{\frac{2m'}{\alpha}} + \sqrt{\frac{\alpha}{2m'}}\right)^2 + \cdots \quad (8.200)$$

or

$$\frac{K_r(T, \omega')}{K_r(0, \omega')} = 1 + \exp\left(-\frac{n^2\hbar\omega_r}{kT}\right)\left(\sqrt[4]{\frac{4\omega'}{n^2\alpha^2\omega_r}} - \sqrt[4]{\frac{n^2\alpha^2\omega_r}{4\omega'}}\right)^2 + \cdots \quad (8.201)$$

From eq. 8.201 the threshold temperature T_c can be determined as

$$T_c \doteq \frac{2n^2\hbar\omega_r/k}{\log(4\omega'/n^2\alpha^2\omega_r)} \quad (8.202)$$

Let us now estimate the isotope effect. For this purpose we use eq. 8.199 and set $\omega' = \omega$. It follows that the ratio of the rates of the deuterated molecule to that of the protonated molecule is given by

$$\frac{K_r(\infty)^D}{K_r(\infty)^H} = \exp\left[\frac{2}{n}\sqrt{\frac{\omega^H}{\omega_r^H}}\left(\sqrt[4]{\frac{\mu_{D-1}}{\mu_H}}\right) + \frac{2}{n}\sqrt{\frac{\omega^H}{\omega_r^H}}\right)$$

$$\times \left\{\left(1 - \sqrt[4]{\frac{\mu_D}{\mu_H}}\right) \times \log\frac{2}{\alpha}\sqrt{\frac{\omega^H}{\omega_r^H}} - \sqrt[4]{\frac{\mu_D}{\mu_H}}\log\sqrt[4]{\frac{\mu_D}{\mu_H}}\right\}\right] \quad (8.203)$$

Noticing that $\alpha \sim 1$ and $n = 4, 6, 8, \ldots$, we can easily estimate the ratio $K_r(\infty)^D/K_r(\infty)^4$ to be much smaller than unity; that is, eq. 8.203 predicts a large inverse deuterium effect (Kittel and Lin).

In concluding this section, it should be noted that in obtaining eq. 8.199, the free rotator approximation has been employed, which is valid for the final state m' but may not be a good approximation for the initial state m (Eyring et al., 1944), that a planar rotator instead of a three-dimensional rotator has been used, and that using either eq. 8.184 or eq. 8.188, one should be able to determine the amount of vibrational energy flowing into the "molecular rotation" and into lattice vibration. Furthermore, a better expression for the "force" F involved in eq. 8.107 should be able to describe the coupling between the molecular vibration and the "molecular rotation," the "translation" of the molecule and the lattice vibration, and remains to be found.

REFERENCES

Abramowitz, M. and I. A. Stegun (1965), *Handbook of Mathematical Function*, Dover, London.

Alfano, R. R. and S. L. Shapiro (1971), *Phys. Rev. Lett.*, **26**, 1247.

Allamandolla, L. J. and J. W. Nibler (1974), *Chem. Phys. Lett.*, **28**, 335.

Amme, R. C. (1975), *Adv. Chem. Phys.*, **28**, 171.

Bates, D. R. (1961), *Quantum Theory. I. Elements*, Academic, New York.

Bates, D. R. (1962), *Atomic and Molecular Processes*, Academic, New York.

Berry, R. S. (1970), in *Molecular Beams and Reaction Kinetics*, C. H. Schlier, Ed., Academic, New York, p. 229.

Brus, L. E. and V. E. Bondybey (1975), *J. Chem. Phys.*, **63**, 786.

Burke, F. P. and G. J. Small (1974), *Chem. Phys.*, **5**, 198; *J. Chem. Phys.*, **61**, 4588.

Child, M. S. (1974), *Molecular Collision Theory*, Academic, New York.

Diestler, D. J. (1974), *J. Chem. Phys.*, **60**, 2692.

Dubost, H. (1976), *Chem. Phys.*, **12**, 139.

Earl, B. L. and R. R. Herm (1974), *J. Chem. Phys.*, **60**, 4568.

Eyring, H., J. Walter, and G. E. Kimball (1944), *Quantum Chemistry*, Wiley, New York.

Fleming, G. R., O. L. J. G-jzman, and S. H. Lin (1974), *J. Chem. Soc. Faraday Trans. II*, **70**, 37.

Flygare, W. H. (1968), *Acc. Chem. Res.*, **1**, 121.

Freed, K. F. and H. Metiu (1977), *Chem. Phys. Lett.*, **48**, 262.

Jackson, J. M. and N. F. Mott (1932), *Proc. Roy. Soc.* (Lond.), **A137**, 703.

Karl, G., P. Kruus, and J. C. Polanyi (1967), *J. Chem. Phys.*, **46**, 224.

Knittel, D. R. and S. H. Lin, to be published.

Lambert, J. D. and R. Salter (1959), *Proc. Roy. Soc.* (Lond.), **A253**, 277.

Landau, L. D. and E. M. Lifshitz (1960), *Mechanics*, Pergamon, Oxford.

Lauberau, A., D. von der Linde, and W. Kaiser (1971), *Phys. Rev. Lett.*, **27**, 802.

Lauberau, A. and W. Kaiser (1975), *Ann. Rev. Phys. Chem.*, **26**, 83.

Legay, F. (1977) in *Chemical and Biological Applications of Lasers*, Vol. III. C. B. Moore, Ed., in press.

Lengyel, B. A. (1971), *Lasers*, Wiley-Interscience, New York.

Levine, R. D. and R. B. Bernstein (1974), *Molecular Reaction Dynamics*, Oxford University Press, Oxford.

Lin, S. H. and R. Bersohn (1968), *J. Chem. Phys.*, **48**, 2732.

Lin, S. H. (1974), *J. Chem. Phys.*, **61**, 3810.

Lin, S. H. and H. Eyring (1974), *Ann. Rev. Phys. Chem.*, **25**, 39.

Lin, S. H. (1976), *J. Chem. Phys.*, **65**, 1053.

Lin, S. H., H. P. Lin, and D. R. Knittel (1976), *J. Chem. Phys.*, **64**, 441.

Lin, S. H. (1978), *Radiationless Transitions*, Academic, New York.

Mahan, B. (1970), *J. Chem. Phys.*, **52**, 5221.

Millikan, R. C. and D. R. White (1963), *J. Chem. Phys.*, **39**, 3209.

Moore, C. B. (1973), *Adv. Chem. Phys.*, **23**, 41.

Montroll, E. W. and K. E. Shuler (1957), *J. Chem. Phys.*, **26**, 454.

Montroll, E. W. (1960), *Lect. Theor. Phys.*, **3**, 221.

Mott, N. F. and H. S. W. Massey (1965), *Theory of Atomic Collisions*, Oxford University Press, Oxford.

Nikitin, E. E. (1960), *Opt. Spectrom.*, **9**, 8.

Nikitin, E. E. (1968), in *Chemische Elementarprozesse*, H. Hartmann, Ed., Springer-Verlag, Berlin, p. 43; *Adv. Quant. Chem.*, **5**, 135.

Nikitin, E. E. (1974), *Physical Chemistry, Advanced Treatise*, 6A187, Academic, New York.

Nitzan, A. and J. Jortner (1973), *Mol. Phys.*, **25**, 713.

Nitzan, A., S. Mukamel, and J. Jortner (1974), *J. Chem. Phys.*, **60**, 3929.

Résibois, P. (1961), *Physica*, **27**, 541.

Schwartz, R. N., Z. I. Slawsly, and K. F. Herzfeld (1952), *J. Chem. Phys.*, **20**, 1591.

Sun, H. Y. and S. A. Rice (1965), *J. Chem. Phys.*, **42**, 3826.

Takayanagi, K. (1963), *Prog. Theor. Phys. Suppl.* (Jap.), **25**, 1.

Tinti, D. S. and G. W. Robinson (1968), *J. Chem. Phys.*, **49**, 3229.

Tolman, R. C. (1938), *The Principles of Statistical Mechanics*, Oxford University Press, New York.

van den Bergh, H. E., M. Faubel, and J. P. Toennies (1973), *Discussion Faraday Soc.*, **55**, 203.

von der Linde, D., A. Laubereau, and W. Kaiser (1971), *Phys. Rev. Lett.*, **26**, 954.

Weitz, E. and G. Flynn (1974), *Ann. Rev. Phys. Chem.*, **25**, 275.

Zwanzig, R. (1961), *J. Chem. Phys.*, **34**, 1931.

Zwanzig, R. (1964), *Physica*, **30**, 1109.

Nine

Reactions in Condensed Phases

CONTENTS

9.1 INTRODUCTION

In this chapter we discuss the reaction kinetics in solids and liquids. In view of the small amount of free space available in a liquid or solid, diffusion effects are often important in condensed-phase reactions. The way in which kinetic rate expressions might be modified so as to take account of the encounter phenomena peculiar to liquids and solids is treated in detail in this chapter.

Although reactions in the gas phase are simpler to deal with theoretically, the fact remains that most reactions encountered in practice occur in solutions. The question then arises as to whether there is any fundamental difference in the kinetics of reactions occurring in condensed media compared to that in the gaseous state. The answer is essentially as follows: When a reaction follows the same mechanism in solution and in the gaseous state, the kinetics is not changed appreciably. However, because of the increased interactions in condensed media, the mechanism is frequently changed completely and the kinetics correspondingly altered. In fact, there is a wide variety of reactions that do not occur in the gas phase at all but that go more or less readily in various solvents. The favored mechanisms in solution are ionic ones, involving formation and interaction of charged particles. Such mechanisms are virtually impossible in the vapor state (excluding wall and surface reactions). Polar solvents in general are the best media for ionic reactions.

In this chapter we also discuss how to apply the transition state theory to the solvent effect and pressure effect on the reaction kinetics in solution, and we present a quantum-statistical mechanical treatment for unimolecular rate processes (e.g., diffusion, dielectric relaxation, electron transfer, etc.) in condensed phases.

9.2 REACTIONS IN LIQUIDS—DIFFUSION-CONTROLLED KINETICS

Most of the recent treatments of reactions in liquid solutions assume the applicability of Fick's macroscopic diffusion laws to the diffusion of molecules in solution. Chandrasekhar (1943) has shown that diffusion of a molecule in solution considered as a random-flight process, after a great number of flights, is described by Fick's laws if the concentration in Fick's equation is replaced by a probability density function. Kirkwood (1946) demonstrated that Fick's laws apply in times as short as 10^{-13} sec after a molecule begins to diffuse. Thus it appears that the diffusion equations derived from Fick's laws are good approximations to the diffusion of molecules in solution (Noyes, 1961; Yguerabide et al., 1964; Scheider, 1972).

The theory of the rate of diffusion-controlled reactions was first formulated by Smoluchowski. The basic idea is that the rate of a reaction can be dominated by the slow diffusive motion required for reacting partners to approach each other, followed by almost instantaneous reaction. Of the many applications of the theory we mention growth of colloids or aerosol particles, precipitation, catalysis, and fluorescence quenching. An application on a more microscopic scale is combustion, in which the diffusing species is the oxidant and the sink a fuel droplet. Smoluchowski's theory was extended by Debye (1942) to include the effect of potential interactions. A quite general mathematical formulation was recently given by Wilemski and Fixman (1973). They have applied their theory to intrachain reactions of polymers (Wilemski and Fixman, 1974).

We consider a system of reactants A and B with concentrations $C_A(r_A, t)$ and $C_B(r_B, t)$. If Fick's laws are applicable, the rates of change of C_B and C_A are given by (Crank, 1956; Yguerabide et al., 1964)

$$\frac{\partial C_B}{\partial t} = D_B \nabla^2 C_B - K_B \qquad (9.1)$$

and

$$\frac{\partial C_A}{\partial t} = D_A \nabla^2 C_A - K_A \qquad (9.2)$$

where D_A and D_B are the diffusion coefficients of A and B, and K_A and K_B denote the rates of change of C_A and C_B caused by the chemical reaction. Equations 9.1 and 9.2 can be simplified by considering B molecules to be stationary, so that the first term on the right-hand side can be set equal to zero, and by taking B molecules as the origin of the system of coordinates of the A molecules. It follows that

$$\frac{\partial C_B}{\partial t} = -K_B \qquad (9.3)$$

and

$$\frac{\partial C_A}{\partial t} = D \nabla^2 C_A - K_A \qquad (9.4)$$

where $D = D_A + D_B$.

The choice of the functional form of K_A and K_B depends on the type of reaction kinetics considered and on the theoretical models used. If it is assumed that the reaction is bimolecular and that the reaction between A and B occurs only on the surface at distance R_c from the center of A or B

with constant probability, then K_B and K_A can be expressed as (Yguerabide et al., 1964; Noyes, 1961)

$$K_B = kC_B(r_B, t) \int C_A(r_A, t)\delta_3(|\mathbf{r}_B - \mathbf{r}_A| - R_c)\, d\mathbf{r}_A \qquad (9.5)$$

and

$$K_A = kC_A(r_A, t) \int C_B(r_B, t)\delta_3(|\mathbf{r}_B - \mathbf{r}_A| - R_c)\, d\mathbf{r}_B \qquad (9.6)$$

where $\delta_3(\chi)$ is related to the delta function $\delta(\chi)$ by $\delta_3(\chi) = \delta(\chi)/4\pi\chi^2$, and k represents the rate constant. Other types of K_A and K_B can be discussed similarly.

The boundary conditions used in solving the differential equations may take a variety of forms. The so-called Smoluchowski boundary condition refers to the case in which the A and B molecules react immediately on collision. For the case in which the A and B molecules may undergo un-reactive collisions, it is commonly assumed that the rate of reaction on the reaction sphere (of radius R_c) is equal to the flux of a molecules through that surface. This is called the *radiation boundary condition* (Carslaw and Jaeger, 1959).

9.2.1 Smoluchowski Boundary Condition

If the system is spherically symmetric, for $r_A > R_c$, eqs. 9.3 and 9.4 can be written

$$\frac{\partial C_B}{\partial t} = -kC_B \int C_A(r_A, t)\delta_3(|\mathbf{r}_A| - R_c)\, d\mathbf{r}_A \qquad (9.7)$$

and

$$\frac{\partial C_A}{\partial t} = \frac{D}{r_A^2} \cdot \frac{\partial}{\partial r_A}\left(r_A^2 \frac{\partial C_A}{\partial r_A}\right) \qquad (9.8)$$

To solve eq. 9.8 we let $C_A = C_{Ar}/r_A$. It follows that

$$\frac{\partial C_{Ar}}{\partial t} = \frac{D\partial^2 C_{Ar}}{\partial r_A^2} \qquad (9.9)$$

Carrying out the Laplace transformation on both sides of eq. 9.9 yields (Appendix One)

$$\frac{D\, d^2\bar{C}_{Ar}}{dr^2} = p\bar{C}_{Ar} - r_A\bar{C}_A^0 \qquad (9.10)$$

where C_A^0 represents the initial concentration of A and $\bar{C}_{Ar}$ is defined by

$$\bar{C}_{Ar} = \int_0^\infty \exp(-pt)C_{Ar}\,dt \tag{9.11}$$

Equation 9.10 can easily be solved and the result is given by

$$\bar{C}_{Ar} = B_1 \exp(-qr_A) + B_2 \exp(qr_A) + \left(\frac{r_A c_A^0}{p}\right) \tag{9.12}$$

where $q^2 = p/D$ and B_1 and B_2 are arbitrary constants to be determined by the boundary conditions. Since C_A is finite as $r_A \to \infty$, we obtain $B_2 = 0$. In addition, the Smoluchowski boundary condition implies that $\bar{C}_{Ar} = 0$ at $r_A = R_c$, which gives us $B_1 = -(R_c C_A^0/p)\exp(qR_c)$. Using these results, we find

$$\bar{C}_{Ar} = \left(\frac{C_A^0}{p}\right)\{r_A - R_c \exp[q(R_c - r_A)]\} \tag{9.13}$$

By inverting the Laplace transformation (Crank, 1956; Carslaw and Jaeger, 1959), we obtain

$$C_{Ar} = C_A^0\left[r_A - R_c \operatorname{erfc}\frac{r_A - R_c}{(4Dt)^{1/2}}\right] \tag{9.14}$$

and

$$C_A = C_a^0\left[1 - \frac{R_c}{r_A}\operatorname{erfc}\frac{r_A - R_c}{(4dt)^{1/2}}\right] \tag{9.15}$$

where $\operatorname{erfc}(x)$ is defined by

$$\operatorname{erfc}(x) = \left(\frac{2}{\pi^{1/2}}\right)\int_x^\infty dZ \exp(-z^2) \tag{9.16}$$

To solve eq. 9.7 we notice that

$$k\int C_A(r_A, t)\delta_3(|\mathbf{r}_A| - R_c)\,d\mathbf{r}_A = 4\pi R_c^2 D\left(\frac{\partial C_A}{\partial r_A}\right)_{r_A = R_c} \tag{9.17}$$

which yields

$$k\int C_A(r_A, t)\delta_3(|\mathbf{r}_A| - R_c)\,d\mathbf{r}_A = 4\pi R_c D C_A^0\left[1 + \frac{R_c}{(\pi Dt)^{1/2}}\right] \tag{9.18}$$

Substituting eq. 9.18 into eq. 9.17 and carrying out the integration with respect to t yields

$$\frac{C_B}{C_B^0} = \exp[-4\pi R_c C_A^0 Dt - 8C_A^0 R_c(\pi Dt)^{1/2}] \tag{9.19}$$

where C_B^0 represents the initial concentration of B. From eq. 9.18 we can calculate the number of collisions between A and B per unit time Z_{AB} as

$$Z_{AB} = 4\pi R_c D C_B C_A^0 \left[1 + \frac{R_c}{(\pi D t)^{1/2}} \right] \qquad (9.20)$$

from which we obtain the diffusion-controlled rate constant k_D as $k_D = 4\pi D R_c$.

9.2.2   Radiation Boundary Condition

The differential equations for this case are exactly the same as those given in eqs. 9.7 and 9.8, and hence the solution for $\bar{C}_{Ar}$ is given by eq. 9.12. To determine B_1 we use the boundary condition

$$4\pi R_c^2 D \left(\frac{\partial \bar{C}_A}{\partial r_A} \right)_{r_A = R_c} = k C_A \qquad (9.21)$$

or

$$(k + 4\pi R_c D) C_{Ar} = 4\pi R_c^2 D \left(\frac{\partial C_{Ar}}{\partial r_A} \right)_{r_A = R_c} \qquad (9.22)$$

Substituting eq. 9.12 into 9.22 yields

$$B_1 = -\frac{k R_c C_A^0 \exp(q R_c)}{p(k + 4\pi R_c D + 4\pi R_c^2 D q)} \qquad (9.23)$$

It follows that the expression for $\bar{C}_{Ar}$ is given by

$$\bar{C}_{Ar} = \frac{C_A^0}{P} \left\{ r_A - \frac{k R_c \exp[-q(r_A - R_c)]}{k + 4\pi R_c D + 4\pi R_c^2 D q} \right\} \qquad (9.24)$$

Inverting the Laplace transformation in eq. 9.24 we obtain

$$\begin{aligned}
C_{Ar} = C_A^0 \Bigg(r_A - & \frac{k R_c}{k + 4\pi R_c D} \\
& \times \left\{ \mathrm{erfc}\left[\frac{r_A - R_c}{(4Dt)^{1/2}} \right] - \exp[h(r_A - R_c) + h^2 D t] \right. \\
& \left. \times \mathrm{erfc}\left[h(Dt)^{1/2} + \frac{r_A - R_c}{(4Dt)^{1/2}} \right] \right\} \Bigg),
\end{aligned} \qquad (9.25)$$

where $h = (1/R_c)[1 + (k/4\pi R_c D)]$. Recalling that $C_A = C_{Ar}/r$, we have

$$\frac{C_A}{C_A^0} = 1 - \frac{kR_c}{r_A(k + 4\pi R_c D)}$$

$$\times \left\{ \text{erfc}\left[\frac{r_A - R_c}{(4Dt)^{1/2}}\right] - \exp[h(r_A - R_c) + h^2 Dt] \right.$$

$$\left. \times \text{erfc}\left[h(Dt)^{1/2} + \frac{r_A - R_c}{(4Dt)^{1/2}}\right]\right\} \tag{9.26}$$

Substituting eq. 9.26 into eq. 9.7 yields

$$\frac{\partial C_B}{\partial t} = -kC_B C_A(R_c, t) \tag{9.27}$$

which can be integrated to give

$$\frac{C_B}{C_B^0} = \exp\left[-k \int_0^t dt C_A(R_c, t)\right] \tag{9.28}$$

or

$$\frac{C_B}{C_B^0} = \exp\left(- \frac{4\pi\gamma t C_A^0 R_c D}{1 + \gamma} - \frac{4\pi\gamma^2 C_A^0}{h^3}\right.$$

$$\left. \times \left\{2h\left(\frac{Dt}{\pi}\right)^{1/2} - 1 + \exp(h^2 Dt)\text{erfc}[h(Dt)^{1/2}]\right\}\right) \tag{9.29}$$

where $\gamma = k/4\pi R_c D = k/k_D$ represents the ratio of the reaction rate constant to the diffusion-controlled rate constant $4\pi R_c D$ and $h = (1 + \gamma)/R_c$.

For the case in which γ is small, to the first-order approximation of γ, eq. 9.29 takes the form

$$\frac{C_B}{C_B^0} = \exp(-ktC_A^0) \tag{9.30}$$

This indicates that when the rate of chemical reaction is slow compared with that of diffusion encounters, the conventional reaction kinetics holds. On the other hand, for $\gamma \to \infty$, eq. 9.29 reduces to eq. 9.20; in this case we obtain the diffusion-controlled kinetics. In other words, the Smoluchowski boundary condition is just a particular case of the radiation boundary condition. From eq. 9.30 we can see that the observed rate constant k_{obs} is given by $k_{\text{obs}} = k/(1 + \gamma)$ (Noyes, 1961).

In the preceding derivation we have only discussed the reactions between spherically symmetric molecules or atoms. The kinetics of diffusion-controlled reaction between chemically asymmetric molecules has recently been discussed by Sole and Stockmayer (1971) and Scheider (1972) and will not be described here.

9.2.3 Reactions Between Interacting Molecules

The considerations described in the preceding sections can be extended readily to reacting molecules that are attracted to or repelled from each other as a result of electrostatic or other forces. In this case, we have to consider the continuity equation resulting from a concentration gradient as well as a potential gradient (Fitts, 1962); that is, instead of eq. 9.8 we have

$$\frac{\partial C_A}{\partial t} = D\left[\nabla^2 C_A + \frac{C_A}{kT}\nabla^2 U(r_A) + \frac{1}{kT}\frac{\partial C_A}{\partial r_A}\frac{\partial U}{\partial r_A}\right] \tag{9.31}$$

where $U(r_A)$ denotes the electrical or any other potential field prevailing between an A and a B molecule.

It is convenient to introduce a new function $\alpha(r_A, t)$ defined by (Montroll, 1946; Weller, 1957, 1961)

$$\alpha(r_A, t) = C_A \exp\left[\frac{U(r_A)}{kT}\right] \tag{9.32}$$

For the spherically symmetric system the differential equation for $\alpha(r_A, t)$ after substituting eq. 9.32 into eq. 9.31 is given by

$$\frac{\partial \alpha}{\partial t} = D\left[\frac{\partial^2 \alpha}{\partial r_A^2} + \frac{\partial \alpha}{\partial r_A}\left(\frac{2}{r_A} - \frac{1}{kT}\frac{\partial U}{\partial r_A}\right)\right] \tag{9.33}$$

This equation has also been derived in a somewhat different manner by Umberger and LaMer (1945) and Chandrasekhar (1943). To solve eq. 9.32 or eq. 9.33 we use the initial condition

$$C_A(r_A, 0) = C_A^0 \exp\left[-\frac{U(r_A)}{kT}\right] \tag{9.34}$$

or

$$\alpha(r_A, 0) = C_A^0 \tag{9.35}$$

Here we only discuss the solution of eq. 9.33 (or eq. 9.31) under the Smoluchowski boundary conditions; the solution of eq. 9.33 under the radiation boundary conditions can be carried out in an analogous fashion and is not reproduced here. The Smoluchowski boundary conditions in this case are given by

$$C_A(\infty, t) = C_A^0 \qquad \text{or} \qquad \alpha(\infty, t) = C_A^0 \tag{9.36}$$

and

$$C_A(R_c, t) = 0 \qquad \text{or} \qquad \alpha(R_c, t) = 0 \qquad (9.37)$$

To obtain the transient solution of eq. 9.33 we introduce new functions $S(r_A)$ and $f(r_A, t)$ defined by

$$S(r_A) = \int_{r_A}^{\infty} \frac{dr}{r^2} \exp\left[\frac{U(r)}{kT}\right] \qquad (9.38)$$

and

$$\frac{\alpha(r_A, t)}{C_A^0} = 1 + \frac{S(r_A)}{S(R_c)}\left[f(r_A, t) - 1\right] \qquad (9.39)$$

By substituting eq. 9.39 into eq. 9.33, we see that $f(r_A, t)$ satisfies

$$\frac{\partial f}{\partial t} = D\frac{\partial 2f}{\partial r_A^2} + \frac{\partial f}{\partial r_A}\left[\frac{2S'(r_A)}{S(r_A)} + \left(\frac{2}{r_A} - \frac{1}{kT}\frac{\partial U}{\partial r_A}\right)\right] \qquad (9.40)$$

$$f(R_c, t) = 0, \qquad f(r_A, 0) \pm 1, \qquad f(\infty, t) < \infty \qquad (9.41)$$

where

$$S'(r_A) = \frac{\partial S}{\partial r_A} = -\frac{1}{r_A^2}\exp\left[\frac{U(r_A)}{kT}\right] \qquad (9.42)$$

Exact analytical solutions of eq. 9.33 have not been carried out. The transformation eq. 9.39 is useful because in the range of large r_A and medium r_A, the coefficient of $\partial f/\partial r_A$ has been shown to be negligible (Montroll, 1946). We shall on this account neglect the term involving $\partial f/\partial r_A$ in eq. 9.40; eq. 9.40 then reduces to

$$\frac{\partial f}{\partial t} = \frac{D\partial^2 f}{\partial r_A^2} \qquad (9.43)$$

The solution of eq. 9.33 has been discussed in the previous section, and under the condition eq. 9.41 is

$$f(r_A, t) = \text{erfc}\,\frac{r_A - R_c}{2(Dt)^{1/2}} \qquad (9.44)$$

Substituting eq. 9.44 into eq. 9.39 yields

$$\frac{\alpha(r_A, t)}{C_A^0} = 1 - \frac{S(r_A)}{S(R_c)}\text{erfc}\,\frac{r_A - R_c}{2(Dt)^{1/2}} \qquad (9.45)$$

Combining eq. 9.45 with eq. 9.32, we obtain

$$C_A(r_A, t) = C_A^0\left[\exp - \frac{U(r_A)}{kT}\right]\left[1 - \frac{S(r_A)}{S(R_c)}\text{erfc}\,\frac{r_A - R_c}{2(Dt)^{1/2}}\right] \qquad (9.46)$$

To solve eq. 9.7 we again use an equation equivalent to eq. 9.17, which in this case is given by

$$k \int C_A(\mathbf{r}_A, t)\delta_3(|\mathbf{r}_A| - R_c) \, d\mathbf{r}_A = 4\pi D R_c^2 \left(\frac{\partial C_A}{\partial r_A} + \frac{C_A}{kT}\frac{U}{r_A}\right)_{r_A = R_c}$$

(9.47)

or

$$k \int C_A(\mathbf{r}_A, t)\delta_3(|\mathbf{r}_A| - R_c) \, d\mathbf{r}_A = 4\pi D R_c^2 \left[\exp\left\{-\frac{U(R_c)}{kT}\right\}\right]\left(\frac{\partial \alpha}{\partial r_A}\right)_{r_A = R_c} \quad (9.48)$$

After substituting eq. 9.45 into eq. 9.48, we find

$$k \int C_A(\mathbf{r}_A, t)\delta_3(|\mathbf{r}_A| - R_c) \, d\mathbf{r}_A$$

$$= 4\pi D C_A^0 \left\{\frac{1}{S(R_c)} + \frac{R_c^2 \exp[-U(R_c)/kT]}{(\pi D t)^{1/2}}\right\} \quad (9.49)$$

Using relation 9.49, we can solve eq. 9.7 easily; the result is

$$\frac{C_B}{C_B^0} = \exp\left[-\frac{4\pi D C_A^0 t}{S(R_c)} - 8C_A^0 R_c^2 (\pi D t)^{1/2} \exp\left\{-\frac{U(R_c)}{kT}\right\}\right] \quad (9.50)$$

which should be compared with eq. 9.19. In this case we obtain the diffusion-controlled rate constant k_D as

$$k_D = \frac{4\pi D}{S(R_c)} = \frac{4\pi D}{\int_{R_c}^{\infty}(dr/r^2)\exp[U(r)/kT]} \quad (9.51)$$

When $U(r) = 0$, eq. 9.51 reduces to $k_D = 4\pi D R_c$, as is to be expected. An important application of eq. 9.51 is to calculate the diffusion-controlled rate constants for reactions between ions. The potential energy of the interaction between ions is given by

$$U(r_A) = \frac{Z_A Z_B e^2}{\varepsilon_0 r_A} \quad (9.52)$$

where ε_0 is the dielectric constant. From eq. 9.38 we have

$$S(R_c) = \int_{R_c}^{\infty} \frac{dr}{r^2} \exp\left(\frac{Z_A Z_B e^2}{\varepsilon_0 r k T}\right) = \int_{R_c}^{\infty} \exp\left(\frac{Z_A Z_B r_c}{r}\right)$$

$$= \frac{\exp(Z_A Z_B r_c / R_c) - 1}{Z_A Z_B r_c} \quad (9.53)$$

or

$$\frac{1}{S(R_c)} = \frac{Z_A Z_B r_c}{\exp(Z_A Z_B r_c / R_c) - 1}$$

where

$$r_c = \frac{e^2}{\varepsilon_0 k T} \tag{9.54}$$

In water at 25°C, r_c is 7.1×10^{-8} cm.

The quantity $1/S(R_c)$ is often called the *ionic reaction radius*. Note that the ionic reaction radius is positive for any combination of signs for the charges. If oppositely charged ions are reacting, the exponential factor is small, because R_c is generally less than r_c, and the ionic reaction radius is approximately $-Z_A Z_B r_c$. If the ions have charges of like sign, the exponential factor is considerably greater than unity and the ionic reaction radius becomes quite small for small R_c. The diffusion-controlled rate constant k_D may be found by inserting eq. 9.53 in eq. 9.51.

$$k_D = \frac{4\pi D Z_A Z_B r_c}{\exp(Z_A Z_B r_c / R_c) - 1} \tag{9.55}$$

Recently the hydrodynamic effect on diffusion-controlled reaction rates has been examined by several authors (see Wolynes and Deutch, 1976). They have noted that the diffusion of the reactant particles toward each other is impeded by a hydrodynamic effect that arises because the particles must force the solvent out of the path of their mutual approach. The resulting drag force on each particle is greater than if the particle were moving by itself; this additional drag is referred to as the *hydrodynamic interaction*. This effect can be taken into account by the use of a relative diffusion constant that depends on the separation of the particles $D(r)$. For the detailed discussion the original papers should be consulted.

9.2.4 Applications

A comparison of some rate constants for rapid reactions with those computed from $k_D = 4\pi R_c D$ is given in Table 9.1. Some of the diffusion coefficients are estimated by comparison with similar stable species. Note that only a few of the reactants are spherically symmetric; this was one of the assumptions of the derivation. The worst case from this point of view is the reaction between two ethyl radicals. The radius chosen for the reaction is related to the radius of a CH_2 group, and not the whole radical. For the other reactions the reduced

Table 9.1
Diffusion-Controlled Reactions[a]

Reaction	Solvent	T (°C)	D_A	D_B	R_c	k_{calc}	k_{obs}	Ref.
$I + I \rightarrow I_2$	CCl_4	25	4.2	—	4	1.3	0.82	b
$OH + OH \rightarrow H_2O_2$	H_2O	24	2.6	—	3	0.6	0.5	c
$OH + C_6H_6 \rightarrow C_6H_6OH$	H_2O	24	2.6	1.1	3	0.8	0.33	c
$C_2H_5 + C_2H_5 \rightarrow C_4H_{10}$	C_2H_6	−177	0.7	—	2	0.1	0.02	c
$H^+ + NH_3 \rightarrow NH_4^+$	H_2O	25	10.0	3.0	4	4.0	4.3	d

[a] D in 10^{-5} cm^2 sec^{-1}, R_c in Å, and k in 10^{10} liters/mole · sec.
[b] Noyes (1961).
[c] Portman and Matheson (1965).
[d] Eigen (1963).

symmetry is not serious and an average van der Waals radius is used. None of the observed rate constants exceeds the diffusion limit by more than experimental error, which is 10% or more. For the reaction $H^+ + NH_3 \rightarrow NH_4^+$, an ion-dipole interaction is present. However, the high dielectric constant of water (78.5 at 25°C) reduces the contribution to the rate constant of the electrostatic potential to about 10%.

Many diffusion-controlled reactions between ions have been studied in aqueous solution and a representative selection of rate constants is given in Table 9.2 together with values computed from eq. 9.55 using a uniform value of 5 × 10^8 cm for R_c (Anbar and Hart, 1968). Note that the excellent agreement between theory and experiment for reactions between oppositely charged ions fades seriously for ions of the same charge. In the case of reactions between ions of the same charge, the computed rate constant k_D is very sensitive to the choice of R_c.

The equation $k_D = 4\pi R_c D$ and eq. 9.55 do not predict that the reactions in Tables 9.1 and 9.2 will be diffusion controlled. They only give the upper limit of the rate constant k_D. The reactions were found to agree reasonably well with the upper limit, and hence they are said to be diffusion controlled. All reactions of hydrogen ions with strong bases that have been studied appear to be diffusion controlled (Anbar and Hart, 1968; Moelwyn-Hughes, 1972; North, 1964; Weston and Schwarz, 1972). These rates are generally measured by relaxation methods. The hydrated electron, which is essentially the conjugate base of the hydrogen ion, is produced in the flash photolysis of many inorganic aqueous solutions and in the pulse radiolysis of water. Most rate constants measured for hydrated electron reactions are also diffusion controlled.

The recombination reactions of ions present interesting examples of the diffusion-controlled reactions of species exerting long-range interactions. Eigen (1954) has discussed the significance of the rate constants for the reactions

$$H^+ + SO_4^{2-} \longrightarrow HSO_4^-, \quad k = 10^{11} \text{ liters/mole} \cdot \text{sec}$$

$$NH_4^+ + OH^- \longrightarrow NH_4OH, \quad k = 4 \times 10^{10} \text{ liters/mole} \cdot \text{sec}$$

$$H^+ + OH^- \longrightarrow H_2O, \quad k = 1.4 \times 10^{11} \text{ liters/mole} \cdot \text{sec}$$

These rate constants agree with the values predicted using eq. 9.55 if it is assumed that ion recombination takes place when the partners approach each other within a distance of 5×10^{-8} cm. This calculated collision separation and the anomalously large mobility of hydrogen and hydroxide ions in aqueous systems are indicative of an exchange mechanism in the hydration shell,

Summers and Burr (1972) have recently examined the Stokes–Einstein equation $D = kT/\sigma \pi R_s N$ by measuring the viscosities of the following two types of solution: (1) aqueous solutions of polyethylene oxide (Union Carbide WSR-301 and WSR-205) and (2) aqueous solutions of glycerine; and the diffusion coefficients of uracil and glucose (standard) in these solutions. Their results are shown in Tables 9.3 and 9.4. These tables list the observed values compared to calculated values using the Stokes–Einstein equation. The data indicated that polymer-water solutions are not classical with respect to the Stokes–Einstein equation but glycerine-water solutions are.

The theory of diffusion-controlled recombination of spherical bodies in a liquid phase has also been applied to the studies of chemically induced dynamic nuclear polarization (CCI DNP) (see, for example, Lawler, 1973).

Table 9.2
Diffusion-Controlled Reactions Between
Ions in Water[a,b]

Reaction	k_{obs}	k_{calc}
$e_{aq} + CO(NH_3)_6^{3+}$	8.2	9.3
$e_{aq} + Cr(en)_3^{3+}$	7.8	9.3
$e_{aq} + Cr(H_2O)_6^{2+}$	6.2	6.5
$H^+ + SO_4^{2-}$	10	6.5
$e_{aq} + Ag^+$	3.2	4.1
$H^+ + OH^-$	14	11
$HS^- + H^+$	7.5	9.0
$e_{aq} + NO_2^-$	0.46	0.97
$e_{aq} + NO_3^-$	0.85	0.97
$e_{aq} + SeO_4^{2-}$	0.10	0.38
$e_{aq} + Ni(CN)_4^{2-}$	0.41	0.38
$e_{aq} + Fe(CN)_6^{3-}$	0.30	0.13
$H^+ + (H_2NC_2H_4)_3NH^{3-}$	0.56	0.26

[a] k in 10^{10} liters/mole · sec.
[b] Anbar and Hart, 1968.

Recent picosecond optical studies of the iodine recombination in CCl_4 (Chuang et al., 1974) have required a more detailed solution to the diffusion-controlled recombination problem. Evans and Fixman (1976) have proposed an improvement of Fick's law that includes an effective spatial dependence in the relative diffusion constant. However, much remains to be done in this area.

Table 9.3
Polarographic Determination of the Diffusion Coefficient
for Uracil Based on the I]kovic Equation

Solvent	T (°C)	Viscosity (cP)	D_{obs}[a]	D_{calc}[a,b]
Water	16	1.19	6.2	5.8
Glycerine-water	16	3.1	1.85	1.8
0.2% WSR-301 (water)	16	7.0	4.9	0.97
0.6% WSR-301 (water)	16	122.0	2.25	0.06

[a] In 10^{-6} cm²/sec.
[b] Based on the Stokes–Einstein equation.

Table 9.4
Direct Determination of Diffusion Coefficient by Tracer Labeling

Solvent	T (°C)	Viscosity (cP)	D_{obs}	D_{calc}	Solute
Water	16	1.19	6.73[a]		[^{14}C] Glucose
			10.7	5.8	[^{14}C] Uracil
0.2% WSR-301 water	16	7.0	4.70		[^{14}C] Glucose
			7.50	0.97	[^{14}C] Uracil
0.6% WSR-301 water	16	122	0.3		[^{14}C] Glucose
			1.77	0.06	[^{14}C] Uracil
0.4% WSR-205 water	16	4.7	9.0	1.4	[^{14}C] Uracil
0.9% WSR-205 water	16	13.5	2.2	0.5	[^{14}C] Uracil

[a] Standard for apparatus, 2.2×10^{-2} M glucose in water, $D = 6.73 \times 10^{-6}$ cm^2/sec.

An aspect of the theory of diffusion-controlled reactions that so far has received surprisingly little attention is the effect of concentration. In the dilute limit it suffices to consider a single pair of reactants, but at higher concentrations the reaction rate will be affected by the competition between neighboring sinks. This effect was first studied by Frisch and Collins (1952) and has begun to attract some attention (Felderhof and Deutch, 1976; Monchick et al., 1957; Waite, 1957).

Recently Monchick (1957) has studied the effect of time delay on the diffusion-controlled reaction by modifying the radiation boundary condition. This effect can be explained as follows. The mechanism of the diffusion-controlled reaction should consist of two fragments or molecular aggregates passing through the surface $r = R_c$. Then, roughly speaking, one may say that there are two fluxes, one entering at time t and another returning after a time delay τ, and a reduction of sticking probability caused by reaction.

The short-time transient rate for the disappearance of electrons in a scavenging reaction has attracted some attention in recent years in the pulse radiolysis of liquids and glasses. A few notable ideas have been proposed for the decay kinetics of electrons. Hamil (1969) has introduced the idea of "dry" (presolvated) electrons to explain the very early ($< 10^{-13}$ sec) disappearance of electrons. Schwarz (1971) and Czapski and Peled (1973), on the other hand, have suggested that the initial rapid decay may be related to the time-dependent reaction rate for electrons that happen to be in the immediate neighborhood of scavengers. Miller (1975), however, suggested that the decay of solvated electron in glassy media involves a direct long-range

(~ 50 Å) electron tunneling from the initial site of solvent trap to the scavenger rather than a diffusion motion going over a number of intermediate sites. It is indeed true that one cannot explain the decay characteristics of solvated electrons in glassy media if one assumes a simple diffusion kinetics of electron scavenging, particularly in the time scale that is comparable to or shorter than the average jump time for the diffusion motion. The failure of the simple diffusion kinetics, however, does not necessarily imply that the concept of a diffusive motion of a localized charge itself is not applicable in these problems. Recently Helman and Funabashi (1977) have calculated time-dependent rates of electron scavenging reactions using the model of continuous-time random walk (CTRW) of Montroll and Weiss (1965) for the hopping time distribution functions of an exponential form and a few nonexponential forms. The result was applied to the charge neutralization reaction using the Laplace transformation technique. It was found that the rate constant in the CTRW model is smaller than the Smoluchowski transient rate but is larger than the time-independent (asymptotic) value at the time scale that is comparable to or shorter than the average jump time. The differences in the decay pattern and the survival fraction of electrons between the CTRW model and the Smoluchowski treatment are much less than those in the rate constants themselves. The results of Helman and Funabashi are, of course, not sufficient to make a definitive statement on the relevance of the dry electron mechanism.

In the course of electrodialysis, the solution near the membrane is depleted, and a concentration gradient arises in the boundary layer close to the membrane—a phenomenon called *polarization of electrolyte*. One might think that as the concentration of electrolyte on the membrane solution interface approaches zero, the voltage current becomes saturated, the current in the system reaching its limit value. Actually after a plateau in the voltage curve, one observes a region of rapid current increase. This is usually said to be due to a progressive participation of dissociated (split) water in charge transfer, as the voltage increases (Spiegler, 1966; Forgues et al., 1975). Together with the voltage-current charges just mentioned one observes a considerable shift of pH in the layers near the membrane. [An excess of H^+ (OH^-) arises near the anion (cation) exchange membrane.] This phenomenon is usually called *water splitting* or *acid-base generation*. It is obvious that an understanding of the water-splitting phenomenon is impossible without a clear knowledge of properties of electrodiffusional ion transfer in the solution layers near the membranes when reversible water dissociation takes place. Recently Rubinstein (1977) has proposed a diffusion model of water dissociation within an unstirred layer in a cross electric field. It is concluded that for realistic values of the bulk salt concentration, the salt ions are potential determining over a wide range of voltages, particularly deeply within the domain of "water splitting." As a corollary, the shift of pH in the unstirred layer is determined by voltage and does not depend on the bulk concentration of the salt. The efficiency of the current depends strongly on the latter. It is shown that the diffusional overvoltage approximation leads to overestimated values of the current carried by the H^+ and OH^- ions, compared with those obtained by numerical solution of the exact problem.

9.3 QUENCHING OF LUMINESCENCE

In system where the distance between the excited donor D^* and the acceptor A is large compared to molecular dimensions and does not change during the lifetime of the former, the transfer of the electronic excitation energy is usually assumed to proceed by resonance transfer as developed by Förster and Dexter (Förster, 1948; Dexter, 1953; Lin, 1971). The Förster–Dexter theory (see Chaper 7) has to be modified when energy transfer takes place in liquid solutions where the distance between the excited donor and acceptor varies during the lifetime of the former as a result of Brownian motion. In any quantitative treatment of nonradiative energy transfer influenced by diffusion, one has to account for the fact that the excited donor molecules D^* that happen to have an acceptor A close by at time of excitation rapidly transfer their energy to the latter. After a short period, only D^* molecules are left, whose vicinity is depleted of A molecules. An average concentration gradient of the latter is thus established, and the Borwnian motion leads to a flux of A molecules toward the excited donors. A transfer of energy in excess of that found in a stationary solution thus takes place (Feitelson, 1966a; Steinberg and Katchalski, 1968).

Let us consider the energy transfer process in solution between two solutes D^* and A, the latter being present in excess. As an approximation, it is assumed that the energy transfer between D^* and A takes place instantaneously whenever the molecules D^* and A approach one another to a distance R_c between their centers of mass. No attractive or repulsive forces are assumed to prevail between the reacting molecules whenever the distance between their centers of mass is greater than R_c.

For a system of excited donors D^* and acceptors A (or quenchers) with concentrations $C_{D^*}(\mathbf{r}_{D^*}, t)$ and $C_A(\mathbf{r}_A, t)$, if Fick's laws are applicable and if τ represents the lifetime of D^* in the absence of acceptors (or quenchers), then the rates of change of concentrations C_{D^*} and C_A are given by

$$\frac{\partial C_{D^*}}{\partial t} = D_{D^*}\nabla^2 C_{D^*} - \left(\frac{C_{D^*}}{\tau}\right) - K_{D^*} \qquad (9.56)$$

and

$$\frac{\partial C_A}{\partial t} = D_A \nabla^2 C_A - K_A \tag{9.57}$$

where D_A and D_{D^*} are the diffusion coefficients of A and D^*. In these equations K_A and K_{D^*} denote the rates of change of C_A and C_{D^*} caused by direct energy transfer (or quenching). The distribution of acceptor (or quencher) molecules around an excited donor molecule is not the same for every D^* molecule in the solution. This effect has been shown to be a high-order effect and not to be measurable in most cases (Yguerabide et al., 1964). If one ignores this effect, then we can proceed to the solution of eqs. 9.53 and 9.57 as in Sect. 9.2. For convenience we discuss the cases of the Smoluchowski and radiation boundary conditions separately.

9.3.1 Smoluchowski Boundary Condition

The differential equations 9.56 and 9.57 with the Smoluchowski boundary condition can be solved in exactly the same manner as that given in Sec. 9.2. The results are given by

$$\frac{C_A}{C_A^0} = 1 - \frac{R_c}{r_A} \operatorname{erfc} \frac{r_A - R_c}{(4Dt)^{1/2}} \tag{9.58}$$

and

$$\frac{C_{D^*}}{C_{D^*}^0} = \phi(t) = \exp\left[-t\left(\frac{1}{\tau} + 4\pi R_c C_A^0 D\right) - 8C_A^0 R_c^2 (\pi Dt)^{1/2} \right] \tag{9.59}$$

where $D = D_A + D_{D^*}$ and C_A^0 and $C_{D^*}^0$ represent the initial concentration of A and D^*, respectively. Equation 9.59 also represents the decay function of donor luminescence for flash excitation. Equation 9.59 is actually a derivation of the average specific rate of reaction of an A molecule with a D^* molecule over all random configurations of A about D^*.

The expressions derived in the previous paragraphs for the overall decay of excited molecules apply directly only to excitation by an instantaneous pulse. In practice, the exciting pulse may have any of a variety of forms, depending on the experimental techniques used, but any type of exciting pulse may be regarded as an infinite sum of instantaneous pulses. If we let a pulse be represented by the function $F(t)$, and if the effect produced by this instantaneous pulse follows a law represented by the function $\phi(t)$, then for the time interval of observation, the response of the system can be expressed by a superposition integral (Yguerabide et al., 1964; Inokuti and Hirayama, 1965):

$$\bar{C}_{D^*}(t) = \int_0^t F(t')\phi(t - t')\, dt' \tag{9.60}$$

where $\bar{C}_{D*}(t)$ represents the number of excited donors at time t. Equation 9.60 is the starting point for considerations of experimental results. If $F(t)$ represents a step function of amplitude F_0—that is, $F(t) = F_0 H(t)$, where $H(t)$ denotes the Heaviside function [i.e., $H(t) = 0, t < 0; H(t) = 1, t \geq 0$]— then eq. 9.60 can be written

$$\bar{C}_D(t) = F_0 \int_0^t \phi(t - t')\, dt' = F_0 \int_0^t \phi(t')\, dt' \tag{9.61}$$

Physically, F_0 represents the number of excited molecules produced per unit time. When $t \to \infty$, $\bar{C}_{D*}(\infty)$ is the number of excited donors in the steady state. Other types of excitation pulses can be discussed similarly.

Substitution of eq. 9.59 into eq. 9.60 gives us an expression for $\bar{C}_{D*}(t)$ when the system is subjected to a step function pulse:

$$\bar{C}_{D*}(t) = F_0 \int_0^t dt' \exp\left[-t'\left(\frac{1}{\tau} + 4\pi R_c C_A^0 D\right) - 8C_A^0 R_c^2 (\pi D t')^{1/2} \right] \tag{9.62}$$

which can easily be integrated:

$$\begin{aligned}
\bar{C}_D(t) = \frac{F_0}{a} &\left(1 - \exp(-at - bt^{1/2}) - \frac{b}{2}\left(\frac{\pi}{2}\right)^{1/2} \right. \\
&\left. \times \left(\exp\frac{b^2}{4a}\right)\left\{ \mathrm{erf}\left[(at)^{1/2} + \frac{b}{2a^{1/2}} \right] - \mathrm{erf}\,\frac{b}{2a^{1/2}} \right\} \right)
\end{aligned} \tag{9.63}$$

where $a = 1/\tau + 4\pi R_c C_A^0 D$, $b = 8C_A^2 R_c^0 (\pi D)^{1/2}$, and

$$\mathrm{erf}(x) = 1 - \mathrm{erfc}(x) = \left(\frac{2}{\sqrt{\pi}}\right) \int_0^t dz \exp(-z^2) \tag{9.64}$$

To obtain the steady-state value of $\bar{C}_{D*}$, we let $t \to \infty$:

$$\bar{C}_{D*}(\infty) = \frac{F_0}{a}\left[1 - \frac{b}{2}\left(\frac{\pi}{a}\right)^{1/2}\left(\exp\frac{b^2}{4a}\right)\mathrm{erfc}\,\frac{b}{2a^{1/2}} \right] \tag{9.65}$$

In the absence of acceptors (or quenchers), $C_A^0 = 0$, $a = 1/\tau$, $b = 0$, and $\bar{C}_{D*}(\infty) = \tau F_0$. Using this relation, we can obtain the Stern–Volmer-type equation

$$\begin{aligned}
\frac{I}{I_0} = \frac{\eta}{\eta_0} &= \frac{1}{a\tau}\left[1 - \frac{b}{2}\left(\frac{\pi}{a}\right)^{1/2}\left(\exp\frac{b^2}{4a}\right)\mathrm{erfc}\,\frac{b}{2a^{1/2}} \right] \\
&= \frac{Y}{1 + 4\pi\tau R_c C_A^0 D} = \frac{Y}{1 + \tau k_D C_A^0}
\end{aligned} \tag{9.66}$$

where I and I_0 represent the luminescent intensities at the steady state in the presence and absence of acceptors (or quenchers), respectively, and

$$Y = 1 - \frac{b}{2}\left(\frac{\pi}{a}\right)^{1/2}\left(\exp\frac{b^2}{4a}\right)\operatorname{erf}\frac{b}{2a^{1/2}} \tag{9.67}$$

Equation 9.66 also represents the relative quantum yield η/η_0. The decay function for steady-state excitation, which may be of practical interest, is given by

$$\phi_S(t) = \frac{\int_t^\infty \phi(t')\,dt'}{\int_0^\infty \phi(t')\,dt'} = \frac{1}{Y}\left\{\exp(-at - bt^{1/2}) - \frac{b}{2}\left(\frac{\pi}{a}\right)^{1/2}\right.$$
$$\left.\times\left(\exp\frac{b^2}{4a}\right)\operatorname{erfc}\left[(at)^{1/2} + \frac{b}{2a^{1/2}}\right]\right\} \tag{9.68}$$

From eqs. 9.66 and 9.67 we can see that the factor Y is a correction factor to the Stern–Volmer relation. In other words, when $Y = 1$, we obtain the conventional Stern–Volmer relation.

9.3.2 Radiation Boundary Condition

Again the solution of the differential equations given in eqs. 9.56 and 9.57 is carried out in the same way as that described in Sect. 9.2.2. The results are given by

$$\frac{C_A}{C_A^0} = 1 - \frac{R_c}{r_A(1 + \gamma)}\left\{\left(\operatorname{erfc}\frac{r_A - R_j}{(4Dt)^{1/2}}\right) - \exp[h(r_A - R_c) + h^2Dt]\right.$$
$$\left.\times\operatorname{erfc}\left[h(Dt)^{1/2} + \frac{r_A - R_e}{(4Dt)^{1/2}}\right]\right\} \tag{9.69}$$

where $k_D = 4\pi R_c D$, $\gamma = k/4\pi R_c D = k/k_D$, and $h = (1 + \gamma)/R_c$.

The quantity γ denotes the ratio of the quenching rate constant to the diffusion-limited rate constant. When $\gamma \to 0$, to the first-order approximation of y, eq. 9.70 reduces to

$$\frac{C_{D^*}}{C_{D^*}^0} = \phi(t) = \exp\left\{-t\left[\left(\frac{1}{\tau}\right) + kC_A^0\right]\right\} \tag{9.71}$$

In other words, when the rate of energy transfer (or quenching) is slow compared with that of diffusion, the conventional reaction kinetics holds and the decay is exponential. When $\gamma \to \infty$, eq. 9.70 reduces to eq. 9.59. Expressions for the relative quantum yields η/η_0 (or I/I_0) and the decay function for steady-state excitation $\phi_s(t)$ can be obtained from eqs. 9.61 and 9.68 by using the decay function for flash excitation $\phi(t)$ given in eq. 9.65 as

$$
\frac{I}{I_0} = \frac{\eta}{\eta_0} = \frac{1}{\tau} \int_0^\infty dt \, \exp\left(-t\left(\frac{1}{\tau} + \frac{\gamma k_D C_A}{1+\gamma}\right) - \frac{4\pi\gamma^2 C_A^0}{h^3}\right.
$$

$$
\times \left. \left\{2h\left(\frac{Dt}{\pi}\right)^{1/2} - 1 + [\exp(h^2 Dt)]\mathrm{erfc}[h(Dt)^{1/2}]\right\}\right) \tag{9.72}
$$

and

$$
\phi_S(t) = \frac{\eta_0}{\tau\eta} \int_t^\infty dt' \, \exp\left(-t'\left(\frac{1}{\tau} + \frac{\gamma k_D C_A^0}{1+\gamma}\right) - \frac{4\pi\gamma^2 C_A^0}{h^3}\right.
$$

$$
\times \left. \left\{2h\left(\frac{Dt}{\pi}\right)^{1/2} - 1 + [\exp(h^2 Dt)]\mathrm{erfc}[h(Dt)^{1/2}]\right\}\right) \tag{9.73}
$$

Exact analytical expressions for η/η_0 (or I/I_0) and $\phi_S(t)$ are not possible in this case. But if the factor

$$
\frac{4\pi\gamma^2 C_A^0}{h^3} \left\{2h\left(\frac{Dt}{\pi}\right)^{1/2} - 1 + [\exp(h^2 Dt)\mathrm{erfc}[h(Dt)^{1/2}]\right\}
$$

is negligible compared with $t\{(1/\tau) + [\gamma k_D C_A^0/(1+\gamma)]\}$, then we obtain the Stern–Volmer relation and after a very short period, the decay rates [both $\phi_S(t)$ and $\phi(t)$] are exponential. It has been shown by Yquerabide et al. (1964) that good approximate expressions for $\phi_S(t)$ and η/η_0 can be obtained by using the following approximate relation:

$$
2\left(\frac{h^2 Dt}{\pi}\right)^{1/2} - 1 + [\exp(h^2 Dt)]\mathrm{erfc}[h(Dt)^{1/2}]
$$

$$
= 2\left[\left(\frac{1}{P} + \frac{h^2 Dt}{\pi}\right)^{1/2} - \left(\frac{1}{P}\right)^{1/2}\right] \tag{9.74}
$$

where $P = 4.93$. Using 9.74 we obtain

$$
\frac{C_{D^*}}{C_{D^*}^0} = \phi(t) = \exp\left\{-t\left(\frac{1}{\tau} + \frac{kC_A^0}{1+\gamma}\right) - \frac{8\pi^2\gamma^2 C_A^0}{h^3}\left[\left(\frac{1}{P} + \frac{h^2 Dt}{\pi}\right)^{1/2}\right.\right.
$$

$$
\left.\left. - \left(\frac{1}{P}\right)^{1/2}\right]\right\} \tag{9.75}
$$

$$
\frac{I}{I_0} = \frac{\eta}{\eta_0} = \frac{Y_R}{1 + (k\tau C_A^0/1 + \gamma)} \tag{9.76}
$$

and

$$\phi_S(t) = \frac{1}{Y_R} \exp\left[-t\left(\frac{1}{\tau} + \frac{kC_A^0}{1+\gamma}\right) - \frac{8\pi\gamma^2 C_A^0}{h^3}\left(\frac{1}{P} + \frac{h^2 Dt}{\pi}\right)^{1/2} + \frac{8\pi\gamma^2 C_A^0}{P^{1/2}h^3}\right]$$

$$\times \exp\left[\left(\frac{\pi}{h^2 PD}\right)^{1/2}\left(\frac{1}{\tau} + \frac{kC_A^0}{1+\gamma}\right)^{1/2}\right.$$

$$\left. + \frac{4\pi\gamma^2 C_A^0}{h^2}\left(\frac{D}{\pi}\right)^{1/2}\left(\frac{1}{\tau} + \frac{kC_A^0}{1+\gamma}\right)^{-1/2}\right]^2$$

$$\times \operatorname{erfc}\left[\frac{4\pi\gamma^2 C_A^0}{h^2}\left(\frac{D}{\pi}\right)^{1/2}\left(\frac{1}{\tau} + \frac{kC_A^0}{1+\gamma}\right)^{-1/2}\right.$$

$$\left. + \left(t + \frac{\pi}{DPh^2}\right)^{1/2}\left(\frac{1}{\tau} + \frac{kC_A^0}{1+\gamma}\right)^{1/2}\right] \tag{9.77}$$

respectively, where

$$Y_R = 1 - \frac{4\pi\gamma^2 C_A^0 D^{1/2}}{h^2}\left(\frac{1}{\tau} + \frac{kC_A^0}{1+\gamma}\right)^{-1/2}$$

$$\times \exp\left[\left(\frac{\pi}{DPh^2}\right)^{1/2}\left(\frac{1}{\tau} + \frac{kC_A^0}{1+\gamma}\right)^{1/2} + \frac{4\pi\gamma^2 C_A^0}{h^2}\left(\frac{D}{\pi}\right)^{1/2}\left(\frac{1}{\tau} + \frac{kC_A^0}{1+\gamma}\right)^{1/2}\right]^2$$

$$\times \operatorname{erfc}\left[\left(\frac{\pi}{DPh^2}\right)^{1/2}\left(\frac{1}{\tau} + \frac{kC_A^0}{1+\gamma}\right)^{1/2} + \frac{4\pi\gamma^2 C_A^0}{h^2}\left(\frac{D}{\pi}\right)^{1/2}\left(\frac{1}{\tau} + \frac{kC_A^0}{1+\gamma}\right)^{-1/2}\right] \tag{9.98}$$

In the discussion of the solution of differential equations 9.56 and 9.57 with the Smoluchowski and radiation boundary conditions, the initial distribution of acceptors (or quenchers) has been assumed constant. This assumption has been discussed by Yquerabide et al. (1964) by using the discrete acceptor (or quencher) distribution. Our results correspond to their first-order approximation. Their results in the second-order approximation differ only slightly from those in the first-order approximation and the experimental data are incapable of showing the effect of high-order terms. Thus the second-order approximation will not be given here.

From the derivation given here, we can see that both $\phi(t)$ and $\phi_S(t)$ are not exponential, although after some time both $\phi(t)$ and $\phi_S(t)$ will approach exponential decay. The exponential decay portion of $\phi(t)$ or $\phi_S(t)$ is often used to determine the Stern–Volmer-type relation $(1/\tau) + k_{\text{obs}} C_A^0$, where C_A^0 represents the acceptor (quencher) concentration, and hence to obtain k_{obs}, the so-called observed bimolecular energy transfer rate constant, which is related to k by $k_{\text{obs}} = k(1 + \gamma)$.

9.3.3 Applications

The first experimentally sound study of the effect of solvent viscosity on the transfer of electronic excitation energy was made by Melhuish (1963). On examining the donor-acceptor combination of 9-methyl-anthracene-perylene as a function of solvent viscosity at a constant temperature, he found that an increase in solvent fluidity brought about an increase in energy transfer efficiency. Elkana et al. (1968) recently studied in quantitative form the effect of diffusion on energy transfer in liquid solution. For this purpose a system was chosen whose energy donor would have a long lifetime in the excited state. This would provide time for diffusion to take place to an appreciable extent, assuring a measurable effect. Naphthalene was found to be an adequate energy donor and anthranilic acid, an acceptor. The lifetime of naphthalene in its lowest excited singlet state is about 10^{-7} sec. A variety of alcohols were used as solvents to test the influence of solvent viscosity (or diffusion of the solute molecules) on energy transfer. As was anticipated, it was found that a decrease in viscosity of the medium and an increase in acceptor concentration enhance markedly the nonradiative energy transfer (see Table 9.5).

Table 9.5

The Quantum Yield of Energy Transfer ϕ_{nr} Between Naphthalene and Anthranilic Acid[a]

Solvent	Solute concentration ($\times 10^{-3}$ M)	ϕ_{nr}	τ_D ($\times 10^9$ sec)	Viscosity (cP)	$D = D_A + D_D$ ($\times 10^5$ cm^2 sec^{-1})
Glycerol	5	0.39	109	1000	0.00
	2	0.28	—	—	—
Cyclohexanol	5	0.53	101	65	0.02
Ethylene glycol	5	0.56	84	17.4	0.08
Amyl alcohol	5	0.77	106	4.3	0.33
t-Butanol	5	0.78	83	4.8	0.30
	2	0.60	—	—	—
	1	0.45	—	—	—
Isobutanol	5	0.76	106	4.1	0.35
n-Butanol	5	0.86	88	2.8	0.51
Isopropanol	5	0.84	89	1.95	0.73
Ethanol	5	0.90	96	1.2	1.20
Methanol	5	0.92	85	0.6	2.30
	2	0.83	—	—	—
	1	0.66	—	—	—

[a] ϕ_D[naphthalene in ethanol] $= 0.23$.

In the Förster–Dexter formulation it is assumed that the Brownian movement of molecules is slow enough that each individual transfer process may be considered to occur at a constant distance. For solutions of high viscosity or solid solutions, this is a valid assumption. However, the average intermolecular distance for a donor–acceptor combination is not necessarily constant over the lifetime of the donor excited state in liquid solutions of low to moderate viscosities. This is particularly true if the donor happens to be a triplet, where the lifetime can range from microseconds to tens of milliseconds or longer. Therefore, even in dilute solutions, diffusion can turn the seemingly inefficient triplet-singlet transfer into a highly efficient process. Its application to photochemistry, photobiology, and chemiluminescence studies may well prove to be an extremely productive photochemical tool.

Vando and Hercules (Vando and Hercules, 1970) studies the phosphorescence quenching of the donor–acceptor pairs benzophenone-perylene and phenanthrene-Rhodamine B in fluid solution at 20°C. No room temperature phosphorescence from either phenanthrene or benzophenone was detectable and the interference of the triplet-triplet absorption bands of perylene with those benzophenone prevented flash photolysis from being used in the benzophenone-perylene system. Thus independent quenching studies on the benzophenone-perylene system and the phenanthrene-Rhodamine B systems were performed using the phosphorescence emission and the triplet-triplet absorption methods, respectively. The bimolecular energy transfer rate constants were obtained from the slope of the plot of observed first-order rate constants versus acceptor concentration. They found that k_{obs} for the benzophenone-perylene system is 7.1×10^9 liters/mole · sec and for the phenanthrene-Rhodamine B system is 3.9×10^7 liters/sec · mole (see Table 9.6). The quenching rate constant for the benzophenone-perylene system is comparable with the diffusion-controlled constant.

The quenching of triplet-state organic molecules by added metal complex ions has been investigated extensively (Binet et al., 1968; Hammond and Foss, 1964; Fry et al., 1966; Bell and Linshitz, 1963; Banfield and Husain, 1969), and two modes of energy deactivation have been proposed:

$$T + M \xrightarrow{\ k_Q\ } S_0 + M \tag{9.79}$$

$$T + M \xrightarrow{\ k_E\ } S_0 + M^* \tag{9.80}$$

where T and S_0 represent the donor triplet and ground singlet states, and M and M^* designate the ground and excited electronic states of the quencher. Equation 9.79 is equivalent to an energy deactivation enhanced by interaction with the quencher and eq. 9.80 represents an intermolecular energy transfer. Binet et al. 1968 reported studies on triplet-state benzil molecules quenched by Cr^{111} complexes in the methanol-water (88 : 12 by volume) at $-113°C$.

At $-113°C$ the solutions were viscous, but fluid. They measured the relative quantum yields η/η_0 and the relative lifetimes τ/τ_0 for donors and the Stern–Volmer relation was used to determine the quenching rate constants $k_Q + k_E$ (see Table 9.6). The sum of quenching constants $k_E + k_Q$ obtained from η_0/η versus C_A and from τ_0/τ versus C_A are in good agreement. They were unable to determine k_E and k_Q separately, but they demonstrated that the quenching at the benzil triplet by $[Cr(NCS)_6]^{3-}$, $Cr(acac)_3$, and $[Cr(en)_3]^{3+}$ involves appreciable energy transfer to the quenching species, by observing the sensitized emission of the Cr^{111} complexes. It should be noticed that the methanol-water solvent at $-113°C$ has a viscosity much like glycerol at room temperature. Rate constants for diffusion-controlled reactions in room temperature glycerol should be about 5×10^6 liters/mole·sec (Wilkinson, 1964) and the k_E and k_Q values of this magnitude (see Table 9.6).

Banfield and Husian (1969) reported experiments on the decay of triplet-state acridine ($C_{13}H_9N$) molecules in the presence of paramagnetic ions complexed with saturated inorganic ligands having no low-lying electronic states. Acridine in the (π, π^*) triplet state was studied in aqueous solution following internal conversion from the first excited singlet state after the application of a high-intensity flash of white light. Triplet-state acridine was monitored by kinetic spectroscopy in absorption following the flash photolysis of acridine solutions in the presence of different concentrations of added metal complex ions. To explain the experimental results, they considered the following mechanism:

$$T \xrightarrow{k_1} S_0, \qquad T + T \xrightarrow{k_2} S_1 + S_0,$$

$$T + Q \xrightarrow{k_Q} S_0 + Q \qquad (9.81)$$

where T, S_0, and S_1 represent the triplet state, the ground singlet, and the first excited singlet state of acridine and Q denotes the quenching species. This mechanism neglects possible deactivation of triplet molecules by ground-state singlet molecules, which was not observed in these experiments. As mentioned before, the conventional kinetics law applies when the rate of reaction is much slower than that of diffusion; in this case the rate of change of $[T]$, the concentration of T, can be written

$$-\frac{d[T]}{dt} = k_1'[T] + k_2[T]^2 \qquad (9.82)$$

where $k_1' = k_1 + k_Q[Q]$. In terms of the optical density A_T, the extinction coefficient ε associated with the triplet-triplet absorption spectrum, and the path length l in the solution, we can rewrite eq. 9.82 as

$$-\frac{d(\log A_T)}{dt} = k_1' + \left(\frac{k_2}{\varepsilon l}\right) A_T \qquad (9.83)$$

Table 9.6
Energy Transfer in Liquids

Donor-acceptor	k (liters/mole · sec)	T (°C)	Solvent	Energy gap ΔE_A	$(\times 10^4 \text{ cm}^{-1})$ ΔE_D	Ref.
Phenanthrene-Rhodamine B	3.9×10^7	20	Ethanol	—	—	a
Benzophenone-perylene	7.1×10^9	20	Freon	2.42	2.25	a
Benzil-[Cr(NCS)$_6$]$^{3-}$	1.1×10^7	−113	Water-methanol	1.88	1.29	b
Benzil-[Cr(CN)$_6$]$^{3-}$	1.8×10^6	−113	Water-methanol	1.88	1.25	b
Benzil-[Cr(en)$_3$]$^{3+}$	1.3×10^6	−113	Water-methanol	1.88	1.50	b
Acridine-Cr(OH)$_6^{3-}$	0	20	Water	1.58	1.72	c
Acridine-Cr(C$_2$O$_4$)$_3^{3-}$	6.1×10^8	20	Water	1.58	1.44	c
Acridine-Mn(H$_2$O)$_6^{2+}$	0	20	Water	1.58	1.89	c
Acridine-Co(H$_2$O)$_6^{2+}$	1.0×10^7	20	Water	1.58	1.13	c
Acridine-Co(NH$_3$)$_6^{2+}$	3.6×10^8	20	Water	1.58	0.9	c
Acridine-Ni(H$_2$O)$_6^{2+}$	3.4×10^7	20	Water	1.58	1.54	c
Acridine-Ni(NH$_3$)$_6^{2+}$	3.3×10^8	20	Water	1.58	1.35	c
Acridine-Cu(H$_2$O)$_6^{2+}$	7.4×10^7	20	Water	1.58	1.26	c
Acridine-Cu(NH$_3$)$_4^{2+}$	4.0×10^8	20	Water	1.58	—	c

[a] Vando and Hercules (1970).
[b] Binet et al. (1968).
[c] Banfield and Husain (1969).

This indicates that a plot of the slope of the first-order decay curve ($\log A_T$ versus t) against A_T is a straight line of intercept k_1' and slope $k_2/\varepsilon l$. The quenching constant k_Q can in turn be determined from the slope of the linear plot, k_1' versus $[Q]$. Using this procedure the results obtained for k_1 and k_2 by Banfield and Husain were $k_1 = 8 \pm 10^3$ sec^{-1} and $k_2 = 9 \pm 5 \times 10^9$ liters mole · sec. These values should be compared with those obtained by Kira and Koizumi (1967) for the triplet decay in the absence of quenching ions in benzene, $k_1 = 2 \times 10^3$ sec^{-1} and $k_2 = 2 \times 10^9$ liters/mole · sec. The values of k_Q obtained by Banfield and Husain and claimed to be generally accurate to approximately 10% and the results for various ions are given in Table 9.6. From Table 9.6 the zero values of k_Q for $Mn(H_2O)_6^{2+}$ and $Cr(OH)_6^{3+}$ should be noticed; this may be due to the fact that electronic excitation of these ions accompanying the quenching of triplet acridine is energetically unfavorable. The rates of quenching of triplet acridine by paramagnetic ions given in Table 9.6 at room temperature are 10^2–10^3 times slower than diffusion-controlled rates. Banfield and Husain are of the opinion that quenching by the paramagnetic ions studied proceeds almost exclusively by an electronic–electronic energy transfer mechanism, since these ions are characterized by excited d-electron states lying below 15,840 cm^{-1}. Thus at least tens or hundreds of encounters are necessary before energy transfer can take place. This indicates that a favorable spatial overlap is necessary between the π orbitals of the donor triplet and the orbitals of the metal ion; any effect that facilitates this overlap will increase the efficiency of quenching. This seems to be consistent with Hoijtink's exchange mechanism of paramagnetic quenching (Hoijtink, 1960; Hill and Lin, 1969, 1970).

Cundall et al. (1969) reported some results obtained on the quenching of biacetyl phosphorescence by a number of species, including free radicals and bisgalvinoxyl, which exists in its ground state as a triplet. Pulse radiolysis has been used to populate the excited states of biacetyl and to follow the subsequent decay of the phosphorescing triplet. Quenching solutes were present in low concentration and direct excitation of the quencher was avoided. Benzene was used as a solvent and all measurements were carried out at 25°C. In their experiments the biacetyl concentration was fixed at 10^{-2} M and added quencher concentrations varied between 5×10^{-7} and 10^{-4} M. The method for measuring the rate constants of the biacetyl phosphorescence was as follows: τ_0, the lifetime of the biacetyl triplet state in the absence of the quencher, and τ, the lifetime in the presence of the quencher at concentration $[Q]$, were measured from the oscilloscope traces. From plots of $1/\tau$ versus $[Q]$, the observed quenching rate constant k_{obs} could be determined from the relation $1/\tau = (1/\tau_0) + k_{obs}[Q]$.

The results are shown in Table 9.7. From eqs. 9.75 and 9.77 we can see that $k_{obs} = k/(1 + \gamma)$. The highest quenching rate constants for DPPH

Table 9.7
Quenching Rate Constants for Biacetyl Phosphorescence

Quencher	$k_{obs} \times 10^{-9}$ (liters/mole · sec)
Diphenylpicrylhydrazil (DPPH)	5.5 ± 0.5

	5.5 ± 0.5

Bisgalvinoxyl

trans-Stilbene	4.0 ± 0.2
Ferric acetylacetonate	3.1 ± 0.5
Galvinoxyl (free radical)	2.9 ± 1.0
Azulene	2.4 ± 0.2
Ferrocene	2.4 ± 0.2
Cyclohexene	$(1.18 \pm 0.1) \times 10^{-4}$
Biacetyl	$(5 \pm 1) \times 10^{-5}$

Source: Cundall et al. (1969).

(diphenylpicrylhydrazil) and bisgalvinoxyl do not exceed other values that have been reported for this type of process. They are less than values expected if the most efficient quenching process in benzene were completely diffusion controlled. The modified Debye equation used by Osborne and Porter (1965), $k_D = 8RT/2000\eta$, with η being the viscosity of the solvent, gives a value of 1.55×10^{10} liters/mole · sec for benzene at 25°C. It has been indicated that more complex expressions, which should presumably be more accurate, increase the predicted value (Wagner and Kochevar, 1968). Spectrofluorimetric studies that were made by Cundall et al. give a value of 4.0×10^{10} liters/mole · sec for the first excited singlet state of benzene by biacetyl. They are of the opinion that their results are consistent with the view that some chemical complex formation involving steric and probably other restrictions is involved in the bimolecular deactivation of the triplet state.

Osborne and Porter (1963) have examined the decay of triplet naphthalene in which the radiative process competes with quenching by 1-iodonaphthalene. Flash photolysis equipment was used to obtain the initial concentration of triplet naphthalene. They compared the experimental results with those computed using a modified equation originally due to Debye, $k_D = 2R_c RT/1000\eta R_S$, in which the diffusion coefficient in the Smoluchowski equation has been replaced by the modified Stokes–Einstein relationship $D = kT/4\pi\eta R_S$. In these two relations R_S is the hydrodynamic radius of the diffusing particle and the constant 4 (relevant for small molecules) replaces the Stokes–Einstein coefficient of 6 (see McGlaughlin, 1959). The comparison of experimental and calculated results is presented in Table 9.8 (Osborne and Porter, 1963; North, 1964). The agreement is very good with propylene glycol, glycerol, and hydrocarbon solvents containing a low percentage of liquid paraffin. In the solvents containing a high percentage of the long-chain paraffin, it is obvious that the macroscopic viscosity is measured for long-chain molecules, having been reported for iodine atom recombinations (Booth and Noyes, 1960) and in polymerization processes (North and Reed, 1961).

Table 9.8
Quenching of Triplet Naphthalene by
1-Iodonaphthalene[a,b]

Solvent	k_{obs}	k_{calc}
Isopropanol	17	50
Propylene glycol	2.8	2.5
Glycerol	0.21	0.20
Liquid paraffin-*n*-hexane		
50:20	16	15
60:10	12	5.0
65:5	5.4	1.9
70:0	2.8	0.6

[a] k_{obs} and k_{calc} are in 10^8 liters/mole · sec.
[b] Osborne and Porter, 1963; North, 1964.

Table 9.9

Values of R_c and D_{AQ} That Give the Best Fits to Decay Curves of 1,2-benzanthracene Quenched by CBr_4 in 1,2-propanediol

CBr_4 Concentration (M)	a	10	15	20	25	30	35
				Temperature (°C)			
0.098	R_c		9.0	8.5	8.5	8.5	8.5
	D_{AQ}		0.23×10^{-6}	0.36×10^{-6}	0.50×10^{-6}	0.57×10^{-6}	0.88×10^{-6}
0.18	R_c	9.0	9.0	9.0	9.0	8.5	8.0
	D_{AQ}	0.17×10^{-6}	0.24×10^{-6}	0.32×10^{-6}	0.44×10^{-6}	0.65×10^{-6}	0.95×10^{-6}
0.29	R_c	8.5	8.0	7.5	7.0	7.5	7.0
	D_{AQ}	0.18×10^{-6}	0.28×10^{-6}	0.43×10^{-6}	0.66×10^{-6}	0.79×10^{-6}	1.1×10^{-6}

[a] R_c in Å, D_{AQ} in cm^2/sec.

The theory of diffusion-controlled reactions, based on the applications of Fick's law to a continuum model, has received considerable attention during the past several decades. The experimental aspects of this subject are by comparison quite underdeveloped. Although suggestions regarding corrections to be applied to kinetic data have been in the literature for some time, detailed experimental tests of the continuum model appear limited.

Recently Ware and Nemzek (1973, 1975) reported observing nonexponential fluorescence decay in the fluorescence quenching of 1,2-benzantharene by CBr_4 in the solvent of 1,2-propanediol, which was attributed to transient effects in diffusion-controlled reactions. It was possible to observe nonexponential decay that followed the decay law $\exp(-at - 2b\sqrt{t})$ as predicted by the continuum model. Analysis of these decay curves gave the encounter radius and pair diffusion coefficient. Reasonable values were obtained for both (see Table 9.9). Evidence was found for theground-state complex formation. When this was taken into account, good agreement between the continuum model and the experimental results was obtained.

9.4 REACTIONS IN LIQUIDS—REACTION-CONTROLLED KINETICS

9.4.1 The Transition State Theory

The majority of reactions that have been studied in solution are considerably slower than the diffusion limit for the reaction. Slow reactions are characterized by the inequality $k \ll k_D$. Many theoretical models have been proposed for the reaction-controlled kinetics (North, 1964), but here we discuss only the transition state theory.

Rates of reaction can be written in general for elementary processes whose rate is characterized by a single potential barrier. For the net rate of the forward crossing of a potential barrier, in or out of the liquid state, we have

$$\text{rate} = \kappa\left(\frac{kT}{h}\right)(C_f^{\ddagger} - C_b^{\ddagger}) \tag{9.84}$$

where $C_f^{\ddagger}$ is the concentration of activated complexes in a length along the reaction coordinate of $\delta^{\ddagger} = h(2\pi m^{\ddagger}kT)^{-1/2}$, which would be in equilibrium with reactants, and $C_b^{\ddagger}$ represents the analogous quantity for activated complexes in equilibrium with products. Here κ is the transmission coefficient (for the detailed discussion of the physical significance of κ, see Lin et al., 1971), and $m^{\ddagger}$ denotes the mass of the activated complex in motion through the distance $\delta^{\ddagger}$ atop the potential energy barrier.

The formation of the activated complex may be expressed as (Eyring et al., 1963; Eyring and Urry, 1963):

$$aA + bB + \cdots \rightarrow C^{\ddagger} + dD + \cdots \leftarrow lL + mM + \cdots \qquad (9.85)$$

where $C^{\ddagger}$ represents the activated complex and D represents a reaction intermediate. It is assumed that there is no interaction between activated complexes, since they are spatially separated in the reaction vessel and their concentrations are extremely dilute. Let λ be the absolute activity, which in the ideal system is defined as the number of molecules in the system divided by the partition function for the system (the general definition is given in eq. 9.89). For the equilibrium number of activated complexes formed from reactants $N_f^{\ddagger}$ we have

$$\lambda_f^{\ddagger} = \frac{N_f^{\ddagger}}{f^{\ddagger}} = \frac{N_f^{\ddagger}/V^{\ddagger}}{f^{\ddagger}/V} = \frac{C_f^{\ddagger}}{F_0^{\ddagger}} \qquad (9.86)$$

where $F_0^{\ddagger}$ is the partition function per unit volume for the activated state with zero reference level being the lowest energy state of the reactants. With an analogous relation for the number of activated complexes arising from products, eq. 9.84 becomes

$$\text{rate} = \kappa\left(\frac{kT}{h}\right)(\lambda_f^{\ddagger} F^{\ddagger} - \lambda_b^{\ddagger} F_0^{\ddagger}) \qquad (9.87)$$

where

$$F_0^{\ddagger} = F^{\ddagger} \exp\left(-\frac{E_a^{\ddagger}}{kT}\right) \qquad (9.88)$$

$E_a^{\ddagger}$ represents the activation energy and $F^{\ddagger}$ is the partition function per unit volume for the activated complex, $F^{\ddagger}$ being defined with the ground state of the complex as the zero of energy.

The absolute activity λ_i can be written in general as (Eyring et al., 1963)

$$\lambda_i = \exp\left(\frac{\mu_i}{kT}\right) \qquad (9.89)$$

where μ_i represents the chemical potential and is related to the Helmholtz free energy A and the Gibbs free energy G by

$$\mu_i = \left(\frac{\partial A}{\partial N_i}\right)_{T,V,N_j} = \left(\frac{\partial G}{\partial N_i}\right)_{T,P,N_j} = \cdots \qquad (9.90)$$

Using equilibrium theory to release reactants to the activated complexes crossing the barrier in the forward direction gives

$$a\mu_A + b\mu_B + \cdots = \mu_f^{\ddagger} + d\mu_D + \cdots \qquad (9.91)$$

which with eq. 9.89 becomes

$$\lambda_A^a \lambda_B^b \cdots = \lambda_b^{\ddagger} \lambda_D^d \cdots \tag{9.92}$$

The analogous relation for the backward direction is

$$\lambda_L^l \lambda_M^m \cdots = \lambda_b^{\ddagger} \lambda_D^d \cdots \tag{9.93}$$

The ground state of the reactants will be taken as the zero reference for all the absolute activities.

Using eqs. 9.91 and 9.92, we can rewrite eq. 9.87 as

$$\text{rate} = \kappa \, \frac{kT}{h} \, \frac{F_0^{\ddagger}}{\lambda_D^d \ldots} \, (\lambda_A^a \lambda_B^b \cdots - \lambda_L^l \lambda_M^m \cdots) \tag{9.94}$$

This expression for the net forward rate is a general expression that is true whether the rate process involves gaseous, liquid, or solid mixtures. With the use of eqs. 9.89 and 9.90 the rate of a reaction takes on two useful forms,

$$\text{rate} = \kappa \, \frac{kT}{h} \, \frac{F_0}{\exp[(d\mu_D + \cdots)/kT]}$$

$$\times \left[\exp\left(\frac{a\mu_A + b\mu_B + \cdots}{kT} \right) - \exp\left(\frac{l\mu_L + m\mu_M + \cdots}{kT} \right) \right] \tag{9.95}$$

and

$$\text{rate} = \kappa \, \frac{kT}{h} \, \frac{F_0^{\ddagger}}{\exp\{(1/kT)[d(\partial A/\partial C_D) + \cdots]\}}$$

$$\times \left\{ \exp\left[\frac{1}{kT} \left(a \frac{\partial A}{\partial C_A} + b \frac{\partial A}{\partial C_B} + \cdots \right) \right] \right.$$

$$\left. - \exp\left[\frac{1}{kT} \left(l \frac{\partial A}{\partial C_L} + m \frac{\partial A}{\partial C_M} + \cdots \right) \right] \right\} \tag{9.96}$$

The rate expressed in terms of chemical potentials has been shown to give rise to Onsager's reciprocal relations in the limit of small displacements from equilibrium. The rate expressions given in the forms of eqs. 9.95 and 9.96 can apply not only to reaction kinetics but also to many other rate processes (Eyring and Urry, 1963; Glasstone et al., 1940).

The rate expressed in terms of the partial derivatives of the Helmholtz free energy with respect to particle number per unit volume becomes the complete expression for reactions in solutions. The significant structure theory of liquids (Eyring and Jhon, 1969) can be applied to calculate the Helmholtz free energy and hence to discuss the reaction kinetics in solutions (Eyring and Urry, 1963).

In most reactions the reverse process in eq. 9.85 and the reaction intermediates D are negligible. In this case the rate expression reduces to

$$\text{rate} = \kappa \frac{kT}{h} F_0^{\ddagger} \lambda_A^a \lambda_B^b \cdots = \kappa \frac{kT}{h} F_0^{\ddagger} \exp \frac{a\mu_A + b\mu_B + \cdots}{kT}$$

$$= \kappa \frac{kT}{h} F_0^{\ddagger} \exp\left[\frac{1}{kT}\left(a\frac{\partial A}{\partial C_A} + b\frac{\partial A}{\partial C_B} + \cdots\right)\right] \tag{9.97}$$

In this equation it has been assumed that the concentration at the activated complexes is so dilute that the activated complexes behave ideally. If the nonideality is pronounced, this equation should be divided by the activity coefficient of the activated complexes $\gamma^{\ddagger}$, that is,

$$\text{rate} = \kappa \frac{kT}{h} \frac{F_0^{\ddagger}}{\gamma^{\ddagger}} \lambda_A^a \lambda_B^b \cdots = \kappa \frac{kT}{h} \frac{F_0^{\ddagger}}{\gamma^{\ddagger}} \exp\left(\frac{a\mu_A + b\mu_B + \cdots}{kT}\right)$$

$$= \kappa \frac{kT}{h} \frac{F_0^{\ddagger}}{\gamma^{\ddagger}} \exp\left[\frac{1}{kT}\left(a\frac{\partial A}{\partial C_A} + b\frac{\partial A}{\partial C_B} + \cdots\right)\right] \tag{9.98}$$

For the real solution, the chemical potentials μ_A, μ_B, ... are related to the activities a_A, a_B, ... (or activity coefficients γ_A, γ_B, ...) by

$$\mu_A = \mu_A^0 + kT \log a_A = \mu_A^0 + kT \log \gamma_A + kT \log C_A \tag{9.99}$$

$$\mu_B = \mu_B^0 + kT \log a_B = \mu_B^0 + kT \log \gamma_B + kT \log C_B \tag{9.100}$$

and so on, where μ_A^0, μ_B^0, ... represent the standard chemical potentials of $A, B \ldots$. Substituting eqs. 9.99 and 9.100 into eq. 9.98, we obtain the rate constant k_r as

$$k_r = \kappa \frac{kT}{h} F_0^{\ddagger} \exp\left(\frac{a\mu_A^0 + b\mu_B^0 + \cdots}{kT}\right) \frac{\gamma_A^a \gamma_B^b \cdots}{\gamma^{\ddagger}} = k_r^0 \frac{\gamma_A^a \gamma_B^b \cdots}{\gamma^{\ddagger}} \tag{9.101}$$

where

$$k_r^0 = \kappa \frac{kT}{h} F_0^{\ddagger} \exp\left(\frac{a\mu_A^0 + b\mu_B^0 + \cdots}{kT}\right) \tag{9.102}$$

which represents the rate constant in the reference state (or standard state).

The activity coefficients are defined as unity in the reference state (or standard state) and the reference state may be chosen to suit the experimental condition. For example, the dilute gas is a convenient reference state in comparing the rates between gas and liquid phases, whereas the hypothetical state of infinite dilution is convenient for ionic reactions. In the activated complex theory, the activity coefficient of the activated complex has the same meaning as the activity coefficient of any ordinary species. However, a particular activated complex is unique to a given reaction, and unlike activity coefficients of ordinary chemical species, $\gamma^{\ddagger}$ cannot be determined in one reaction and used in another. The activity coefficient of the activated complex $\gamma^{\ddagger}$ cannot be found by the usual experimental techniques such as osmotic pressure, boiling point elevation, melting point depression, or potentiometric measurements. In some cases, however, $\gamma^{\ddagger}$ can be predicted by analogy between the activated complex and stable compounds or by the use of thermodynamic theories (Weston and Schwartz, 1972; Moelwyn-Hughes, 1972).

Notice that in general from eq, 9.97 we have $k_r = \kappa(kT/h)\exp(-\Delta G^{\ddagger}/RT)$ and $\Delta G^{\ddagger} = \Delta H^{\ddagger} - T\Delta S^{\ddagger}$. Negative or very small values of $\Delta H^{\ddagger}$ are rare. They obviously cannot be associated with a single step, and they give overwhelming evidence for a multistep process that includes a preequilibrium. For example, negative or near-zero values for $\Delta H^{\ddagger}$ for a few inner-sphere and outer-sphere redox reactions indicate the importance of complexes as intermediates and rule out in these cases a single step, with a single activated complex (Wilkens, 1974).

Values for $\Delta S^{\ddagger}$ can be positive or negative. The same types of arguments used in considering the magnitude and the sign of ΔS values for a reaction can be used in interpreting the values of $\Delta S^{\ddagger}$ in the formation of the activated complex from the reactants. From general considerations, one might expect that $\Delta S^{\ddagger}$ would be more positive for reactions accompanied by topological change for a similar series of reactions that proceed with retention of configuration. This was first pointed out for the aquation of several *cis* and *trans* cobalt(III) complexes (Table 9.10). Subsequent studied have generally supported this behavior for octahedral Co(III) complexes, although there have been a couple of interesting exceptions (see Table 9.10). The relationship means that the steric course is determined in the rate-determining step and this idea can be accommodated by a dissociative type of mechanism. When the five-coordinated intermediate is trigonal-bipyramidal, marked stereochemical change can be expected, leading to the positive $\Delta S^{\ddagger}$ values. With a square-pyramidal intermediate, which differs only slightly from the original octahedron, retention of configuration and more negative $\Delta S^{\ddagger}$ values can be expected.

9.4.2 Solvent Effects

It is difficult to discuss the solvent effects on reaction rates in general; the significant structure theory of liquids may actually serve this purpose (Eyring and Jhon, 1969). Here we discuss first the effect of ionic strength on reactions between ions. According to the transition-state theory, reaction rate constants for reactions between ions vary with ionic strength in a

Table 9.10
Entropies for Activation and Stereochemical Change for Aquation
of Co^{III} Complexes at 25°

Complex	$\Delta S^{\ddagger}$	% Stereochemical change	Ref.
trans-$Coen_2Cl(OH)^+$	+20	75	c
trans-$Coen_2Cl_2^+$	+14	35	c
trans-$Co(NH_3)_4Cl_2^+$	+9	55	a
trans-$Co(RR,SS-2,3,2-tet)Cl_2^+$	+9	100	c
cis- and *trans*-$Coen_2(NO_2)Cl^+$	−2	0	c
trans-$Co(cyclam)Cl_2^+$	−3	0	c
cis- and *trans*-$Coen_2(NH_3)PO_4H_3^{3+}$	−3	0	c
cis-$Coen_2Cl_2^+$	−5	0	c
cis-$CoetrenF_2^+$	−7	0	b
trans-$Co(tet\ a)Cl_2^+$	+17	0	d
trans-$Co(3,2,3-tet)Cl_2^+$	+4	0	d

[a] R. G. Linck, *Inorg. Chem.*, **8**, 1016 (1969).
[b] K. W. Kuo and S. K. Madan, *Inorg. Chem.*, **10**, 229 (1971).
[c] M. L. Tobe, *Inorg. Chem.*, **7**, 1260 (1968).
[d] M. D. Alexander and H. G. Hamilton, Jr., *Inorg. Chem.*, **9**, 1504 (1970).

manner resembling the dependence of equilibrium constants on ionic strength. For this purpose we use the Debye–Hückel theory; for the more sophisticated models of electrolyte solutions, the specialized references should be consulted (Harned and Owen, 1958; Eyring et al., 1963; Rice and Nagasawa, 1961).

The activity coefficient of an ion depends on the extent to which the free energy G of the ion differs from some assumed ideal behavior G_{ideal}:

$$\Delta G = G - G_{ideal} = kT \log \gamma \qquad (9.103)$$

The part of the free energy of an ion with radius a that may be attributed to its charge ze may be found from the work done to charge the ion from zero to ze. The contribution to G from an element of charge dq is

$$dG = dqV(q, a) \qquad (9.104)$$

where $V(q, a)$ represents the potential function acting on the ion. The potential function must satisfy Poisson's equation, which is, for a spherically symmetric charge distribution:

$$\frac{1}{r^2}\frac{d}{dr}\left[r^2\frac{dV(r)}{dr}\right] = -\frac{4\pi\delta}{\varepsilon_0} \qquad (9.105)$$

where ε_0 is the dielectric constant and δ is the charge density at r. For an isolated ion A, δ is zero in the space surrounding the ion and eq. 9.105 reduces to Coulomb's law. When other ions are present, δ will not be zero in the space surrounding the ion A because there will be a tendency for ions of opposite sign to concentrate near A. In the Debye–Hückel theory it is assumed that all ions other than the ion A are present as a structureless continuum around A, which allows the system to preserve spherical symmetry. The concentration of any type of ion at r from A can be determined from Boltzmann's law

$$C_i = C_i^0 \exp\left[\frac{-z_i e V(r)}{kT}\right] \tag{9.106}$$

where C_i is the number of ions per unit volume at r, C_i^0 is the concentration of the ith ion at large r, where $V(r)$ is zero, and $z_i e$ is the charge on the ion. The charge density δ is obtained by summing $C_i z_i e$ over all ions present:

$$\delta = \delta(r) = \sum_i C_i z_i e = \sum_i C_i^0 z_i e \exp\left[\frac{-z_i e V(r)}{kT}\right] \tag{9.107}$$

Substituting eq. 9.107 into eq. 9.105 yields

$$\frac{1}{r^2}\frac{d}{dr}\left[r^2\frac{dV(r)}{dr}\right] = -\frac{4\pi e}{\varepsilon_0}\sum_i C_i^0 z_i \exp\left(-\frac{z_i e V(r)}{kT}\right) \tag{9.108}$$

To facilitate the solution of eq. 9.108, Debye and Hückel made use of the approximation $\exp(-x) \doteq 1 - x$. Using this relation, eq. 9.108 becomes

$$\frac{1}{r^2}\frac{d}{dr}\left[r^2\frac{dV(r)}{dr}\right] = -\frac{4\pi e}{\varepsilon_0}\left[\sum_i C_i z_i - \sum_i \frac{C_i^0 z_i^2 e}{kT}V(r)\right] \tag{9.109}$$

The first sum on the right-hand side of eq. 9.109 is zero because the total number of positive and negative charges must balance. Thus

$$\frac{1}{r^2}\frac{d}{dr}\left[r^2\frac{dV(r)}{dr}\right] = \frac{V(r)}{r_0^2} \tag{9.110}$$

where

$$\frac{1}{r_0^2} = \frac{4\pi e^2}{\varepsilon_0 kT}\sum_i C_i^0 z_i^2 \tag{9.111}$$

The constant r_0 has the dimensions of length and is sometimes referred to as the radius of the ionic atmosphere.

Now the ionic strength μ of the solution is defined as

$$\mu = \frac{1}{2}\sum_i m_i z_i^2 \tag{9.112}$$

where m_i is the molar concentration of ion i. Consequently,

$$\sum_i C_i^0 z_i^2 = \frac{2N_A \, d\mu}{1000} \qquad (9.113)$$

where N_A is Avogadro's number and d is the density of the solution. Substituting eq. 9.113 into 9.111, we obtain

$$\frac{1}{r_0^2} = \frac{8\pi e^2 N_A \, d\mu}{1000\varepsilon_0 kT} \qquad (9.114)$$

For example, in water at 25°C, $r_0^{-2} = 1.08 \times 10^{15} \, \mu \, \text{cm}^{-2}$.

The differential equation 9.110 can be solved easily; the general solution is given by

$$V(r) = \frac{A_1}{r} \exp\left(-\frac{r}{r_0}\right) + \frac{A_2}{r} \exp\left(\frac{r}{r_0}\right) \qquad (9.115)$$

where A_1 and A_2 are integration constants. The constant A_2 is zero because $V(r)$ approaches zero as $r \to \infty$. If the ion A is considered to be a point charge, A_1 may be evaluated by noting that $V(r)$ must approach the value computed for an isolated charge (Coulomb's law) as r approaches zero, that is,

$$\lim_{r \to 0} V(r) = \lim_{r \to 0} \frac{A_1}{r} \exp\left(-\frac{r}{r_0}\right) = \frac{z_A e}{\varepsilon_0 r} \qquad (9.116)$$

which yields $A_1 = z_A e/\varepsilon_0$ and the potential $V(r)$ is

$$V(r) = \left(\frac{z_A e}{\varepsilon_0 r}\right) \exp\left(-\frac{r}{r_0}\right) \qquad (9.117)$$

If the ion A is of finite size with radius a, the integration constant A in eq. 9.115 should be evaluated at the surface of the ion A (Harned and Owen, 1958), in which case it is

$$A_1 = \frac{z_A e}{\varepsilon_0} \frac{\exp(a/r_0)}{1 + (a/r_0)} \qquad (9.118)$$

It should be noted that if a, the radius of the ion A, is 3 Å, then $a/r_0 = \sqrt{\mu}$, in water at 25°C.

Equation 9.117 is valid only for large r_0, so nothing is lost by expanding the exponential to give

$$V(r) = \left(\frac{z_A e}{\varepsilon_0 r}\right) - \left(\frac{z_A e}{\varepsilon_0 r_0}\right) \qquad (9.119)$$

Thus we obtain an approximate expression for the potential $V(r)$ describing an ion and its ionic atmosphere. The first term on the right-hand side of eq.

9.119 is simply the potential of the isolated ion. The contribution to $V(r)$ from the ionic atmosphere is represented as a depression of the potential by a constant amount, independent of the radius. Of course this cannot be correct at large r, since $V(\infty)$ must be zero. Fortunately, however, we are only concerned with the immediate neighborhood of the ion A.

Substituting eq. 9.119 into eq. 9.104, we obtain

$$dG = q\left(\frac{1}{\varepsilon_0 a} - \frac{1}{\varepsilon_0 r_0}\right) dq \qquad (9.120)$$

so that

$$G = G_0 + \int_0^{zj} q \, dq \left(\frac{1}{\varepsilon_0 a} - \frac{1}{\varepsilon_0 r_0}\right) = G_0 + \frac{z^2 e^2}{2\varepsilon_0 a} - \frac{z^2 e^2}{2\varepsilon_0 r_0} \qquad (9.121)$$

where G_0 represents all parts of the free energy not due to the interaction of the charge with the medium. If we choose the reference state as an infinitely dilute solution of ions, then $1/r_0 = 0$ and we have

$$G_{\text{ideal}} = G_0 + \left(\frac{z^2 e^2}{2\varepsilon_0 a}\right) \qquad (9.122)$$

or

$$\Delta G = -\frac{z^2 e^2}{2\varepsilon_0 r_0} \qquad (9.123)$$

Combining eq. 9.123 with eq. 9.103, we obtain

$$\log \gamma = \frac{-z^2 e^2}{2\varepsilon_0 r_0 kT} = -\frac{z^2 2^2}{2\varepsilon_0 kT}\left(\frac{8\pi N_A e^2 d}{1000\varepsilon_0 kT}\right)^{1/2} \sqrt{\mu} \qquad (9.124)$$

In other words, $\log \gamma$ varies linearly with $\sqrt{\mu}$. In aqueous solution at 25°C, eq. 9.124 can be written

$$\log \gamma = -0.51 z^2 \sqrt{\mu} \qquad (9.125)$$

Equation 9.124 or eq. 9.125 is true for each ion in dilute electrolyte solution, that is,

$$\log \gamma_i = -\beta z_i^2 \sqrt{\mu} \qquad (9.126)$$

where $\beta = (e^2/2\varepsilon_0 kT)(8\pi N_A e^2 d/1000\varepsilon_0 kT)^{1/2}$. The effect of ionic strength on reactions between ions can be seen by substituting eq. 9.126 into eq. 9.101, that is,

$$\log k_r = \log k_r^0 + a \log \gamma_A + b \log \gamma_B + \cdots - \log \gamma^{\ddagger}$$
$$= \log k_r^0 + \beta\sqrt{\mu}(-1)[az_A^2 + bz_B^2 + \cdots - (z_A + z_B + \cdots)^2]$$

$$(9.127)$$

In particular, for second-order reactions, $a = 1$, $b = 1$, and eq. 9.127 reduces to

$$\log k_r = \log k_r^0 + 2\beta z_A z_B \sqrt{\mu} \qquad (9.128)$$

Notice that in water at 25°C, $\beta = 0.51$. A somewhat more precise form of eq. 9.128 for ions of finite size is

$$\log k_r = \log k_r^0 + 2\beta_A z_B \left(\frac{\sqrt{\mu}}{1 + \sqrt{\mu}}\right) \qquad (9.129)$$

by using eq. 9.118, $a/r_0 \div \sqrt{\mu}$, and repeating the foregoing derivation.

From eq. 9.128 or eq. 9.129 we can see that the effect of ionic strength on rate constants depends on both the magnitude of the charges on the reactants and the signs of the charges. Rate constants increase with ionic strength for reactions between ions of the same sign, decrease with ionic strength for reactions between ions of opposite sign, and are unaffected by ionic strength for reactions in which one species is uncharged. To demonstrate quantitatively the effect of ionic strength on rate constants, Fig. 9.1 shows the variation of some rate constants with μ (Weston and Schwarz, 1972). The lines drawn in the figure are the theoretical slopes for the different reactions pairs as given by eq. 9.129. Most of the reactions in Fig. 9.1 are much slower than the diffusion limit; however, one diffusion-controlled reaction, e_{aq} and NO_2^-, is included. From Fig. 9.1 we can see that the agreement with the theory is quite good, particularly at low ionic strength; there is some specific effect of SO_4^{2-} on the $OH\text{-}CO(NH_3)_5Br^{2+}$ reaction; and the ionic strength effect on the inversion of sucrose is not quite nil, as is expected from eq. 9.130.

The specific effect of SO_4^{2-} ions probably results from the formation of an ion-pair complex, $CO(NH_3)_5Br^{2+}SO_4^{2-}$, which reacts less readily with the OH^- ion because it is uncharged (Davies, 1961). The formation of this complex is highly favored, for both ions are double charged. Such specific ion effects are a constant danger in ionic strength experiments. Whenever possible, one should avoid using multiple charged ions to vary the ionic strength and should avoid using ionic strengths above 0.5 μ. At high concentrations ionic strength effects may disappear altogether or may even be reversed (Moelwyn-Hughes, 1972).

The effect of ionic strength on the sucrose inversion is an order of magnitude smaller than would be found if sucrose had a negative charge; still, it is real and is due to the large dipole moment of sucrose. Dipolar interactions were not considered in the derivation of the Debye–Hückel theory; the effect of dipolar interactions on reaction rates is discussed in detail by Moelwyn-Hughes (1972).

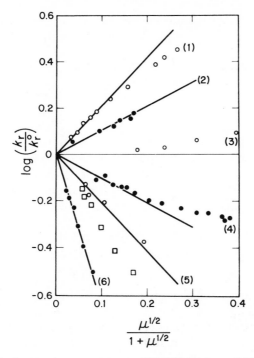

Fig. 9.1 Effect of ionic strength on rate constants. (1) $BrCH_2COO^- + S_2O_3^{2-}$; (2) $e_{aq} + NO_2^-$; (3) $H_3O^+ + C_{12}H_{22}O_{11}$ (inversion of sucrose); (4) $H_3O^+ + Br^- + H_2O_2$; (5) $OH^- + Co(NH_3)_5Br^{2+}$ [ionic strength varied with NaBr (circles) and Na_2SO_4 (squares)]; (6) $Fe(H_2O)_6^{2+} + Co(C_2O_4)_3^{3-}$. Data from LaMer and Fessenden (1932), Czapski and Schwarz (1962), Kautz and Robinson (1928), Livingston (1926), Olson and Simonson (1949), Barnett and Baxendale (1956).

Next we consider the effect of solvent on reactions between ions at negligibly low ionic strength. In this case, eq. 9.121 reduces to

$$G = G_0 + \left(\frac{z^2 e^2}{2\varepsilon_0 a} \right) \qquad (9.130)$$

as $1/r_0 = 0$ because of $\mu = 0$. Choosing the reference state as a solvent of infinite dielectric constant, that is, $G_0 = G_{ideal}$, we have

$$\Delta G = \frac{z^2 e^2}{2\varepsilon_0 a} \qquad (9.131)$$

and the corresponding activity coefficient is given by

$$\log \gamma = \frac{z^2 e^2}{2\varepsilon_0 a k T} \qquad (9.132)$$

Thus the effect of solvent on reactions between ions at negligibly low ionic strength can be seen by substituting eq. 9.132 into eq. 9.101:

$$\log k_r = \log k_r^0 + \frac{e^2}{2\varepsilon_0 kT}\left[\frac{az_A^2}{a_A} + \frac{bz_B^2}{a_B} + \cdots - \frac{(z_A + z_B)^2}{a^\ddagger}\right] \quad (9.133)$$

For second-order reactions, $a = b = 1$, and eq. 9.133 becomes

$$\log k_r = \log k_r^0 + \frac{e^2}{2\varepsilon_0 kT}\left[\frac{z_A^2}{a_A} + \frac{z_B^2}{a_B} - \frac{(z_A + z_B)^2}{a^\ddagger}\right] \quad (9.134)$$

In other words, $\log k_r$ varies linearly with respect to $1/\varepsilon_0$.

We cannot make an a priori estimate of k_r^0 for any actual reaction, but eq. 9.133 does predict the dependence of k_r on the dielectric constant of the medium if it is assumed that k_r^0 is independent of the properties of the solvent. If we can replace a_A, a_B, and $a^\ddagger$ in eq. 9.134 by an average value R, then eq. 9.134 becomes

$$\log\left(\frac{k_r}{k_r^0}\right) = -\frac{z_A z_B e^2}{\varepsilon_0 kTR} \quad (9.135)$$

The term R is often called the reaction radius. The dielectric constant can be adjusted by changing the solvent composition. For example, the dielectric constant of a mixture of water and methanol can be varied continuously from 78 in pure water to 32 in pure methanol. Equation 9.135 indicates that the logarithm of the rate constant of an ionic reaction should vary inversely with dielectric constant, and indeed it often does over a limited range of composition (Amis, 1966). The slope of the line should yield a value for R that is of the magnitude of atomic or molecular dimensions. Some typical data are given in Table 9.11 (Amis, 1966).

Equation 9.133 is only valid over a limited range of solvent composition, because its derivation neglects all but electrostatic interactions and the molecular nature of the solvent. Both of these are oversimplifications of the actual ion-solvent interaction. For example, the proton is a much different

Table 9.11
Reaction Radii for Ionic Reactions

Reaction	Solvent	$R(\text{Å})$
Tetrabromophenolphthalein + OH$^-$	H$_2$O + EtOH	1.2
	H$_2$O + MeOH	1.5
BrCH$_3$COO$^-$ + S$_2$O$_3^{2-}$	H$_2$O + sucrose	5.1
(CH$_3$)$_3$S$^+$ + OH$^-$	H$_2$O + EtOH	1.4

species in water than it is in methanol. In aqueous solution, the proton is usually formulated as H_3O^+ to indicate intimate association with at least one solvent molecule, whereas in methanol it forms $CH_3OH_2^+$. The free energy difference between H_3O^+ and $CH_3OH_2^+$ cannot be accounted for by classical considerations like Coulomb's law. Similar, though weaker, associations take place between other ions and solvents.

Next we discuss how to apply the significant structure theory of liquids, to study the solvent effects on reaction rates. In the significant structure theory of liquids, liquids are regarded as having an excess volume that for simple liquids is proportional to the difference in molar volumes of the liquid and the solid, that is, the excess volume is $V - V_s$. Whereas solids have a limited number of static vacancies, liquids have many fluidized vacancies, resulting in soft space in which there is a potential hole of molecular size. Dynamic vacancies of molecular size are most likely to occur, since they leave all but nearest neighbors of the vacancy relatively undisturbed. In order to utilize the potential hole of molecular size, a nearest neighbor must have sufficient kinetic energy to push back the other nearest neighbors; that is, it must have enough kinetic energy to equal or exceed the mean kinetic energy that the other nearest neighbors gained by the presence of the hole, except that some slight potential energy decrease aids the change. In the soft spaces of the liquid a dynamic vacancy should move through the liquid about as freely as a molecule moves in the gas. Hence one expects the hole to have associated with it about the same heat and entropy as does a vapor molecule.

In view of the foregoing discussion, the partition function f_N for a mole of pure liquid can be written as (Eyring and Jhon, 1969)

$$f_N = (f_s')^{NV_s/V}(f_g)^{N(V - V_s)/V} \tag{9.136}$$

where N is Avogadro's number and f_s' and f_g are the partition functions of solidlike and gaslike degrees of freedom, respectively. In deriving f_s' we have to consider a positional degeneracy factor that multiplies the usual partition function for a solid. As stated earlier, if a molecule is to have access to fluidized vacancies, it must push the competing neighboring molecules aside. When the molecule has the required energy, the additional site becomes available to it and there is a degeneracy factor equal to the number of such sites made available plus the original site. The number of additional sites will be equal to the number of vacancies around a solidlike molecule multiplied by the probability that the molecule has the required energy E_h/N to move into a site. Thus the number of additional sites is $n_h \exp(-E_h/RT)$. Here n_h represents the number of vacancies, $n_h = n(V - V_s)/V_s$; n must be chosen so that n_h is equal to the number of neighboring vacancies at the melting point (e.g., for close packing, the number of nearest neighbors is 12, and $V/V_s = 1.12$

at the melting, which leads us to $n = 12 \times 1.12 = 10.7$), and E_h should be inversely proportional to the number of vacancies and directly proportional to the energy of sublimation of the solid E_s, that is, $E_h = aE_s V_s/(V - V_s)$, where a is a proportionality constant that can be calculated theoretically (Eyring and Jhon, 1969).

The total number of positions available to a given molecule is $1 + n_h \exp(-E_h/RT)$. If we assume that an Einstein oscillator model is an adequate representation of the lattice vibrational degree of freedom of the solidlike molecules at the temperature of interest, we can write f'_s for a liquid as follows:

$$f'_s = \frac{\exp(E_s/RT]}{[1 - \exp(-\theta/T]^3} f^{(s)}_{rot} \left[\prod_i \frac{1}{1 - \exp(-hv_i/kT)} \right] (1 + n_h e^{-E_h/RT}) \tag{9.137}$$

where $f^{(s)}_{rot}$ represents the molecular rotational partition function of the molecule in the solid; it could be free rotation or hindered rotation. Here θ is the Einstein characteristic temperature, E_s is the sublimation energy, and the v_i is the normal vibrational frequencies of the molecules.

For the partition function of the gaslike degrees of freedom f_g we use the nonlocalized independent ideal gas partition function for the $N(V - V_s)/V$ gaslike molecules moving in the excess volume $V - V_s$. This gives us

$$f_g^{N(V - V_s)/V} = \left[f^{(g)}_{rot} f_{vib} \left(\frac{2\pi mkT}{h^2} \right)^{3/2} \frac{eV}{N} \right]^{N(V - V_s)/V} \tag{9.138}$$

Here Stirling's approximation $n! = (n/e)^n$ has been used.

In terms of the canonical partition function f_N, we can calculate the Helmholtz free energy from

$$A = -kT \log f_N = -kT \left[\frac{NV_s}{V} \log f'_s + \frac{N(V - V_s)}{V} \log f_g \right] \tag{9.139}$$

Other thermodynamic quantities can be calculated by using the thermodynamic relations. For example, the absolute activity (or the chemical potential) of the pure liquid can be computed as

$$\log \lambda = \frac{\mu}{kT} = \frac{1}{kT} \left(\frac{\partial A}{\partial N} \right)_{T,V} = - \left[\frac{V_s}{V} \log f'_s + \frac{V - V_s}{V} \log \frac{f_g}{e} \right] \tag{9.140}$$

The foregoing method can be generalized to liquid mixtures as is required in calculating chemical potentials (or absolute activities) for reaction rates in solution. For this purpose we discuss a binary mixture as an example and the multicomponent mixtures can be discussed similarly. Notice that according to the significant structure theory of liquids, there are three significant structures in a liquid: (1) molecules with solidlike degrees of freedom, (2) positional degeneracy in the solidlike structure, and (3) molecules with gaslike degrees of freedom. In other words, eq. 9.136 can be written as

$$f_N = (f_s e^{E_s/RT})^{NV_s/V} (f_{deg})^{NV_s/V} (f_g)^{N(V - V_s)V}$$ (9.141)

where

$$f_s = \frac{1}{(1 - e^{-\theta/T})^3} f_{rot}^{(s)} f_{vib}$$ (9.142)

and

$$f_{deg} = 1 + n_h \exp\left(-\frac{E_h}{RT}\right)$$ (9.143)

To extend the theory to binary mixtures we must consider the concentration dependence of the parameters. Thus the following assumptions are made: (1) Nonrandom mixing is negligible; (2) the same characteristic temperatures of vibration θ are retained for mixtures that are used for the pure substances; (3) molecules of both components continue to possess their gaslike translational degrees of freedom; (4) the degeneracy term has the same form as that for a pure liquid; and (5) the parameters E_s, V_s, n, and a may be taken as suitable averages of the parameters for the pure components. With these assumptions, the partition function for a binary mixture $f_{N_1 N_2}$ takes the following form:

$$f_{N_1 N_2} = \frac{(N_1 + N_2)!}{N_1! N_2} (f_{s1}^{N_1} f_{s2}^{N_2})^{V_s/V} (f_{deg} e^{E_s/RT})^{(N_1 + N_2)V_s/V} (f_{g1}^{N_1} f_{g2}^{N_2})^{(V - V_s)/V}$$ (9.144)

where

$$f_{si} = \frac{1}{[1 - \exp(-\theta_i/T)]^3} f_{rot, i}^{(s)} f_{vib, i}$$ (9.145)

$$f_{gi} = \left(\frac{2\pi M_i kT}{h^2}\right)^{3/2} f_{rot, i}^{(g)} f_{vib, i} \frac{eV}{N_1 + N_2}$$ (9.146)

and

$$f_{deg} = 1 + n_h e^{-E^h/RT}$$ (9.147)

$$E_s = X_1^2 E_{st} + X_2^2 E_{s_2} + 2(1 + \sqrt{E}) X_1 X_2 (E_{s_1} E_{s_2})^{1/2}$$ (9.148)

$$V_s = X_1 V_{st} + X_2 V_{s_2} + X_1 X_2 \sqrt{V} (V_{st} V_{s_2})^{1/2}$$ (9.149)

$$n = X_1 n_1 + X_2 n_2$$ (9.150)

$$a = X_1 a_1 + X_2 a_2$$ (9.151)

The X_i represents the mole fractions. The quantities $\sqrt{E}$ and $\sqrt{V}$ are the only parameters in the mixture partition function that are not evaluated from the pure liquids. The quantities $\sqrt{E}$ and $\sqrt{V}$ are the correction parameters used in the cross terms and in the terms containing higher powers of the concentrations. In general, these values are very small, and reasonable results are obtained when they are taken to be zero.

Using eq. 9.144, the chemical potential for component 1 can be calculated as follows:

$$
\begin{aligned}
\mu_1 = \left(\frac{\partial A}{\partial N_1}\right)_{T,V,N} &= kT \log \lambda_1 = -kT\bigg\{ - \log X_1 \\
&+ \left[\frac{V_s}{V} \log\left(f_{s_1} f_{\text{deg}} \exp \frac{E_s}{RT}\right) + \frac{V - V_s}{V} \log \frac{f_{g_1}}{e}\right] \\
&+ \frac{(N_1 + N_2)}{V} \frac{\partial V_s}{\partial N_1}\left(x_1 \log \frac{f_{s_1}}{f_{g_1}} + X_2 \log \frac{f_{s_2}}{f_{g_2}} + \log f_{\text{deg}} \exp \frac{E_s}{RT}\right) \\
&+ \frac{V_s}{V}(N_1 + N_2) \frac{\partial}{\partial N_1}\left(\log f_{\text{deg}} \exp \frac{E_s}{RT}\right)\bigg\}
\end{aligned}
\tag{9.152}
$$

Equation 9.152 can be written as

$$
\mu_1 = \mu_1^0 + kT \log X_1 + kT \log Q_1
\tag{9.153}
$$

where μ_1^0 represents the chemical potential of pure component 1,

$$
\mu_1^0 = -kT\left[\frac{V_s}{V} \log\left(f_s f_{\text{deg}} \exp \frac{E_s}{RT}\right) + \frac{V - V_s}{V} \log \frac{f_{g_1}}{e}\right]
\tag{9.154}
$$

and Q_1 is defined by

$$
\begin{aligned}
\log Q_1 = &-\frac{(N_1 + N_2)}{V} \frac{\partial V_s}{\partial N_1} \\
&\times \left(X_1 \log \frac{f_{s_1}}{f_{g_1}} + X_2 \log \frac{f_{s_2}}{f_{g_2}} + \log f_{\text{deg}} \exp\left(\frac{E_s}{RT}\right)\right) \\
&- (N_1 + N_2) \frac{V_s}{V} \frac{\partial}{\partial N_1}\left(\log f_{\text{deg}} \exp \frac{E_s}{RT}\right)
\end{aligned}
\tag{9.155}
$$

or

$$
\begin{aligned}
Q_1 = &\left\{\exp\left[-(N_1 + N_2) \frac{V_s}{V} \frac{\partial}{\partial N_1}\left(\log f_{\text{deg}} \exp\left(\frac{E_s}{RT}\right)\right)\right]\right\} \\
&\times \left\{\frac{[f_{s_1} f_{\text{deg}} \exp(E_s/RT)]^{(N_1/V)(\partial V_s/\partial N_1)}}{f_{g_1}}\right. \\
&\times \left.\frac{[f_{s_2} f_{\text{deg}} \exp(E_s/RT)]^{(N_2/V)(\partial V_s/\partial N_2)}}{f_{g_2}}\right\}^{-1}
\end{aligned}
\tag{9.156}
$$

To calculate Q_1 the following relations can be used:

$$\frac{\partial V_s}{\partial N_1} = \frac{X_2}{N_1 + N_2} [V_{s_1} - V_{s_2} + (X_2 - X_1)\sqrt{V}(V_{s_1}V_{s_2})^{1/2}] \quad (9.157)$$

$$\frac{\partial n}{\partial N_1} = \frac{X_2}{N_1 + N_2} (n_1 - n_2); \quad \frac{\partial a}{\partial N_1} = \frac{X_2}{N_1 + N_2} (a_1 - a_2) \quad (9.158)$$

and

$$\frac{\partial E_s}{\partial N_1} = \frac{2X_2}{N_1 + N_2} [(X_1 E_{s_1} - X_2 E_{s_2}) + (1 + \sqrt{E})(X_2 - X_1)(E_{s_1}E_{s_2})^{1/2}]$$

$$(9.159)$$

A similar equation for the chemical potential of component 2 can be obtained. Equation 9.153 indicates that if we choose the pure substance as the reference state, then Q_1 plays the role of the activity coefficient for component 1. This derivation can be generalized to multicomponent systems. There has been no application of the significant structure theory of liquids to the effect of solvents on reaction rates quantitatively.

We can arbitrarily divide solvents into three categories: protic, including both proton donors and acceptors; dipolar aprotic, solvent with dielectric constant larger than 15 but without hydrogen capable of forming hydrogen bonds; and aprotic, having neither acidic nor basic properties—for example, CCl_4. These may be expected to interact in widely different ways with complex ions containing large internal charges. There are basically two ways in which the solvent may be regarded, although assessing their distinction and relative importance is very difficult.

1. The solvent may be regarded as an "inert" medium. In this case, the dielectric constant of the solvent is the important parameter, and the effect can be semiquantitatively evaluated for ion-ion or ion-dipolar reactant mixtures, where electrostatic considerations dominate.

2. The solvent may act as a nucleophile and an active participator in the reaction. It is extremely difficult to assess the function of the solvent in solvolysis reactions. Some attempts to define the mechanism for the replacement of ligand by solvent in octahedral complexes have been made using mixed solvents and the solvating power concept.

The rate of solvolysis (k_s) of tert-butyl chloride in a particular solvent S compared with the rate (k_0) in 80% v/v aqueous ethanol is used as a measure of that solvent's ionizing power Y_s:

$$Y_s = \log \frac{k_s}{k_0}$$

For any other substrate acting by an $S_N 1$ mechanism, it might be expected that

$$\log \frac{k_s}{k_0} = mY_s$$

where m depends on the substrate and equals 1.0 for t-BuCl. Such an expression holds for the aquation of a number of cobalt(III) complexes in a variety of mixed aqueous solvents. The value of m is in the range 0.23–0.36; and although this much reduced value, compared with the value of t-BuCl, can be rationalized in terms of a dissociative mechanism, it is apparent that more data and patterns of behavior are required before the concept is likely to be diagnostic of the mechanism. The Y_s parameter has been used in drawing conclusions about the mechanisms of isomerization of *trans*-$Co(C_2O_4)_2$-$(H_2O)_2$ — in aqueous mixtures with a number of solvents (Wilkins, 1974).

9.4.3   Pressure Effect

From eq. 9.101 the thermodynamic formulation of the transition-state theory may be written as follows (Glasstone et al., 1940):

$$k_r = \kappa \left(\frac{kT}{h}\right)\exp\left(\frac{-\Delta G^{\ddagger}}{kT}\right) \qquad (9.160)$$

where $\Delta G^{\ddagger} = -a\mu_A^0 - b\mu_B^0 - \cdots + \mu^{\ddagger 0}$. The effect of pressure on k_r may be obtained by taking the partial derivative of $\log k_r$ with respect to pressure P at constant temperature T, that is,

$$\left(\frac{\partial \log kr}{\partial P}\right)_T = -\frac{1}{kT}\left(\frac{\partial \Delta G^{\ddagger}}{\partial P}\right)_T \qquad (9.161)$$

Since $(\partial \Delta G / \partial P)_T = \Delta V$, we have

$$\left(\frac{\partial \log k_r}{\partial P}\right)_T = -\frac{\Delta V^{\ddagger}}{RT} \qquad (9.162)$$

where $\Delta V^{\ddagger}$ represents the difference in molar volume between the reactants and the activated complex, $\Delta V^{\ddagger} = -(aV_A + bV_B + \cdots) + V^{\ddagger}$

To demonstrate the dependence of k_r on pressure, Fig. 9.2 shows the plot of $\log k_r$ versus P for two reactions (LeNoble, 1967).

$$t\text{-}C_5H_{11}(CH_3)_3N^+ + OH^- \longrightarrow (CH_3)_3N + H_2O + CH_3C_4H_8$$

and

$$t\text{-}C_5H_{11}Cl + C_2H_5OH \longrightarrow t\text{-}C_5H_{11}OH + C_2H_5Cl$$

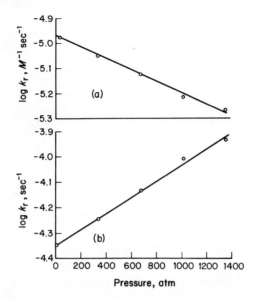

Fig. 9.2 Pressure effect on rate constants in solution. (*a*) (t-C_5H_{11})(CH_3)$_3$N$^+$ + OH$^-$ → (CH_3)$_3$N + H_2O + methylbutene, in ethanol at 85°C, $\Delta V^\ddagger = 15$ cm^3/mole (*b*) t-C_5H_{11}Cl + C_2H_5OH → t-C_5H_{11}OH + C_2H_5Cl, in 80% aqueous ethanol at 34°C, $\Delta V^\ddagger = -18$ cm^3/mole. (Data from Brower and Chen, 1965.)

where $CH_3C_4H_8$ is methylbutene. The linear dependence of log k_r on pressure is good in these examples, but often there is a pronounced curvature; in that case, it means that $\Delta V^\ddagger$ depends on pressure significantly and the slope of the log k_r versus P plot must be estimated at various pressures.

The volume of activation can be a useful quantity in determining the reaction mechanism (LeNoble, 1967). For example, consider a unimolecular reaction proceeding by the breaking of a bond. If the bond is represented as a cylinder with a volume $\pi R^2 l$, where R is a van der Waals radius of about 2 Å and l is the bond length, then in the activated complex in which the bond is stretched by about 1 Å, the volume increase $\Delta V^\ddagger$ on activation would be about 12 Å^3/molecule or 7 cm^3/mole. This is demonstrated in Table 9.12 by some bond cleavage reactions (LeNoble, 1967).

Volume changes of the solvent resulting from the reaction are also included in $\Delta V^\ddagger$. In the process of ionization there is a charge separation in the activated complex that is not present in the parent molecule. The presence of a charge induces a contraction of the solvent called electrostriction, so that the volume of activation should be negative. Typical magnitudes are -10 to -20 cm^3/mole (LeNoble, 1967). Typically, $\Delta V^\ddagger$ for bond formation is

Table 9.12
Volume of Activation $\Delta V^{\ddagger}$

Reaction	Solvent	$\Delta V^{\ddagger}$ (cm^3/mole)
$(CH_3)_3COOC(CH_3)_3 \rightarrow 2(CH_3)_3CO$	Cyclohexane	$+6.7$
$(CH_3)_2C-N=N-C(CH_3)_2 \rightarrow 2(CH_3)CN + N_2$	Toluene	$+3.8$
$C_6H_5COO \cdot OOCC_6H_5 \rightarrow 2C_6H_5COO$	CCl_4	$+11$
$2C_5H_6(\text{cyclopentadiene}) \rightarrow C_{10}H_{12}$	n-C_4H_9Cl	-22
$OH^- + (C_2H_5)_3S^+ \rightarrow C_2H_5OH + (C_2H_5)_2S$	Water	$+10$
$OH^- + N(C_2H_5)_4^+ \rightarrow CH_3OC_2H_5 + N(C_2H_5)_3$	Methanol	$+20$

-10 to -20 cm^3/mole and for neutralization is $+10$ to $+20$ cm^3/mole (cf. Table 9.12). It should be noted that volume changes for the reverse process may be found by noticing that the same activated complex is reached as in the forward reaction. Thus, by conservation of volume, we obtain a useful relation

$$\Delta V_{\text{forward}}^{\ddagger} - \Delta V_{\text{reverse}}^{\ddagger} = \Sigma V_{\text{products}} - \Sigma V_{\text{reactants}} \qquad (9.163)$$

The molar volumes of the products and reactants are generally known from density measurements.

The volume of activation $\Delta V^{\ddagger}$ can be considered as made up of two parts, (1) the direct volume change when reactants are transferred into the transition complex and (2) the attendant volume changes in the surrounding solvent when the transition state is formed (electrostriction of solvent). Component 1 is less important in reactions in which the charge types of reactant and product are the same, as in exchange reactions. It is usually positive and small in dissociative mechanisms.

Consider the effect of pressure on the exchange rate constant k, for the reaction (Stranks and Swaddle, 1971)

$$Cr(H_2O)_6^{3+} + H_2^{18}O \longrightarrow Cr(H_2O)_5(H_2^{18}O)^{3+} + H_2O$$

The value of $\Delta V^{\ddagger}$ is -9.3 ± 0.3 cm^3/mole. The markedly smaller volume of the reactants is interpreted in terms of an associative intercharge mechanism in which water in the second coordination sphere or solvation sheath is interchanged with water in the first coordination sphere via an activated process of this type

$$Cr(H_2O)_6 \cdot (H_2O)_x^{3+} \longrightarrow [Cr(H_2O)_7 \cdot (H_2O)_{x-1}^{3+}]^{\ddagger}$$

The species $Cr(H_2O)_6^{3+}$ and $Cr(H_2O)_7^{3+}$ are considered to have the same volume, so that the vacancy created in the solvation sheath leads to the negative $\Delta V^{\ddagger}$ value. A direct associative process appears to be ruled out because bulk and coordinated waters differ in their compressibility and therefore a pressure-dependent $\Delta V^{\ddagger}$ would be expected. Both first and second coordination sphere water would be expected to be relatively incompressible and therefore the $\Delta V^{\ddagger}$ value would likely be pressure independent, as is observed.

Hunt and Taube (1958) in an early study found $\Delta V^{\ddagger} = +1.2 \text{ cm}^3/\text{mole}$ for

$$Co(NH_3)_5H_2O^{3+} + H_2^{18}O \longrightarrow Co(NH_3)_5H_2^{18}O^{3+} + H_2O$$

Since coordinated water occupies less volume than bulk water, this result can be interpreted in terms of a dissociative mechanism. This is now well established (Stranks and Swaddle 1971).

Comparison of some parameter of the activated complex with the same parameter of reactants or products has been a continuous theme in this section. Thus if $\Delta V^{\ddagger}$ can be compared with ΔV for the overall reaction, some estimate may be possible of the relative structures of product and activated complex. Such a comparison has been made with rewarding results for the equation of a series of Co(III) complexes (Table 9.13). Such a result is entirely consistent with the dissociative mechanism proposed for the reactions. The final products have almost been formed in reaching the activated complex. Water from the solution sheath simply has to transfer to the first coordination sphere of the Co(III), and the volume change associated with this is quite small. For the corresponding reaction of $Cr(NH_3)_5X^{n+}$, $X = Cl^-$, Br^-,

Table 9.13
Volumes of Activation and Volumes of
Reaction for the Aquation of
$Co(NH_3)_5X^{(3-n)+}$ at 25°, $= 0.1 M^a$

X^{n-}	$\Delta V_0^{\ddagger}$ cm^3/mole	ΔV cm^3/mole
NO_3^-	-5.9	-7.2
Br^-	-8.7	-10.8
Cl^-	-9.9	-11.6
SO_4^{2-}	-17.0	-19.2
H_2O^a	$+1.2$	0.0

[a] W. E. Jones, L. R. Carey, and T. W. Swaddle (1973), *Can. J. Chem.*, **51**, 821.

I^-, and H_2O, the slope of the $\Delta V_0^{\ddagger}/\Delta V$ plot is only 0.59, indicating that separation of X^{n-} and provision of its solvation are only about half complete in the transition state. The mechanisms of aquation of Co(III) and Cr(III) pentammine complexes are clearly different. Since a neat method is available for the determination of ΔV, this approach promises to be a useful one.

Although volumes and entropies of activation often parallel one another, $\Delta V^{\ddagger}$ is the more easily understood and calculable parameter, and it is therefore likely to be the more sensitive criterion of mechanism.

9.5 REACTIONS IN SOLIDS

Gas-solid reactions accompanied by a phase change are the subject of the kinetic studies to be discussed here. Experimentally, the gross overall mechanism is usually being studied in order to identify those regimes where diffusion plays a prominent role in the mechanism and where other factors such as reaction at interphase boundaries or at reaction centers or ordering processes or nucleation-growth influence the limiting rate of the reaction.

For example, in the experiments to study the kinetic behavior of the phase reactions of rare earth oxide systems, the phase reaction in reduction is interpreted to have occurred with some combinations of (1) monophasic reduction of the reactant phase, (2) nucleation, and (3) growth of the product phase. With oxidation, however, it is often observed to be either diffusion controlled or reaction controlled, or both. Thus for convenience of discussion, these two cases are treated separately.

9.5.1 Heterogeneous Reactions

Kinetics of physicochemical transitions or reactions in solids may consist of several simple physical and chemical "elementary" processes involving movement of the particles of matter, changes in structure, physical state, and chemical composition. For example, for the oxidation of metals, these processes manifest themselves in such a way that depending on experimental conditions and reactants, various rate laws (such as logarithmic, inverse logarithmic, linear, and parabolic and cubic) have been observed to occur. It must be understood that these rate laws only describe limiting cases, and since they have been well studied, the conditions under which these equations apply have been quite well understood and documented (Knittel et al., 1977).

In the experimental measurements to follow the kinetics of gas-solid reactions, only information about the weight change of the sample as a function of time at a particular temperature or pressure is usually obtained. Thus in order to find meaningful rate constants of the elementary processes associated with the phase reaction of, say, praseodymium oxide, the experimental data must be fit into a particular kinetic model. Many of these have been proposed by various workers. All the existing kinetic models in solid-state reactions have been compiled and solved by Knittel et al. (1977) and Lin et al. (1978). Some of them are presented in this section.

First we consider the simple diffusion model. According to this model, the weight change of the sample is simply controlled by the diffusion of an absorbed species. In this case, the diffusion equation

$$\frac{\partial C}{\partial t} = D \nabla^2 C \tag{9.163}$$

must be solved subject to the following conditions:

$$C = C^*, \quad r = r_0, \quad t \geq 0,$$
$$C = C_0, \quad 0 < r < r_0, \quad t = 0$$

The solution of eq. 9.163 can be obtained by using the Laplace transformation method. For a spherically symmetric system, we find

$$\bar{C} = \frac{C_0}{p} + \frac{r_0}{rp}(C^* - C_0)\frac{\sinh qr}{\sinh qr_0} \tag{9.164}$$

where $q = \sqrt{p/D}$. The inverse Laplace transform of eq. 9.164 gives (Norwick and Burton, 1975; Crank, 1956)

$$C = C^* - \frac{2r_0}{\pi r}(C^* - C_0)\sum_{n=0}^{\infty}\frac{(-1)^n}{n+1}\sin\frac{(n+1)\pi r}{r_0}\exp[-(n+1)^2 D^* t] \tag{9.165}$$

where $D^* = \pi^2 D/r_0^2$. How to derive eqs. 9.164 and 9.165 is left as an exercise.

If we let M_t be the amount diffused at time t into a sphere of radius r_0, then

$$M_t = \int_0^{r_0} 4\pi r^2 C \, dr \tag{9.166}$$

Substituting eq. 9.165 for C in eq. 9.166 and integrating yields

$$M_t = \tfrac{4}{3}\pi r_0^3 C^* - \frac{8r_0^3(C^* - C_0)}{\pi}\sum_{n=0}^{\infty}\frac{e^{-(n+1)^2 D^* t}}{(n+1)^2} \tag{9.167}$$

If we let

$$M_0 = \tfrac{4}{3}\pi r_0^3 C_0$$

and

$$M^* = \tfrac{4}{3}\pi r_0^3 C^*$$

then eq. 9.167 becomes

$$M_t = M^* - (M^* - M_0)\frac{6}{\pi^2}\sum_{n=0}^{\infty}\frac{e^{-(n+1)^2 D^* t}}{(n+1)^2} \tag{9.168}$$

Equation 9.168 can be written as

$$F = \frac{M_t - M_0}{M_\infty - M_0} = 1 - \frac{6}{\pi^2}\sum_{n=0}^{\infty}\frac{e^{-(n+1)^2 D^* t}}{(n+1)^2} \tag{9.169}$$

From eq. 9.168 we observe that $M_\infty = M^*$ thus eq. 9.169 represents the fraction reacted.

Next we consider the model that leads to the so-called parabolic law. This is a one-dimensional model where the concentration gradient of the diffusing gas in the product is assumed to be constant. Consider a diffusion-controlled reaction

$$A_{\text{solid}} + B_{\text{gas}} \rightarrow AB_{\text{solid}} \tag{9.170}$$

in one dimension with boundary conditions

$$C = C^* \quad \text{at } x = 0 \tag{9.171}$$

and

$$C = C_1 \quad \text{at } x = l$$

The change dl of the gas boundary in the solid can be related to the change in time dt by (Booth, 1948)

$$-D\frac{\partial C}{\partial x}\bigg|_{x=l}\, dt = \frac{m_B}{m_{AB}}\rho\, dl \tag{9.172}$$

where m_B and m_{AB} are the molecular weights of B and AB, respectively, ρ is the density of AB, C is the concentration of the gas, and D is the diffusion constant.

The parabolic law arises when the concentration gradient of B in AB is assumed to be constant, that is,

$$\frac{\partial C}{\partial x} = k \tag{9.173}$$

The constant k in eq. 9.173 can be obtained by the use of the boundary conditions eq. 9.171. Equation 9.173 becomes

$$\frac{\partial C}{\partial x} = \frac{C_1 - C^*}{l} \tag{9.174}$$

Substituting eq. 9.174 into eq. 9.172 and solving the differential equation yields the parabolic law

$$l^2 = \frac{2Dm_{AB}}{m_B \rho}(C^* - C_1)t \tag{9.175}$$

Other derivations of the parabolic law can be found in the literature (Douglas, 1971).

Now we consider the moving boundary model. According to this model, the reaction takes place on the boundary between the product and the reactant and the thickness of the product layer is changing with time. In other words, the absorbed gaseous species must diffuse through the product layer in order to react at the boundary. For a planar system the exact solution of the model has been outlined by Crank but the detailed numerical calculation has not been carried out. For a spherically symmetric system, the model has been solved only when the steady-state approximation has been used; this type of solution has been widely used in the kinetic study of the reduction of iron oxides, the oxidation of nickel, and so on (Seth and Ross, 1965; Carter, 1960).

Consider a system defined by the boundary conditions

$$C = C^* \quad \text{at } r = r_0, \quad t \geq 0$$

$$C = C_1 \quad \text{at } r = r_1, \quad t > 0$$

At steady state the concentration does not change with time because the rate of diffusion through $r_0 - r_1$ is balanced by the reaction rate at r_1. The steady-state condition is represented by

$$D\nabla^2 C = 0 \tag{9.176}$$

where D is the diffusion constant. For a spherical system eq. 9.176 can be solved to yield

$$C = \frac{A}{r} + B \tag{9.177}$$

where A and B are constants determined from the boundary conditions. Applying the boundary conditions, eq. 9.177 becomes

$$C = \frac{r_0 r_1 (C^* - C_1)}{r(r_1 - r_0)} + \frac{r_1 C_1 - r_0 C^*}{r_1 - r_0} \tag{9.178}$$

The rate of diffusion into the sphere of radius r_1 is

$$R_{\text{diff}} = 4\pi r_1^2 D \left(\frac{dC}{dr}\right) r = r_1 \tag{9.179}$$

If the concentration is given by eq. 9.178, eq. 9.179 becomes

$$R_{\text{diff}} = -4\pi r_1^2 D \frac{r_0(C^* - C_1)}{r_1(r_1 - r_0)} \tag{9.180}$$

Equating the diffusion rate with the reaction rate

$$R_{Re} = 4\pi r_1^2 k(C_1 - C_{eq}) \tag{9.181}$$

where k is a rate constant yields

$$C_1 = \frac{Dr_0 C^* + kr_1(r_0 - r_1)C_{eq}}{Dr_0 + kr_1(r_0 - r_1)} \tag{9.182}$$

The reaction rate can be related to the weight change. If W is the weight of the product, then

$$W = \tfrac{4}{3}\pi C_0(r_0^3 - r_1^3) \tag{9.183}$$

where C_0 is the concentration of the reactant. The fraction remaining y is represented by

$$y = \frac{\tfrac{4}{3}\pi r_1^3 C_0}{\tfrac{4}{3}\pi r_0^3 C_0} \tag{9.184}$$

Substituting for r_1^3 in eq. 9.183 yields

$$W = \tfrac{4}{3}\pi r_0^3 C_0(1 - y) \tag{9.185}$$

The reaction rate becomes

$$\frac{dW}{dt} = -\tfrac{4}{3}\pi r_0^3 C_0 \frac{dy}{dt} \tag{9.186}$$

Equating eq. 9.186 with eq. 9.181 yields

$$\frac{dy}{dt} = \frac{-3kD[C^* - C_{eq}]y^{2/3}}{C_0 r_0[D + kr_0 y^{1/3}(1 - y^{1/3}]} \tag{9.187}$$

The solution of eq. 9.187 yields

$$\frac{kD}{C_0 r_0}(C^* - C_{eq})t = D\{-(1 - F)^{1/3} + 1\}$$

$$+ \frac{kr_0}{2}\{-(1 - F)^{2/3} + 1\} - \tfrac{1}{3}kr_0 F \tag{9.188}$$

where $F = 1 - y$ the fraction reacted. In particular, if $D/kr_0 \to 0$, then eq. 9.188 reduces to

$$-\tfrac{1}{3}F + \tfrac{1}{2}\{-(1 - F)^{2/3} + 1\} = \frac{DC^*}{C_0 r_0^2} t \tag{9.189}$$

In the preceding three models, either the reaction rate process is completely ignored (because of its rapidity) or the reaction is assumed to take place at the boundary between the reactant and product. Another model is considered that takes into account the reaction rate process in an entirely different manner, that is,

$$\frac{\partial C}{\partial t} = D\nabla^2 C - \frac{\partial S}{\partial t} \tag{9.190}$$

and

$$\frac{\partial S}{\partial t} = kC - \mu S \tag{9.191}$$

Both forward and reverse reactions have been considered. We shall impose the boundary conditions of eq. 9.190 given by $C = C^*$ at $r = r_0$ and $t \geq 0$ and $S = S_0$ and $C = C_0$ at $0 < r < r_0$ and $t = 0$. The solution of eq. 9.190 for a sphere system is obtained by taking the Laplace transform of both sides, yielding (left as an exercise; $\beta = p(p + \lambda + \mu)/D(p + \mu)$)

$$\bar{C} = \frac{C_0}{p} + \frac{\mu S_0 - \lambda C_0}{p(p + \mu + \lambda)} + \frac{r_0}{rp}\left(C^* - C_0 - \frac{\mu S_0 - \lambda C_0}{p + \mu + \lambda}\right)\frac{\sinh r\sqrt{\beta}}{\sinh r_0\sqrt{\beta}} \tag{9.192}$$

If we let $A_n = (\lambda^* + \mu^* + n^2)^2 - 4n^2\mu^*$, then carrying out the inverse Laplace transform of eq. 9.192 yields

$$C(t) = C^* + \frac{2r_0}{r}(C^* - C_0)\sum_{n=1}^{\infty}\frac{(-1)^n}{\pi n}\quad\sin\frac{n\pi r}{r_0}$$

$$\times\left\{\left(1 + \frac{\mu^* + \lambda^* - n^2}{\sqrt{A_n}}\right)\exp\left(\sqrt{A_n}\frac{D^*t}{2}\right)\right.$$

$$+ \left(1 - \frac{\mu^* + \lambda^* - n^2}{\sqrt{A_n}}\right)\exp\left(-\sqrt{A_n}\frac{D^*t}{2}\right)\right\}$$

$$\times\exp\left[-(\lambda^* + \mu^* + n^2)\frac{D^*t}{2}\right] + \frac{2r_0}{r}(\mu^*S_0 - \lambda^*C_0)$$

$$\times\sum_{n=1}^{\infty}\frac{(-1)^n}{\pi n}\sin\frac{n\pi r}{r_0}\left\{\exp\left[\sqrt{A_n}\frac{D^*t}{2}\right] - \exp\left[-\sqrt{A_n}\frac{D^*t}{2}\right]\right\}$$

$$\times\frac{\exp[-(\lambda^* + \mu^* + n^2)(D^*t)/2]}{\sqrt{A_n}} \tag{9.193}$$

To be able to consider the amount of C in a sphere of radius r we define a term M_t that is the amount of C diffused at time t given by

$$M_t = \int 4\pi r^2 C(t)\, dr \tag{9.194}$$

Substituting eq. 9.193 for $C(t)$ in eq. 9.194 yields

$$M_t = M^* - (M^* - M_0)\frac{6}{\pi^2}\sum_{n=1}^{\infty}\frac{\exp[-(\lambda^* + \mu^* + n^2)(D^*t)/2]}{2n^2}$$

$$\times \left\{\left(1 + \frac{\mu^* + \lambda^* - n^2}{\sqrt{A_n}}\right)\exp\left[\sqrt{A_n}\frac{D^*t}{2}\right]\right.$$

$$+ \left(1 - \frac{\mu^* + \lambda^* - n^2}{\sqrt{A_n}}\right)\exp\left(-\sqrt{A_n}\frac{D^*t}{2}\right)\right\}$$

$$+ (\mu^*M_0^S - \lambda^*M_0)\frac{6}{\pi^2}\sum_{n=1}^{\infty}\frac{\exp[-(\lambda^* + \mu^* + n^2)(D^*t)/2]}{n^2\sqrt{A_n}}$$

$$\times \left\{\exp\left(\sqrt{A_n}\frac{D^*t}{2}\right) - \exp\left(-\sqrt{A_n}\frac{D^*t}{2}\right)\right\} \qquad (9.195)$$

where $M^* = \frac{4}{3}\pi r_0^3 C^*$, $M_0 = \frac{4}{3}\pi r_0^3 C_0$, and $M_0^S = \frac{4}{3}\pi r_0^3 S_0$. Equation 9.195 can also be written in terms of the hyperbolic functions as

$$M_t = M^* - (M^* - M_0)\frac{6}{\pi^2}\sum_{n=1}^{\infty}\frac{\exp[-(\lambda^* + \mu^* + n^2)(D^*t)/2]}{n^2}$$

$$\times \left\{\cosh\sqrt{A_n}\frac{D^*t}{2} + \frac{\mu^* + \lambda^* - n^2}{\sqrt{A_n}}\sinh\sqrt{A_n}\frac{D^*t}{2}\right\}$$

$$+ (\mu^*M_0^S - \lambda^*M)\frac{6}{\pi^2}\sum_{n=1}^{\infty}\frac{2\exp[-(\lambda^* + \mu^* + n^2)(D^*t)/2]}{n^2\sqrt{A_n}}$$

$$\times \sinh\sqrt{A_n}\frac{D^*t}{2} \qquad (9.196)$$

The fraction reacted F can be obtained from eq. 9.196 as

$$F = \frac{M_t - M_0}{M_\infty - M_0} = 1 - \frac{6}{\pi^2}\sum_{n=1}^{\infty}\frac{\exp[-(\lambda^*\mu^* + n^2)(D^*t)/2]}{n^2}$$

$$\times \left\{\cosh\sqrt{A_n}\frac{D^*t}{2} + \frac{\mu^* + \lambda^* - n^2}{\sqrt{A_n}}\sinh\sqrt{A_n}\frac{D^*t}{2}\right.$$

$$+ \frac{\mu^*M_0^S/M_\infty - \lambda^*M_0/M_\infty}{1 - M_0/M_\infty}\frac{6}{\pi^2}$$

$$\times \sum_{n=1}^{\infty}\frac{2\exp[-(\lambda^* + \mu^* + n^2)(D^*t)/2]}{n^2\sqrt{A_n}}\sinh\sqrt{A_n}\frac{D^*t}{2} \qquad (9.197)$$

In all the kinetic models of solid-state reactions discussed so far, the diffusion plays a very important role. Needless to say, the diffusion-controlled kinetic models discussed in Sec. 9.2 can also be included here.

In the oxidation of some metal oxides (or the polymerization), because of the fact that the oxygen diffusion is fast compared with the chemical reaction, the oxidation mechanism may be described as follows. The oxidation reaction initiates at nucleus-forming sites formed by oxidation and the oxidation product nuclei grow by the chemical reaction taking place at the interface between the nuclei and the reaction matrix. This type of the oxidation mechanism has been treated by Lin et al. (1978) and is briefly discussed in the following.

If we let A represent the reactant unit in the solid and B_n represent the product unit in the solid, then the reactive nucleation and growth model may be described by

$$A \xrightarrow{k_0} B_1, A + B_1 \xrightarrow{k_1} B_2, \ldots, A + B_{i-1} \xrightarrow{k_{i-1}} B_i, \ldots,$$

$$A + B_{n-1} \xrightarrow{k_{n-1}} B_n$$

Because of the fact that the oxygen transport is fast compared with the chemical reaction, the rate constants k_{i-1} should depend on the oxygen partial pressure. The reverse processes in Eq. 9.198 that are slow in the oxidation of solids have been ignored but they can be included; in that case, the perturbation method may be employed to treat these reverse processes.

From eq. 9.198 the rate expression for B_i and A can be written as

$$\frac{dB_1}{dt} = k_0 A - k_1 A B_1 \tag{9.200}$$

$$\frac{dB_i}{dt} = k_{i-1} A B_{i-1} - k_i A B_i \tag{9.201}$$

$$\frac{dB_n}{dt} = B_{n-1} A B_{n-1} \tag{9.202}$$

and

$$\frac{dA}{dt} = -k_0 A - \sum_{i-1}^{n=1} k_i A B_i \tag{9.203}$$

We assume that at $t = 0$, $A = A_0$ and $B_i = 0$. Notice that

$$A_0 = A + \sum_{i=1}^{n-1} i B_i \tag{9.204}$$

To solve eqs. 9.200–9.203 we introduce the following dimensionless variables:

$$A^* = \frac{A}{A_0}, \qquad B_i^* = \frac{B_i}{A_0} \tag{9.205}$$

$$k_i^* = \frac{k_i}{k_0} A_0 \tag{9.206}$$

$$d\tau = k_0 A^* \, dt, \qquad k_0 t = \int_0^\tau \frac{d\tau}{A^*} \tag{9.207}$$

Using eqs. 9.205–9.207, eqs. 9.200–9.203 can be rewritten as

$$\frac{dB_1}{d\tau} = 1 - k_1^* B_1^* \tag{9.208}$$

$$\frac{dB_i^*}{d\tau} = k_{i-1}^* B_{i-1}^* - k_i^* B_i^* \tag{9.209}$$

$$\frac{dB_n^*}{d\tau} = -1 - \sum_{i=1}^{n-1} k_i^* B_i^* \tag{9.211}$$

In eq. 9.198 we have limited the size of B to be n; n depends on k_{n-1}, which in turn depends on the oxygen pressure. In other words, we have restricted ourselves to consider the closed-end problem. The open-end problem (i.e., $n \to \infty$) can be treated similarly. One way to solve eqs. 9.208–9.211 is to solve eq. 9.208 first for B_1^*; after finding $B_1^*(\tau)$, we may substitute it into the next equation to solve for $B_2^*(\tau)$. This process may be repeated until we obtain all of $B_i^*(\tau)$'s.

Another more systematic way to solve eqs. 9.208–9.211 is to employ the Laplace transformation method. Notice that the Laplace transform of $B_i^*(\tau)$ is defined by

$$\overline{B_i^*}(p) = \int_0^\infty e^{-p\tau} B_i^*(\tau) \, d\tau \tag{9.212}$$

Carrying out the Laplace transformation of eqs. 9.208–9.210 yields

$$\overline{B_i^*}(p) = \frac{1}{p(p + k_1^*)} \tag{9.213}$$

$$\overline{B_i^*}(p) = \frac{k_{i-1}^*}{p + k_i^*} \, \overline{B_{i-1}^*}(p) \tag{9.214}$$

$$\overline{B_n^*}(p) = \frac{k_{n-1}^*}{p} \, \overline{B_{n-1}^*}(p) \tag{9.215}$$

from which we find

$$\overline{B_i^*}(p) = \prod_{j=1}^{i} \frac{k_j^*}{(p + k_j^*)} \cdot \frac{1}{k_i^* p} \tag{9.216}$$

for $i \neq n$, and

$$\overline{B_n^*}(p) = \frac{1}{p^2} \prod_{j=1}^{n-1} \frac{k_j^*}{p + k_j^*} \tag{9.217}$$

Carrying out the inverse Laplace transformation of eqs. 9.213–9.217 yields

$$B_1^*(\tau) = \frac{1}{k_1^*} - \frac{e^{-k_1^* \tau}}{k_1^*} \tag{9.218}$$

$$B_i^*(\tau) = \prod_{j=1}^{i=1} k_j^* \left[\prod_{j=1}^{i} \frac{1}{k_j^*} - \sum_{l=1}^{i} \frac{e^{-k_l^* \tau}}{k_l^* \prod_j^{'i} (k_j^* - k_l^*)} \right] \tag{9.219}$$

and

$$B_n^*(\tau) = \sum_{l=1}^{n-1} \left(\tau + \frac{e^{k_l^* \tau} - 1}{k_l^*} \right) \prod_j^{'} \frac{k_j^*}{k_j^* - k_l^*} \tag{9.220}$$

The term $A^*(\tau)$ can be calculated from either eq. 9.204 or eq. 9.211,

$$A^*(\tau) = 1 - n\tau + \frac{1 - e^{k_1^* \tau}}{k_1^*} + \sum_{i=2}^{n-1} \sum_{l=1}^{i} \frac{(1 - e^{k_l^* \tau})}{k_l^*} \prod_j^{'} \frac{k_j^*}{k_j^* - k_l^*} \tag{9.221}$$

An interesting particular case is $k_i^* = k^*$, that is, $k_1 = k_2 = \cdots = k_{n-1}$; this case has also been solved by Lin et al. (1978) but is not presented here.

9.5.2 Decomposition Reactions

It is generally accepted that the nucleation and growth processes play important roles in the decomposition of a solid to yield a second product phase. In this case, the chemical reaction initiates at certain discrete points in the reactant, called nucleus-forming sites, which result in the formation of submicroscopic particles of the solid product phase called nuclei. These nuclei are believed to distribute over the surface of, or embedded in the bulk of, the reactant and further chemical reaction is localized at the interface between the nuclei and the reactant matrix so that the nuclei grow in size as the reaction proceeds. The theory of the decomposition of solids through the nucleation and growth of the product phase has been well established (Allnatt and Jacobs, 1968; Hannay, 1976).

The differential equation controlling the number of nuclei n_i that contain i product atoms is

$$\frac{dn_i}{dt} = k_{i-1} n_{i-1} - k_i n_i \tag{9.222}$$

which corresponds to

$$A_{i-1} \xrightarrow{k_{i+1}} A_i, A_i \xrightarrow{k_{i-1}} A_{i+1} \tag{9.223}$$

Equation 9.222 is to be solved subject to the boundary conditions $n_i = 0$ for $i > 0$ at $t = 0$ and $n_i = n_0(0)$, for $i = 0$ at $t = 0$.

By using the Laplace-transformation method, we find

$$\bar{n}_i(p) = \frac{1}{k_i} n_0(0) \prod_{l=0}^{i} \frac{k_l}{p + l_l} \tag{9.224}$$

Carrying out the inverse transformation of eq. 9.224 yields

$$n_i(t) = n_0(0) k_0 k_1 k_2 \cdots k_{i-1} \sum_{l=0}^{i} \frac{e^{-k_i t}}{\prod_{j=0}^{i} (k_j - k_l)} \tag{9.225}$$

According to Bagdassarian (1945), who introduced the concept of multistep nucleation, in general, r successive molecular decompositions at a single site are required to form a stable growth nucleus and nuclei containing less than r product atoms are not growth nuclei, but germ nuclei, which may become growth nuclei by acquiring the requisite number of product atoms. In other words, Bagdassarian considered only the special case

$$\begin{aligned} k_0 = k_1 = k_2 \cdots = k_{r-1} \\ k_r = k_{r+1} = k_{r+2} = \cdots = k \end{aligned} \tag{9.226}$$

In this case, we have for $i < r$

$$n_i(p) = \frac{k_0^i}{(p + k_0)^{i+1}} n_0(0) \tag{9.227}$$

and

$$n_i(t) = n_0(0) \frac{(k_0 t)^i}{i!} \exp(-k_0 t) \tag{9.228}$$

and for $i > r$

$$\bar{n}_i(p) = \frac{k^{i-r}}{(p + k)^{i-r+1}} \cdot \frac{k_0^r}{(p + k_0)^r} n_0(0) \tag{9.229}$$

and

$$n_i(t) = k^{i-r} k_0^r n_0(0) I_i(t) \tag{9.230}$$

where

$$I_i(t) = \int_0^t d\tau \frac{(t - \tau)^{i-r}}{(i - r)!} \frac{\tau^{r-1}}{(r - 1)!} \exp[-k(t - \tau) - k_0 \tau] \tag{9.231}$$

The integration of eq. 9.231 can be carried out by integration by parts.

Germ nuclei are converted into growth nuclei in the manner described, but growth nuclei formed from germ nuclei, which happen to be near one another in the reactant, will impinge together and overlap during growth. This ingestion and this overlap result in the fractional decomposition α being smaller than the extended fractional decomposition α_{ex}, which Avrami (1941) defines as the fractional decomposition that would have occurred without ingestion and overlap. If we assume that decomposition is a purely random phenomenon, then

$$d\alpha = d\alpha_{ex}(1 - \alpha) \tag{9.232}$$

Equation 9.232 is the simplest expression that satisfies the required boundary conditions, $d\alpha/dt = d\alpha_{ex}/dt$ at $\alpha = 0$, $d\alpha/dt = 0$ but $d\alpha_{ex}/dt$ finite at $\alpha = 1$. It follows that

$$\alpha_{ex} = -\log(1 - \alpha) \tag{9.233}$$

Suppose that nucleation is a multistep process characterized by rate constants $k_0, k_1, \ldots, k_r$ and that once a nucleus containing r atoms of product is formed it becomes an active growth nucleus. The rate of formation of active growth nuclei is therefore

$$\frac{dn_r}{dt} = k_{r-1}n_{r-1} = n_0(0) \sum_{l=0}^{r-1} A_{rl} \exp(-k_l t) \tag{9.234}$$

where

$$A_{rl} = k_l \prod_j^{r-1}{}' \frac{k_j}{k_j - k_l} \tag{9.235}$$

The extended fractional decomposition α_{ex} is then given by (Erofeev, 1946)

$$\alpha_{ex} = \frac{\sigma}{V(\infty)} \int_0^t d\tau \left(\frac{dn_r}{dt}\right)_{t=\tau} [G(t - \tau)]^3 \tag{9.236}$$

where G is the constant radial growth rate, σ is a shape factor ($\sigma = 4\pi/3$ for spherical nuclei), and $V(\infty)$ is the total volume of product when reaction is complete. Substituting eq. 9.234 into eq. 9.236, integrating and using eq. 9.233, we obtain (Allnatt and Jacobs, 1968)

$$-\log(1 - \alpha) = \frac{6\sigma}{V(\infty)} n_0(0)G^3 \sum_{l=0}^{r-1} \frac{A_{rl}}{k_l^4} [\exp(-k_l t) - 1 + k_l t$$
$$- \tfrac{1}{2}(k_l t) + \tfrac{1}{6}(k_l t)^3] \tag{9.237}$$

Equation 9.237 gives the general solution to the problem of multistep nucleation followed by a constant rate of growth in three dimensions. Many applications of eq. 9.237 can be found in the literature (Hannay, 1976).

9.6 QUANTUM-STATISTICAL MECHANICAL THEORY
OF RATE PROCESSES

Many rate processes, like diffusion (Glasstone et al., 1940), dielectric relaxation (Fong and Diestler, 1972), electron transfer reactions (Schmidt, 1972;
Levich, 1956; Brent and Trone, 1971), nonradiative decay (Lin, 1966, 1970,
1972; Lin and Bersohn, 1968), resonance transfer of electronic excitation
(Lin, 1971), thermally activated motion (Sewell, 1963; Landauer and Swanson, 1961); and so on, in dense media seem to be closely related to one another
theoretically and can be treated from a unified quantum-mechanical viewpoint. In this section, a general theoretical approach is presented that will treat
these rate processes by properly choosing the perturbation associated with
each rate process (Lin and Eyring, 1972). In the foregoing unimolecular rate
processes, two potential surfaces (or curves) are involved, and it is assumed
that when the two potential surfaces (or curves) cross, the resonance interaction is small (Fong, 1976).

9.6.1 General Considerations

We consider any rate process for which the transition probability for the
transition $bv' \rightarrow av''$ can be described by the golden rule expression

$$k_{bv'av''} = \left(\frac{2\pi}{\hbar}\right)|\langle bv'|\hat{H}'|av''\rangle|^2\delta(E^a_{v''} - E_{bv'}) \qquad (9.238)$$

where (v', v'') represent the quantum states of nuclear motion and (a, b)
represent the quantum states of the two potential surfaces. Here (bv') is the
initial state and (av'') is the final state.

If the rate of energy equilibrium is much faster than the rate process under
consideration, then the transitions always originate from a Boltzmann
distribution of the energy levels of nuclear motion and the rate constant is
given by

$$k_{ba}(\beta) = \left(\frac{2\pi}{\hbar}\right)\sum_{v'}\sum_{v''} P_{bv'}|\langle bv'|\hat{H}'|av''\rangle|^2\delta(E_{av''} - E_{bv'}) \qquad (9.238)$$

where $P_{bv'}$ represents the Boltzmann factor.

Using the adiabatic approximation, we have

$$|bv'\rangle = \phi_b(q, Q)\theta_{bv'}(Q) \qquad \text{and} \qquad |av''\rangle = \phi_a(q, Q)\theta_{av''}(Q) \qquad (9.239)$$

It follows that

$$k_{ba}(\beta) = \left(\frac{2\pi}{\hbar}\right)\sum_{v'}\sum_{v''} P_{bv'}|\langle\theta_{bv'}|H'_{ba}|\theta_{av''}\rangle|^2\delta(E_{av''} - E_{bv'}) \qquad (9.240)$$

where $\hat{H}'_{ba} = \langle \phi_b | H' | \phi_a \rangle$. The matrix element H'_{ba} in general depends on the nuclear coordinates. We have been concerned with the rate process taking place in a dense medium. Thus the intermolecular motion is approximately treated as vibrations. In this section we attempt to simplify eq. 9.240 without specifying explicitly the perturbation $\hat{H}'$.

Introducing the integral expression for the delta function, eq. 9.240 becomes

$$k_{ba}(\beta) = \left(\frac{1}{\hbar^2}\right) \sum_{v'} \sum_{v''} \int_{-\infty}^{\infty} dt \, \exp(it\omega_{av'', bv'}) P_{bv'} |\langle \theta_{bv'} | H'_{ba} | \theta_{av''} \rangle|^2 \quad (9.241)$$

where $E_{av''} - E_{bv'', bv'}$. If the matrix element H'_{ba} does not vary rapidly with normal coordinates Q'_i, we can expand H'_{ba} in power series of Q'_i,

$$H'_{ba} = H'^{(0)}_{ba} + \sum_i \left(\frac{\partial H'_{ba}}{\partial Q'_i}\right)_0 Q'_i + \cdots \quad (9.242)$$

In this section we shall discuss two general cases; case I refers to the processes in which H'_{ba} is independent of normal coordinates and case II represents the processes in which the matrix element varies linearly with respect to Q'_i, that is,

$$H'_{ba} = \left(\frac{\partial H'_{ba}}{\partial Q'_i}\right)_0 Q'_i$$

For case I the rate constant $k_{ba}(\beta)$ can be written as

$$k_{ba}(\beta) = \left(\frac{|H'_{ba}|^2}{\hbar^2}\right) \sum_{v'} \sum_{v''} \int_{-\infty}^{\infty} dt \, \exp(it\omega_{av''bv'}) P_{bv'} |\langle \theta_{bv'} | \theta_{av''} \rangle|^2 \quad (9.243)$$

and similarly for case II, we have

$$k_{ba}(\beta) = \left(\frac{1}{\hbar^2}\right) \left| \left(\frac{\partial H'_{ba}}{\partial Q'_i}\right)_0 \right|^2 \sum_{v''} \sum_{v'} \int_{-\infty}^{\infty} dt$$

$$\times \exp(it\omega_{av''bv'}) P_{bv'} |\langle \theta_{bv'} | Q'_i | \theta_{av''} \rangle|^2 \quad (9.244)$$

These two cases cover a great number of rate processes, depending on the choice of H'_{ab}. Writing the wave functions of nuclear motion as a product of wave functions of normal vibrations is valid only for the rate processes in the solid phase; for the rate processes in the liquid, we regard the intermolecular motion as approximating vibrations; eqs. 9.243 and 9.244 become

$$k_{ba}(\beta) = \left(\frac{|H'_{ab}|^2}{\hbar^2}\right) \int_{-\infty}^{\infty} dt \, \exp(it\omega_{ab}) \prod_j G_j(t) \quad (9.245)$$

where $G_j(t)$ is defined by (Lin, 1966)

$$G_j = \sum_{v'_j} \sum_{v''_j} P_{bv'_j} |\langle X_{v''_j} | X_{v_j} \rangle|^2 \exp\{it[(v''_j + \tfrac{1}{2})\omega''_j - (v'_j + \tfrac{1}{2})\omega'_j]\} \quad (9.246)$$

and

$$k_{ba}(\beta) = \left[\frac{R_i(ab)}{\hbar^2}\right] \int_{-\infty}^{\infty} dt [\exp(it\omega_{ab})] K_i(t) \prod_j' G_j(t) \qquad (9.247)$$

respectively, where $R_i(ab) = |\langle \phi_b | (\partial \hat{H}' / \partial Q_i')_0 | \phi_a \rangle|^2$, and

$$K_i(t) = \sum_{v_i'} \sum_{v_i''} P_{bv_i} |\langle X_{bv_i} | Q_i' | X_{av_i''} \rangle|^2$$

$$\times \exp\{it[(v_i'' + \tfrac{1}{2})\omega_i'' - (v_i' + \tfrac{1}{2})\omega_i']\} \qquad (9.248)$$

$X_{bv'}$, $X_{av''}$, and so on, represent the wave functions of normal vibrations. In the following sections we show how to simplify the rate constants of these two important classes of rate processes and to obtain the temperature dependence of these rate constants so that the rate processes of these two classes can be distinguished from the measurements of the temperature dependence of the rate constants (also see Chapter 7).

9.6.2   Case I

To simplify eq. 9.244 we assume that both intermolecular and intramolecular vibrations are harmonic. In terms of the changes in normal coordinates and frequencies, $d_i'' - d_i'$ and ϕ_i, defined by

$$Q_i'' - Q_i' = d_i' - d_i'', \qquad \omega_i'' = \omega_i'(1 - \phi_i) \qquad (9.249)$$

the expression for the rate constant $k_{ba}(\beta)$ to the first-order approximation of ϕ_i can be shown to be (Lin, 1966; Lin and Bersohn, 1968)

$$k_{ba}(\beta) = \frac{|H_{ab}'|^2}{\hbar^2} \exp\left(-\frac{1}{2}\sum_j \Delta_j^2 \coth\frac{\beta\hbar\omega_j'}{2}\right)$$

$$\times \int_{-\infty}^{\infty} dt \exp\left[it\omega_{ab} - \frac{it}{2}\sum_j \phi_j\omega_j' \coth\frac{\beta\hbar\omega_j'}{2} + \frac{1}{2}\sum_j \Delta_j^2 \operatorname{csch}\frac{\beta\hbar\omega_j'}{2}\right]$$

$$\times \cos\left(\omega_j't - \frac{i\beta\hbar\omega_j'}{2}\right)\bigg] \qquad (9.250)$$

where $\beta = 1/kT$, and $\Delta_j^2 = \beta_j'^2(d_j' - d_j'')^2$, with $\beta_j'^2 = \omega_j'/\hbar$. If we define $\bar{n}_j = [\exp(\beta\hbar\omega_j') - 1]^{-1}$, $S = \frac{1}{2}\sum_j \Delta_j^2$, and $\omega_{ab}' = \omega_{ab} - \frac{1}{2}\sum_j \phi_j\omega_j'$, eq. 9.250 can be rewritten as

$$k_{ba}(\beta) = \left(\frac{|H_{ab}'|^2}{\hbar^2}\right)\exp(-S) \int_{-\infty}^{\infty} dt \exp\left[it\omega_{ab}' - it\sum_j \phi_j\omega_j'\bar{n}_j\right.$$

$$\left. + \frac{1}{2}\sum_j \Delta_j^2 \exp(it\omega_j') + \sum_j \bar{n}_j\Delta_j^2(\cos \omega_j't - 1)\right] \qquad (9.251)$$

The derivation used here is similar to that in radiationless transitions (Lin, 1966, 1972).

First we consider the weak coupling case (Fischer, 1970; Engleman and Jortner, 1970), that is, $S \leq 1$. In this case, we apply the method of steepest descent to evaluate the integral in eq. 9.251. For this purpose we set $T = 0$ in eq. 9.251:

$$k_{ba}(\beta) = \left(\frac{|H'_{ab}|^2}{\hbar^2} \right) \exp(-S) \int_{-\infty}^{\infty} dt \, \exp[f(t)] \qquad (9.252)$$

where

$$f(t) = it\omega'_{ab} + \frac{1}{2} \sum_j \Delta_j^2 \exp(it\omega'_j) \qquad (9.253)$$

It follows that

$$f'(t) = i\omega'_{ab} + \frac{1}{2} \sum_j i\Delta_j^2 \omega'_j \exp(i\omega'_j t)$$

$$f''(t) = -\frac{1}{2} \sum_j \Delta_j^2 \omega'^2_j \exp(i\omega'_j t) \qquad (9.254)$$

The saddle point of t is determined by $f'(t^*) = 0$, or

$$\omega'_{ba} = \frac{1}{2} \sum_j \Delta_j^2 \omega'_j \exp(i\omega'_j t^*) = S\bar{\omega} \exp(it^* \bar{\omega}) \qquad (9.255)$$

where $\bar{\omega}$ represents an average frequency defined by this equation. Applying the method of steepest descent to eq. 9.252 yields (Lin, 1973)

$$k_{ba}(\infty) = \frac{|H_{ab}|^2}{\hbar^2} \exp\left\{ -\frac{1}{2} \sum_j \Delta_j^2 [1 - \exp(it^* \omega'_j)] + it^* \omega'_{ab} \right\}$$

$$\times \left[\frac{4\pi}{\sum_j \Delta_j^2 \omega'^2_j \exp(it^* \omega'_j)} \right]^{1/2} \qquad (9.256)$$

where it^* is to be determined from eq. 9.255.

Expanding the integrand in eq. 9.251 around t^* and applying the method of steepest descent, we obtain (Lin, 1973)

$$k_{ba}(\beta) = k_{ba}(\infty) \exp\left(\sum_j \bar{n}_j A'_j \right) \qquad (9.257)$$

where

$$A'_j = \frac{\Delta_j^2}{2} \left[\left(\frac{\omega'_{ba}}{S\bar{\omega}} \right)^{\omega'_j/2\bar{\omega}} - \left(\frac{\omega'_{ba}}{S\bar{\omega}} \right)^{-\omega'_j/2\bar{\omega}} \right]^2 - \frac{\phi_j \omega'_j}{\bar{\omega}} \log \frac{\omega'_{ba}}{S\bar{\omega}}$$

$k_{ba}(\infty)$ represents the rate constant at $T = 0$. When T is low, eq. 9.257 can be written as (Lin, 1972)

$$k_{ba}(\beta) = k_{ba}(\infty)\left[1 + \sum_j A'_j \exp(-\beta\hbar\omega'_j) + \cdots\right] \qquad (9.258)$$

The process of nonradiative energy transfer of electronic excitation in the Condon approximation belongs to this category (Lin, 1971).

Next we turn to the strong coupling case (Fischer, 1970; Engleman and Jortner, 1970), $S \gg 1$. In this case we can expand the exponential term in the integrand of eq. 9.254 in power series of t. Retaining only up to the quadratic term in t, and carrying out the integration, we obtain

$$k_{ba}(\beta) = \frac{|H'_{ab}|^2}{\hbar^2}\left[\frac{2\pi}{\sum_j(\frac{1}{2} + \bar{n}_j)\Delta_j^2\omega_j'^2}\right]$$

$$\times \exp\left[-\frac{(\omega'_{ab} - \sum_j \phi_j\omega'_j\bar{n}_j + \frac{1}{2}\sum_j \Delta_j^2\omega'_j)^2}{2\sum_j(\frac{1}{2} + \bar{n}_j)\Delta_j^2\omega_j'^2}\right] \qquad (9.259)$$

At $T = 0$, eq. 9.259 reduces to

$$k_{ba}(\infty) = \frac{|H_{ab}|^2}{\hbar^2}\left[\frac{4\pi}{\sum_j \Delta_j^2\omega_j'^2}\right]^{1/2}\exp\left[-\frac{(\omega'_{ab'} + \frac{1}{2}\sum_j \Delta_j^2\omega'_j)^2}{\sum_j \Delta_j^2\omega_j'^2}\right] \qquad (9.260)$$

When T is high but not high enough so that the harmonic oscillator approximation breaks down, we have $\bar{n}_j = kT/\hbar\omega'_j$, and eq. 9.259 becomes

$$k_{ba}(\beta) = \frac{|H'_{ab}|^2}{\hbar^2}\left[\frac{2\pi\hbar}{kT\sum_j \Delta_j^2\omega'_j}\right]^{1/2}$$

$$\times \exp\left[-\frac{\hbar(\omega'_{ab} - (kT/\hbar)\sum_j \phi_j + \frac{1}{2}\sum_j \Delta_j^2\omega'_j)^2}{2kT\sum_j \Delta_j^2\omega'_j}\right] \qquad (9.261)$$

In particular, if the ϕ_j are zero and $\omega'_{ab} = 0$, we can simplify eq. 9.261 as

$$k_{ba}(\beta) = \frac{|H'_{ab}|^2}{\hbar^2}\left[\frac{\pi}{4kT\Delta E}\right]^{1/2}\exp\left(-\frac{\Delta E}{kT}\right) \qquad (9.262)$$

where $\Delta E = \frac{1}{8}(\hbar \sum_j \Delta_j^2\omega'_j)$, which represents the minimum potential crossing in the multidimensional potential surfaces. The electron transfer reactions and some thermally activated processes belong to this case.

9.6.3   Case II

For small ϕ_i values, $K_i(t)$ in eq. 9.248 can be simplified to (Lin, 1966)

$$K_i(t) = \frac{\hbar}{4\omega'_i}\left[\left(1 + \coth\frac{\beta\hbar\omega'_i}{2}\right)\exp(it\omega'_i)\right.$$

$$\left. + \left(\cot h\frac{\beta\hbar\omega'_i}{2} - 1\right)\exp(-it\omega'_i)]G_i(t)\right. \qquad (9.263)$$

Substituting eq. 9.248 into eq. 9.247, we obtain

$$k_{na}(\beta) = \frac{C_{i(ab)}}{\hbar^2} \int_{-\infty}^{\infty} dt \, \exp(it\omega_{ab}) \left[\frac{1}{2}\left(\coth \frac{\beta\hbar\omega_{i'}}{2} + 1 \right) \exp(it\omega_{i'}) \right.$$

$$\left. + \frac{1}{2}\left(\coth \frac{\beta\hbar\omega_{i'}}{2} - 1 \right) \exp(-it\omega_{i'}) \prod_j G_j(t) \right] \qquad (9.264)$$

or

$$k_{ba}(\beta) = \left(\frac{C_{i(ab)}}{\hbar^2} \right) \exp(-S) \int_{-\infty}^{\infty} dt \{ \exp(it\omega_i') + \bar{n}_i[\exp(it\omega_i') + \exp(-it\omega_i')] \}$$

$$\times \exp\left[i\omega_{ab}'t - it \sum_j \phi_j\omega_j'\bar{n}_j + \frac{1}{2}\sum_j \Delta_j^2 \exp(it\omega_i') \right.$$

$$\left. + \sum_j \bar{n}_j\Delta_j^2(\cos \omega_j't - 1) \right] \qquad (9.265)$$

where $C_{i(ab)} = (\hbar/2\omega_i')R_i(ab)$. Equation 9.265 should be compared with eq. 9.251.

For the weak coupling case $S \leq 1$, the expression for $k_{ba}(\beta)$ in eq. 9.265 is exactly the same as that for the nonradiative decay rate constant of large molecules (Lin, 1970). By the method of steepest descent, eq. 9.265 can be expressed as

$$k_{ba}(\beta) = k_{ba}(\infty)\{1 + \bar{n}_i[1 + \exp(-2it^*\omega_i')]\}$$

$$\times \exp\left[\sum_j \bar{n}_j\{\Delta_j^2(\cos \omega_j't^* - 1) - it^*\phi_j\omega_j' \right] \qquad (9.266)$$

where the saddle point value t^* is determined by

$$\omega_{ba}' - \omega_i' = \frac{1}{2}\sum_j \Delta_j^2\omega_j' \exp(it^*\omega_j') \qquad (9.267)$$

Again eq. 9.267 can be solved for t^* by introducing an average frequency (cf. eq. 9.255), and substituting this t^* value into eq. 9.266 yields (Lin, 1973)

$$k_{ba}(\beta) = k_{ba}(\infty)\left[1 + \bar{n}_i\left\{ 1 + \left(\frac{S\bar{\omega}}{\Delta\omega} \right)^{2\omega_i'/\bar{\omega}} \right\} \right]\exp\left(\sum_j \bar{n}_j A_j \right) \qquad (9.268)$$

where $\Delta\omega = \omega_{ba}' - \omega_i'$.

$k_{ba}(\infty)$ is given by

$$k_{ba}(\infty) = \frac{C_{i(ab)}}{\hbar^2} \exp\left\{ -\frac{1}{2}\sum_j \Delta_j^2[1 - \exp(i\omega_j't^*)] - it^*\Delta\omega \right\}$$

$$\times \left[\frac{4\pi}{\sum_j \Delta_j^2\omega_j'^2 \exp(it^*\omega_j')} \right]^{1/2} \qquad (9.269)$$

where t^* is to be determined from eq. 9.193. When T is low, eq. 9.268 can be reduced to an equation similar to eq. 9.258 with A'_j replaced by A_j.

Next we turn to the strong coupling case $S \gg 1$. In this case we also expand the exponential term in eq. 9.265 in power series of t and retain only up to the quadratic term in t. It follows that

$$k_{ba}(\beta) = \frac{C_{i(ab)}}{\hbar^2} \left[\frac{2\pi}{\sum_j (\frac{1}{2} + \bar{n}_j)\Delta_j^2 \omega_j'^2} \right]^{1/2}$$

$$\times \exp\left[-\frac{(\omega'_{ab} - \sum_j \phi_j \bar{n}_j \omega'_j + \frac{1}{2}\sum_j \Delta_j^2 \omega_j')^2}{2\sum_j (\frac{1}{2} + \bar{n}_j)\Delta_j^2 \omega_j'^2} \right] \quad (9.270)$$

$$\times \{\exp(it^*\omega'_i) + \bar{n}_i[\exp(it^*\omega'_i) + \exp(-it^*\omega'_i)]\}$$

where t^* is determined by

$$t^* = \frac{i \sum \Delta_j^2 \omega'_j}{2\sum_j (\frac{1}{2} + \bar{n}_j)\Delta_j^2 \omega_j'^2} \quad (9.271)$$

When $T = 0$, eq. 9.270 reduces to

$$k_{ab}(\infty) = \frac{C_{i(ab)}}{\hbar^2} \left[\frac{4\pi}{\sum_j \Delta_j^2 \omega_j'^2} \right]^{1/2} \exp\left[-\frac{(\omega'_{ab} - \frac{1}{2}\sum_j \Delta_j^2 \omega'_j)^2}{\sum_j \Delta_j^2 \omega_j'^2} \right]$$

$$\times \exp\left(-\frac{\sum_j \Delta_j^2 \omega'_j \omega'_i}{\sum_j \Delta_j^2 \omega_j'^2} \right) \quad (9.272)$$

When T is high so that $\bar{n}_j = kT/\hbar\omega'_j > 1$, we obtain $k_{ba}(\beta)$ as

$$k_{ba}(\beta) = \frac{C_{i(ab)}}{\hbar^2} \frac{kT}{\hbar\omega'_i} \left[\frac{8\pi\hbar}{kT \sum_j \Delta_j^2 \omega'_j} \right]^{1/2}$$

$$\times \exp\left[-\frac{\hbar(\omega'_{ab} - kT/\hbar \sum_j \phi_j + \frac{1}{2}\sum_j \Delta_j^2 \omega'_j)}{2kT \sum_j \Delta_j^2 \omega'_j} \right] \quad (9.273)$$

For the case in which $\phi_j = 0$, eq. 9.273 can be written

$$k_{ba}(\beta) \doteq \frac{C_{i(ab)}}{\hbar} \left[\frac{\pi kT}{(\Delta E)\hbar^2 \omega_i'^2} \right]^{1/2} \exp\left(-\frac{\Delta E}{kT} \right) \quad (9.274)$$

where $\Delta E = \frac{1}{8}\hbar \sum_j \Delta_j^2 \omega'_j$ represents the minimum potential crossing. An equation similar to eq. 9.274 has been derived by Fong and Diestler for dielectric relaxation by using the correlation function method (Kubo, 1957; Yamamoto, 1960). It is well known that the correlation function expression for a rate constant can be reduced to the golden rule expression. Fong and Diestler have also compared the expression for $k_{ba}(\beta)$ given in eq. (9.274) with that obtained from the absolute reaction rate theory. Finally, we should notice the difference in the temperature dependence of the rate constants between case I and case II (see eqs. 9.274 and 9.262).

It should be noted that in eq. 9.255 if $\omega'_{ba} = 0$, the saddle point method fails. A similar situation exists in eq. 9.267. In a recent study the Arrhenius type of the temperature dependence of the rate constant has been demonstrated by numerical calculation and by other theoretical methods (Ma et al., 1978). Ma et al have also studied the damping effect on the rate constant. They consider the temperature effect on the rate constant for the case in which $\omega'_{ab} = 0$ and $\phi_j = 0$. In this case, using the model, eq. 9.251 becomes

$$k_{ab}(\beta) = \left(\frac{2\pi}{\hbar^2}\right)|H'_{ab}|^2 \exp[-S(1 + 2\bar{n})] \sum_{m=0}^{\infty} \sum_{m'=0}^{\infty}$$

$$\times \frac{(S\bar{n})^m}{m!} \frac{[S(\bar{n} + 1)]^{m'}}{m'!} \delta(m\omega - m'\omega) \tag{9.275}$$

where $\delta(m\omega - m'\omega)$ represents the delta function. To eliminate the delta function, notice that the density of states in the Einstein model is $(\hbar\omega)^{-1}$. It follows that

$$k_{ba}(\beta) = \left(\frac{2\pi}{\hbar^2\omega}\right)|H_{ab'}|^2 \exp[-S(1 + 2\bar{n})] \sum_{m=0}^{\infty} \frac{[S^2\bar{n}(\bar{n} + 1)]^m}{(m!)^2} \tag{9.276}$$

or

$$\frac{k_{ba}(\beta)}{k_{ba}(\infty)} = \exp(-2S\bar{n}) \sum_{m=0}^{\infty} \frac{[S^2\bar{n}(\bar{n} + 1)]^m}{(m!)^2} \tag{9.277}$$

where $k_{ba}(\infty)$ is the rate constant at $T = 0$. For convenience of displaying the temperature effect on the rate constant, they define

$$k_{ba}^* = \frac{k_{ba}(\beta)}{k_{ba}(\infty)}, \qquad T^* = \frac{kT}{\hbar\omega} \tag{9.278}$$

From the numerical calculation they show that for $S = 1$, k_{ba}^* decreases with T^*; for $S = 5$, k_{ba}^* first decreases with T^* and then increases with T^*; and for $S = 10$ and $S = 15$, k_{ba}^* increases steadily with T^*. Notice the existence of the threshold in the temperature effect on the rate constant. For large S values we can see that the rate constant shows the Arrhenius temperature dependence with the activation energy $\frac{1}{4}S$, which is the potential energy in the minimum potential crossing in the multidimensional surfaces.

The Arrhenius behavior of the rate constant is usually shown by using the so-called strong coupling approximation. It can be shown by first expressing $1/m!$ in eq. 9.277 by the contour integral

$$k_{ba}(T^*) = \exp(-2S\bar{n})(2\pi i)^{-1} \int_{\gamma} dz z^{-1} \exp\left[\frac{z + S\bar{n}(n + 1)}{z}\right] \tag{9.279}$$

and by evaluating the resulting contour integral by the saddle point method

$$k_{ba}(T^*) = \exp\{-2S\bar{n} + [1 + 4S^2\bar{n}(n + 1)^{1/2}\}\left\{2\pi\left[\frac{1 + 2S^2\bar{n}(\bar{n} + 1)}{z^*}\right]\right\}^{-1/2}$$

$$(9.280)$$

where $z^* = \frac{1}{2} + \frac{1}{2}[1 + 4S^2\bar{n}(\bar{n} + 1)]^{1/2}$. In the high T^* range, eq. 9.280 reduces to

$$k_{ba}(T^*) = \exp\left(\frac{S - S/4\bar{n} + 1}{4S\bar{n}}\right)\{4\pi[S^2\bar{n}(\bar{n} + 1)]^{1/2}\}^{-1/2} \qquad (9.281)$$

To consider the damping effect on the rate constant, Ma et al. replaced the delta function in the rate constant expression by the corresponding Lorentzian:

$$k_{ba}(\beta) = \left(\frac{|H'_{ab}|^2}{\hbar^2}\right)\exp(-S)\int_{-\infty}^{\infty}dt$$

$$\times\exp\left[it\omega_{ab'} - it\sum_j\phi_j\omega'_j\bar{n}_j + \frac{1}{2}\sum\Delta_j^2\exp(it\omega'_j)\right.$$

$$\left.+ \sum_j\bar{n}_j\Delta_j^2(\cos\omega'_jt - 1) - \hbar^{-1}|t|\Delta_{ab}\right] \qquad (9.282)$$

where Δ_{ab} represents the average damping constant. For the case $\omega_{ab'} = 0$ and $\phi_j = 0$, for the Einstein model eq. 9.282 reduces to

$$k_{ba}(\beta) = \left(\frac{|H'_{ab}|^2}{\hbar^2}\right)\exp[-S(2\bar{n} + 1)]\sum_{m=0}\sum_{m'=0}$$

$$\times\frac{S^{m+m'}\bar{n}^m(1 + \bar{n})^{m'}}{m!m'!}\frac{2\Delta_{ab}/\hbar}{\Delta_{ab}/\hbar^2 + (m - m')^2\omega^2} \qquad (9.283)$$

It follows that

$$k_{ba}(T^*) = \frac{\exp(-2S\bar{n})}{I(\Delta)}\sum_{m=0}\sum_{m'=0}\frac{S^{m+m'}\bar{n}^{m(1+\bar{n})^{m'}}}{m!m'!}\frac{\Delta^2}{\Delta^2 + (m - m')^2} \qquad (9.284)$$

where $\Delta = \Delta_{ab}/\hbar\omega$ and

$$I(\Delta) = \sum_{m=0}^{\infty}\frac{(S^m/m!)\Delta^2}{m^2 + \Delta^2} \qquad (9.285)$$

As $\Delta \to 0$, eq. 9.284 reduces to eq. 9.277.

For $\Delta = 0.01$, the damping effect is negligible except for the $S = 15$ case. For $\Delta = 0.1$, however, the damping effect is significant for both $S = 10$ and $S = 15$.

Table 9.14
Numerical Comparison[a]

T^*	$\ln k_{ba}^*(T^*)$exact	$\ln k_{ba}^*(T^*)_1$	$\ln k_{ba}^*(T^*)_2$	T^*	$\ln k_{ba}^*(T^*)$exact	$\ln k_{ba}^*(T^*)_1$	$\ln k_{ba}^*(T^*)_2$
$S=1$				$S=5$			
0.8	−0.30	−0.21	−0.07	0.3	0.43	0.52	0.14
1.2	−0.47	−0.38	−0.16	0.6	1.30	1.36	0.60
1.6	−0.60	−0.52	−0.25	0.9	1.67	1.71	0.90
2.0	−0.69	−0.62	−0.33	1.2	1.84	1.87	1.05
2.4	−0.77	−0.71	−0.39	1.6	1.94	1.96	1.15
2.8	−0.84	−0.79	−0.46	2.0	1.97	1.99	1.18
				2.8	1.98	1.99	1.19
$S=10$				$S=15$			
0.3	1.61	1.69	0.31	0.3	2.98	3.04	0.48
0.6	3.97	4.00	1.40	0.6	6.78	6.81	2.19
1.2	5.46	5.47	2.56	1.2	9.23	9.24	4.08
2.0	6.00	6.01	3.07	2.0	10.18	10.18	4.96
2.8	6.18	6.19	3.25	2.8	10.53	10.54	5.31

[a] $\ln k_{ba}^*(T^*)_1$ refers to that given by eq. 9.280; $\ln k_{ba}^*(T^*)_2$ refers to that given by the use of the strong coupling approximation.

In Table 9.14 we compare the performance of $k_{ba}^*(T^*)$ given by eq. 9.80 and that given by the strong coupling approximation with the exact results. The rate constant in the strong coupling approximation can only provide the order of magnitude agreement, whereas the results given by eq. 9.280 are in good agreement with exact results in the whole temperature range.

The successful applications of the Arrhenius form of rate constants in rate processes are well known (Glasstone et al., 1940); numerous other sphere electron transfer reactions were also found to agree quite well with correlation made on the basis of the transition state theory (Cohen and Marcus, 1968; Meisel, 1975). However, in polaron migration experiments (Burshtein and Williams, 1978; Schein, 1977), it is found that the mobility is very weakly dependent on temperature, and it cannot be represented by the Arrhenius form. Measurements of exciton migration in anthracene (Ern et al., 1972) have found that the diffusion coefficient decreases with temperature whereas the Arrhenius law predicts an increase. Other spectroscopic experimental work (Wolf and Hoover, 1970) tends to verify this trend. Recent numerical calculations of the temperature dependence of the rate constant based on the theory described in this section showed that for the weak coupling ease ($S \sim 1$), the rate constant for the motion between equivalent sites does indeed decrease with temperature (Ma et al., 1978).

REFERENCES

Allnatt, A. R. and P. W. M. Jacobs (1968), *Can. J. Chem.*, **46**, 111.

Amis, E. S. (1966), "Solvent Effects on Reaction Rates and Mechanisms," Academic Press, New York.

Anbar, M. and E. J. Hart (1969), *in Radiation Chemistry*, Vol. I, E. J. Hart, Ed., American Chemical Society, p. 79.

Avrami, M. (1941), *J. Chem. Phys.*, **9**, 177.

Bagdassarian, C. H. (1945), *Acta Physicochim. U.R.S.S.*, **20**, 441.

Banfield, T. L. and D. Husain (1969), *Trans. Faraday Doc.*, **65**, 1985.

Barnett, J. and J. H. Baxendale (1956), *Trans. Faraday Doc.*, **52**, 210.

Bell, J. A. and H. Linschitz (1963), *J. Am. Chem. Soc.*, **85**, 528.

Binet, D. J., E. J. Goldberg, and L. S. Foster (1968), *J. Phys. Chem.*, **72**, 3017.

Booth, D. and R. M. Noyes (1961), *J. Am. Chem. Soc.*, **57**, 859.

Booth, F. (1948), *Trans. Faraday Soc.*, **44**, 796.

Brent, J. P. and K. Trone (1971), *Transfer Coefficients in Electrochemical Kinetics*, Academic.

Brower, K. R. and J. S. Chen (1965), *J. Am. Chem. Soc.*, **87**, 3396.

Burschtein, Z. and D. F. Williams (1978), *J. Chem. Phys.*, **69**, 983.

Carslaw, H. S. and J. C. Jaeger (1959), *Conduction of Heat in Solids*, Oxford University Press.

Carter, R. (1960), *J. Chem. Phys.*, **34**, 2010.

Chandrasekhar, S. (1943), *Rev. Mod. Phys.*, **15**, 1.

Chuang, T. J., G. W. Hoffman, and K. B. Eisenthel (1974), *Chem. Phys. Lett.*, **25**, 201.

Cohen, A. Ö. and R. A. Marcus (1968), *J. Phys. Chem.*, **72**, 891, 4249.

Crank, J. (1956), *The Mathematics of Diffusion*, Oxford University Press.

Cundall, R. B., G. B. Evans, and E. J. Land (1969), *J. Phys. Chem.*, **73**, 3892.

Czapski, G. and E. Peled (1973), *J. Phys. Chem.*, **77**, 893.

Czapski, G. and H. A. Schwarz (1962), *J. Phys. Chem.*, **66**, 471.

Danckwerts, D. (1951), *Trans. Faraday Soc.*, **42**, 1014.

Davies, C. W. (1961), *Prog. React. Kinet.*, **1**, 161.

Debye, P. (1942), *Trans. Electrochem. Soc.*, **82**, 265.

Dexter, D. L. (1953), *J. Chem. Phys.*, **21**, 836.

Douglas, D. L., Ed. (1971), *Oxidation of Metals and Alloys*, American Society of Metals, 1971.

Eigen, M. (1954), *Discuss. Faraday Soc.*, **17**, 194.

Eigen, M. (1963), *Prog. React. Kinet.*, **2**, 285.

Elkana, Y., J. Feitelson, and E. Katchalski (1968), *J. Chem. Phys.*, **48**, 2399.

Engleman, R. and J. Jortner (1970), *Mol. Phys.*, **18**, 145.

Ern, V., A. Suna, Y. Tomkiewiecz, P. Avakian, and R. P. Groff (1972), *Phys. Rev.*, **B5**, 3222.

Erofeev, B. V. (1946), *Compt. Rend. Acad. Sci., U.R.S.S.*, **52**, 511.

Evans, G. T. and M. Fixman (1976), *J. Phys. Chem.*, **80**, 1544.

Eyring, H. and M. S. Jhon (1969), *Significant Liquid Structures*, Wiley-Interscience.

Eyring, H. and D. Urry (1963), *Z. Elektrochem.*, **67**, 731.

Eyring, H., D. Henderson, B. Stover, and E. M. Eyring (1964), *Statistical Mechanics and Dynamics*, Wiley-Interscience.

Feitelson, J. (1966a), *J. Chem. Phys.*, **44**, 1497.

Feitelson, J. (1966b), *J. Chem. Phys.*, **44**, 1500.

Felderhof, B. U. and J. M. Deutch (1976), *J. Chem. Phys.*, **64**, 4551, 4559.

Fischer, S. (1970), *J. Chem. Phys.*, **53**, 3195.

Fitts, D. D. (1962), *Nonequilibrium Thermodynamics*, McGraw-Hill.

Fong, F. K. and D. J. Diestler (1972), *J. Chem. Phys.*, **57**, 4953.

Fong, F. K. (1976), *Theory of Molecular Relaxation*, Wiley-Interscience.

Forgues, C., J. Leibovitz, R. O'Brien, and K. S. Spiegler (1975), *Electrochimica*, **20**, 555.

Förster, Th. (1948), *Ann. Phys.*, **2**, 55.

Fry, A. J., R. S. H. Liu, and G. S. Hammond (1966), *J. Am. Chem. Soc.*, **88**, 4781.

Glasstone, S., K. J. Laidler, and H. Eyring (1940), *Theory of Rate Processes*, McGraw-Hill.

Hamil, W. H. (1969), *J. Phys. Chem.*, **73**, 1341.

Hammond, G. S. and R. P. Foss (1964), *J. Phys. Chem.*, **63**, 3739.

Hannay, N. B., Ed. (1976), *Treatise on Solid State Chemistry*, Vol. 4, Plenum.

Harned, J. S. and B. B. Owens (1958), *The Physical Chemistry of Electrolyte Solutions*, Van Nostrand Reinhold.

Helman, W. P. and K. Funabashi (1977), *J. Chem. Phys.*, **66**, 5790.

Hill, C. O. and S. H. Lin (1969), *Int. J. Quantum Chem.*, **3S**, 315.

Hill, C. O. and S. H. Lin (1970), *J. Chem. Phys.*, **53**, 608.

Hoijtink, G. J. (1960), *Mol. Phys.*, **3**, 67.

Hunt, H. R. and H. Taube (1958), *J. Am. Chem. Soc.*, **80**, 2642.

Inokuti, M., and F. Hirayama (1965), *J. Chem. Phys.*, **43**, 1978.

Kautz, C. F. and A. S. Robinson (1928), *J. Am. Chem. Soc.*, **50**, 1022.

Kira, A. and Koizumi, M. (1967), *Bull. Chem. Soc. Jap.*, **40**, 2486.

Kirkwood, J. G. (1946), *J. Chem. Phys.*, **14**, 180.

Knittel, D. R., S. Pack, H. Inaba, S. H. Lin, and L. Eyring (1977), *Kinetic Models of Solid State Reactions. I. Technical Report*, Arizona State University.

Kubo, R. (1957), *J. Phys. Soc. Jap.*, **12**, 570.

LaMer, V. K. and R. W. Fessenden (1932), *J. Am. Chem. Soc.*, **54**, 2351.

Landauer, R. and J. A. Swanson (1961), *Phys. Rev.*, **121**, 1668.

Lawler, R. G. (1973), *Prog. Nucl. Magn. Resonance Spectrom.*, **9**, 3.

LeNoble, W. J. (1967), *Prog. Phys. Org. Chem.*, **5**, 207.

Levich, V. G. (1965), *Adv. Electrochem. Electrochem. Eng.*, **4**, 249.

Lin, S. H. (1966), *J. Chem. Phys.*, **44**, 3759.

Lin, S. H. (1970), *J. Chem. Phys.*, **53**, 3766.

Lin, S. H. (1971), *Mol. Phys.*, **21**, 853.

Lin, S. H. (1972), *J. Chem. Phys.*, **56**, 2648.

Lin, S. H. (1973), *J. Chem. Phys.*, **58**, 5760.

Lin, S. H. and R. Bersohn (1968), *J. Chem. Phys.*, **48**, 2732.

Lin, S. H. and H. Eyring (1972), *Proc. Nat. Acad. Sci. U.S.*, **69**, 3192.

Lin, S. H., T. Groy, F. Carney, H. Inaba, and L. Eyring (1978), *Kinetic Models of Solid State Reactions*: II. Technical Report, Arizona State University.

Lin, S. H., K. H. Lau, and H. Eyring (1971), *J. Chem. Phys.*, **55**, 5657.

Livingston, R. S. (1926), *J. Am. Chem. Soc.*, **48**, 53.

Ma, S. M., S. H. Lin, D. Wutz, Y. Fujimura, and H. Eyring (1978), *Chem. Phys. Lett.*, **58**, 159.

McGlaughlin, E. (1959), *Trans. Faraday Soc.*, **55**, 28.

Meisel, D. (1975), *Chem. Phys. Lett.*, **34**, 263.

Melhuish, W. H. (1963), *J. Phys. Chem.*, **67**, 1681.

Miller, J. R. (1975), *J. Phys. Chem.*, **79**, 1071.

Moelwyn-Hughes, E. A. (1972), *Chemical Statics and Kinetics of Solutions*, Academic.

Monchick, L., J. L. Magee, and A. H. Samuel (1957), *J. Chem. Phys.*, **26**, 935.

Monchick, L. (1975), *J. Chem. Phys.*, **62**, 1907.

Montroll, E. W. (1946), *J. Chem. Phys.*, **14**, 202.

Montroll, E. W. and G. H. Weiss (1965), *J. Math. Phys.*, **6**, 167.

North, A. M. (1964), *The Collision Theory of Chemical Reactions in Liquids*, Methuen.

North, A. M. and G. A. Reed (1961), *Trans. Faraday Soc.*, **57**, 859.

Nowick, A. S. and J. J. Burton, Eds. (1975), *Diffusion in Solids*, Academic; Tanaka, S., J. D. Clewley, and T. B. Flanagan (1977), *J. Phys. Chem.*, **81**, 1684.

Noyes, R. M. (1961), *Prog. React. Kinet.*, **1**. 129.

Olson, A. R. and T. R. Simonson (1949), *J. Chem. Phys.*, **17**, 1167.

Osborne, A. D. and G. Porter (1963), *Int. Symp. Free Radicals*, Cambridge.

Osborne, A. D. and G. Porter (1965), *Proc. Roy. Soc.* (Lond.), **A284**, 9.

Portman, L. M. and M. S. Matheson (1965), *Prog. React. Kinet.*, **3**, 237.

Rise, S. A. and M. Nagasawa (1961), *Polyelectrolyte Solutions*, Academic.

Scheider, W. (1972), *J. Phys. Chem.*, **76**, 349.

Schein, L. B. (1977), *Phys. Rev.*, **15**, 1024.

Schmitt, P. P. (1972), *J. Chem. Phys.*, **56**, 2775.

Schwarz, H. A. (1971), *J. Chem.*, **55**, 3647.

Seth, B. and H. Ross (1965), *Trans. TMS-AIME*, **233**, 180.

Sole, K. and W. H. Stockmayer (1971), *J. Chem. Phys.*, **54**, 2981.

Spiegler, K. S. (1966), *Principles of Desalination*, Academic.

Steinberg, I. Z. and E. Katchalski (1968), *J. Chem. Phys.*, **48**, 2404.

Stranks, D. R. and T. W. Swaddle (1971), *J. Am. Chem. Soc.*, **93**, 2783.

Summer, W. A., Jr., and J. G. Burr (1972), *J. Phys. Chem.*, **76**, 3137.

Swell, G. L. (1963), *Phys. Rev.*, **129**, 597.

Umberger, J. Q. and V. K. LaMer (1945), *J. Am. Chem. Soc.*, **67**, 1099.

Vando, A. F. and D. M. Hercules (1970), *J. Am. Chem. Soc.*, **92**, 3573.

Wagner, P. J. and I. Kochevar (1968), *J. Am. Chem. Soc.*, **90**, 2232.

Waite, T. R. (1957), *J. Chem. Phys.*, **58**, 4009 (1973).

Ware, W. R. and T. L. Nemzek (1973), *Chem. Phys. Lett.*, **23**, 557.

Ware, W. R. and T. L. Nemzek (1975), *J. Chem. Phys.*, **62**, 477.

Weller, A. (1957), *Z. Phys. Chem.*, **13**, 335.

Weller, A. (1961), *Prog. React. Kinet.*, **1**, 187.

Weston, R., Jr., and H. A. Schwarz (1972), *Chemical Kinetics*, Prentice-Hall.

Wilemski, G. and M. Fixman (1973), *J. Chem. Phys.*, **58**, 4009.

Wilemski, G. and M. Fixman (1974), *J. Chem. Phys.*, **60**, 866, 878.

Wilkins, R. G. (1974), *The Study of Kinetics and Mechanisms of Reactions of Transition Metal Complexes*, Allyn and Bacon.

Wolf, H. C. and D. Haarer (1970), *Mol. Cryst. Liq. Cryst.*, **10**, 359.

Wolynes, P. G. and J. M. Deutch (1976), *J. Chem. Phys.*, **65**, 450; and the references given therein.

Yamamoto, T. (1960), *J. Chem. Phys.*, **33**, 281.

Yguerabide, J., M. A. Dillon, and M. Burton (1964), *J. Chem. Phys.*, **40**, 3040.

Appendix One

Laplace Transformation

The Laplace transformation is defined by

$$\bar{x}(p) = L[x(t)] = \int_0^\infty e^{-pt} x(t)\, dt \tag{1}$$

and $\bar{x}(p)$ or $L[x(t)]$ is called the Laplace transform of $x(t)$. The term p in eq. 1 is a real positive number large enough to make the integral converge. For example, if $x(t) = e^{at}$, we must have $p > a$. In the following discussion we list some important theorems on the Laplace transformation that can be proved easily (see, for example, Carslaw and Jaeger, 1963; Crank, 1956; Morse and Feshbach, 1953).

Theorem I. If $\bar{x}_1(p) = L[x_1(t)]$ and $\bar{x}_2(p) = L[x_2(t)]$, then

$$\bar{x}_1(p) \pm \bar{x}_2(p) = L[x_1(t) \pm x_2(t)]$$

Theorem II. If $x(t)$ is differentiable for all $t > 0$ and $x(t)$ and dx/dt both have Laplace transforms, and if also $x(t) \to x_0$ as $t \to 0$ and $e^{-pt} x(t) \to 0$, as $t \to \infty$, then

$$L[dx/dt] = p[x(t)] - x_0 = p\bar{x}(p) - x_0. \tag{3}$$

Theorem III. If $\lim_{t \to \infty} (e^{-pt} \int_0^t x(r)\, dr) = 0$, then

$$L\left[\int_0^t x(r)\, dr \right] = \frac{1}{p} \bar{x}(p) \tag{4}$$

Theorem IV. If a is a constant, then

$$L[e^{-at} x(t)] = \bar{x}(p + a) \tag{5}$$

Theorem V. If $a > 0$, then

$$L[x(t - a)H(t - a)] = e^{-ap}\bar{x}(p) \tag{6}$$

where $H(t - a)$ represents the Heaviside function

$$\left.\begin{array}{ll} H(t - a) = 0, & t \leq a \\ H(t - a) = 1, & t > a \end{array}\right\} \tag{7}$$

Theorem VI. If $\bar{x}_1(p) = L[x_1(t)]$ and $\bar{x}_2(p) = L[x_2(t)]$, then

$$\bar{x}_1(p)\bar{x}_2(p) = L\left[\int_0^t x_1(r)x_2(t - r)\, dr\right] = L\left[\int_0^t x_1(t - r)x_2(r)\, dr\right] \tag{8}$$

Theorem VII. If two continuous functions $x_1(t)$ and $\chi_2(t)$ both have the same Laplace transform $\bar{x}(p)$, then they are identically equal.

Theorem VIII. If ω is a real number, then

$$L[x(\omega t)] = \frac{1}{\omega}\,\bar{x}\left(\frac{p}{\omega}\right) \tag{9}$$

Theorem IX. If $x(t)$ is periodic with period T, then

$$\bar{x}(p) = \frac{1}{1 - e^{-pT}} \int_0^T e^{-pt}x(t)\, dt \tag{10}$$

Theorem X. If $p \to \infty$ through real values, $\bar{x}(p) \to 0$, and if, in addition, the conditions of Theorem II are satisfied,

$$\lim_{p \to \infty} p\bar{x}(p) = \lim_{p \to 0} x(t) \tag{11}$$

Theorem XI. If $\bar{x}(p_0)$ exists and $p \to p_0 + 0$ through real values,

$$\bar{x}(p) \to \bar{x}(p_0) \tag{12}$$

In particular

$$\lim_{p \to 0} \bar{x}(p) = \int_0^\infty x(t)\, dt \tag{13}$$

and

$$\lim_{p \to 0} p\bar{x}(p) = \lim_{t \to \infty} x(t) \tag{14}$$

provided that the quantities on the right-hand sides of eqs. 13 and 14 exist, and for eq. 14, that in addition the conditions of Theorem II are satisfied.

Theorem XII. *If* $\bar{x}(p) = L[x(t)]$ and $L[K(t, u)] = \phi(p)\exp[-u\psi(p)]$, where $\phi(p)$ and $\psi(p)$ are independent of u, then

$$L\left[\int_0^\infty K(t, u)x(u)\,du\right] = \phi(p)\bar{x}[\psi(p)] \tag{15}$$

Theorem XIII. If $x(t)$ has a continuous derivative, $|x(t)| < Ke^{ct}$, where K and c are constants, and

$$\bar{x}(p) = L[x(t)] = \int_0^\infty e^{-pt}x(t)\,dt,\ \text{Re}(p) > c \tag{16}$$

then

$$x(t) = \frac{1}{2\pi i}\lim_{\omega \to \infty}\int_{\gamma - i\omega}^{\gamma + i\omega} e^{pt}\bar{x}(p)\,dp = L^{-1}[\bar{x}(p)] \tag{17}$$

where $\gamma > c$. This is, of course, the inversion theorem.

Theorem XIV. If $\bar{x}(p)$ can be expanded in the power series

$$\bar{x}(p) = \sum_{n=1}^\infty a_n p^{-n} \tag{18}$$

which has a finite radius of convergence, then

$$x(t) = \sum_{n=1}^\infty \frac{a_n t^{n-1}}{(n-1)!} \tag{19}$$

Theorem XV. If we let $\bar{x}(p)$ be the Laplace transform of the real function $x(t)$, then

$$\frac{1}{2\pi}\int_{-\infty}^\infty |\bar{x}(\gamma + iy)|^2\,dy = \int_0^\infty e^{-2\gamma t}[x(t)]^2\,dt \tag{20}$$

Next we give some examples to show the applications of the Laplace transformation.

Example 1. If $\bar{x}(p) = p + 8/p^2 + 4p + 5$, find $x(t)$. The term $\bar{x}(p)$ can be written as

$$\bar{x}(p) = \frac{(p + 2) + 6}{(p + 2)^2 + 1} = \frac{p + 2}{(p + 2)^2 + 1} + \frac{6}{(p + 2)^2 + 1} \tag{21}$$

By using Theorem IV, we obtain

$$x(t) = e^{-2t}(\cos t + 6 \sin t) \tag{22}$$

Example 2. This example is concerned with the application of Theorem XII. Important special cases can be written down by choosing functions $K(t, u)$ from the Table of Laplace Transforms whose Laplace transforms have the required form. For example,

$$K(t, u) = \frac{1}{\sqrt{\pi t}} e^{-u^2/4t}, \quad \bar{K}(p, u) = \frac{e^{-u\sqrt{p}}}{\sqrt{p}} \tag{23}$$

Theorem XII then gives the result that

$$L\left[\frac{1}{\sqrt{\pi t}} \int_0^\infty e^{-u^2/4t} x(u) \, du\right] = \frac{1}{\sqrt{p}} \bar{x}(\sqrt{p}) \tag{24}$$

Example 3. Suppose we have to solve

$$\frac{d^2 x}{dt^2} + 2\frac{dx}{dt} + 2x = F(t), \quad t > 0 \tag{25}$$

with x, and dx/dt equal to x_0 and x_1 where $t = 0$. We carry out the Laplace transformation of eq. 25,

$$\int_0^\infty e^{-pt} \frac{d^2 x}{dt^2} \, dt + 2 \int_0^\infty e^{-pt} \frac{dx}{dt} \, dx + 2\bar{x}(p) = \bar{F}(p) \tag{26}$$

Using Theorem II we obtain

$$[p\{p\bar{x}(p) - x_0\} - x_1] + 2[p\bar{x}(p) - x_0] + 2\bar{x}(p) = \bar{F}(p) \tag{27}$$

or

$$\bar{x}(p) = \frac{px_0 + 2x_0 + x_1}{p^2 + 2p + 2} + \frac{\bar{F}(p)}{p^2 + 2p + 2} \tag{28}$$

The first term can be inverted easily. To find the function whose transform is the second term, we take $\bar{x}_1(p) = \bar{F}(p)$ and $\bar{x}_2(p) = (p^2 + 2p + 2)^{-1}$ in Theorem VI. Then $x_1(t) = F(t), x_2(t) = e^{-t} \sin t$, and we obtain finally

$$x(t) = x_0 e^{-t} \cos t + (x_0 + x_1)e^{-t} \sin t + \int_0^t F(r)e^{-(t-r)} \sin(t - r) \, dr \tag{29}$$

Example 4. Suppose we want to deduce the function $x(t)$ whose Laplace transform $\bar{x}(p)$ is given by

$$\bar{x}(p) = \frac{f(p)}{g(p)} \tag{30}$$

We first put $\bar{x}(p)$ into partial fractions,

$$\frac{f(p)}{g(p)} = \sum_{r=1}^n \frac{Ar}{p - a_r} \tag{31}$$

Then

$$f(p) \equiv \sum_{r=1}^{n} A_r(p - a_1) \cdots (p - a_{r-1})(p - a_{r+1}) \cdots (p - a_n) \qquad (32)$$

and putting $p = a_r$ in this yields

$$f(a_r) = A_r(a_r - a_1) \cdots (a_r - a_{r-1})(a_r - a_{r+1}) \cdots (a_r - a_n) \qquad (33)$$

Substituting for A_r from eq. 33 into eq. 31 gives

$$\bar{x}(p) = \sum_{r=1}^{n} \frac{1}{p - a_r} \cdot \frac{f(a_r)}{(a_r - a_1) \cdots (a_r - a_{r-1})(a_r - a_{r+1}) \cdots (a_r - a_n)} \qquad (34)$$

Now since

$$g(p) = (p - a_1)(p - a_2) \cdots (p - a_n) \qquad (35)$$

we have,

$$g'(p) = \sum_{r=1}^{n} (p - a_1) \cdots (p - a_{r-1})(p - a_{r+1}) \cdots (p - a_n) \qquad (36)$$

Setting $p = a_r$ in eq. 36 yields

$$g'(a_r) = (a_r - a_1) \cdots (a_r - a_{r-1})(a_r - a_{r+1}) \cdots (a_r - a_n) \qquad (37)$$

and using eq. 37 in eq. 34, $\bar{x}(p)$ can be written

$$\bar{x}(p) = \sum_{r=1}^{n} \frac{f(a_r)}{(p - a_r)g'(a_r)} \qquad (38)$$

On carrying out the inverse transformation of eq. 38 we obtain immediately

$$x(t) = \sum_{r=1}^{n} \frac{f(a_r)}{g'(a_r)} \exp(a_r t) \qquad (39)$$

This result applies only to the case in which $g(p)$ has no repeated zeros, but it can readily be generalized for the case of repeated factors (Crank, 1956; Carslaw and Jaeger, 1963). Thus eq. 39 implies that to each linear factor $p - a_r$ of the denominator of $\bar{x}(p)$ there corresponds a term $f(a_r)/g'(a_r)\exp(a_r t)$ in the solution. The generalization is that, to each squared factor $(p - b)^2$ of the denominator of $^1(p)$ there corresponds a term

$$t\left[\frac{(p - b)^2 f(p)}{g(p)}\right]_{p=b} \exp(bt) + \left[\frac{d}{dp}\left\{\frac{(p - b)^2 f(p)}{g(p)}\right\}\right]_{p=b} \exp(bt) \qquad (40)$$

in the solution. To each multiple factor $(p - c)^m$ of the denominator of $\bar{x}(p)$ there corresponds a term

$$\sum_{s=0}^{m-1} \left[\frac{d^s}{dp^s} \left\{ \frac{(p - c)^m f(p)}{g(p)} \right\} \right]_{p=c} \cdot \frac{t^{m-s-1}}{s!(m - s - 1)!} \exp(ct) \qquad (41)$$

in the solution.

The Laplace transform method has been widely applied to kinetics of heterogeneous chemical reactions (Lin and Eyring, 1970, 1971a) and to the solution of the time-dependent Schrödinger equation (Lin and Eyring, 1971b; Sasakawa, 1966).

REFERENCES

Carslaw, H. S. and J. C. Jaeger (1963), *Operational Methods in Applied Mathematics*, Dover.

Crank, J. (1956), *The Mathematics of Diffusion*, Oxford University Press.

Lin, S. H. and H. Eyring (1970), *Proc. Natl. Acad. Sci. U.S.*, **65**, 47.

Lin S. H. and H. Eyring (1971a), *Proc. Natl. Acad. Sci. U.S.*, **68**, 777.

Lin S. H. and H. Eyring (1971b), *Proc. Natl. Acad. Sci. U.S.*, **68**, 76.

Morse, P. M. and H. Feshbach (1953). *Methods of Theoretical Physics*, Vol. I, McGraw-Hill.

Sasakawa, T. (1966), *J. Math. Phys.*, **7**, 721.

Appendix Two

The Method of Steepest Descent (Saddle Point Method)

To discuss the method of steepest descent we follow the concise treatment given by Kubo (Kubo, 1965; Kittel, 1958) by considering the complex integral

$$F(E) = \frac{1}{2\pi i} \int_{\gamma - i\infty}^{\gamma + i\infty} Q(z) \exp(zE) \, dz \tag{1}$$

Here we should notice that the inverse Laplace transform is of the same form as eq. 1. If there are no singularities of the integral in the right half-plane, the path of integration may thus be taken as an arbitrary line parallel to the imaginary axis, and we can write the preceding integral as

$$F(E) = \frac{1}{2\pi} \int_{-\infty}^{\infty} \exp[f(z)] \, d\alpha \tag{2}$$

Here the relation $z = \gamma + i\alpha$ has been used. The term $f(z)$ in eq. 2 is defined by

$$f(z) = \log Q(z) + zE \tag{3}$$

Expanding $f(z)$ about the point γ in the imaginary direction yields

$$f(z) = \log Q(\gamma) + \gamma E + \left[E + \left(\frac{\partial \log Q}{\partial z} \right)_{z = \gamma} \right] (i\alpha) + \sum_{n=2}^{\infty} \frac{1}{n!} \left(\frac{\partial^n \log Q}{\partial z^n} \right)_{z = \gamma} (i\alpha)^n \tag{4}$$

458

If the function $f(\gamma)$ has a minimum $[\partial^2 \log Q/\partial z^2)_{z=\gamma} > 0]$ at a point $\beta^* = (\beta^*, 0)$ given by the condition

$$f'(\beta^*) = 0 \quad \text{or} \quad \left(\frac{\partial \log Q}{\partial z}\right)_{z=\beta^*} = -E \tag{5}$$

this must be a saddle point in the complex plane. Since $f''(\beta^*)$ is real, it follows that the steepest descent of $f(z)$ and hence the whole integrand is in the imaginary direction. Choosing γ to have the saddle point value β^* given by eq. 5, eq. 4 can be written

$$f(z) = \log Q(\beta^*) + \beta^* E + \sum_{n=2}^{\infty} \frac{1}{n!} \left(\frac{\partial^n \log Q}{\partial z^n}\right)_{z=\beta^*} (i\alpha)^n \tag{6}$$

Substituting eq. 6 into eq. 2, we obtain,

$$F(E) = \frac{Q(\beta^*)}{2\pi} e^{\beta^* E} \int_{-\infty}^{\infty} d\alpha \exp\left[\sum_{n=2}^{\infty} \frac{1}{n!} \left(\frac{\partial^n \log Q}{\partial z^n}\right)_{z=\beta^*} (i\alpha)^n\right] \tag{7}$$

Introducing the notation

$$b_n = \frac{1}{n!} \left(\frac{\partial^n \log Q}{\partial z^n}\right)_{z=\beta^*} \tag{8}$$

we can write $F(E)$ as

$$F(E) = \frac{Q(\beta^*)}{2\pi} e^{\beta^* E} \int_{-\infty}^{\infty} d\alpha \, e^{-b_2\alpha^2} \sum_{n=0}^{\infty} B_n \alpha^n \tag{9}$$

where the B_n are coefficients of α^n in the following product expression:

$$\prod_{k=3}^{\infty} \exp[b_k(i\alpha)^k] = \sum_{n=0}^{\infty} B_n \alpha^n \tag{10}$$

It can easily be shown that $B_0 = 1$. Thus to the lowest order of approximation, eq 9 can be integrated to give

$$F_1(E) = \frac{Q(\beta^*)}{2\pi} e^{\beta^* E} \int_{-\infty}^{\infty} d\alpha e^{-b_2\alpha^2} = \frac{Q(\beta^*)}{2\pi} e^{\beta^* E} \sqrt{\frac{\pi}{b_2}}$$

$$= \frac{Q(\beta^*) e^{\beta^* E}}{[2\pi(\partial^2 \log Q/\partial z^2)_{z=\beta^*}]^{1/2}} \tag{11}$$

Now the integrals in eq. 9 for odd n vanish, and since it is the coefficients of α^n alone that contain a factor of i, $F(E)$ is correctly exhibited as a real function. Carrying out the integrations in eq. 9 and reindexing the summation, we obtain

$$F(E) = \frac{Q(\beta^*)}{2\pi} e^{\beta^* E} \left[\sum_{k=1}^{\infty} B_2 k \cdot \frac{1 \cdot 3 \cdot 5 \cdots (2k-1)}{2^k} \sqrt{\frac{\pi}{b_2^{2k+1}}} + \sqrt{\frac{\pi}{b_2}} \right] \quad (12)$$

where the relation

$$\int_{\infty}^{\infty} d\alpha \, \alpha^{2k} e^{-b_2\alpha^2} = \frac{1 \cdot 3 \cdot 5 \cdots (2k-1)}{2^k} \sqrt{\frac{\pi}{b_2^{2k+1}}} \quad (13)$$

has been used. Equation 12 can be rewritten as

$$F(E) = F_1(E) \left[1 + \sum_{k=1}^{\infty} B_{2k} \cdot \frac{1 \, 3 \, 5 \cdots (2k-1)}{2k \cdot b_2^k} \right]$$

$$= F_1(E) \left[1 + \sum_{k=1}^{\infty} B_{2k} \cdot \frac{(2k)!}{2^{k \cdot} k!} \left(\frac{\partial^2 \log Q}{\partial z^2} \right)^{-k}_{z=\beta^*} \right] \quad (14)$$

The coefficients B_n are related to the derivatives b_k (Hoare, 1970; Lin and Eyring, 1971) and the first few even coefficients B_{2k} are given in the following:

$$B_2 = 0, \qquad B_4 = b_4, \qquad B_6 = -(b_6 + \tfrac{1}{2}b_3^2), \qquad B_8 = b_8 + \tfrac{1}{2}b_4^2 + b_3 b_5$$

$$B_{10} = -(b_{10} + \tfrac{1}{2}b_5^2 + b_4 b_5 + b_3 b_7 + \tfrac{1}{2}b_3^2 b_4)$$

$$B_{12} = b_{12} + \tfrac{1}{2}b_8^2 + b_5 b_7 + b_4 b_8 + \tfrac{1}{6}b_4^3 + b_3 b_9 + b_3 b_4 b_5 + \tfrac{1}{2}b_3^2 b_6 + \tfrac{1}{24}b_3^4$$

and so on. Their structure is of the form

$$B_{2k} = \sum_{\{i_n, j_n\}} (-1)^k \left[\frac{b_{i_1}^{j_1} b_{i_2}^{j_2} \cdots}{(j_1!)(j_2!)\cdots} \right] \quad (15)$$

with the summation over all possible integers i_n, j_n satisfying the condition

$$i_1 j_1 + i_2 j_2 + \cdots = 2k \quad (16)$$

The use of eq. 14 is actually more complicated than might appear since the terms as written do not as yet constitute a correct asymptotic expansion in the successive orders of approximation. The terms beyond the first in eq. 14 represent the high-order approximations in the method of steepest descent. The high-order approximations of the steepest-descent method have been obtained (Hoare, 1970; Lin and Eyring, 1971; Lau and Lin, 1971). In most cases the first-order approximation is sufficient. We now give the expression for $F(E)$ up to the second order of approximation of the method of steepest descent:

$$F_2(E) = F_1(E) \left[1 + \left(\frac{3}{4} \cdot \frac{b_4}{b_2^2} - \frac{15}{16} \cdot \frac{b_3^2}{b_2^3} \right) + \cdots \right] \quad (17)$$

Now let us show the application of the method of steepest descent by a couple of examples.

Example 1. Suppose we want to derive the Stirling formula. For this purpose we notice that the gamma function $\Gamma(n + 1)$, where $\Gamma(n + 1) = n!$ if n is an integer, can be expressed by the contour integral

$$\frac{1}{\Gamma(n + 1)} = \frac{1}{2\pi i} \oint \frac{dz}{z^{n+1}} e^z \tag{18}$$

To apply the method of steepest descent to eq. 18, we compare eq. 18 with eq. 1. Thus

$$f(z) = z(n + 1)\log z, \qquad \log Q(z) = -(n + 1)\log z \tag{19}$$

To find the saddle point we set $f'(\beta^*) = 0$, to obtain

$$\beta^* = n + 1 \tag{20}$$

Substituting eqs. 19 and 20 into eq. 11 yields

$$\frac{1}{\Gamma_1(n + 1)} = \frac{\beta^{*-(n+1)} e^{\beta^*}}{[2\pi(n + 1)/\beta^{*2}]^{1/2}} = \frac{e^{(n+1)}}{(n + 1)^n \sqrt{2\pi(n + 1)}} \tag{21}$$

or

$$\Gamma_1(n + 1) = \sqrt{2\pi}(n + 1)^{n+1/2} e^{-(n+1)} \tag{22}$$

Example 2. Suppose we are concerned with the evaluation of the following contour integral appearing in the Darwin–Fowler method (Fowler, 1955; Eyring et al., 1964):

$$F(E) = \frac{1}{2\pi i} \oint g(z) [\phi(z)]^E \frac{dz}{z} \tag{23}$$

where the contour is circulating once counterclockwise around $z = 0$, and both $g(z)$ and $\phi(z)$ are regular analytic functions except perhaps at $z = 0$. Furthermore, $\phi(z)$ is expressible in the form

$$\phi(z) = z^{-\alpha_0} [f_1(z)]^{\alpha_1} [f_2(z)]^{\alpha_2} \cdots \tag{24}$$

where the α_n's are positive constants, integral after multiplication by E, and $f_n(z)$'s are power series in z that start with nonzero constant terms and have real positive integral coefficients and radii of convergence unity.

To apply the method of steepest descent to eq. 23, we proceed by making the contour γ pass through the saddle point of $\phi(z)$ in such a direction that the value of the integrand falls off along γ from a maximum value at the saddle point at the greatest possible rate. This is here achieved by taking for the contour γ the circle $|z| = \theta$, where θ represents the root of $\phi'(z) = 0$. It is easy to show rigorously that when E is large all parts of the contour except that in the immediate neighborhood of $z = \theta$ make contributions exponentially small compared to this critical region (Fowler, 1955).

If we set $z = \theta e^{i\alpha}$, then when α is small

$$\log \phi(z) = \log \phi(\theta) - \frac{\alpha^2 \theta^2}{2} \cdot \frac{\phi''(\theta)}{\phi(\theta)} + \cdots \tag{25}$$

and

$$g(z) = g(\theta) + i\alpha g'(\theta) + \cdots \tag{26}$$

where $g'(\theta)$, $\phi''(\theta)$, and so on, represent the derivatives of $g(z)$ or $\phi(z)$ with respect to z evaluated at $z = \theta$. Substituting eqs. 25 and 26 into eq. 23 and neglecting the higher-order terms, we obtain

$$F_1(E) = \frac{[\phi(\theta)]^E}{2\pi} \int_{-\infty}^{\infty} [F(\theta) + i\alpha F'(\theta)] \exp\left[\frac{-\alpha^2 \theta^2}{2} E \cdot \frac{\phi''(\theta)}{\phi(\theta)} \right] d\alpha \tag{27}$$

The error in taking the range of integration with respect to α infinity instead of some small number is exponentially small. The odd term in eq. 27 vanishes and eq. 27 becomes

$$F_1(E) = \frac{F(\theta)[\phi(\theta)]^E}{[2\pi E \theta^2 (\phi''(\theta)/\phi(\theta))]^{1/2}} \tag{28}$$

REFERENCES

Eyring, H., D. Henderson, E. M. Eyring, and B. Stover (1964), *Statistical Mechanics and Dynamics*, Wiley-Interscience.

Fowler, R. H. (1955), *Statistical Mechanics*, Cambridge University Press.

Hoare, M. R. (1970), *J. Chem. Phys.*, **52**, 5695.

Kittel, C. (1958), *Elementary Statistical Physics*, Wiley-Interscience.

Kubo, R. (1965), *Statistical Mechanics*, North-Holland.

Lau, K. H. and S. H. Lin (1971), *J. Phys. Chem.*, **75**, 981.

Lin, S. H. and H. Eyring (1971), *Proc. Natl. Acad. Sci. U.S.*, **68**, 402.

Appendix Three

Stochastic Model
of the RRKM Theory

The RRKM theory is based on the assumption of the internal energy equilibration of the energized molecule and on the steady-state approximation of the distribution of energized molecules. In this appendix we derive the RRKM theory by using a stochastic model in which the vibrational relaxation and chemical processes of internal quantum states of the molecule are considered explicitly (Lin, 1972; Lin et al., 1972; Lin and Eyring, 1974; Widom, 1971).

In terms of detailed rate constants W_i and $W_{ij'}$ we rewrite the Lindemann scheme as

$$A_i + M \; \underset{w_{ji'}}{\overset{w_{ji'}}{\rightleftharpoons}} \; A_j, \qquad A_i \; \xrightarrow{w_i} \; P_i \tag{1}$$

where $i = 1, 2, 3, \ldots, N$, M is the collision partner and A_i represents the reactant molecule in the ith state. The molecule is regarded as energized when $i \geq n$, that is, $W_i = 0$ for $i = 1, 2, \ldots, n - 1$. From eq. 1 the rate of disappearance of A_i can be expressed as

$$-\frac{dA_i}{dt} = \sum_j{}' W_{ji} A_i - \sum_j{}' W_{ij} A_j + W_i A_i \tag{2}$$

From the operational viewpoint the unimolecular rate constant $k^{(1)}$ is defined by

$$\frac{dP}{dt} = k^{(1)} A \tag{3}$$

where $P = \sum_i P_i$ and $A = \sum_i A_i$. From eqs. 2 and 3 we obtain

$$k^{(1)} = \frac{\sum_i W_i A_i}{\sum A_i} \tag{4}$$

Applying the steady-state approximation to the energized molecules A_i ($i \geq n$) yields

$$A_i = \frac{\sum_j' W_{ij} A_j}{W_i + \sum_j' W_{ji}} \qquad (5)$$

Substituting eq. 5 into eq. 4, we obtain

$$k^{(1)} = \sum_i W_i \frac{\sum_j' W_{ij} f_j}{W_i + \sum_j' W_{ji}} \qquad (6)$$

where $f_j = A_j/\sum_i A_i$. If f_j is approximated by the normalized equilibrium thermal distribution function P_j, then, by the principle of detailed balance, $W_{ij} P_j = W_{ji} P_i$ and we obtain the unimolecular rate constant of the RRKM theory

$$k^{(1)} = \sum_i \frac{W_i P_i}{1 + (W_i/\sum_j' W_{ji})} \qquad (7)$$

in which the transition state theory is applied to calculate W_i. Replacing the summation in eq. 7 by integration, we obtain eq. 5.97 or eq. 5.98.

From eq. 6 we can see that $k^{(1)}$ depends on f_j, which in turn is determined by A_j; A_j should be determined from the solution of the master equation given by eq. 2. Thus in the short time range, $k^{(1)}$ may vary with time and become constant after a certain time (Lin et al., 1972); in view of the two approximations used in the RRKM theory, one has to examine the validity ond limitation of the RRKM theory. This has been carried out based on the solution of eqs. 1 and 2 (Lin et al., 1972).

REFERENCES

Lin, S. H. (1972), *J. Chem. Phys.*, **56**, 4155.

Lin, S. H., K. H. Lau, W. Richardson, L. Volk, and H. Eyring (1972), *Proc. Natl. Acad. Sci. U.S.*, **69**, 2778.

Lin, S. H. and H. Eyring (1974), *Ann. Rev. Phys. Chem.*, **25**, 39.

Widom, B. (1971), *J. Chem. Phys.*, **55**, 44.

Appendix Four

Problems

1. Use the l'Hospital rule to show that as $a \to 1$

$$\frac{1}{a-1}\left(\frac{1}{C_A^{a-1}} - \frac{1}{C_{A_0}^{a-1}}\right) = kt$$

and

$$t_{1/2} = \frac{2^{a-1} - 1}{(a-1)kC_{A_0}^{a-1}}$$

can be reduced to

$$\log \frac{C_A}{C_{A_0}} = -kt$$

and

$$t_{1/2} = \frac{\log 2}{k}$$

respectively. Similarly show that

$$\frac{1}{C_{B_0} - C_{A_0}} \log \frac{C_B}{C_A} = kt + \frac{1}{C_{B_0} - C_{A_0}} \log \frac{C_{B_0}}{C_{A_0}}$$

can be reduced to

$$\frac{1}{C_A} - \frac{1}{C_{A_0}} = kt$$

for $C_{A_0} = C_{B_0}$.

2. (a) Find the integrated expression for the reaction

$$A + B + C \to \text{products} \qquad (C_{A_0} \neq C_{B_0} \neq C_{C_0})$$

(b) Find the integrated expression for the reaction

$$2A + B \to \text{products} \qquad (C_{A_0} \neq 2C_{B_0})$$

3. (a) A second-order reaction (single component) is found to be 75% completed in 92 min when the initial concentration of reactant is 0.24 M. Find how long it will take for the concentration to reach 0.16 M under the same conditions.

   (b) The reaction $CH_3CH_2NO_2 + OH^- \rightarrow CH_3CHNO_2^- + H_2O$ was carried out at 273°K with an initial concentration of each reactant of 5.00×10^{-3} mol/l. The OH^- concentration fell to 2.00×10^{-3} after 5 min, 1.70×10^{-3} after 10 min, and 1.30×10^{-3} after 15 min. Show that the reaction was second order and determine k_2.

4. (a) Apply the method of separation of variables to

   $$A \underset{k_b}{\overset{k_f}{\rightleftharpoons}} B$$

   to find the integrated expressions for C_A and C_B.

   (b) The racemization of an optically active halide in solution is first order with respect to the reactant in each direction and the rate constants are equal, $R_1R_2R_3C \times$ (dextro) $\rightleftharpoons R_1R_2R_3C \times$ (laevo). If the initial reactant is pure dextro and the rate constant is 1.90×10^{-6} sec^{-1}, find (i) the time to 10% reaction and (ii) the percent reaction after 24 hr.

5. Apply the determinant method to the reactions

   $$A \xrightarrow{k_1} B \xrightarrow{k_2} C$$

   assuming the initial concentrations $C_{A_0} \neq 0$, and $C_{B_0} = C_{C_0} = 0$.

   (a) Write down the rate equations.

   (b) Find the eigenvalues of the solution.

   (c) Show that the solutions are given by

   $$C_A = C_{A_0} e^{-k_1 t}, \qquad C_B = \frac{k_1 A_0}{k_2 - k_1}(e^{-k_1 t} - e^{-k_2 t})$$

   $$C_C = C_{A_0} + \frac{A_0}{k_2 - k_1}(k_1 e^{-k_2 t} - k_2 e^{-k_1 t})$$

6. A gaseous reaction takes place between A and B. A is in great excess. The half-life $t_{1/2}$ as a function of initial pressure at 50°C is as follows:

P_A (mm)	500	125	250	250
P_B (mm)	10	15	10	20
$t_{1/2}$ (min)	80	213	160	80

   (a) Show that the rate expression is

   $$\text{rate} = kP_A P_B^2$$

(b) Evaluate the rate constant for concentration units of moles/liter and time in seconds.

7. Apply the determinant method to the reactions

(a) Write down the rate equations.
(b) Find the eigenvalues of the solution.
(c) Obtain the expressions for C_A, C_B, and C_C in terms of C_{A_0}, C_{B_0}, C_{C_0}, and time t.
(d) Find the equilibrium conditions and equilibrium concentrations of C_A, C_B, and C_C.

8. (a) Find the Laplace transforms of t^n, $\sin at$, and $(t/2a)\sin at$.
   (b) If $\bar{\chi}(p)$ is the Laplace transform of $\chi(t)$ and a is any constant, then $\bar{\chi}(p + a)$ is the Laplace transform of $e^{-at}\chi(t)$.
   (c) Find the Laplace transforms of $e^{-bt}t^n$, $e^{-bt}\sin at$, and $(t/2a)e^{-bt}\sin at$.

9. Suppose we consider the reaction that goes to completion

$$aA + bB \longrightarrow cC + dD$$

If a physical quantity D (e.g., pressure, optical density or absorbance) is linearly related to concentrations, then at any time we have

$$D = D_A + D_B + D_C + D_D = \varepsilon_A C_A + \varepsilon_B C_B + \varepsilon_C C_C + \varepsilon_D C_D$$

where ε_A, ε_B, ε_C, ε_D are proportional constants. Noticing that $C_A = C_{A_0} - ax$ and so on, show that

$$D - D_0 = x \, \Delta\varepsilon$$

where $\Delta\varepsilon = c\varepsilon_C + d\varepsilon_D - a\varepsilon_A - b\varepsilon_B$ and D_0 is the D value at $t = 0$. If at the completion of reaction, A disappears, show that

$$D_\infty - D_0 = \frac{C_{A_0}}{a} \Delta\varepsilon$$

$$\frac{D - D_0}{D_\infty - D_0} = \frac{ax}{C_{A_0}} \; ; \qquad \frac{C_A}{C_{A_0}} = \frac{D_\infty - D}{D_\infty - D_0}$$

and

$$C_B = C_{B_0} - C_{A_0} \cdot \frac{b}{a} \cdot \frac{D - D_0}{D_\infty - D_0}$$

10. (a) Starting from $\mathbf{J} = -D\nabla C$ for the three-dimensional case, derive the diffusion equation

$$\frac{\partial C}{\partial t} = D\nabla^2 C \tag{1}$$

(b) Notice that

$$\nabla^2 \equiv \frac{1}{r^2}\frac{\partial}{\partial r}\left(r^2\frac{\partial}{\partial r}\right) + \frac{1}{r^2\sin\theta}\frac{\partial}{\partial\theta}\left(\sin\theta\frac{\partial}{\partial\theta}\right) + \frac{1}{r^2\sin^2\phi}\frac{\partial^2}{\partial\phi^2} \tag{2}$$

Show that for the spherical-symmetric system, eq. 1 reduces to

$$\frac{\partial C}{\partial t} = \frac{D}{r^2}\frac{\partial}{\partial r}\left(r^2\frac{\partial C}{\partial r}\right) \tag{3}$$

(c) Solve eq. 3 by first letting $C = C_r/r$ and then applying the Laplace transformation with the boundary conditions as follows:

$$C \text{ finite as } r \to \infty$$

and

$$C = 0 \text{ at } r = R.$$

Show that the solution can be expressed as

$$C = C_0\left[1 - \frac{R}{r}\,\text{erfc}\,\frac{(r-R)}{\sqrt[2]{Dt}}\right]$$

where C_0 represents the initial concentration.

11. Consider the reactions

(a) Write down the rate equations.
(b) Apply the Laplace transformation to find $\bar{C}_A(P)$, $\bar{C}_B(P)$, and $\bar{C}_C(P)$.
(c) Obtain $C_A(t)$, $C_B(t)$, and $C_C(t)$. Let C_{A_0}, C_{B_0}, and C_{C_0} be the initial concentrations of A, B, and C, respectively.

12. The kinetics of the catalytic decomposition of NH_3 into the elements on a hot tungsten filament at $1100°C$ was found that the times $t_{1/2}$ required for

half of NH_3 to decompose with no N_2 or H_2 present at the start depended on the initial NH_3 pressures as follows:

P (torr)	$t_{1\,2}$ (min)
265	7.6
130	3.7
58	1.7

Determine (a) the order of the reaction, and (b) the rate constant.

13. Consider the reaction

$$A \xrightarrow{k_1} B \xrightarrow{k_2} C.$$

Solve the rate equations for this reaction by the Laplace transform method.

14. A hypothetical reaction is given by $\frac{1}{2}A + B \xrightarrow{k}$ products.
    (a) Find the expression for dC_A/dt.
    (b) Solve the rate equation.

15. A graduate student has synthesized an alkyl halide, RX, which solvolyzes in water according to the equation

$$RX + H_2O \longrightarrow ROH + HX$$

He determined the rate of this reaction (assumed to be first order with respect to RX) by measuring the conductivity of the solution, which is proportional to the concentration of HX. The following data are obtained:

Time (hr)	0	0.1	0.2	0.3	0.4	0.5	0.6	0.7	0.8	0.9
Λ^a	0	89.4	144.6	181.4	206.0	223.0	235.0	244.4	251.4	257.2

Time (hr)	1.0	1.2	1.4	1.6	1.8	2.0	∞
Λ^a	261.8	269.6	275.2	279.8	283.6	286.6	300.0

[a] Conductivity in arbitrary units.

(a) Determine if RX is a pure compound.
(b) If not, state what you can about the relative amounts of its components and their relative solvolysis rates.

Suggestion: Plot the data on semilogarithmic graph paper in the same way they would be plotted for a first-order reaction. How does this look when the elapsed time is large?

16. The rate of a gaseous reaction $A + B \rightarrow C$ is determined by measuring total pressure. The following data are obtained at $500°K$:

P_A (torr)	P_B (torr)	Initial rate (torr/min)
50	100	0.262
100	100	1.05
100	147	1.60
100	203	2.15
153	100	2.50
198	100	4.10

(a) Find the reaction order with respect to A and B.

(b) Determine the rate constant in the customary units.

17. (a) Suppose that

$$A \underset{k_b}{\overset{k_f}{\rightleftharpoons}} B + C.$$

Find the integrated expression for C_A assuming that C_B and C_C are zero at $t = 0$.

(b) Consider

$$A + B \underset{k_b}{\overset{k_f}{\rightleftharpoons}} C + D.$$

Find the integrated expression for C_A if only A and B are present initially and at concentrations C_{A_0} and C_{B_0}.

18. Use the steady-state approximation to set up the rate expressions for the following reactions. The intermediates whose concentrations can be considered very small are indicated in each case.

(a) Overall reaction: $H_2 + Br_2 \rightarrow 2\,HBr$. Intermediates: H, Br. Mechanism:

$$
\begin{array}{lll}
Br_2 + M & \longrightarrow\;\; 2\,Br + M & k_1 \\
Br + H_2 & \longrightarrow\;\; HBr + H & k_2 \\
H + Br_2 & \longrightarrow\;\; HBr + Br & k_3 \\
H + HBr & \longrightarrow\;\; H_2 + Br & k_4 \\
2Br + M & \longrightarrow\;\; Br_2 + M & k_{-1}
\end{array}
$$

(b) Overall reaction- $N_2O_5 \rightarrow 2\,NO_2 + \frac{1}{2}O_2$. Intermediates: NO, NO_3. Mechanism:

$$
\begin{array}{lll}
N_2O_5 & \longrightarrow\;\; NO_2 + NO_3 & k_1 \\
NO_2 + NO_3 & \longrightarrow\;\; N_2O_5 & k_{-1} \\
NO_2 + NO_3 & \longrightarrow\;\; NO + O_2 + NO_2 & k_2 \\
NO + NO_3 & \longrightarrow\;\; 2\,NO_2 & k_3
\end{array}
$$

(c) Overall reaction: $(CH_3)_2CO + X_2 \rightarrow CH_2XCOCH_3 + HX$. Intermediates: $(CH_3)_2COH^+$, $CH_2C(OH)CH_3$, $CH_2XC(OH)CH_3^+$. Mechanism (HA represents any protonic acid, X_2 a halogen):

$$(CH_3)_2CO + HA \longrightarrow (CH_3)_2COH^+ + A^- \qquad k_1$$
$$(CH_3)_2COH^+ + A^- \longrightarrow (CH_3)_2CO + HA \qquad k_{-1}$$
$$(CH_3)_2COH^+ + A^- \longrightarrow CH_2 = C(OH)CH_3 + HA \qquad k_2$$
$$CH_2 = C(OH)CH_3 + X_2 \longrightarrow CH_2XC(OH)CH_3^+ + X^- \qquad k_3$$
$$CH_2XC(OH)CH_3^+ + A^- \longrightarrow CH_2XCOCH_3 + HA \qquad k_4$$

19. The major path for thermal decomposition of di-t-butyl peroxide is $(CH_3)_3COOC(CH_3)_3 \rightarrow 2(CH_3)_2CO + C_2H_6$ [J. R. Raley, F. F. Rust, and W. E. Vaughan, *J. Amer. Chem. Soc.*, **70**, 88 (1948)]. The rate is followed by measuring the *total pressure* (P_t) as a function of time. Use the data given in the following, obtained at 147.2°C., to find the reaction order with respect to peroxide, the half-life, and the rate constant.

Time (min)	P_t (torr)^a	Time	P_t
0	179.5	26	252.5
2	187.4	30	262.1
6	198.6	34	271.3
10	210.5	38	280.2
14	221.2	40	284.9
18	231.9	42	288.9
20	237.3	46	297.1
22	242.3		

^{a} Corrected for pressure of N_2 (3.1 torr) used to force peroxide into reaction vessel.

20. Consider the isotropic exchange reaction

$$AX + BX^* \longrightarrow AX^* + BX$$

with $(AX^*) + (BX^*) \ll (AX) + (BX)$. Show that the exchange is a first-order reaction, regardless of the mechanism, and that

$$-\log(1 - F) = \frac{R[(AX) + (BX)]t}{(AX)(BX)}$$

where $F = (AX^*)_{t=t}/(AX^*)_{t=\infty}$ and R is the gross rate of exchange of X atoms (i.e., regardless of labeling) between AX and BX. [This expression was originally derived by H. A. C. McKay., *Nature*, **142**, 997 (1938) and *J. Amer. Chem. Soc.*, **65**, 702 (1943).]

Suggestion: The rate at which X^* in BX^* can exchange with X in AX is proportional to the fraction $(BX^*)/[(BX^*) + (BX)]$ and to the fraction $(AX)/[(AX^*) + (AX)]$. Similarly for the reverse exchange.

21. The expression derived in the preceding problem is applicable to the ferrous-ferric exchange reaction [J. Silverman and R. W. Dodson, *J. Phys. Chem.*, **56**, 846 (1952)].
    (a) Find R from the following data:

$$(Fe(III)) = 3.11 \times 10^{-4} \, M, \qquad (Fe(II)) = 3.95 \times 10^{-4} \, M,$$

$$(HClO_4) = 0.547 \, M$$

t (sec)	0	80	330	580	890	1180	1450
F	0.00	0.07	0.34	0.43	0.57	0.67	0.74

(b) Find the order of reaction with respect to ferrous ion and ferric ion from the following data [$(HClO_4) = 0.547 \, M$]:

$(Fe^{III}) \times 10^4 \, M$	$(Fe^{II}) \times 10^4 \, M$	$R \times 10^8 \, M/\text{sec}$
1.03	0.99	1.38
1.05	1.97	2.84
1.06	2.94	4.11
1.08	3.92	5.61
2.10	3.94	11.2
3.11	3.95	16.5
4.24	3.85	21.9
4.18	3.92	23.1

22. Consider the modified Lindemann scheme

$$A \underset{k_{-1}}{\overset{k_1}{\rightleftarrows}} B \longrightarrow C$$

Solve the rate equations for this reactions by the Laplace transform method. Show that

$$C_A^* = \frac{1}{\lambda_1^* - \lambda_2^*} [(1 - \lambda_2^*)e^{-\lambda_1^* t^*} + (\lambda_1^* - 1)e^{-\lambda_2^* t^*}]$$

$$C_B^* = \frac{(\lambda_1^* - 1)(1 - \lambda_2^*)}{k_{-1}^*(\lambda_1^* - \lambda_2^*)} (e^{-\lambda_2^* t^*} - e^{-\lambda_1^* t^*})$$

where

$$C_A^* = \frac{C_A}{C_{A_0}}, \qquad C_B^* = \frac{C_B}{C_{A_0}}, \qquad t^* = k_1 t, \qquad k_{-1}^* = \frac{k_{-1}}{k_1}$$

$$\lambda_1^* = \tfrac{1}{2}(1 + k_{-1}^* + k_2^* + \sqrt{(1 + k_{-1}^* + k_2^*)^2 - 4k_2^*})$$

$$\lambda_2^* = \tfrac{1}{2}(1 - k_{-1}^* + k_2^* - \sqrt{(1 + k_{-1}^* + k_2^*)^2 - 4k_2^*}$$

Here it is assumed that at $t = 0$, $C_B = C_C = 0$ and $C_A = C_A$.

23. Construct a potential energy surface for H_3 by using the LEP method.

24. Construct a potential energy surface for H_3 by using the LEPS method.

25. Verify eq. 2.55.

26. Derive eq. 2.65 and discuss its validity.

27. Derive eq. 2.68 and make a plot of Δ_{ab} versus R_{ab}.

28. Prove that the effective potential for the Lennard–Jones (6, 12) potential (Fig. 3.3) has a point of inflection at $E^* = U^* = 0.8$, $L^* = 1.569$, and $r^* = 1.3077$.

29. Consider an elastic collision of a N_2 molecule with a CO molecule. Before collision each molecule has a speed of 4×10^4 cm/sec and the two velocities are perpendicular to each other. Determine the maximum and minimum values of laboratory speed that the N_2 molecule could possibly reach. Find the collision energies in the CM and LAB coordinates initially.

30. Consider an elastic collision of two particles with particle 2 initially at rest.
    (a) Calculate the speed of particle 1 after the collision.
    (b) Show that for head-on collision, the change in kinetic energy of particle 1 is given by

    $$E_1 = \frac{-4m_1 m_2 E_1}{(m_1 + m_2)^2}$$

    (c) Find the maximum possible laboratory angle through which particle 1 can be scattered.
    (d) Show that v_2', the final velocity of particle 2, can be expressed as

    $$v_2' = \frac{2m_1 v_1 \sin(\chi/2)}{(m_1 + m_2)}$$

    Use this expression to give a general expression for the LAB energy change of particle 1.

31. Derive the general expression for the angle of deflection if the collision is subject to a square potential well. Derive the angle of deflection, differential cross section, and the total elastic cross section for hard sphere collisions.

32. Consider the elastic scattering from a potential of the form $V(r) = C/r^2$, where $C > 0$.

    (a) Find the deflection function as a function of impact parameter and initial collision energy.

    (b) Find simple approximate expressions for χ in the limit of large b and E, and in the opposite limit of small b.

    (c) Find the differential cross section and show that the total cross section is infinite.

    (c) Find the differential cross section and show that the total cross section is infinite.

33. Suppose you are instructed to do a laboratory exercise on the subject of the Rutherford scattering governed by the Coulombic potential $V(r) = C/r$. The main exercise is to measure the angular distribution of α-particles scattered from a gold foil; here, α-particles of 5.77 MeV are prepared from the americium isotope $_{95}AM^{239}$.

    (a) Derive the expression for the CM differential cross section.

    (b) Derive an approximate expression for your measured angular distribution of α-particles from a and evaluate the intensity of ratio at $\chi = 2°$ and $20°$. If the angular resolution of your detector is not better than $0.5°$, explain why your measured angular distribution falls off faster than the equation indicates.

    (c) Determine how much energy α-particles would lose at the laboratory angle $30°$ if gold foil were replaced by aluminum foil.

34. Calculate the elastic scattering of the ion-molecule system, with the potential governed by the ion-induced dipole interaction $V(r) = Cr^{-4}$ where $C = -e\alpha^2/2$ and where α is the polarizability of the molecules. Assume that an initial kinetic energy of the Ar^+ ion beam is 20 eV and that it is scattered by N_2 gas at $300°K$.

    (a) Determine the laboratory speeds of the Ar^+ ions and the N_2 molecules. Find the relative collision energy?

    (b) For an angular resolution of an ion detector of one degree, find the corresponding angular resolution in the CM coordinate.

    (c) Determine the effective total elastic cross section if $\alpha_{N_2} = 2A^3$?

35. Prove the equivalence of the Jeffreys–Born approximation for the phase shift (eq. 3.52) and the high energy approximation for the angle of deflection (eq. 3.15).

36. According to the Born approximation (eq. 3.50), a large deflection may be produced if the interaction potential $V(r)$ varies rapidly with r whether or not it is a strong or weak potential. State the reason for this.

37. Consider a potential of the form $V(r) = C_m r^{-m} + C_n r^{-n}$ where $C_n < 0 < C_m$ and $m > n > 1$.

    (a) Use the Jeffreys–Born approximation to show that the rainbow angle

    $$\chi_r = \frac{\text{constant}}{K}; \quad \text{where } K = \frac{E}{\varepsilon}$$

    (b) Apply the results of part a to a Lennard–Jones (6, 12) potential.

    (c) Use the Jeffreys–Born approximation to show that the maximum in the phase shift

    $$\eta_{\max} = \text{constant} \frac{B^{1/2}}{K^{1/2}}; \quad B = \frac{2\mu\varepsilon\sigma^2}{h^2}$$

    Evaluate η_m as a function of K for a Lennard–Jones potential.

38. Consider the electron scattering off an atom target in which the interaction may be approximated by the shielded coulombic potential $V(r) = A \exp(-\alpha r)/r$.

    (a) Obtain an expression for the differential cross section from the Born approximation.

    (b) Employ the condition for the low energy limit to show the scattering is isotropic.

    (c) Show that in the high energy limit, the Born approximation is equivalent to the Rutherford formula.

39. Suppose you want to measure the total elastic cross section of Ar–Ar scattering by introducing an argon beam into a small scattering chamber containing argon target gas. The Ar beam is prepared in a large vacuum chamber which is pumped by a diffusion pump and a cryogenic pump with overall pumping speed of 10^5 liters/sec.

    (a) If the beam speed of Ar is 4×10^4 cm/sec and you want to maintain the pressure of the vacuum chamber below 5×10^{-7} mm Hg to avoid background scattering due to ambient gas, find the maximum permissible source pressure of argon when you have an orifice of 0.1 cm^2.

    (*Suggestion*: The working formula $SP_2 = \frac{1}{4}(P_1 - P_2)\langle v \rangle A$; $S =$ pumping speed; $P_1 =$ source pressure; $P_2 =$ quasi-equilibrium pressure; $\langle v \rangle =$ averaged speed of beam particles; $A =$ orifice area).

(b) If this beam is traveling through the scattering chamber 2 cm in length with an argon density of 10^{12} #/cc, find the total cross section if the detector records a 5% attenuation.

40. Consider a wire along the z-axis with radius R maintained at a positive potential V. Thus a gaseous positive ion approaching the surface must overcome an energy barrier eV at an infinitesimal distance from the surface before being bound by the short range surface forces. You may assume that the wire is straight and of infinite length in order to eliminate the edge effects.
    (a) Show that the equation of motion of a positive ion moving under the influence of this potential separates into a straight line motion along the z coordinate and a motion in the xy-plane under this potential energy.
    (b) Determine the critical impact parameter b_c in the xy-plane for collision of a positive ion with the surface of the wire.
    (c) Calculate the cross section for collisions between unit length L_0 of the wire and an ion of speed g_{xy} in the xy-plane. Note that the interaction potential is no longer spherically symmetric.

41. Calculate $\omega(F)$ and $\delta(E)$ as a function E for the water molecule (see Table 5.1) by using the inversion of the partition function method.

42. Calculate $\omega(E)$ as a function of E for the water molecule by using the Darwin–Fowler method.

43. Derive the QET rate constant $k(E)$ for the classical harmonic oscillator model (cf. eqs. 5.20 and 5.21).

44. Derive the RRKM rate constant $k^{(1)}$ by treating the molecular vibrations as the classical harmonic oscillators and by approximating $k_{-2}M$ by the elastic collision.

45. Assuming that the reactant velocities are distributed according to a Maxwell–Boltman distribution, obtain the expression for the rate constant as a function of temperature if
    (a) $Q(E) = \pi d^2$
    (b) $Q(E) = \pi d^2 \cdot H(E - E_0)$ where $H(x) = 0$ for $x < 0$
                                                                  $= 1$ $x \geq 0$

    (c) $Q(E) = \pi d^2 \left(1 - \dfrac{E_0}{E}\right) H(E - E_0)$

    The expression for case c is identical to eq. 2.18, which is obtained by postulating that (1) only the relative energy along the line of centers of the colliding particles at the instant of contact is available for reaction and (2) this effective energy has to exceed the threshold energy E_0. Prove these statements.

46. A Sutherland potential is defined as

$$V(r) = -Cr^{-n} \quad \text{for } r > d$$
$$= -\varepsilon \quad r = d$$
$$= \infty \quad r < d$$

(a) If $n > 2$, orbiting collision is possible if the relative collision energy is low enough. Find the maximum kinetic energy at which orbiting can occur.

(b) Sketch the qualitative behavior of $I_n(E,)$ versus (eq. 2.3 or 2.41) if the reaction probability $P(E, b)$ is unity for some distance of closest approach $r > d$.

47. For an electron impact ionization process, the electron temperature is typically much higher than the atom temperature. Suppose the velocity distributions of the electron and of the atomic species are in thermal equilibrium and the effective temperature is given by

$$T^* = \frac{m_e T_a + m_a T_e}{m_a + m_e}$$

where the subscripts a and e denote the parameters for the atom and the electron, respectively.

(a) Derive a general expression for the rate constant as a function of the effective temperature T^*.

(b) Assume that the cross section for single ionization is given as

$$Q(E) = 0 \quad \text{for } E \leq I$$
$$= Q_0(E - I) \quad E > I$$

where I is the ionization potential of atom, Q_0 a constant, and E the relative collision energy. Find the rate constant temperature condition that $I > kT_e$.

48. Consider the reaction $K + HBr \rightarrow KBr + H$ with $\Delta D_0 = 4.2$ kcal/mole carried out in the crossed beam experiment. Assume the K beam with a velocity of 7.3×10^4 cm/sec was intercepted at $90°$ by HBr with a velocity 2.3×10^4 cm/sec. Find the range of the LAB angles at which the product KBr would be confined if $\varepsilon = 0$ or $\varepsilon = \Delta D_0$. You may assume no internal energy, $W = 0$, for HBr in eq. 2.21.

49. A spectator stripping model is often used to describe the reaction dynamics for a number of ion-molecule reactions. For $A + BC \rightarrow AB + C$

reaction, this model postulates that the velocity vector of atom C remains constant before and after the reaction. Show that this model predicts a negative exothermicity (eqs. 2.31 and 2.32).

50. In our discussion of the reaction cross section (Sections 2.1.2 and 2.1.3), the hard sphere model defined by $V(r) = 0$ at $r = a$ and $V(r) = E_0$ at $r = a$ gives (eq. 2.18)

$$Q(E) = \pi a^2 \left(1 - \frac{E_0}{E}\right) \qquad \text{for } E > E_0$$

$$= 0 \qquad\qquad\qquad E \le E_0$$

In addition, we have also showed that the cross section from the orbiting model with an attractive potential $V(r) = -E_0(a/r)^n$ is given by

$$Q(E) = \pi a^2 \left(\frac{n}{n-2}\right)\left(\frac{E_0(n-2)}{2E}\right)^{2/N} \qquad \text{for } n > 2$$

(a) Sketch a qualitative behavior for these energy dependent cross sections for $n = 4$ and compare your results with the measured cross sections for $K + CH_3I$ and $K + HCl$ reactions (Fig. 2.13).

(b) An empirical function of $Q(E)$ has been proposed to fit the measured cross section for a $K + CH_3I$ reaction,

$$Q(E) = \pi a^2 \left(\frac{q - E_0}{E}\right) \qquad \text{for } E > q$$

$$= \pi a^2 \left(1 - \frac{E_0}{E}\right) \qquad \text{for } E_0 \le E < q$$

51. Prove that the total kinetic energy for a three-body collinear configuration of $A + BC \rightarrow ABC \rightarrow AB + C$ is simply given as

$$E = \tfrac{1}{2}(\dot{q}_1^2 + \dot{q}_2^2)$$

Here q_1 and q_2 are the skewed coordinates defined in eqs. 2.36 and 2.37.

52. Consider the reaction

$$
\begin{array}{lll}
H^+ + H^- & \longrightarrow \quad H(1S) + H(2S) & \quad (1) \\
& \longrightarrow \quad H(1S) + H(3S) & \quad (2) \\
& \longrightarrow \quad H(1S) + H(4S) & \quad (3)
\end{array}
$$

carried out with relative collision velocity of 10^6 cm/sec. One of these processes is more important than the others if we have

Process	ΔH_{12} (erg)	R_c (Å)	ΔE (erg)
Process	ΔH_{12} (erg)	R_c (Å)	ΔE (erg)
(1)	8×10^{-13}	6.0	4.24×10^{-12}
(2)	1.6×10^{-14}	19.0	1.22×10^{-12}
(3)	1.6×10^{-17}	143.0	1.6×10^{-13}

Here ΔE is the overall energy change, R_c the crossing radius of ionic-covalent curves and ΔH_{12} the interaction matrix element. Calculate the transition probabilities for these processes and identify the dominant reaction products.

53. Derive eqs. 7.59 and 7.60 from eqs. 7.56 and 7.57 by using the Slater sum.

54. Derive eqs. 7.61 and 7.62.

55. Derive eq. 7.67 from eq. 7.66.

56. Calculate the Franck–Condon factors $|\langle X_{u''}(Q'')|X_0(Q')\rangle|^2$ for distorted and displaced oscillators.

57. Calculate the radiative lifetime of the $2p$ state of the H-atom.

58. If Φ_a and Φ_b represent the singlet electronic states of D^*A and A^*D, respectively, show that by using the molecular orbital theory

$$\left\langle \Phi_a \left| \sum_{i<j} \frac{e^2}{\gamma_{ij}} \right| \Phi_b \right\rangle = 2\left\langle \chi_p^*\chi_A \left| \frac{e^2}{12} \right| \chi_p \chi_A^* \right\rangle - \left\langle \chi_p^*\chi_A \left| \frac{e^2}{\gamma_{12}} \right| \chi_A^*\chi_p \right\rangle$$

where χ_A, χ_A^*, χ_p, χ_p^* are molecular orbitals. If the second term, which is the exchange interaction, is negligible, show how the above expression can reduce to eq. 7.121. Similarly if Φ_b nad Φ_a represent the triplet states of D^*A and A^*D, respectively, show that, using the molecular orbital theory,

$$\left\langle \Phi_a \left| \sum_{i<j} \frac{e^2}{\gamma_{ij}} \right| \Phi_b \right\rangle = -\left\langle \chi_p^*\chi_A \left| \frac{e^2}{\gamma_{12}} \right| \chi_A^*\chi_D \right\rangle$$

59. Derive the decay function $\phi(t)$ of D^* for the triplet-triplet transfer using eq. 7.147 Mitio Inokuti and Fumis Hirayama [See *M*. Inokuti and F. Hirayama, *J. Chem. Phys.*, **43**, 1978 (1965)].

60. The molecular orbitals of the π and π^* states of the formaldehyde molecules are given by [J. W. Sidman, *J. Chem. Phys.*, **29**, 644 (1958)].

$$\chi_\pi = 0.6\,(2P_x)_c + 0.8\,(2P_x)_0$$

and

$$\chi_\pi^* = 0.8\,(2P_x)_2 - 0.6\,(2P_x)_0$$

respectively.

(a) Calculate the transition moment for the $\pi \rightarrow \pi^*$ transition if the C—O distance is 1.26 Å.

(b) Calculate the radiative lifetime of the excited state $(\pi \rightarrow \pi^*)$, if the energy difference between the excited and ground states is 8.0 eV.

61. The temperature dependence of the triplet-state lifetime of *n*-dedecylbenzene between 4.2 and 100°K has been reported by Leubner and Hodgkins [*J. Phys. Chem.*, **73**, 2545 (1969)].

(a) Show that their data can be correlated by the equation

$$\frac{1}{\tau} = \frac{1}{\tau_0} + B \exp\left(-\frac{\theta}{T}\right)$$

(b) Find τ_0, B, and θ.

(c) If we can assume that the temperature dependence of the lifetime τ can be attributed to the nonradiative process, discuss the physical meaning of B and θ.

62. A simple classical model for V-V transfer between two oscillators A and B assumes the force acting on the target B as

$$F(t) = F_0 \cos \omega_A t \, \exp(-\alpha^2 t^2)$$

where ω_A is the circular frequency of the projectile oscillator A. Evaluate the energy transferred $\langle \Delta E \rangle$ assuming a random phase approximation and discuss the physical significance of the dependence of $\langle \Delta E \rangle$ on α, ω_A, and the circular frequency of the target oscillator ω_B.

63. Consider an oscillator fixed in space pointing along the y-axis. An ion of constant velocity g is projected into the z-axis and past the oscillator with an impact parameter b. The oscillator, which has an internal displacement coordinate r, has a dipole moment μ given by $\mu = \mu_0 + \alpha r$ and a vibration frequency ω. Work out the following problems by assuming a straight-line trajectory, $R(t) = (g^2 t^2 + b^2)^{1/2}$ where R is the separation of the collision partners.

(a) Evaluate $^1\Delta E^2$ in the high energy limit where $e^{i\omega t}$ is constant.

   (b) Evaluate $^1\Delta E^2$ for lower values of g. Since the actual Fourier transfer of R^{-3} is rather complicated, you may try to solve the problem by setting the solution between the limits of $b^{-1}R^{-2}$ and bR^{-4}.

   (c) Find inelastic cross section as a function of impact parameter b for parts a) and b).

64. Consider the DWA approximation for evaluating the transition probability of inelastic scattering

$$C_n(+\infty) = -\frac{i}{\hbar g}\int_{-\infty}^{\infty} V_{n0}(b, z)\exp\left(\frac{i}{\hbar g}\int_0^z \omega'_{nc}\,dz'\right)dz$$

   (a) Evaluate this integral by arguing that its main contributions come from Z values near Z_0 where the argument of the exponential is stationary. Expand $\omega'_{n0}(Z)$ in a Taylor series about Z_0 and show that the resulting probability of an inelastic collision from first-order approximation agrees with the Landau–Zener formula for the case of small transition probability.

   (b) From the result of part a, explain why the Landau–Zener formula becomes invalid at sufficiently high velocities.

65. Consider a collision of spin exchange of $A(\uparrow) + B(\downarrow) + A(\downarrow) + B(\uparrow)$ process. Calculate the energy dependence of the total inelastic cross section by postulating that a trajectories which fail to surmount the barrier of effective potential lead to elastic scattering and b those which surmount the rotational barrier are equally likely to lead to either elastic or inelastic scattering.

66. Consider a fast collision between an electron and a target molecule.

   (a) Evaluate the probability of electron excitation of the target molecule by assuming a straight-line trajectory. Explain qualitatively why the subsequent emission from this excited target is highly, linearly polarized.

   (b) Use the expressions for the ion-dipole and ion-quadrupole interactions to derive the transition probabilities for dipole and quadrupole target excitations and explain why the former is more important.

   (c) Assume that the target is being ionized. Describe qualitatively the angular distribution of the emitted electron.

67. With the squared transition dipole $|\mu_a|^2 = 10^{-34}$ (esu · cm)^2 for Na($^2S_{1/2} \to {}^2P_{1/2}$) transition, calculate the following.

   (a) The transition probability for a stationary Na atom collided with an electron at 1 keV with $b = 10$ Å.

   (b) Same as part (a) except with 10 keV proton.

68. In the semi-classical treatment of the electronic excitation of an atom by a heavy charged particle, certain selection rules are derived, that is, only electric dipole transitions would be allowed and only those in which the transition dipole is perpendicular to the direction of the trajectory.

    (a) Determine the effect if the collision velocity is lowered so that the excitation energy ΔE is an appreciable fraction of the incident kinetic energy.

    (b) State what will happen if the actual trajectory is curved but large impact parameter still applies.

    (c) Determine the result when the impact parameter b approaches the atomic dimensions.

69. For a ground state atom excited by an identical atom via the quadrupole transition, estimate the total cross section for $g = 10^5$ cm/sec and the interaction matrix $|Q|^2 = 10^{-55}$ erg $\cdot$ cm^5. Decide if the estimated value can be considered reliable and why.

70. Prove eqs. 8.88 and 8.89. Derive eqs. 8.90 and 8.91.

71. Solve eq. 8.98 by using the initial condition $\rho_{nm}^{(1)} = 0$ at $t = 0$.

72. Solve eq. 8.109 for the initial condition that the oscillator is initially prepared in the u_ith state [cf. E. Montroll and K. Shuler, *J. Chem. Phys.*, **29**, 454 (1957)].

73. Solve eq. 8.115 by the Laplace-transform method for the initial condition that the oscillator is initially populated in the u_ith state.

74. Derive eqs. 8.131 and 8.132 from eq. 8.130.

75. Derive eqs. 8.196, 8.198, and 8.199 from eq. 8.194.

76. Derive the golden rule expression for the transition probability for the perturbation H' being independent of time (cf. Chapters 7, 8, and 9).

77. Derive eq. 9.20 from eq. 9.14. Also derive eq. 9.22 from eq. 9.14.

78. Show that $\Delta E = \frac{1}{8}\hbar \sum_j \Delta_j^2 \omega_j^1$ in eq. 9.25 represents the minimum potential crossing the multidimensional potential surfaces.

79. Check the calculated results of the diffusion controlled reaction rate constants given in Tables 9.1 and 9.2.

Author Index

Subject Index